INTERNATIONAL MINING FORUM

This book aims to present an alternative based on natural processes and an environmental approach to post-excavation site management, e.g., post-coal mining heaps. These sites are places where various mineral excavation by-products are collected. Nevertheless, some post-mineral excavation sites are oligotrophic, terrestrial, wetland, and water habitat islands, providing unique biodiversity enrichment in the landscape. These oligotrophic mineral habitats are essential in over-fertilized, eutrophic, agricultural and urban-industry surroundings. Some post-mineral excavation sites are places where the wildlife can develop and support the functional processes of novel ecosystems. Implementing the newest biogeochemical and comprehensive knowledge into urban-industry landscape management will help to establish the ecosystem's processes and environmental functioning.

There are several post-industrial sites in Europe where the wildlife areas developed due to natural processes, are becoming wildlife hotspots in densely populated urban-industry areas. In this respect, many of the oligotrophic mineral terrestrial, wetland, and water habitats of anthropogenic origin should not be categorized as environmentally dangerous and undergo economic utility-focused reclamation.

Facing the actual environmental constraints of the Anthropocene Epoch, the book's chapters presenting the natural basics and perquisites of the environmental ecosystem mosaics, will be interesting for a broad range of environmentalists (scientists and students), miners, economists, and sociologists.

International Mining Forum

Green Scenarios: Mining Industry Responses to Environmental Challenges of the Anthropocene Epoch

Edited by

Artur Dyczko
Polish Academy of Sciences, Mineral and Energy Economy Research Institute, Kraków, Poland

Andrzej M. Jagodziński
Polish Academy of Sciences, Institute of Dendrology, Kórnik, Poland

Gabriela Woźniak
University of Silesia in Katowice, Faculty of Natural Sciences, Katowice, Poland

CRC Press
Taylor & Francis Group
Boca Raton London New York Leiden

CRC Press is an imprint of the
Taylor & Francis Group, an **informa** business
A BALKEMA BOOK

CRC Press/Balkema is an imprint of the Taylor & Francis Group, an informa business

Typeset by Krzysztof Stachurski, Jacek Jarosz
Printed and bound in Poland

Published by: CRC Press/Balkema
Schipholweg 107C, 2316 XC Leiden, The Netherlands
e-mail: enquiries@taylorandfrancis.com
www.routledge.com – www.taylorandfrancis.com

ISBN: 978-1-032-22212-7 (Hbk)
ISBN: 978-1-032-22215-8 (Pbk)
ISBN: 978-1-003-27160-4 (eBook)
DOI: 10.1201/9781003271604

Table of Contents

Preface

Global change significantly influences the natural environment, human society and economy. The environmental transformation includes complex eco-evolutionary consequences for organisms, mutual relationships among organisms, ecosystem alternation and the biotic versus abiotic environmental elements. The range and intensity of global change proves that the current methods of environmental change mitigation are not efficient enough.

This book aims to present an alternative environmental approach to the management of the post-excavation, e.g., post-coal mining heaps, areas of subsidence filled with groundwater, and salty sedimentation pools, all of which are side-effects of mineral resource exploitation. The rocks and stones and other mineral elements accompanying the exploited mineral resources are deposited on heaps. From the economic point of view, mining heaps are the sites where varied types of mineral excavation by-products are collected.

From the environmental point of view some of the post-mining sites are oligotrophic terrestrial, wetland, and water habitat islands, which can provide unique enrichments of environmental biodiversity on the landscape. The oligotrophic mineral habitats are essential within the surrounding over-fertilized, eutrophic, built up and designed garden areas of urban-industrial landscapes. Some of the post-mineral excavation sites are places where the wildlife can develop and support the improvement of environmental functioning. The implementation of the newest biogeochemical and comprehensive holistic environmental knowledge to urban-industrial landscape management procedures will enable gradual improvement of the ecosystems and environmental functioning.

There are several examples of post-industrial sites in European industrial centers where the wildlife areas developed as results of natural processes. The wildlife places are composed of the primary producer's (plants) vegetation mosaic, animals (e.g. butterflies, birds), and microorganisms such as fungi and bacteria. Such sites are wildlife hotspots in densely populated urban-industrial areas. Unmanaged lands are becoming scarce in urban post-industrial sites. In such circumstances, each opportunity to encourage development of open wildlife areas needs support. The presence of public wildlife areas is particularly crucial for development of new ecosystem functioning and as the prerequisite for ecosystem services based on support of environmental processes. The wildlife areas have scientifically proven ecological significance all over Europe. In this respect, all of the oligotrophic mineral terrestrial, wetland, and water habitats of anthropogenic origin should not be frequently or obligatorily categorized as environmentally dangerous, and undergo economic utility-focused reclamation.

Many aspects of human activity are damaging environmentally desired ecosystem functions, leading to landscape transformation and establishing many types of post-industrial sites. Unfortunately, according to current law, not updated with the newest environmental scientific knowledge, all of these sites are labeled as wasteland. As wasteland sites, they should undergo reclamation, which frequently brings damage to the possible spontaneous re-establishment of natural ecosystems and environmental functioning. Ecosystem and environmental functioning are the essential prerequisites of ecosystem services provision and human survival. At a time of worldwide global change, all environmental management decisions have to be made based on the newest scientific (biological, chemical, geological, and interdisciplinary environmental) knowledge. The usefulness and economic benefit should not be the only criteria for land assessment. Providing large open areas for wildlife colonization and open ecosystem development should also be taken into consideration.

The theory of novel ecosystems and natural capital concept offer a framework for a new environmental management approach, including the possibilities of ecosystem services provided by the ecological and ecosystem functioning focused on managed sites. Application of this new approach requires knowledge about the links between the novel ecosystem operating in urban-industrial environments and ecosystem services provision. A wide range of environmental science methods at scales from molecular to geographical, and from cells of living organisms to landscapes, has to be used for this research. Ecosystem services are essential for human survival. The availability and quality of ecosystem services depend on habitat mosaics available in the environment to support biodiversity and a variety of ecosystem functions.

The aim of this book is to present and link the complex fundamental interdisciplinary biogeochemical aspects of dynamic environmental functioning:

1. The characteristics of natural geological, geochemical and geo-hydrological conditions, which make the mining activity to be conducted possible. In the same way, the geological, geochemical and geo-hydrological conditions are prerequisites for colonization and community assembly by living organisms and ecosystem development. The same natural rules apply to the transformed and newly established sites with respect to, e.g., the moisture, texture, and nutrient gradients of the side-effect mineral material deposited throughout European urban–industrial centers;
2. The unusual *de novo* habitat conditions established, provided by post-excavation heaps of mineral resources, support the colonization and development processes of vegetation and ecosystems composed of specific, unknown species assemblages. The characteristics of some of the novel ecosystems and their functional aspects, including the living organisms that colonize the mineral material deposited on heaps, the vegetation species composition and diversity, as well as some aspects of the biological processes (e.g. the moss species diversity, seed bank biology) are presented;
3. The application of advanced biogeochemical methods is essential to understand the functioning of novel ecosystems. The characteristics of some of the methods that should be additionally applied to conduct comprehensive, holistic research on functioning of novel ecosystems are presented. The proper understanding of ecosystem functioning is basic for the improvement of environmental functioning. The selected methods present a range that should be used at varied organizational levels of the studied ecosystems, from the identification of antibody proteins at the cell level, up to the geo-informatics methods used to study ecosystem processes and spatial-temporal chronosequences of landscapes;
4. The presentation includes some of the abilities and possibilities of post-industrial, post-excavation sites to provide ecosystem function and services, based on the appropriate, best-adapted organisms. The development of vegetation as the basis of each ecosystem, has to be adapted to the unusual habitat conditions provided by harsh post-excavation mineral sites. Adaptation is the natural process that enables the acquisition and maintenance of stress tolerance, which is utilized as a service during phytoremediation – a remediation method for post-industrial areas. The fundamental idea is the natural capital concept, that enables the possibility to consider the ecosystem services;
5. Presenting some analyses of the ultimate solutions for environmental management practices on post-industrial, post-coal mining heaps in the Anthropocene Epoch, by using the natural resources deposited on post-excavation sites, education perspectives, Corporation Social Responsibility (CSR) approaches, and the possibilities for science-business cooperation. Management practices in the Anthropocene have to include the latest scientific and environmental knowledge in order to apply each opportunity to improve the ecosystem and environmental function in urban and industrial landscapes for better provision and enhancement of ecosystem services.

This book provides consideration and arguments for the idea that the underlying biogeochemical processes continue and become adjusted to the habitat conditions of novel ecosystems established due to human activity. The natural phenomena are ongoing, regardless of changes caused by alterations of the global environment, including for example alterations of nutrients or water conditions. The ongoing natural evolutionary adaptation and adjustment mechanisms acting at different scales cannot be ignored and should be incorporated and implemented into management practices. Nature is providing humans with solutions best for the environment and all living organisms, including humans.

Chapters of the **first part** characterize the overview of the geological aspects of mining activity as a factor shaping the urban and industrial landscapes of the Anthropocene. They present the reasons and consequences of the Anthropocene Epoch, including the geological prerequisites of the mineral excavation and mining industry, particularly hard coal exploitation, and the geological, mineral and chemical composition of the side-effect mineral materials deposited on, e.g., the post-coal mine heaps. Global environmental alterations of the Anthropocene challenge the demand for modification of manager attitudes towards environmental management. Also, the role of the newest scientific knowledge about development of ecosystem function on the new habitats that emerged due to human activity. The chapters of this part present the geological and hydrological background that made the excavation of mineral resources possible. Insight into the geological and hydrological basis of the mineral excavation process also gives context to some of the environmental side effects connected with mining activity. The results of the conducted research provide data for proper understanding of the environmental geo-hydrological and chemical feedbacks that can help in making the right management decisions. For example, to hinder from practice the exclusive economic utility approach and agricultural, hydro-technical focused activity. A broad interdisciplinary, holistic vision should be introduced based on the environmental, feedback linkages between all biotic and abiotic elements.

Chapters of the **second part** focus on the natural processes observed on the habitats which emerged because of mining activity. In this part the biology of colonizing organisms, interactions between living organisms and their habitats, ecological processes and ecosystem development join with the appearance of sites providing habitats of new mineral soil substrata in the environment. The biological and ecosystem processes of the first colonizing organisms, the primary producers – particularly vascular plants and mosses – are discussed. Apart from the species diversity, also the variety of vegetation types, plant assemblages occurring spontaneously during primary succession and complex biochemistry of seed bank studies on post-coalmine heaps are presented. This part implements the results of detailed case studies conducted on a range of sites in terrestrial, wetland and water ecosystems. The chapters of this part of the book describe the results obtained from research focused on the mechanisms involved and relationships among different groups of organisms which are established in the novel ecosystems developing on post-excavation mineral soil substrata. Some of the chapters of this part also underline the rarity and uniqueness of plants, fungi, bacteria, and mutualistic organisms.

The **third part** presents the applicability of different methods that have been used, and should be introduced in research focused on revealing the mechanisms of functioning of novel ecosystems. The methods discussed include the geo-informatics modeling approach used for spectral identification and calibration studies solving problems at the landscape geographical scale. For the calibration and assessment part of this study, the methods used for the resolution of vegetation patch, population and individual plant scales are also introduced. At the molecular level immunocytochemistry methods can be applied for identifying the differences in presence of proteins in plant tissues, e.g. the cell walls of individuals of the same species growing in different post-excavation habitat conditions. For studying the belowground soil, soil substratum and root ecosystem processes, the BIOLOG method (identifying the functional diversity of bacteria) and the methods used for identifying and assessing the activity of soil/soil substratum enzymes are introduced.

The **fourth part** presents the potential and recorded possibilities of the post-industrial, post-excavation sites to provide ecosystem services. The first chapter of this "New Approach" part presents the idea and concept of the natural capital framework with respect to the environmental circumstances that have emerged along with the gradient of post-industrial and post-excavation habitats. The following chapters present the results of studies dealing with species composition and diversity of communities in relationship to different groups of organisms on various mineral soil substratum habitats. The characterized elements of novel ecosystem functioning are presented with respect to the possible provision of ecosystem services. The novel ecosystems are self-sustaining, and in the face of human introduced environmental changes, they are becoming resistant to the impacts of global change factors. One of the chapters sums up the knowledge gained from various *in situ* experiments providing the data about the molecular background of plant adaptation mechanisms and plant functioning in transformed, harsh, post-industrial habitats. In this part some examples of post-industrial sites for which the particular ecosystem services have been recognized and identified in England were presented.

The **fifth part** consists of chapters that are focused on various aspects of current and future possible and recommended understanding, education and applicability of the ideas of natural capital, novel ecosystems and environmental health, and the ecosystem services inevitably linked to human conditions of life. The education activity should help to spread information about the potential environmental value of post-excavation sites. Novel ecosystems are particularly important because of their regulation of ecosystem services. The possible educational initiatives that can be undertaken, for example by coal mine enterprises, are presented. The education possibilities are, in part, linked with the social responsibilities of business; there are also additional comments on corporate social responsibility (CSR). The wise, scientifically based management of the transformed terrestrial, wetland and water environment elements that appeared after the exploitation of mineral resources are essential for corporate social responsibility. Apart from the present and future education and CSR activities, the cooperation and /or collaboration between science and business to develop the new approach in management practices for post-excavation sites, is analyzed. It is crucial that the activities presented in the fifth group of chapters will be followed by real, conscious activities and not only written declarations. The environment is mentioned in almost every official document. Many business strategies include separate environmental protection declarations. The Anthropocene Epoch is the time when humans have to start being responsible for the ecosystems and environments on which their quality of life and survival depend.

The complexity of current, accelerating environmental changes poses a major challenge for society and science. As a fundament of the evaluation of management options, scientific approaches need to cover the consequences of global change for organisms, populations, communities, ecosystems, and landscapes. The framework and concept of ecological novelty should enhance the urgent need for cross-disciplinary cooperation, for many reasons including:

1. The possibility to focus both on sites and on organisms, and thus the coexistence of complementary, comprehensive ecological and evolutionary aspects of novelty can be considered;
2. The cross-disciplinary approach facilitates a "common language" and new joint methods to perform the assessment of novelty, and thus will enhance methods, concepts and knowledge

exchange between researchers, helping to understand the significance of the proper use of novelty concept ideas;

3. The integration of research and business efforts on the consequences of global change for all living organisms, their relationships and ecosystem functioning should be taken into consideration. This allows the development of a conceptual common language. The joint concept should include aspects within and beyond biodiversity science, incorporating paleoecology, evolutionary ecology, and microbial ecology. The connection of research lines that are currently largely separated will benefit from an understanding of how global change effects should be managed.

We therefore believe that the inter-disciplinary approach presented in the book will help to work out a framework on ecological and ecosystem novelty, and will also help in the construction of an integrative basis for the overall better understanding of the actions necessary to properly deal with the ecological and evolutionary consequences of global change.

From the perspective of global change, the complex eco-evolutionary effects for organisms and ecosystems (e.g., novel ecosystems concept) need to be included in environmental management decisions. The comprehensive concept of "ecological novelty" should include (i) a site-specific approach, (ii) an individual organism-focused, eco-evolutionary perspective, and (iii) the maintenance of landscape ecosystem mosaics. This allows researchers and managers to address adequate ecological novelty questions from different perspectives. The questions should be derived from defined theoretical basics, e.g., (a) categorical or continuous diversity and functional measures, (b) the reference habitat conditions for occurrences of organisms, and (c) types and intensities of human activities.

The inspiration to write this monograph comes from interdisciplinary scientific sessions organized during the International Mining Forum in 2019 and 2021. In 2019 more than 20 presentations were shown within the "Green scenarios – responsible in action" session. The discussion after the end of this scientific meeting opened our eyes to the importance of presenting the role of post-mining areas to a wide audience, not only in relation to their reasonable and consistent with the current state of knowledge management. The discussion also showed the necessity of a new approach in the management of post-mineral resources areas. Thus, we developed the idea of combining our experiences from various perspectives. In 2021, within the scientific session "Green Scenarios: Mining industry responses to environmental challenges of the Anthropocene Epoch", organized as a part of International Mining Forum 2021, we have presented some of almost ready chapters of this book and discussed them with the conference participants. We believe that our monograph will answer many questions related to post-industrial habitats and their functioning.

Artur Dyczko, Andrzej M. Jagodziński, Gabriela Woźniak

Part One

Geological Background of the Mineral Resources Exploitation

Green Scenarios..., Dyczko, Jagodziński & Woźniak (eds.)

1.1. Geological Origin as the Basis for the Systematics of Deposits

Alicja Kicińska
Department of Environmental Protection,
Faculty of Geology, Geophysics and Environmental Protection,
AGH University of Science and Technology;
30 Mickiewicza Av., 30-059 Kraków, Poland;
kicinska@agh.edu.pl; ORCID: 0000-0003-2725-7319

ABSTRACT: The paper presents the most important concepts related to the origin and systematics of deposits, which undoubtedly include the following definitions: minerals, deposits of ore and metal-ore. The factors determining the economic value of deposits are discussed, dividing them into two most important groups: natural conditions and technical and economic conditions. An important part of the work is to present and discuss the modern classification of deposits, based on the main geological processes that contributed to the formation of deposits. The application of this criterion made it possible to distinguish 3 main groups of deposits (endogenous, exogenous and metamorphic), within which 9 categories, 18 classes and 21 types of deposits were distinguished. Other criteria for the classification of deposits are also presented, i.e. geotectonic, tectonic, morphological, genetic, mineralogical, chemical, technological, which may be helpful in assessing the industrial suitability or the abundance of mineral accumulations.

KEYWORDS: Deposit, metal-ore, mineral, deposit origins, modern deposit classification

INTRODUCTION

The basic concept in the systematics and the origin of deposits is the definition of a deposit and the accompanying description of a useful mineral. A deposit is an accumulation of a useful mineral (or several minerals simultaneously), in such quantity and under such conditions which economically justify its extraction and use (Jaroszewski *et al.* 1985; Candela 2003). A useful mineral is a geological formation within or on the surface of the Earth's crust that may find profitable economic use (Kucherenko *et al.* 2016).

The useful minerals are a diverse group; they can have the form of solids, the so-called solid fossil fuels (i.e. hard coal, lignite, building stones, clay raw materials, etc.) and they can also be in liquid form (i.e. crude oil, drinking water, medicinal water, mineral water). There are also gaseous minerals (e.g. natural gas). However, in order to be obtained, they must meet specific quantitative, qualitative, technical and functional requirements. The aspect of the industrial use of minerals, in the form of profitability and profitability of the undertaken mining activities related to the extraction of useful minerals (Fig. 1), is the basis for distinguishing a useful mineral which we will call a deposit, from an accumulation called an occurrence, i.e. it does not meet the requirements for the acquired minerals and the economic account. Terms are also used in relation to deposits, such as metal-ore and ore deposit. Ores are useful minerals from which metals are obtained. According to Krajewski (1954), ores are minerals that are the source of e.g. Ag, Au and platinum groups.

Geological processes are very complex and diverse. They have taken place on our planet during the last 4.5 billion years. All of this, combined with the continuous development of technology and research methods, have resulted in the mineral extraction of not only one (primary) mineral, but economic reasons also favour obtaining so-called accompanying minerals or associated minerals. The latter occurs with the main useful mineral, but they have different technological and functional properties, which determine the other method of their exploitation and processing (Gruszczyk 1984; Kullerud 2003).

Figure 1. "Szczakowa" mining plant
(Photo: Alicja Kicińska, September 2014)

The factors determining the value of the accumulation of a useful mineral and its recognition as a deposit are:

a) natural conditions, i.e.:
- geological environment of occurrence of the deposits,
- its geotectonic relationships, stratigraphic, lithological, and tectonic conditions,
- form and structure of the deposit,
- the type and quality of the useful mineral,
- its physical, chemical and structural properties,
- abundance of other deposits and their universality;

b) technical and economic conditions, i.e.:
- geographic and climatic conditions,
- mining availability (depth of the deposit),
- hydrogeological, heat and gas conditions, and geotechnical conditions,
- the level of mining technology,
- mechanical processing and technology (Gruszczyk 1984).

THE ORIGIN OF THE DEPOSIT

Deposits can be created in all conditions that prevail on our planet. They are the result of igneous, metamorphic processes (which usually take place deep in the Earth's crust) and may also be formed as a result of weathering and sedimentation processes taking place on the Earth's surface or in the subsurface zone of the crust. Krajewski (1954) emphasized the need to take into account two processes in consideration of the genesis of deposits: 1) the formation of a useful mineral (so-called minerogenesis) and 2) the extraordinary accumulation of a useful mineral (so-called concentration). The first of them usually takes place in all three states of aggregation (solid, liquid and gas). Their pace depends on the dynamics of changes in individual phases. The second type of processes is related to the phase change (generally towards the formation of the most durable minerals that can be formed under given conditions). The complex of all minerals present in one deposit is called the paragenetic complex of minerals. Characteristic paragenetic complexes which are constantly found in various deposits are called formations.

Minerals included in the paragenetic assemblages of magmatic deposits usually crystallize at different times, the so-called mineral succession (Brown 2014; Cooke *et al.* 2014). The fact that a mineral is built up several times during the deposit formation process is called different generations of the mineral. The analysis of the internal structure of minerals and their mutual relationships allows the time of formation of the entire paragenetic complex to be determined. Information on the processes of growth and displacement of younger minerals into older ones, the presence of pseudomorphoses and the occurrence of mineral erosions is helpful. Succession in sedimentary deposits is determined by analysing the litho-stratigraphic layer arrangement. It is based on the principle that in the absence of tectonic disturbances, the younger layers overlap with the older ones. Chemical sedimentation deposits are a special group. According to Krajewski (1954), they are similar to igneous deposits, and usually contain different generations of minerals, as well as crystallization cycles characterized by the so-called crystallization order.

All deposit processes can be classified as rock-forming processes, i.e. processes that play an important role in the formation and/or construction of rocks. Therefore, we can divide them in a relative order to the 4 basic types of rock formation processes (Jaroszewski *et al.* 1985; Heinrich & Candela 2014):

1) magmatism – the processes of creating and displacing magma and the formation of magma bodies in the depths of the Earth and in its surface zone;
2) weathering – mechanical and chemical decomposition of rocks under the influence of exogenous factors, takes place mainly on the surface and in the subsurface zone of the Earth;
3) sedimentation – deposition of clastic material from alimentation areas and material of organic origin, precipitated during natural physicochemical processes from solutions, under pressure and temperature conditions on the Earth's surface;
4) metamorphism – the processes of rock transformation, involving the action of temperature and pressure, and diagenesis, most often with the participation of chemical factors.

Observation of the deposit formation process is practically impossible in most cases. We can only register them in a few cases, such as the formation of rock salt, peat, sand and gravel accumulations, clay deposits, as well as certain effects of volcanic or weathering phenomena. In other cases, the limitation of these registrations is the time scale assigned to geological processes. It is measured in hundreds of thousands or even millions of years. Therefore, in the vast majority of cases, when determining the genesis of deposits, the method of deduction is used, based on the compilation of measurement data (mainly drilling in the rock mass, geophysical soundings and reservoir mapping), and then their in-depth interpretation. These data, being the results of geological research, provide the basis for formulating only a hypothesis of the geological genesis of the deposit, based on the conditions of its occurrence and development. This hypothesis is constantly verified and refined in the course of mining operations. The ambiguity and difficulty in formulating the genesis of the deposits is largely due to the complexity of geological processes taking place in a given area and the fact of the overlapping of several, most often different geological processes, most often separated in time.

The moment of deposit formation, i.e. its relative age (measured in relation to the surrounding rocks), can be the basis for distinguishing two types of deposits: syngenetic and epigenetic (Gruszczyk 1984). Syngenetic deposits formed simultaneously and the surrounding rocks are of the same age. They usually have isometric and plate forms. Examples of such deposits may be salt, anhydrite or coal deposits found among sedimentary rocks (Fig. 2). Epigenetic deposits, on the other hand, arose later in relation to the surrounding rocks. Their example can be veins formed by the contact of igneous rocks and surrounding rocks (e.g. vein mineralization in the Świętokrzyskie Mountains).

The formation of a deposit and the concentration of minerals are most often caused by a change in the physical and chemical conditions of the environment. The precipitation and concentration of minerals occurs as a result of the crystallization of magma or lava, crystallization from solutions and gases, gas-gas reactions, reactions of gases with solutions and solid mineral substances, as well as a result of the reactions of solutions with solutions and solid mineral substances (Gruszczyk 1984). The cause may also be the reducing or oxidizing effect of solid substances, solutions and gases, catalytic and sorption effects of solid substances, the activity of bacteria, plant and animal organisms, recrystallization of a mineral substance during diagenesis, catagenesis or metamorphism, mechanical, chemical and weathering concentration.

A separate important issue related to the origin of the deposits is the determination of the main mineralization processes from the later transformations, the transformations which the deposit underwent, which could have changed its primary character.

Figure 2. Plate coals in the Miocene sediment profile of the Sądecka Basin
(Photo: Alicja Kicińska, November 2010)

CONTEMPORARY DEPOSIT CLASSIFICATION

The modern division of deposits is based on genetic differentiation and is related to the moment of deposit formation. This allows them to be grouped into natural compounds based on the principle of the minerogenesis and concentration processes. On this basis, we distinguish three main groups of deposits: endogenous, exogenous and metamorphogenic (Tab. 1).

Table 1. Genetic classification of deposits (after Gruszczyk (1984))

USEFUL MINERAL DEPOSITS			
GROUP	**CATEGORY**	**CLASS**	**TYPE**
ENDOGENOUS	INTRUSIVE IGNEOUS	CRYSTALLIZATION LIQUIDATION	MODERN EARLY PROPER INJECTION
	PEGMATITE	GRANITE ALKALINE	PROPER MIXED
	CARBONATITE		
	SKARN		
	PNEUMOHYDROTHER- MAL	PNEUMATOLICS	ALBITITE GREYSENE
		HYDROTHERMAL	PLUTONIC SUB-VOLCANIC
	EXTRUSIVE IGNEOUS	PYRITE EXHALATION-SEDIMEN- TARY EFFUSIVE PYROCLASTIC GAS EXHALATIONS ASSOCIATED WITH HOT SPRINGS	
EXOGENOUS	WEATHERING	PLACER	ELUVIAL DELUVIAL TERRACE PLACER
		CHEMICAL-RESIDUAL	
		INFILTRATION	
	SEDIMENTARY	PLACER	AEOLIAN, POSTGLACIAL DELTA, SPIT, LAKE, SEA AND OCEAN MODERN AND FOSSIL SURFACE AND COVERED
		CHEMICAL	
		BIOCHEMICAL	CAUSTOBIOLITHS
		EVAPORATION	
METAMORPHO- GENIC	METAMORPHIZED METAMORPHIC POLY- AND ULTRAMETA- MORPHIC EXUDATION	PARA- AND ORTHOMETA- MORPHIC REGIONAL CONTACT	

ENDOGENOUS DEPOSITS

Endogenous deposits are formed under the influence of factors acting from the inside of the Earth, i.e. thermal energy from magma. In this group of deposits, we distinguish 6 categories (separated on the basis of various types of geological processes, based mainly on the physical parameters in which they occur). These are: intrusive igneous, pegmatite, carbonatite, skarn, pneumohydrothermal and extrusive igneous deposits (Gruszczyk 1984).

Intrusive igneous deposits are associated with igneous (ultra-basic, alkaline and alkaline) deposits. They are divided into two classes: crystallization and liquidation beds. The basis for this division is the process that led to the formation of the deposit (Gruszczyk 1984). The first of them – crystallization beds – were formed as a result of magma crystallization. When this process took place in the initial stages of crystallization, deposits called early (segregating or accumulative) were formed. The ore minerals occurring in them are characterized by idiomorphism, and the remaining minerals, mainly silicate ones, act as a crystalline binder. Deposits from the so-called later stage were formed after the formation of the surrounding rocks, as evidenced by the nature of ore aggregates and the xenomorphic formation of ore-bearing minerals. Examples of such deposits are copper and nickel sulphide ores, titanium-magnetite ores, apatites with nepheline and rare Earth elements, diamond deposits and others (Gruszczyk 1984; Huff & Owen 2013; 2015).

Pegmatite deposits are associated with igneous deposits of geosynclinal-fold areas. They have a mineral composition similar to the surrounding rocks, but they are formed as numerous mineral concentrations in pegmatites. Due to the dominant mineral composition, granite and alkaline pegmatites are distinguished, and due to their relationship to the surrounding rocks, they are divided into: proper deposits (with a similar mineral composition) and mixed deposits (with a different composition).

Carbonatite deposits are formed by carbonate rocks (called carbonatites) with coexisting ultrabasic alkaline rocks. Carbonatites are usually younger in age than ultra-alkaline and alkaline rocks. They are often found in the craters of volcanic cones. Mineralogically, they consist of carbonates (80–90%), and accessory minerals such as sulphates, phosphates, silicates, oxides and sulphides occur sporadically (Gruszczyk 1984; Mungall 2005; 2014).

Skarn deposits are formed at the contact of igneous rocks and carbonate deposits transformed into skarn (tactites). Skarn is a limestone-silicate rock. These deposits are formed at low or medium depths and form layers, lenses, veins, nests and complex forms.

Pneumohydrothermal deposits formed as a result of the action of gases and solutions. They include concentrations of almost all metals and metalloids: As, Au, Be, Co, Fe, Hg, Li, Mo, Ni, Pb, Sb, Th, W, Zn, Zr and rare Earth elements. They occur mainly in the form of veins, nests and in exceptional conditions, beds. They are located in the upper parts of igneous massifs and are formed at great depths. We distinguish two classes of deposits: pneumatolics, in which the gas phase played a fundamental role (types: albitite – formed earlier from high-temperature alkaline solutions, and greysene – formed later, at lower temperatures from acidic solutions), and hydrothermal deposits (plutonic – formed at great depths and sub-volcanic – formed at shallow depths). In the past, they played an important role as a source of useful minerals (Gruszczyk 1984; Klein 2003).

Extrusive igneous deposits are associated with undersea magma eruptions or lava flows onto the Earth's surface. The first results are pyrite deposits in which iron sulphides are the dominant component, as well as chalcopyrite, bornite, sphalerite, galena and others. They form elongated forms with lengths of thousands of meters. This class includes specific mineral formations: pyrite, copper-pyrite and polymetallic-pyrite (Gruszczyk 1984). Other classes of deposits assigned to this category include: exhalation-sedimentary (in which the mineral substance that builds the rocks was brought into the basin by submarine exhalations, but they have a form and structure typical of sedimentary rocks), effusive rocks (related to openings formed in the initial and final stages of geosynclinal development, diversified in terms of minerals and chemicals, including basalts, andesites, dacites), pyroclastic (made of pyroclastic deposits, i.e. blocks and volcanic bombs, lapilles, sands, and volcanic ashes), associated with gas exhalations (e.g. some native sulphur deposits) and associated with hot springs (Hg, Sb, Mn, W, Au and Ag ore deposits).

EXOGENOUS DEPOSITS

Exogenous deposits are formed under the influence of factors from outside of the Earth's crust, that is, those that develop on or near the Earth's surface. Minerals and rocks are transformed under the influence of weathering. Two further categories of deposits are created: weathering and sediment. Weathering processes are largely conditioned by factors (temperature, pressure) on the Earth's surface. The rate of weathering depends on the type of weathered rocks, climate, their physical and chemical properties, as well as geological and hydrogeological conditions (Gruszczyk 1984).

Weathering deposits are formed as a result of weathering processes, but the resulting materials are not transported. Weathered materials remain in situ and can originate from all types of rocks (igneous, sedimentary, metamorphic) and cover previously formed deposits located on the Earth's surface. The topography determines the possibility of deposit formation. They generally have irregular forms and variable thickness. These deposits differ significantly from igneous or metamorphic deposits in their place of occurrence, morphological features and mineral composition (Gruszczyk 1984). Within the group of weathering deposits we distinguish the following classes of deposits: placer (made of hard and weather-resistant minerals, of high density), chemical-residual (made of minerals with a high degree of oxidation) and infiltration (whose mineral composition is similar or different from the primary).

Within the placer class the following types, among others, stand out: – eluvial deposits – rocks which were the source of the mineral substance are located in the place of their original occurrence; – deluvial deposits – rocks which were the source of the mineral substance are located on the slope of the hill; – terrace placer – occurs at the foot of the hill (Gruszczyk 1984). Often, all three types are referred to as placer deposits of (detrital/clastic sediment) weathering origin. Common placers include sand and gravel deposits, as well as block fields. Due to the time of formation, there are modern and fossil deposits, and due to the way they occur, they can be divided into surface and buried. If the object of exploitation is one mineral, we speak of a monomineral deposit, and if there are more, then we have a poly-mineral deposit (Gruszczyk 1984; Vaughan 2005).

The class of chemical-residual deposits includes deposits formed as a result of chemical weathering, the mineral composition of which is a remnant of weathered rocks. The term residual refers to accumulations that remain in situ (i.e. have not been displaced). Additionally, they can be classified as surface or buried deposits, and due to their age, modern or fossil deposits are distinguished.

Infiltration beds (also known as metathetic or descending beds) are formed from a substance that leaches out from the weathering zone and is transported by descending waters and then precipitated in adjacent rocks (Gruszczyk 1984). The locations of such deposits are below the weathering zone, and most often these are fissures and cracks located at considerable depths.

Weathering is a common process on the Earth's surface and in its subsurface zone. In the existing deposits, there are also subsequent weathering processes that can significantly change its original structure and mineral composition. Zones and sub-zones may arise: oxidation, secondary enrichment or cementation, and primary ores. Three subzones occur within the oxidation zone of the deposit itself, the so-called subsurface, ores oxidation subzone, oxide ores leaching subzone and rich oxide ores subzone (Gruszczyk 1984). Due to mineralogical changes, the ore deposits can be divided into those in which the minerals are resistant to oxidation and no significant mineralogical changes are noted, into those in which there is a change of ore minerals (without and with metal discharge) and into those in which the oxidation zone takes place due to concentration of metals forming admixtures in primary deposits. This group of deposits is associated with the formation of an aureole of dispersion. These are the areas in the vicinity of the deposits, characterized by an increased content of minerals present in the deposit. They can be primary (formed during the deposit formation) and secondary (formed as a result of weathering and destruction of the deposit and its primary halo).

The second category of exogenous deposits are sedimentary deposits. The sedimentary deposits are located in sedimentary rocks, are arranged in accordance with them, have a seam, pseudo-seam or lenticular form and are associated with specific stratigraphic and lithological levels (Gruszczyk 1984). They occur in specific facies and paleogeographic conditions and are syngenetic deposits. This group includes deposits of hard coal, lignite, peat, crude oil, natural gas, oil shale, bauxite, iron ore, manganese, copper, zinc, lead, uranium and many others. They are characterized by a simple structure, large resources, simple mineral composition, and can be formed in both terrestrial and marine environments. The useful minerals obtained from this type of deposit can be loose, plastic or compact. The formation of this type of deposit depends on the source of the component useful in the maintenance area and its abundance, climate, the development of the organic world, the possibility and method of transporting the mineral substance, the physical and chemical conditions of the sedimentation environment and the nature of diagenesis (Gruszczyk 1984). In this category of deposits, we can distinguish two classes: placer and chemical, and the latter are divided into: chemical in the strict sense, biochemical and evaporative.

Placer deposits include sedimentary deposits of common rocks and concentrations of ores in contemporary and fossil deposits, formed from minerals resistant to weathering. In this group, further divisions of deposits are used in terms of their mineral composition, commonness in the environment, age or specific physical properties. The fine material is transported (mainly by water, but also by wind – the so-called aeolian deposits, or glaciers – the so-called postglacial deposits – moraine and fluvioglacial), and then deposited outside the area of weathered formations. The place of deposition also determines the name of the later deposit. This classification includes delta, spit, lake, sea and ocean deposits (Gruszczyk 1984). Due to the mineral composition, there are also poly- and monomineral deposits, and due to age – modern and fossil deposits, and due to their occurrence – there are surface and covered deposits. An important factor is the common occurrence in the Earth's crust, which is why we distinguish deposits of common rocks (here we distinguish contemporary loose rock deposits: boulders, rubble, pebbles, gravel, sand, silt, clay) and clastic deposits of rarer minerals (we also divide them into modern and fossil: native gold, platinum, ilmenite, magnetite, zircon, rutile and many others).

Chemical sedimentary deposits are formed from a mineral substance transported in the form of solutions (liquid or colloidal) and deposited as a result of various chemical, biogenic and physical processes. As already mentioned, they can be divided into chemical sensu stricto, biochemical and evaporative (Gruszczyk 1984). This group of deposits is the most numerous and diverse, and it also has the greatest industrial importance. Therefore, the individual types of deposits will only be mentioned briefly.

The deposits of solid caustobioliths are divided into humoliths, humosapropelliths and liptobioliths. Together with acaustobiolites, they form a group of biogenic deposits called biolites. Caustobiolites include humolites, i.e. peat and coal deposits, and similarly to gyttja, putrid sludge or bituminous shale, they also belong to sapropelites. Well-known examples of liptobiolites are amber, ozocerite and resins. Acaustobiolites are chalk formations, coral limestones, diatomaceous Earth, diatomites, radiolarites, spongiolites and phosphates.

Another type of deposit in the group are humolites (peat, brown coal, hard coal). Peat deposits represent the youngest geological formations, i.e. the Pleistocene or the Holocene. They create accumulations called peat bogs (which can be low, transitional and high). They are characterized by the presence of plant residues and admixtures of inorganic material. The carbon content of peats is usually below 60%. They are present in boreal, temperate and tropical climates, on almost all continents. Deposits are formed by seams (beds) lying on the Earth's surface. Depending on the plant (peat-forming) communities, homogeneous and heterogeneous peats are distinguished. They can also be of planktonic, megaplankton, rush, aphid-sedge and forest origin. The peat substance is slightly firm and shows a different degree of decomposition, therefore we are dealing with fresh, boggy, fatty and tar peat (Gruszczyk 1984). Lignite deposits are located in younger geological formations, mainly Neogene, and less often Cretaceous, Jurassic and Triassic. They usually lie horizontally and are composed of loose or weakly compact clastic rocks and clay minerals. Due to the small thickness of the overlays, they are exploited mainly by opencast methods. The basic form of coal occurrence is seam, although these deposits are usually composed of several or a dozen seams. Their thickness ranges from several to several dozen (rarely several dozen) meters. Lignite is yellow-brown or black in colour, it is a weakly compact material, brittle, earthy, containing large amounts of water (even 40–60%). Depending on the chemical composition, brown coal can be divided into: humic, soft, hard, lignite and sapropelic. The content of elemental carbon in lignite varies between 60 and 76%, while the content of mineral admixtures, which are mainly clay minerals, carbonates and sulfides, can be from 3 to 36%. Lignite deposits are known in at least 30 countries. The last of the humolites is hard coal. Hard coal deposits are accumulations of organic deposits and contain from 76 to 96% carbon by weight. They form large units called coal basins. Three types of coal deposits are distinguished: geosynchinal, platform and transitional (Gruszczyk 1984). Hard coal deposits were formed in different geological periods, i.e. Carboniferous, Permian, Jurassic and Neogene. Carboniferous deposits constitute approx. 22% of the world's resources and are mainly found in the eastern part of North America, Europe, northern Africa and western Asia. Permian and Jurassic deposits are found in eastern Asia, Australia, the polar regions, south-eastern Africa and South America and they cover about 17% of the world's total resources. The Upper Cretaceous and Neogene deposits are mainly found in eastern Asia and Australia, on the islands of the Pacific Ocean, in the western part of North and South America, and they account for about 50% of the world's hard coal resources. Hard coal deposits are mainly associated with sedimentary formations (clastic, i.e. sandstones, clay shales, and less often conglomerates), they usually form seams or lenses with a thickness of several centimetres to several meters. Deposits of sapropelites are made of putrid sludge (gyttja), which is obtained from inland water reservoirs in a temperate climate. However, it does not have much raw material importance. The deposit class also includes boghead and cannel coal deposits as well as oil shale deposits. The deposits of caustobiolites are closed by liptobiolites, represented by amber, tasmanite and piropissite.

The deposits of crude oil and natural gas are the accumulation of liquid and gaseous caustobiolites, meeting the requirements of industrial exploitation (Gruszczyk 1984). Most of the crude oil is concentrated in the marginal parts of mountain ranges, in premountain and inter-mountain sinkholes and molasses basins. For centuries they have been associated with all geological formations, from the Cambrian to the Pliocene. There are platform, geosynclinal-fold and subgeosynclinal deposits. In the first type, crude oil is usually found in carbonate rocks, with a constant level and forms large units with indistinct contours. In the second type, it is associated with large complexes of mainly clay-clastic deposits. The deposits of this type are highly tectonically disturbed, characterized by folded structures and variable fall angles. The third type of deposit has some features of the first and second types. Crude oil and gas deposits can form within or outside of the parent series.

The group of coagulant deposits includes bauxite, iron and manganese ores. They have a similar environment, form and structure of ore bodies (Gruszczyk 1984). They can be associated with platform, sinkhole and geosynclinal formations. The form of the deposits, on the other hand, can be onboard, lenticular, socket or pocket. The bauxite deposits contain aluminium hydroxides, diasporas, hydragilite and bemite. However, the content of aluminium oxides should be greater than 45 wt.%. In iron ore deposits there are: oxide ores (i.e. hydrogoethite, goethite, magnetite), carbonate ores (siderite, pyrite, iron silicates) and silicate ores (ferric chlorites: shamosite and thuringite), and the content of Fe compounds should not be less than 25 wt.%. The manganese ores contain hydroxide ores Mn and Fe as well as psylomelan, goethite, pyrolusite, oxide (manganite) and carbonate ores (rhodochrosite, pyrite, and glauconite).

Other important sedimentary deposits are deposits of: phosphorus minerals (guano), limestone, dolomite and related rocks, silica rocks, iron and sulphide-polymetallic ores, uranium, radium and vanadium ores, stratoid deposits of Cu, Ag, Zn ores, Pb and Ba, F and Sr minerals, native sulphur and evaporative deposits (formed in dry climate conditions as a result of water evaporation).

METAMORPHOGENIC DEPOSITS

Metamorphogenic deposits are associated with metamorphic forms. As Gruszczyk (1984) writes, they are genetically, spatially and age-related with transformed bodies. They are characterized by a typical mineral composition, as well as structural and textural features. The following deposit categories were distinguished in this group: metamorphized, metamorphic, poly- and ultrametamorphic, and exudation. Mineral substances found in this type of deposits generally have dense structures, ranging from micro to coarse crystalline. Textures are typically slate, parallel and laminated. Relic structures can be found occasionally.

The metamorphized deposits were formed by transformation, as well as the accumulation of a useful component. Under the influence of metamorphic agents, some properties of minerals, such as structure and texture, have changed. Depending on the primary deposits, para- and orthometamorphic deposits can be formed, and regionally or contact-metamorphised deposits can also be distinguished. Their forms are similar to primal forms, they appear in the form of decks, lenses, sockets or veins. The sizes of these deposits can be considerable and due to multiple folds, deformations or dislocations and their structure is usually complicated. Examples of such deposits are the formations of metamorphized iron ore deposits (quartzite and mica deposits of ferruginous shales and banded ferruginous jaspilites) and manganese, quartzites and manganese shales, metamorphized deposits of gold-bearing and uranium-bearing conglomerates, copper ore, lead and copper ore, metamorphized ore and copper ore formations, zinc, pyrite and apatite.

Metamorphic deposits are deposits in which the useful minerals were completely or partially formed as a result of the metamorphism process which the primary rocks underwent. Examples of such deposits include graphite, corundum, garnet or asbestos. They are important due to the type of minerals obtained from them (iron, manganese, gold, zinc, lead, and subordinate apatite ores). The form of these deposits is usually on-board and their dimensions are considerable. They are of complicated and disturbed structure. The mineral composition of these deposits depends on the type of primary minerals, and the structure and texture depend on the degree of metamorphism. Examples of such deposits are: graphite, marble, quartzite, cyanite slate, corundum and emery, garnet, amphibole asbestos, talcum, titanium ores and slate. This group also includes deposits of metamorphic rocks such as: gneisses, amphibolites, and serpentinites used in road construction and as cut stones (Gruszczyk 1984).

Poly- and ultrametamorphic deposits are the result of deep, often multiple metamorphic transformations. They are characterized by varied formations. Mineral substances underwent multiple transformations and precipitations under frequently changing agents. They occur in the area of precambrian crystalline shields, and they are mainly associated with skarns, limestones and quartzites. They have a complex shape and belong to deposits rich in ore minerals. Their structures are extremely diverse. They include leptite deposits of iron ores, skarns of manganese and sulphide ores, as well as copper, gold and zinc ores (Gruszczyk 1984).

Exudation deposits represent accumulations of a mineral in a metamorphic series that are postmetamorphic in origin. They are genetically epigenetic and form veins, sockets, and lenses. In terms of mineral composition, as well as structures and textures, they are similar to hydrothermal deposits. Examples of this type of deposit are the alpine veins (Gruszczyk 1984).

The deposits can also be classified using other criteria, such as: geotectonic, geological, tectonic, morphological, genetic, mineralogical, chemical, formation, technological, industrial suitability or wealth (Gruszczyk 1984).

REFERENCES

Brown A.C. 2014: Low-temperature sediment-hosted copper deposits. [In:] Holland H.D. & Turekian K.K. (eds.): Treatise on geochemistry. 2nd ed. Elsevier, Pergamon, 251–271.

Candela P.A. 2003: Ores in the Earth's crust. [In:] Holland H.D. & Turekian K.K. (eds.): Treatise on geochemistry. Elsevier, Pergamon, 411–431.

Cooke D.R., Hollings P., Wilkinson J.J. & Tosdal R.M. 2014: Geochemistry of porphyry deposits. [In:] Holland H.D. & Turekian K.K. (eds.): Treatise on geochemistry. 2nd ed. Elsevier, Pergamon, 357–381.

Gruszczyk H. 1984: Nauka o złożach. Wydawnictwa Geologiczne, Warszawa.

Heinrich C.A. & Candela P.A. 2014: Fluids and ore formation in the Earth's crust. [In:] Holland H.D. & Turekian K.K. (eds.): Treatise on geochemistry. 2nd ed. Elsevier, Pergamon, 1–28.

Huff W.D. & Owen L.A. 2013: Volcanic landforms and hazards. [In:] Shroder J. & Owen L.A. (eds.): Treatise on geomorphology. Volume 5. Tectonic geomorphology. Academic Press, San Diego, 148–192.

Huff W.D. & Owen L.A. 2015: Volcanic landforms and hazards. Reference Module in Earth Systems and Environmental Sciences. Elsevier.

Jaroszewski W., Marks L. & Radomski A. 1985: Słownik geologii dynamicznej. Wydawnictwa Geologiczne, Warszawa.

Klein E.M. 2003: Geochemistry of the igneous oceanic crust. [In:] Holland H.D. & Turekian K.K. (eds.): Treatise on geochemistry. Elsevier, Pergamon, 433–463.

Krajewski R. 1954: Zarys nauki o złożach surowców mineralnych. Państwowe Wydawnictwo Naukowe, Kraków.

Kucherenko I., Sinkina E. & Abramova R. 2016: Genetic classification of ore-forming processes. IOP Conf. Series: Earth and Environmental Science, 43, 012077.

Kullerud G. 2003: Ore petrology. [In:] Meyers R.A. (ed.): Encyclopedia of Physical Science and Technology. 3rd ed. Academic Press, 411–433.

Mungall J.E. 2005: Mining geology – magmatic ores. [In:] Selley R.C., Cocks L.R.M. & Plimer I.R. (eds.): Encyclopedia of geology. Elsevier, Amsterdam, 637–645.

Mungall J.E. 2014: Geochemistry of magmatic ore deposits. [In:] Holland H.D. & Turekian K.K. (eds.): Treatise on geochemistry. 2nd ed. Elsevier, Pergamon, 195–218.

Vaughan D.J. 2005: Minerals – sulphides. [In:] Selley R.C., Cocks L.R.M. & Plimer I.R. (eds.): Encyclopedia of geology. Elsevier, Amsterdam, 574–586.

1.2. Geological Conditions Determining the Type of Mining Activity

Radosław Pomykała
Department of Environmental Engineering,
Faculty of Civil Engineering and Resource Management,
AGH University of Science and Technology;
30 Mickiewicza Av., 30-059 Kraków, Poland;
rpomyk@agh.edu.pl; ORCID: 0000-0003-2238-241X

Justyna Adamczyk
Department of Civil & Geotechnical Engineering and Geomechanics,
Faculty of Civil Engineering and Resource Management,
AGH University of Science and Technology;
30 Mickiewicza Av., 30-059 Kraków, Poland;
jadamcz@agh.edu.pl; ORCID: 0000-0002-3274-1001

ABSTRACT: The chapter identifies the most important groups of environmental impacts of resource exploitation, as well as the most important factors affecting their type, range, and intensity. The dominant are geological and mining factors, including the type and properties of rocks and mined raw materials, the shape of the deposits, their disturbance, depth, as well as the method of access, type of mining system, methods of processing, and enrichment, transport and many others. The size and perception of the impacts also depends on geographic factors, including the location of the mine, topography, vegetation and the degree of urbanization, as well as on broadly understood social, economic and legal environmental conditions. The most important ranges of the environmental impacts characteristic for open pit and underground mining are presented.

KEYWORDS: Mining impacts, environment, land deformation, pollutants

INTRODUCTION

Mining activities related to solid raw materials can take very different forms. The size, scope and type of mining activities related to substantial resources depend primarily on the location of the deposit, its size, shape, and type of mineral. The method of exploitation results from the current state of knowledge and technology. The set of these criteria is called geological and mining factors. However, these are not all issues important for mining activity, including its type and scope of environmental impact. Other factors are also important, such as the condition and availability of technology, legal conditions and requirements, geography, including location and topography, climate, as well as broadly understood environmental and social aspects.

The important geological factors include the shape and size of the deposit, the type and features of the raw material excavated. The location of the deposit is significant, including the depth, type of rock mass and the properties of rocks and rock formations that form the resource deposit, deposit tectonics, seismic activity and others. The rock characteristics are valid concerns both in the vicinity of the deposit and adjacent areas. Hydrogeological features are also important, especially the location, type and size of aquifers and the properties of the waters themselves. The possible earlier environmentally-focused operation is also essential.

The mining factors result from the available mining technologies, which include mining methods and means, devices and machines used for this purpose, the mining systems used and their parameter efficiency. The selected exploitation system should ensure the possibility of valid selection and use of the mineral resources. The method used should ensure safety for people and the environment.

Geographical factors include atmospheric conditions, natural terrain conditions including relief and socio-economic, cultural, social and environmental factors. Here, the most important factors are the degree/method of land development, its urbanisation and population, as well as the presence of valuable natural, historical and cultural areas on the surface. This group of factors is very diverse, and depends on the region of the world where exploitation occurs. They are directly related to the permissible range of mining influence on various elements of the environment.

The above factors affect the choice of a specific method and system of operation. They are also elements that directly or indirectly shape the costs of mining activities at all its stages (from exploration, through reaching the deposit – its availability, its exploitation and processing of raw materials, to the decommissioning of the plant and land reclamation). The primary mining methods are opencast and underground mining (Piechota 2008; Spitz & Trudinger 2008). Such a distinction is obvious and results from the location of the raw materials of interest within the rock mass, and thus the place of drilling of mining excavations. This division is also the most obvious and visible to all interested parties. Other types of mining are borehole mining based on methods of extracting solid raw materials previously subjected to changes in physical and chemical properties such as a sulphur smelting, salt leaching, or underground coal gasification methods. There are also many other types of mining activities, such as bio-mining and (sub) sea-mining (Fig. 1).

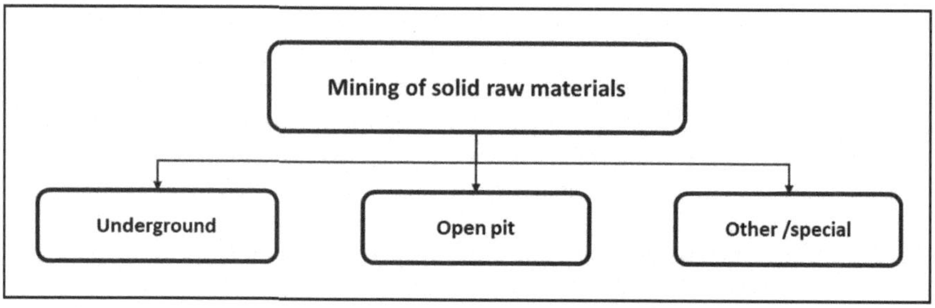

Figure 1. Main types of mining

While sticking to this first stage of mining division, it is worth getting acquainted with a slightly more detailed description of the types.

SPECIFICITY OF OPENCAST MINING

Opencast mining is historically the oldest method of mining. Almost every type of mineral resource is or has been mined in the past by opencast methods. Over time, very large mines began to emerge, and the term "open-pit mine" is more appropriate for them. To avoid misunderstandings, in this chapter the term "opencast" will mean any exploration of resources directly from the surface.

The most easily accessible deposits, i.e. the shallowest ones, are depleted. Due to technical progress the opencast method enables exploitation of both very shallow, small deposits and those located at very great depths and/or that occur over vast areas.

For deposits located next to the surface, exploitation may start directly from an outcrop on the surface. The outcrop approach applies in particular to the exploitation of rock deposits. Quarries are frequent on all continents. However, many deposits of raw materials lie beneath a layer of other rocks known as overburden. The overburden earth or rock layers lying over the deposit are removed to a designated place, outside the mining excavation. At a later stage of operation, when the pit reaches a sufficient size, it is possible to deposit earth masses from the overburden within the mining excavation (the so-called internal dump). The cut-and-cover method itself consists of making excavation on the scale (at the bottom) of the deposit to be exploited. Near the surface, the hole is more massive due to the need to maintain a safe slope angle. The greater the depth, the more significant are the differences. The essential feature of opencast mining is the targeted coverage of almost the entire surface of the mining area for which mining activities have obtained the concession.

Many types of raw materials are mined using the opencast method. In Poland, these are mainly rock materials (the most number of mines) and lignite (the largest mines). The scope and type of exploitation obviously depend on the type of raw materials, the size and value of the deposit and its availability. The more valuable the deposit, the greater the thickness of the overburden rocks, it is profitable to remove it in order to reach the deposit. In the case of common raw materials, such as sand or rock raw materials, deposits with a small overburden are strongly preferred.

SPECIFICITY OF UNDERGROUND MINING

Underground mining is the exploitation of raw materials from underground workings located under the overburden. In underground mining, the area occupied by the mining plant and its infrastructure is relatively small compared to the area of underground mining activity. Directly on the surface, there is a complex of technological buildings, shaft towers, processing plants, material and product warehouses and many other elements of the mine's infrastructure. The entire main plant is usually concentrated, fenced and takes up from several to several dozen hectares. Around it, from a few to several kilometres away, there are other, smaller clusters of infrastructure – around the ventilation shafts allowing for active air exchange.

Depending on the geological conditions, the topography and the depth of the deposit, it is sometimes possible to reach the deposit with the use of excavations such as tunnels (horizontal excavation in a mountain slope) or sloping excavation (inclined excavation). Shafts, i.e. vertical pits (possibly inclined at an angle greater than 45°), is the most expensive type of excavation. Shafts are drilled when there is no other way to reach the deposit Exploitation takes place in underground workings, the shape, size, location and quantity of which depend, among others, on the function, shape and size of the deposit and the adopted system of exploitation. In Poland, essential raw materials exploited underground are hard coal, copper ores, rock salt and zinc and lead ores (Hartman 1992; Piechota 2008).

OTHER METHODS OF MINING EXPLOITATION

There are many methods for the exploitation of solid raw materials, different then opencast pits and underground mining. Among them, the so-called borehole methods allow for extraction with the use of drilling methods, as a result of liquefaction of raw materials with physical or chemical processes. For example, the borehole method is used to extract sulphur. Superheated water melts sulphur in the underground bed and the liquefied sulphur is transported to the surface. Another example is dissolving salt with water. After saturation with salt, the brine is transported to the surface where the salt is recovered by an evaporation process.

This group of exploitation methods also includes metal leaching. Metal leaching is carried out from deposits (*in situ*) and heap leaching methods. These methods of exploitation have different effects on the environment. Other methods of exploitation include the use of hydro-mining, bioleaching or thermal processes such as coal gasification (Hartman 1992; Spitz & Trudinger 2008).

RAW MATERIALS AND THE SCOPE OF THEIR PROCESSING

The type of exploited raw materials is an essential criterion for distinguishing type of mining activities. Many mining techniques and technologies are universal for the exploitation of various kinds of raw materials. The variety of exploited natural resources has an impact on the range of processes that allow separation of unusual raw materials from the gangue. The nature of the utilised geological elements influences the type, properties and quantity of by-products produced in these processes. The most complex processes concern metal ores. Metals or their minerals often appear as fine grains in the rock mass of the geological substrate. Flotation, enrichment in the water environment, is an activity that allows the separation of economically valuable mineral geological elements. As a result of flotation, hydrated fine-grained tailings remain as a by-product.

LAND SURFACE DEFORMATION CAUSED BY MINING ACTIVITIES

Mining is an industry for which many types of impact can be identified, practically on all elements of the environment (Spitz & Trudinger 2008; Popiołek 2009; Korkmaz *et al.* 2011; Jain *et al.* 2016; Mancini & Sala 2018; Boldy *et al.* 2021):

- land – both surface and rock mass: soil stability, natural hazard, land pattern, landscape changes;
- atmosphere – including air quality as a result of gases and dust emission, as well as noise;
- water environment – both underground and surface;
- biosphere – including flora and fauna life conditions, habitats, biodiversity and ecosystem services;
- people – including their economic and occupational status, social amenities and relationships, social patterns and lifestyle, health, personal security, legal conditions.

The scope, type and magnitude of impacts on individual elements of the environment depends on the type of mining activity and broadly understood geological and mining factors, including the size, shape and type of the deposit, methods of exploitation, including mining, transport, processing and enrichment of raw materials.

The scope of environmental impacts depends not only on the activities/operations related to mining but also on the condition, type and quality of the environment in which this activity is conducted. This affects both the scope and nuisance of impacts, as well as the possibility of their evaluation and analysis.

The legal environment imposing greater or lesser care to avoid adverse impacts is also important here. Currently, it is standard to use environmental impact assessment tools already at the mine design and construction stages. Even then, operations are planned at all locations of the mine's operation, including the scene after its liquidation.

Other methods of classifying the environmental impacts of mining have also been used. One of the most popular is the division into direct and indirect influences. The first ones include transformations related to the exploitation itself, such as the removal of plant and soil layers dug in the trenches (in the opencast method), or excavation of workings and the deformation of the land caused by them (in the underground process). This group also includes activities related to the emission of pollutants as a result of the operation of machines, noise, etc. All other impacts are indirect. Some of them may be sudden, such as geomechanical phenomena: mining tremors, landslides and related damage to buildings, structures and surface infrastructure, as well as various types of accidents and disasters. However, a significant part of indirect impacts is changes of a long-term nature, often difficult to quantify. Changes in the aquatic environment are incredibly important, including lowering the groundwater level, the formation of floodplains and flooding, changes in water quality as a result of underground water discharge or leachate from landfills, or secondary deformation of the area related to, for example, drainage or ground watering. Very important, although often difficult to measure, are various types of environmental pollution, as well as a whole range of phenomena related to changes in the natural environment caused by mining activity or its effects. A separate, significant issue is the social aspects of the impact of mining. The scope and intensity of the different types of impacts can vary greatly. As a rule, however, indirect effects are usually felt in a much larger area than the operation itself, their effects are felt for a long time, and their prevention and possible repair or compensation are significantly difficult. Furthermore, we usually deal with complex phenomena; various types of interactions are intertwined, creating difficulties both for their analysis and, above all, for the methods of limiting unfavourable influences or their compensation (Spitz & Trudinger 2008; Popiołek 2009; Korkmaz *et al.* 2011).

The stage of mining activity is crucial for the analysis of the environmental impact, as well as for the application of methods of limiting unfavourable effects. Traditionally, there are the design and construction phases of mines (reaching the target capacity), the operational phase, the closure and liquidation of the mine, and the post-exploitation phase related to reclamation and possible revitalization of converted areas.

The transformed areas on the surface are in a direct or indirect way the result of mining activities, affecting the natural environment. This applies in particular to areas where reclamation processes are planned or commenced, as well as those that have not been rehabilitated or the effects have proved insufficient.

IMPACTS OF OPENCAST MINING ON THE SURFACE

Opencast mining means the transformation of the entire area of land in the vicinity of the selected deposit. Before the start of opencast mining, it is necessary to remove all surface objects, plant cover, the soil layer, and finally the remaining rocks of the overburden. Due to the need to maintain a safe slope and shape of the slopes, the surface of the excavation is obviously larger than the surface of the deposit to be exploited. The removed earth or rock masses are first deposited outside the exploitation area, creating an external dump.

In the next stages, if it is possible (e.g. when exploiting seam deposits), the removed overburden may be deposited in the excavation, thus reducing its volume. Such action means a complete change of the landscape, often associated with the liquidation of localities and infrastructure, as well as natural forms and habitats, sometimes shaped for thousands of years. After the end of exploitation, the area of the exploitation itself and the dumps require reclamation activities aimed at geotechnical stabilization, ensuring safety, as well as reconstruction of the biological layer and environmental functions (depending on the adopted reclamation direction).

During opencast mining, gas and dust emissions to the atmosphere must also be taken into account. The type of pollution and its intensity depends both on the adopted mining system, mining method, and means of transport, processing methods and installations, as well as operations related to transport and tipping of earth or rock masses. The most important sources of such contamination include blasting works, mechanical mining and grinding processes, as well as transport and secondary emission from dumps or temporary storage of materials (Chaulya 2004; Jiang *et al.* 2017; Trechera *et al.* 2021). The most important types of dust pollutants are fine dust, the content of which in the air is expressed by the PM2.5 and PM10 indicators. The essential gaseous pollutants include nitrogen oxides, sulphur dioxide, carbon monoxide, and volatile aromatic hydrocarbons. These types of emissions to the atmosphere, although they also occur during underground mining, have a much greater range in the case of opencast mines, depending mainly on the topography and wind strength.

The impacts of opencast mining on the aquatic environment are particularly important. Very often, especially in the case of deposits located over a large area (such as Polish lignite deposits), it is necessary to drain, i.e. continuously pump out water from aquifers. This process is essential for safe operation. It creates an area of water scarcity with an area much larger than the surface of the workings. This harms both the agricultural economy and the natural environment at a distance of up to several dozen kilometres from the site of exploitation.

IMPACTS OF UNDERGROUND MINING ON THE SURFACE

As a result of underground mining, the land surface is also transformed, although in a way different form that in the case of opencast mining. Three types of such transformations can be distinguished: continuous deformations, discontinuous deformations and various types of earth structures erected to deposit waste resulting from the operation or processing and enrichment of raw materials. They can progress slowly, automatically, with the progress of exploitation – such as continuous deformations (subsidence trough), or they can be sudden – as in the case of landslides or sinkholes. Sometimes there is more than one cause of deformation, in which case we are talking about indirect influences, the range of which may be greater than the above-mentioned direct deformations. The factors that strengthen and initiate deformations to the greatest extent are changes in water relations. These include, first of all: drainage and waterlogging as a result of pumping groundwater, particularly severe in the case of opencast mining, as well as leaching and washing – creating underground voids and local water reservoirs. Dehydration deformations can be both continuous and discontinuous. Tremors of the rock mass caused by mining activities, including blasting works, or those related to the rapid discharge of accumulated stresses in the rock mass resulting from the violation of the original state of equilibrium, can also be a very important initiator and strengthening factor (Hartman 1992; Spitz & Trudinger 2008; Popiołek 2009; Jianjun *et al.* 2012; Jiang *et al.* 2017; Adelsohn *et al.* 2020).

Continuous deformations occur where the exploitation of deposits is carried out with large horizontal dimensions (areas) and small vertical dimensions concerning the depth of exploitation. Continuous deformations are mild depressions of the terrain in the form of troughs (load trough) with a range significantly exceeding the contours of exploitation (Popiołek 2009). They are characteristic of large-scale exploitation – most often of seam deposits. They occur in all regions of underground coal and copper ore mining in Poland. The size, range and shape of the load troughs depend on the thickness of the selected seam, the depth of mining, the type and properties of the overlying rocks (rock mass), and the progress of mining. It is characteristic that deformations result from the cumulative exploitation of the seams lying one under the other. As a result, in regions with a long mining tradition, e.g. in Upper Silesia in Poland, the cumulative surface losses in many places exceed 20 meters, reaching up to 40 m locally (Ostrowski 2001). Such deformations can cause various types of damage to buildings, structures and surface infrastructure, especially continuous ones (e.g. pipelines). Continuous deformations, however, are predictable. The deformation parameters allow determining the so-called category of terrain deformation, as well as carrying out not only repair but also preventive actions. Continuous deformations may also cause disturbance of surface watercourses, including local flooding and changes in wetland ecosystems.

Another type of rock mass deformation is the so-called discontinuous deformation of the terrain surface in the form of cracks, crevices or sills (linear deformations). Various kinds of sinkholes (point deformations) or landslides (surface deformations) can occur, although the latter is much more common in opencast mining. Discontinuous deformations, although associated primarily with shallow underground mining, also appear as a result of mining even at considerable depths, currently characteristic of Polish mining (Kotyrba & Kortas 2016; Malinowska *et al.* 2019).

Land surface deformations are often the first herald or initiator of a series of impacts that result from mining activities, listed at the beginning of this chapter.

REFERENCES

Adelsohn E., Iannacchione A. & Winn R. 2020: Investigations on longwall mining subsidence impacts on Pennsylvania highway alignments. International Journal of Mining Science and Technology, 30 (1), 85–92.

Boldy R., Santini T., Annandale M., Erskine P.D. & Sonter L.J. 2021: Understanding the impacts of mining on ecosystem services through a systematic review. The Extractive Industries and Society, 8 (1): 457–466.

Chaulya S.K. 2004: Assessment and management of air quality for an opencast coal mining area. Journal of Environmental Management, 70 (1), 1–14.

Hartman H.L. (ed.) 1992: SME mining engineers handbook. Society for Mining, Metallurgy & Exploitation, Littleton.

Jain R.K., Cui Z. & Domen J.K. 2016: Environmental impact of mining and mineral processing. Management, monitoring and auditing strategies. Elsevier, Butterworth-Heinemann.

Jiang N., Zhou C., Lu S. & Zhang Z. 2017: Propagation and prediction of blasting vibration on slope in an open pit during underground mining. Tunnelling and Underground Space Technology, 70, 409–421.

Jianjun S., Chunjian H., Ping L., Junwei Z., Deyuan L., Minde J., Lin Z., Jingkai Z. & Jianying S. 2012: Quantitative prediction of mining subsidence and its impact on the environment. International Journal of Mining Science and Technology, 22 (1), 69–73.

Korkmaz H., Çetin B., Ege İ., Karataş A., Bom A. & Özşahin E. 2011: Environmental effects of stone pits in Hatay (Turkey). Procedia Social and Behavioral Sciences, 19, 504–510.

Kotyrba A. & Kortas Ł. 2016: Sinkhole hazard assessment in the area of abandoned mining shaft basing on microgravity survey and modelling – Case study from the Upper Silesia Coal Basin in Poland. Journal of Applied Geophysics, 130, 62–70.

Malinowska A.A., Witkowski W.T., Hejmanowski R., Chang L., van Leijen F.J. & Hanssen R.F. 2019: Sinkhole occurrence monitoring over shallow abandoned coal mines with satellite-based persistent scatterer interferometry. Engineering Geology, 262, 105336.

Mancini L. & Sala S. 2018: Social impact assessment in the mining sector: Review and comparison of indicators frameworks. Resources Policy, 57, 98–111.

Ostrowski J. (ed.) 2001: Ochrona środowiska na terenach górniczych. Wydawnictwo Instytutu Gospodarowania Surowcami Mineralnymi i Energią PAN, Kraków.

Piechota S. 2008: Technika podziemnej eksploatacji złóż i likwidacji kopalń. Uczelniane Wydawnictwa Naukowo-Dydaktyczne AGH, Kraków.

Popiołek E. 2009: Ochrona terenów górniczych. Wydawnictwa AGH, Kraków.

Spitz K. & Trudinger J. 2008: Mining and the Environment. From Ore to Metal. Taylor & Francis Group, London.

Trechera P., Moreno T., Córdoba P., Moreno N., Zhuang X., Li B., Li J., Shangguan Y., Dominguez A.O., Kelly F. & Querol X. 2021: Comprehensive evaluation of potential coal mine dust emissions in an open-pit coal mine in Northwest China. International Journal of Coal Geology, 235, 103677.

1.3. Hydrological and Hydrochemical Conditions for the Reclamation of Anthropogenic Water Bodies and Wetlands in Opencast Mines

Joanna Kidawa
Institute of Earth Sciences,
Faculty of Natural Sciences, University of Silesia in Katowice;
60 Będzińska Str., 41-200 Sosnowiec, Poland;
joanna.kidawa@us.edu.pl; ORCID: 0000-0001-6479-3108

Damian Chmura
Institute of Environmental Protection and Engineering,
Faculty of Materials, Civil and Environmental Engineering, University of Bielsko-Biala;
2 Willowa Str., 43-309 Bielsko-Biała, Poland;
dchmura@ath.bielsko.pl; ORCID: 0000-0002-0242-8962

Tadeusz Molenda
Institute of Earth Sciences,
Faculty of Natural Sciences, University of Silesia in Katowice;
60 Będzińska Str., 41-200 Sosnowiec, Poland;
tedimolenda@interia.pl; ORCID: 0000-0003-3000-8460

ABSTRACT: Studies on waters and a botanical survey were carried out in 2014 in the territory and in the vicinity of the three opencast mines: "Górażdże" (a limestone quarry), "Nowogród Bobrzański" (a sand mine) and "Wójcice" (a gravel mine). It was observed that the chemical composition of waters in post-exploited reservoirs and wetlands is determined by the type of rock that is being exploited. The hydrological conditions (the degree of hydration of the excavation, the morphometric parameters of water body, etc.) also contribute to the type of vegetation and the direction of spontaneous succession in post-exploited areas. The proper choice of reclamation should be performed using a modified succession that is adapted to specific hydrological-hydrochemical conditions. In the case of flow-through-containers, it should provide a good quality of water using special technical solutions, i.e. an Olszewski tube. The rehabilitation of slurry clarifiers in which sand is a waste requires a modified succession and the use of stress-tolerating plants that can grow under conditions in which there is a lack of water and low fertility of the substratum such as a forecrop. The vicinity near post-exploited water bodies and excavations on limestone ought to undergo reclamation into wetlands or artificial water bodies due to the flooding of the floor of excavation. This enables the creation of biotopes for fens and aquatic plant communities.

KEYWORDS: Quarry, sand pit, primary succession, spontaneous vegetation, opencast mine

INTRODUCTION

The geological background is the main element influencing the course, pace, dynamics, composition and structural effects of natural processes on sites which have emerged as the by-products of mineral resources exploitation. The habitats established on these sites have varied conditions in terms of moisture. Apart from the terrestrial sites like heaps, there are also wetland and water habitats provided during the processes of post-industrial, post-excavation establishment of sites. In this way the geology determines the hydrological and hydrochemical conditions of anthropogenic water bodies and wetlands in opencast mines.

The studies that were conducted earlier indicated that the direction of natural processes, colonization of post-excavation mineral pits by plant species and vegetation, is mainly determined by the type of rock and mineral resources that were exploited in a particular mine (Nowak 2005; Nejfeld 2007; Molenda *et al.* 2013; Kasprowska-Nowak & Beczała 2014; Beczała & Dorda 2015). Flooded excavations in which the waters are rich in calcium and magnesium ions have the appropriate conditions for forming unique mires with calciphilous plant species. Some of the calciphilous species are protected plant species (Stebel & Błońska 2010; König 2017; Błońska *et al.* 2020).

In sand pits and gravel excavations in which the hydrological-hydrochemical conditions are different or lack of water occurs, other type of vegetation do well (Woźniak & Sierka 2004; Nowak & Hebda 2006; Chmura *et al.* 2013). Temperate sand grasslands and therophyte swards thrive in dry places (Czylok & Szymczyk 2009; Rahmonov & Szymczyk 2010). On wetter sites reed beds and tall-sedge vegetation develop (Chmura & Molenda 2007). The choice of the appropriate type of reclamation should be preceded by a detailed understanding of the hydrological and hydrochemical conditions in an excavation area (Molenda & Kidawa 2020). The kind of soil-substratum in post-exploited areas is also essential. The spontaneous process of succession under natural conditions can proceed very slowly and does not always occur over the entire area of excavation. Therefore, the reclamation of these sites is necessary. The introduction of plant species and the creation of new biotopes must conform to the habitat requirements of these species and the abiotic conditions of a specific site (Řehounková *et al.* 2011). Even when rehabilitation of a given site meets these criteria, it is still possible that the course of succession will be modified compared to succession under natural conditions. We chose three types of opencast mines that differed in a variety of exploited materials. We had water bodies that were created during natural subsidence succession or were formerly existing water reservoirs. The aim was to check whether there were links between abiotic properties of water bodies and characteristics of post-mining vegetation.

OBJECTIVES

The main aim of the research was to answer the following questions:

- What are the hydrographic and hydrochemical relations of excavations in limestone, sand and gravel mines?
- What are the morphometric attributes of the excavations in limestone, sand and gravel mines?
- What is the direction of the natural primary succession on these sites?
- What are the best methods for reclamation of the specific excavations in terms of their hydrological, hydrochemical and vegetation conditions?

STUDY SITES

Three different opencast mines, all of which are located in Poland, were chosen for the purpose of the study: "Górażdże" limestone quarry, "Wójcice" gravel mine; "Nowogród Bobrzański" sand mine (Fig. 1). In the "Górażdże" mine the reservoirs and wetlands that were situated on the floor of the excavation were studied. In the "Nowogród Bobrzański" mine the study focused on the post-exploitation water bodies and their surroundings and in the "Wójcice" mine sedimentation pools and slurry heaps were the subject of research.

METHODS

The underlying physical and chemical properties of the waters (temperature, pH, electrical conductivity, oxygenation and redox potential) were measured. Measurement in the field was performed by using an EDS 6600 multiparameter probe from the US company YSI. Before the measurements, the probe was calibrated using standard solutions. The water parameters studied were determined in water columns at 0.5 m ("Nowogród Bobrzański"). The profile was located in the deepest parts of the reservoir. In the remaining water bodies ("Górażdże", "Wójcice"), the measurements were conducted in the coastal zone. Additionally, water transparency was determined using a Secchi disk (SD).

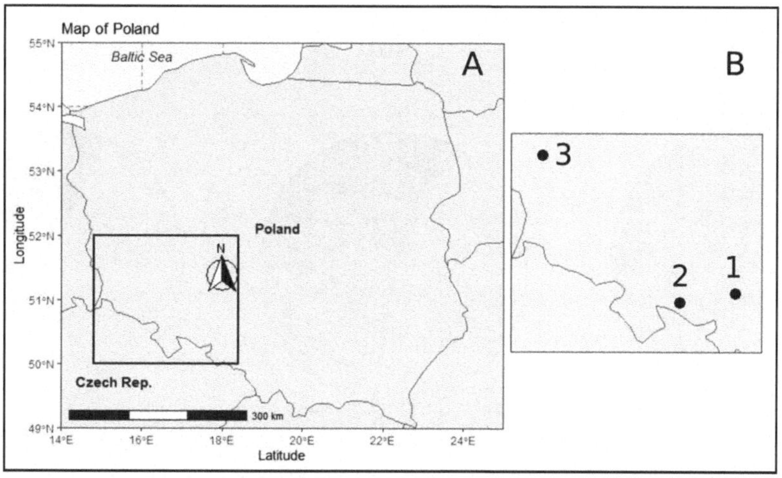

Figure 1. Situation of the studied opencast mines "Górażdże" (1), "Wójcice" (2) and "Nowogród Bobrzański", (3) in the western part (B) of Poland (A)

During field research, water samples were collected in polyethene bottles (0.5 L) for chemical analyses in the laboratory. Samples that were collected represented the surface layer of water (0.5 m). The water surface layer was collected using a telescopic boom. Transport of water samples to the laboratory was conducted at a temperature of +4°C. The samples were filtered using a 0.45 mm filter (Millipore) before analyses. Laboratory analyses included the determination of the major cations and anions in the samples: Ca^{2+}, Mg^{2+}, Na^{+}, K^{+}, NH_4^{+}, HCO_3^{-}, SO_4^{2-}, Cl^{-}, NO_3^{-} and PO_4^{2-}. The analyses were performed using a Metrohm 850 Professional IC ion chromatograph. The hydrochemical type was determined according to the classification by Szczukariew-Prikłoński (Macioszczyk & Dobrzyński 2002).

The botanical survey included mapping of vegetation in the vicinity of the post-exploited reservoirs or sedimentation pools and slurry heaps, using the Braun-Blanquet approach (Westhoff & Van Der Maarel 1978). The distribution of rare and protected species in relation to the geological characteristics of the mineral excavated within the phytocoenoses was studied. The description of succession was done using the chronosequence method (Johnson & Miyanishi 2008). Information about the hydrochemical and hydrological conditions and vegetation composition was collected. The ana lyses enabled us to propose reclamation practices for re-establishment of wetland ecosystems for sites on which opencast mines are operating.

RESULTS AND DISCUSSION

Hydrological conditions of excavations

In limestone excavations, the waters are rich in calcic, sulphate and bicarbonate ions (Fig. 2). The ponds that are located in the vicinity of the "Górażdże" mines are examples of these types of water bodies. The post-exploitation reservoirs that formed in mineral aggregate excavations (the "Nowogród Bobrzański" mine and the pool in "Wójcice") are distinct in terms of the physical-chemical properties of waters. Their waters are three or four ionic and more similar to each other, characterized by higher participation of bicarbonate ions magnesium and sodium (Fig. 2).

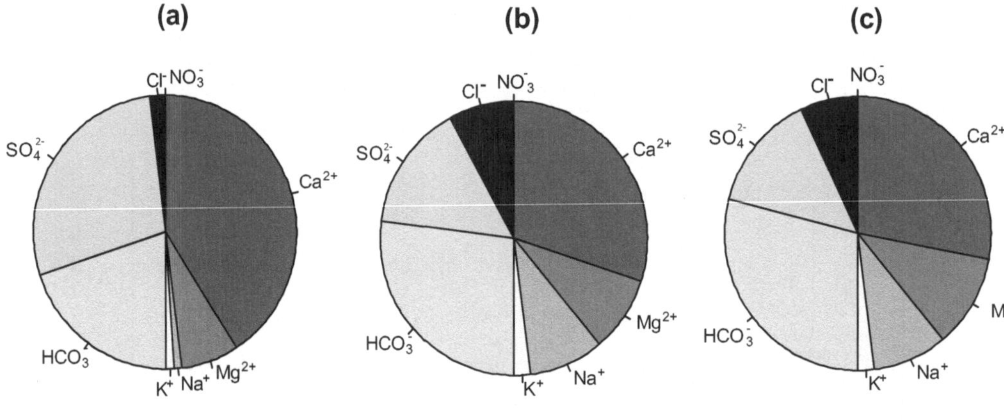

Figure 2. Chemical composition of the waters in the post-exploitation reservoirs in "Górażdże" (a),"Wójcice" (b), "Nowogród Bobrzański" (c)

In "Górażdże" the water in these reservoirs is representative of a bicarbonate-sulphate-calcic-magnesium type of water (HCO_3^- - SO_4^{2-} - Ca^{2+} - Mg^{2+}). Thus, they are four ionic-type waters, and they differ from the waters that were observed in the sump of the "Górażdże" mine in which the waters are representative of a sulphate-bicarbonate-calcic type of water (SO_4^{2-} - HCO_3^- - Ca^{2+}). Within one excavation (mine) site, profound differences in the hydrochemistry of water were recorded. The mineralization of the water can also be significantly different. There are significant differences in mineralization of the water among the ponds that were situated in the vicinity of the "Górażdże" opencast mine, which results from differentiation in the water supply of these sites. The ponds that were mainly supplied by underground waters were characterized by higher mineralization (> 0.6 g/L). In contrast, the shallow water bodies that have precipitation as the dominant type of water supply were typified by lower mineralization of the waters (< 0.6 g/L). The highest mineralization was found in the underground waters of the sump (> 0.7 g/L). In "Wójcice" and "Nowogród Bobrzański" the waters of these reservoirs were characterized by lower mineralization (0.2 g/L).

Vegetation in the vicinity of reservoirs

The chemistry of water bodies has a major impact on riparian vegetation. In the immediate vicinity of the reservoirs of each of the mines, reed beds of the *Phragmition* alliance, i.e. patches of plant associations: *Phragmitetum australis*, *Typhetum latifoliae*, *Scirpetum lacustris*, *Sparganietum erecti*; however, the surrounding vegetation differed significantly. In the area of the "Górażdże" mine, in spite of cultivation of Scots pine *Pinus sylvestris* and larch *Larix decidua*, species of broad-leaved forest prevail on the forest floor (the *Fagetalia* alliance) such as: *Anemone nemorosa*, *Viola reichenbachiana*, and *Primula elatior*. However, the tree stands are dominated by coniferous trees and birch *Betula pendula* due to previously conducted land reclamation. Moreover, the amount of invasive alien boxelder *Acer negundo* and black alder *Alnus glutinosa* is still increasing (Kacprzak & Bruchal 2011). These forests grew in beech forest habitats. Potential natural vegetation in this area is best illustrated by the vegetation in the "Kamień Śląski" reserve (Spałek 2003) which is located in the surrounding area. Close to the bank zone in an adjacent area of rushes there are patches of initial fens with *Epipactis palustris* and *Centaurium erytharea* of the *Scheuzcherio-Caricetea nigrae* (Fig. 3).

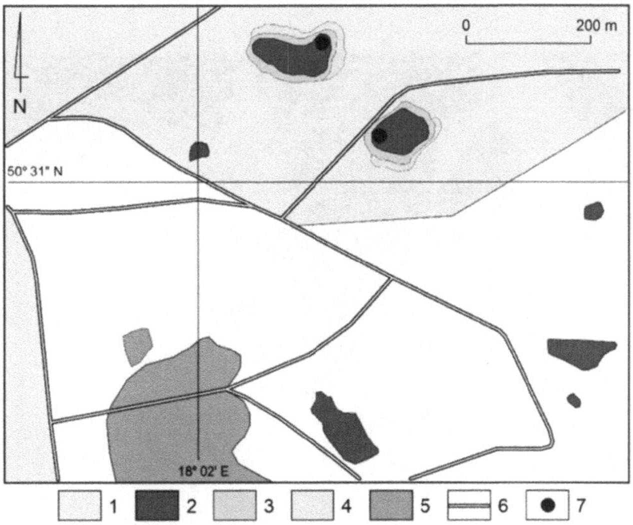

Figure 3. Localization of investigated objects and vegetation in the limestone quarry "Górażdże". Explanations: 1 – woodland/thickets, 2 – reservoirs, 3 – rushes (*Phragmition*), 4 – initial fen communities (the *Scheuzcherio-Caricetea nigrae* class), 5 – forest, 6 – roads, 7 – sampling points

In the "Wójcice" mine area, the vegetation developed on two different sites: at the bottom of the sedimentation pools and slurry heaps, mainly built by sand. Along the banks of water bodies rushes of the *Phragmition* alliance grew. The only vegetation with high contribution of trees was at the border of the territory of the mine. These are disturbed patches of species-poor oak-hornbeam forest *Tilio-Carpinetum* (Fig. 4). In this forest, apart from *Q. robur*, the codominant tree was *Robinia pseudoacacia* of North American origin. Several species found within the area of the mine originated from this forest. However, a majority of the area of the mine is almost non vegetated or with only initial development of sand grasslands. The most frequent plants on sand heaps were *Saponaria officinalis*, *Oenothera* spp., and *Senecio viscosus*. The lack of water and nutrients in the soil substratum seemed to be limiting factors for growth of plants on slurry heaps. In remaining areas on the flat sites *Salix* scrub developed (thickets of *Salix purpurea*, *S. fragilis*).

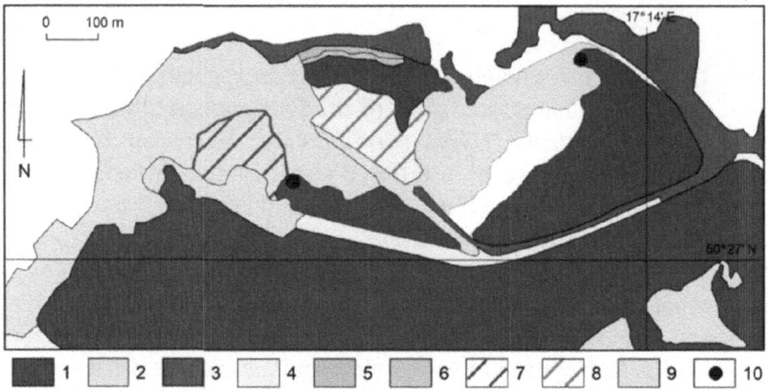

Figure 4. Localization of investigated objects and vegetation in the gravel mine "Wójcice". Explanations: 1 – reservoirs and the slurry clarifiers, 2 – old slurry heaps, 3 – degenerated oak-hornbeam forest, 4 – the contemporary dry slurry clarifier, 5 – meadow, 6 – ecotone, 7 – the area of succession with *Saponaria officinalis*, *Oenothera* spp., 8 – the area of succession with willow (*Salix* spp.), 9 – rushes (the *Phragmition* alliance), 10 – sampling points of water

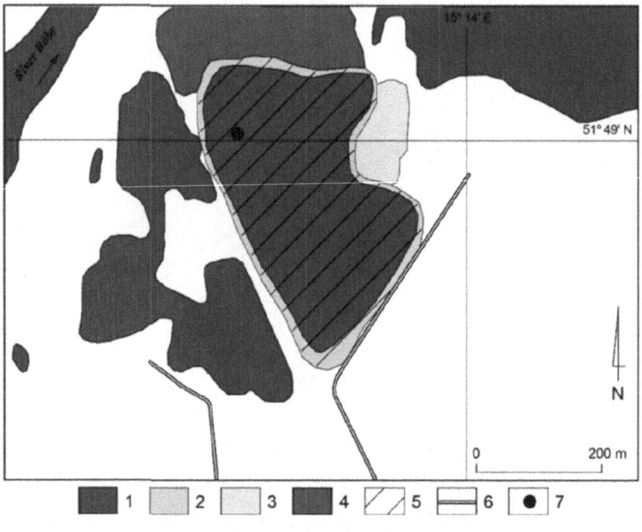

Figure 5. Localization of investigated objects and vegetation in the sand mine "Nowogród Bobrzański". Explanations: 1 – reservoirs, 2 – rushes (*Phragmition*), 3 – sand grasslands and therophyte swards (the *Koelerio-Corynephoretea* class), 4 – floodplain forest (the *Alno-Ulmion* alliance), 5 – reservoir, 6 – roads, 7 – sites of profile studies and sampling points of water

In the area of the "Nowogród Bobrzański" mine, there were patches of temperate sand grasslands and therophyte swards (the *Koelerio-Corynephoretea* class) in the sandy sites close to two main artificial water reservoirs (Fig. 5). In a further zone of the reservoirs there was a degenerated floodplain forest of the *Alno-Ulmion* alliance with abundant populations of the invasive alien species *Impatiens parviflora* and *I. glandulifera*, however, other dominant species included native *Iris pseudacorus*, *Urtica dioica*, and *Chaerophyllum aromatcium*. The causeway was planted with trees, mostly monumental individuals of oaks: red oak *Quercus rubra* of alien origin and native European oak *Q. robur*. In this area more than 120 vascular plant species were found including the Red List species *Potentilla rupestris* (Burda *et al.* 2016).

Management of excavations after the exploitation of limestones and dolomites (a study of "Górażdże" mine)

The most interesting ecosystems are associated with the transitional zones between the land and a water body. There were several occurrences of protected and rare species, including *Epipactis helleborine* and (not previously registered) *Centaurium erythraea*. There were conditions for shaping of wetland ecosystems (Fig. 6). The occurrence of species was closely associated with the hydrogeochemical conditions. The floor of the excavation pit is permanently wet (the water table during the year is no lower than 0.5 m from the surface area). The waters are rich in calcium ions and magnesium and therefore they are preferred by calciphilous plants.

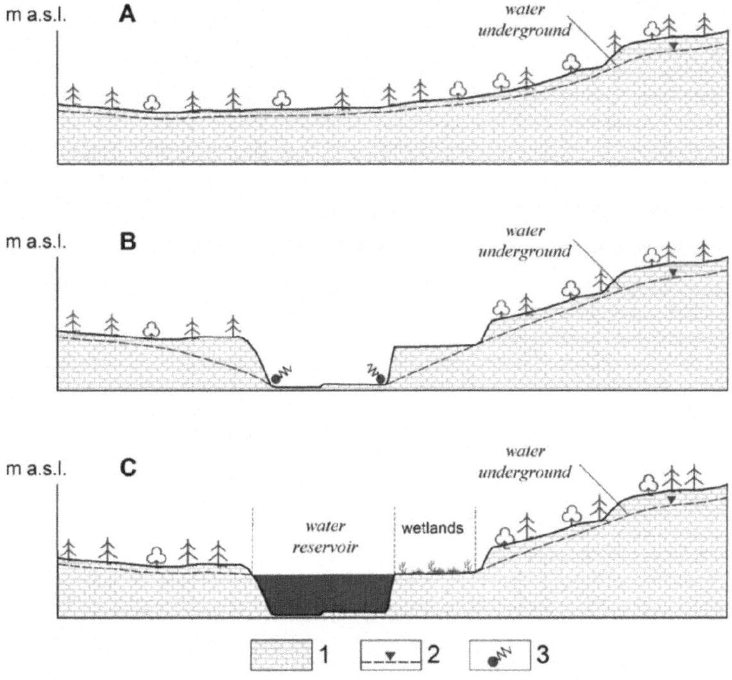

Figure 6. The scheme of water reservoirs forming in excavation of limestones and dolomites. Explanations: A – the state before exploitation, B – the state during exploitation, C – the state after exploitation cessation; 1 – limestones or dolomites, 2 – level of underground waters, 3 – outflows of underground waters

Possible solution – creation of wetland ecosystems

The example of the "Górażdże" mine showed that the reclamation of excavations after the cessation of the exploitation of dolomites and limestones should include the creation of wetland ecosystems. This is possible in excavation pits in which the floor or the excavation overlap with the underground water table. The deeper parts of excavations after the discontinuation of exploitation that are subjected to flooding should be preserved as water bodies, which can provide habitats for precious rare species such as algae (*Charophyta*). Wetland ecosystems can also be created in sand mines.

RECLAMATION OF SLUDGE (THE CASE OF THE "WÓJCICE" MINE)

The exploitation of gravel deposits creates considerable amounts of wastes. These wastes are the fine-grained mineral fractions (loams, dusts, sands), which are formed during the screening and washing of spoil. The wastes are transported in the form of water pulp into the sedimentation pool. After the sedimentation pool is completely filled, it is desiccated and closed. Such an object was situated in the vicinity of the "Wójcice" mine. Fine-grained material is blown around by winds and contributes to an increase in the dusting area. Land reclamation of this type of site is very difficult. Conditions are unsuitable for vegetation development in old, unexploited sedimentation pools. These conditions are connected with:

- a lack of organic material and nutrients;
- extremely harsh water conditions (a lack of groundwater).

Colonization by pioneer plants is ongoing in the vicinity of the "Wójcice" mine in spite of harsh conditions for organisms in the sedimentation pool. The succession has proceeded along two pathways and direction was connected with two things – access to water and relief (Fig. 7). In dry biotopes in the initial stage *Saponaria officinalis* and *Oenothera* spp. prevailed. During succession *Robinia pseudoacacia* starts to appear and finally it can form monospecific stands. In wet, temporarily flooded flat terrain, the successional process of vegetation is similar to those described for the "Błędów" desert (Rahmonov 2007), which resulted in willow scrub. In turn, on the slopes of the "dunes", the final stage of the succession is a black locust forest.

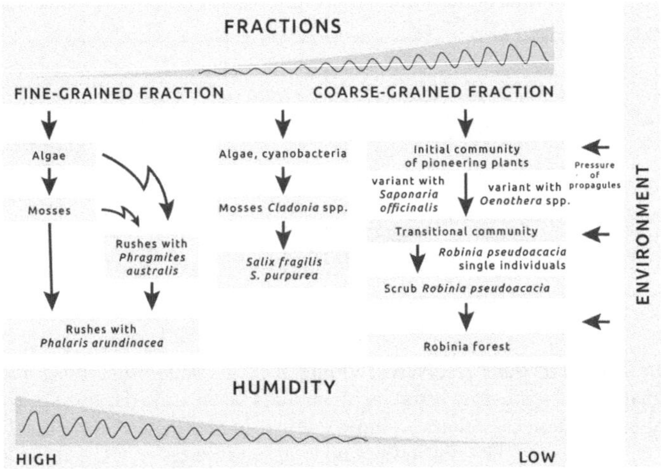

Figure 7. Model of vegetation succession in the "Wójcice" opencast mine and its dependence on moisture of the substratum (according to Kidawa *et al.* (2021))

Solution to the problem of material blown away by wind

The estimated time for the formation of this forest is about 25 years (based on observations by the mine staff members). Some plants that can tolerate extreme stress conditions, i.e. a lack of water and nutrients, can be used as forecrop species during land reclamation. These include: the common soapwort *Saponaria officinalis* and evening primrose *Oenothera* spp.; the first one is especially easy to cultivate and can be sown in sandy habitats. This plant should be used for the rehabilitation of sand heaps. In flat places, it is recommended that willow shrubs, e.g. *Salix purpurea* be planted. Although the aforementioned plants would appear naturally on the site, this process can be speeded up and extended for the whole area of the excavation using the modified succession land reclamation method. Moreover, sowing potential pioneer plants as "forecrop plants" can prevent the so-called arrested succession in which unexpected conditions delay or even stop ecological processes (Boyes *et al.* 2011).

THE PROBLEM OF WATER EUTROPHICATION OF POST-EXPLOITED RESERVOIRS EXEMPLIFIED BY GRAVEL EXCAVATION – "NOWOGRÓD BOBRZAŃSKI"

The contemporary exploitation of resources uses equipment and ladling methods from significant depths, which often go to depths of 45 meters below the water level (Craig *et al.* 2003). As a consequence, deep post-exploited water bodies are formed after the cessation of exploitation. A few good examples of the morphometric diversity of reservoirs are sites in the vicinity of the "Nowogród Bobrzański" mine. One of two water bodies was created as a result of the use of old methods of exploitation. Its maximum depth was 5.8 m and the bottom was quite diversified. In turn, the second one, which was created using contemporary methods, is different. Its maximum depth was 21.0 m and the morphometry of its bottom was weakly diversified. The underwater slopes of the reservoir were very steep (Molenda 2015).

The morphometric differentiation of post-exploitation water bodies has an impact on the structure of aquatic vegetation. In the case of older water bodies, hydrophytes, aquatic submerged macrophytes such as *Myriophyllum spicatum* and *Ceratophyllum demersum* occurred in many areas of the reservoir and formed underwater meadows. In the case of newer water body, the zone of vegetation that was submerged was limited to the narrow coastal belt while in the case of the quarries of solid rocks (granites, limestones); the depth close to the bank was so deep that it prevented any growth and development of plants rooted on the bottom. Taking into account bathymetric plans (Molenda 2014; 2015) and vegetation structure, it can be concluded that depth is the main factor that determines the development of aquatic plants. In the last phase of any exploitation, it is advisable to form the bottom in such way so as to enable the creation of vegetation zones that are characteristic for natural lakes in a separate section of the reservoir, i.e.:

- a water and marsh zone;
- rushes;
- macrophytes with floating leaves;
- macrophytes with submerged leaves.

Deep post-exploitation reservoirs such as new water bodies were vulnerable to the formation of anaerobic zones in the summer. This especially concerns water bodies that are supplied with poor quality water. The problem of water eutrophication arises very quickly in such reservoirs. Eutrophication limits the utilization of these reservoirs for various purposes, mainly recreation. An example of such a site was the newer reservoir in the vicinity of the "Nowogród Bobrzański" mine. The oxygen profile that was depicted indicated that in summer the waters of the hypolimnion were deoxygenated while toxic hydrogen sulfide and ammonia occurred in the demersal zone (Fig. 8).

In the top-most layer (epilimnion), the waters were characterized by a high degree of supersaturation of oxygen (160%), which was the result of the intense processes of photosynthesis, which led to the alkalinisation of the waters. In the summer the water reaction exceeded pH = 9.1, which was a very unfavourable phenomenon because it may have caused a fish kill. Moreover, water transparency as measured using a Secchi disk was low and varied between 1.8 and 1.2 m. Such a low degree of transparency is the result of the intense development of phytoplankton (water bloom). Increasing electrolytic conductivity (EC) was also observed with increasing depth (Fig. 8).

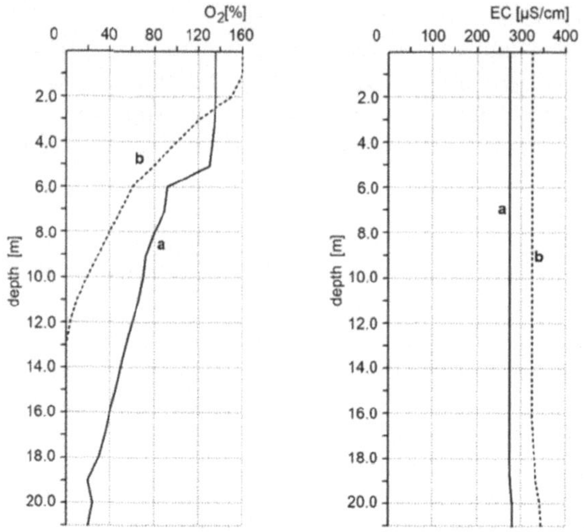

Figure 8. The vertical distribution of electric conductivity (EC) and oxygen saturation in the waters in "Nowogród Bobrzański": a – spring, b – summer

Methods of water quality improvement in post-exploitation reservoirs

In each reservoir that is a flow-through container, the most rational (and inexpensive) method is the installation of an Olszewski tube (Olszewski 1959; 1961). There are numerous examples demonstrating that this solution enables export of nutrients from the deepest layers and prevents euthrophication (Klapper 2003; Kostecki 2012). Such a solution can only be applied in flow-through containers such as in the newer reservoir in the vicinity of the "Nowogród Bobrzański" mine. This type of tube provides leverage. As a result of the separation of the outflow through the installation of a weir, the hydrostatic pressure increases, which pushes the benthic waters. The deoxygenated waters from the demersal zone are removed, which results in a decrease in the anaerobic zones; thus the water quality improves. A permanent improvement in water quality in a reservoir that has a low capacity can occur in as little as one year after the installation of an Olszewski tube. The Olszewski tube should be a permanent element of the management of flow-through containers and it should be installed immediately after the exploitation has ceased.

FINAL REMARKS

The results obtained in the presented studies of three different objects and review of the literature, lead to the conclusion that there is no one universal method of management of opencast water bodies. The high differentiation of hydrological and hydrochemical conditions of the water and wetland sites is the reason for the lack of one universal management method. Therefore, the authors propose various directions of reclamation depending on the type of raw material being mined (Fig. 9).

The results demonstrated that the direction of natural succession of vegetation in opencast mines is ruled by two factors: type of exploited rock and hydrological conditions. They determine the hydrochemistry of waters, which in turn determine the dominant species of aquatic and riparian plants colonising the area and subsequent vegetation development. In limestone and dolomite excavations the most ecologically valuable plants that can appear are species of the genus *Chara* (Owsianny & Gąbka 2007). The plant cover includes species that represent the vegetation type that is protected under the NATURA 2000 Habitat Directive (code 3140). Close to water bodies, in riparian habitats, fens with orchids and other rare plants can occur. In flooded subsidences after gypsum exploitation rare and Red List species, such as common mare's-tail *Hippuris vulgaris* found habitat conditions optimal for growth. This plant has wide-tolerance for calcium ions typical for such type of waters (Gałczyńska 2006) which was proven by transplantation experiments of this species (Gałczyńska & Malas 2010). Depending on local conditions, reclamation can be conducted either for the maintenance of aquatic vegetation and/or wetlands.

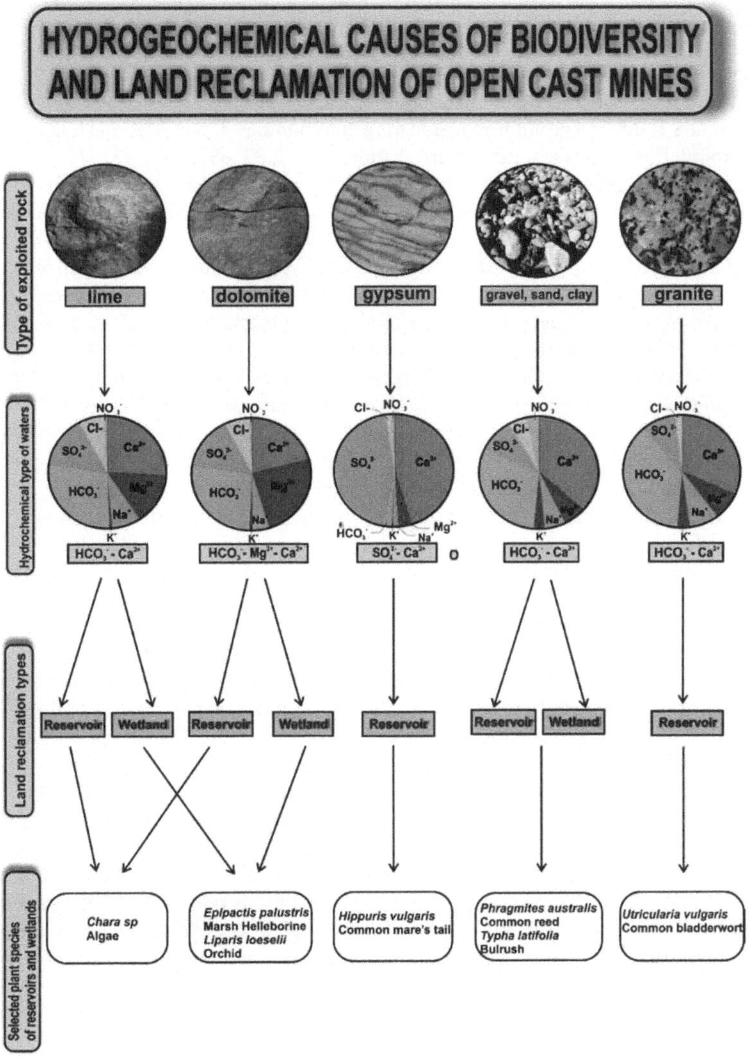

Figure 9. Hydrological-hydrochemical conditions of the reclamation of various opencast mines

REFERENCES

Beczała T. & Dorda A. 2015: Tereny po eksploatacji surowców skalnych ostoją bioróżnorodności – na przykładzie wybranych wyrobisk na Pogórzu Cieszyńskim. Bezpieczeństwo Pracy i Ochrona Środowiska w Górnictwie, (9), 17–25.

Błońska A., Kidawa J., Molenda T. & Chmura D. 2020: Hydrogeochemical conditions of the development of anthropogenic carbonate swamps: A case study of an abandoned Polish sandpit. Polish Journal of Environmental Studies, 29 (1), 561–569.

Boyes L.J., Gunton R.M., Griffiths M.E. & Lawes M.J. 2011: Causes of arrested succession in coastal dune forest. Plant Ecology, 212 (1), 21–32.

Burda E., Jerzak L., Najbar B., Wieczorek M., Walińska K., Nowacka-Chiari E. & Nowacki R. (eds.) 2016: A measures to optimize the habitat conditions within the KSM "Nowogród Bobrzański". Quarry Life Award Project, Zielona Góra. https://www.quarrylifeaward.pl/download-final-report/30026/pl-52_en-optimize_habitat_conditions.pdf, access: 06.02.2020.

Chmura D. & Molenda T. 2007: The anthropogenic mire communities of the Silesian Upland (S Poland): a case of selected exploitation hollows. Nature Conservation, 64 (7), 57–63.

Chmura D., Błońska A. & Molenda T. 2013: Hydrographic properties and vegetation differentiation in selected anthropogenic wetlands. 13th International Multidisciplinary Scientific GeoConference. SGEM 2013 Conference Proceedings, June 16–22, 2013, Volume 1, 555–562.

Craig J.R., Vaughan D.J. & Skinner B.J. 2003: Zasoby Ziemi. Wydawnictwo Naukowe PWN, Warszawa.

Czylok A. & Szymczyk A. 2009: Sand quarries as biotopes of rare and critically endangered plant species. [In:] Mirek Z. & Nikel A. (eds.): Rare, relict and endangered plants and fungi in Poland. W. Szafer Institute of Botany, Polish Academy of Sciences, Kraków, 187–192.

Gałczyńska M. & Malas A. 2010: Ocena redukcji wartości stężeń potasu, wapnia, magnezu i sodu przez przęstkę pospolitą (*Hippuris vulgaris* L.) w roztworach zanieczyszczonych ściekiem miejskim i osadem ściekowym. Zeszyty Problemowe Postępów Nauk Rolniczych, 547, 101–109.

Gałczyńska M. 2006: Impact of chemical properties of water in small reservoirs on development of *Hippuris vulgaris*. Polish Journal of Environmental Studies, 15 (5), 567–570.

Johnson E.A. & Miyanishi K. 2008: Testing the assumptions of chronosequences in succession. Ecology Letters, 11 (5), 419–431.

Kacprzak M. & Bruchal M. 2011: Procesy rekultywacji terenów pogórniczych na przykładzie Kopalni Wapienia "Górażdże". Inżynieria i Ochrona Środowiska, 14, 49–58.

Kasprowska-Nowak K. & Beczała T. 2014: Przemiany krajobrazu w obszarze i otoczeniu piaskowni „Krasna-Bielowiec" (Pogórze Cieszyńskie). Prace Komisji Krajobrazu Kulturowego, 26, 139–154.

Kidawa J., Chmura D. & Molenda T. 2021: The hydrological-hydrochemical factors that control the invasion of the black locust (*Robinia pseudoacacia* L.) in succession in areas with opencast mines. Plants, 10, 40.

Klapper H. 2003: Technologies for lake restoration. Journal of Limnology, 62 (Suppl. 1), 73–90.

König P. 2017: Plant diversity and dynamics in chalk quarries on the islands of Rügen and Wolin (Western Pomerania/Germany and Poland). Biodiversity: Research and Conservation, 47, 23–39.

Kostecki M. 2012: Rekultywacja zbiornika antropogenicznego metodą usuwania hypolimnionu (południowo-zachodnia Polska). Inżynieria i Ochrona Środowiska, 15 (2), 101–117.

Macioszczyk A. & Dobrzyński D. 2002: Hydrogeochemia. Wydawnictwo Naukowe PWN, Warszawa.

Molenda T. & Kidawa J. 2020: Natural and anthropogenic conditions of the chemical composition of pit lake waters (based on example pit lakes from Central Europe). Mine Water and the Environment, 39, 473–480.

Molenda T. 2014: „Górka" – najgłębszy w Polsce (41,7 m) zbiornik poeksploatacyjny surowców budowlanych – charakterystyka morfometryczna. Górnictwo Odkrywkowe, 55 (3), 126–129.

Molenda T. 2015: Conditions for development of anthropogenic meromictic reservoirs in the workings of crystalline rocks (based on the examples of the quarries of the Žulovská pahorkatina, NE Czech Republic). Environmental Earth Sciences, 74 (3), 2259–2271.

Molenda T., Błońska A. & Chmura D. 2013: Hydrochemical diversity of selected anthropogenic wetlands developed in disused sandpits. 13th International Multidisciplinary Scientific GeoConference. SGEM 2013 Conference Proceedings, June 16–22, 2013, Volume 1, 547–554.

Nejfeld P. 2007: Chronione i zagrożone gatunki roślin naczyniowych w nieczynnych kamieniołomach wapienia w Kotlinie Żywieckiej (zewnętrzne Karpaty Zachodnie, Południowa Polska). [In:] Kasza H. & Klama H. (eds.): Zapobieganie zanieczyszczeniu, przekształcaniu i degradacji środowiska, 14, 121–131.

Nowak A. & Hebda G. (eds.) 2006: Biodiversity of quarries and pits. Opole Scientific Society, Opole-Górażdże.

Nowak A. 2005: Anthropogenic habitats as sites of occurrence of endangered, rare and protected plants on the example of Opole Silesia, SW Poland. Thaiszia – Journal of Botany, 15, 155–172.
Olszewski P. 1959: Usuwanie hypolimnionu jezior. Wyniki pierwszego roku eksperymentu na Jeziorze Kortowskim. Zeszyty Naukowe Wyższej Szkoły Rolniczej w Olsztynie, 19 (81), 331–339.
Olszewski P. 1961: Versuch einer Ableitung des hypolimnischen Wassers an einem See. Ergebnisse des ersten Versuchsjahres. Internationale Vereinigung für Theoretische und Angewandte Limnologie: Verhandlungen, 14, 855–861.
Owsianny P.M. & Gąbka M. 2007: Zbiorniki ramienicowe i dystroficzne – cechy diagnostyczne w świetle programu Natura 2000 i przykładów z Lasów Pilskich. Studia i Materiały Centrum Edukacji Przyrodniczo-Leśnej, 9 (2–3), 584–600.
Rahmonov O. & Szymczyk A. 2010: Relations between vegetation and soil in initial succession phases in post-sand excavations. Ekológia (Bratislava), 29 (4), 412–429.
Rahmonov O. 2007: Relacje między roślinnością i glebą w inicjalnej fazie sukcesji na obszarach piaszczystych. Wydawnictwo Uniwersytetu Śląskiego, Katowice.
Řehounková K., Řehounek J. & Prach K. 2011: Near-natural restoration vs. technical reclamation of mining sites in the Czech Republic. Faculty of Science, University of South Bohemia in České Budějovice.
Spałek K. 2003: Flora i roślinność rezerwatu „Kamień Śląski" (Śląsk Opolski). Parki Narodowe i Rezerwaty Przyrody, 22 (1), 71–89.
Stebel A. & Błońska A. 2010: *Moerckia hibernica* (Marchantiophyta) in anthropogenic habitats in southern Poland. Herzogia, 25 (1), 113–117.
Westhoff V. & Van Der Maarel E. 1978: The Braun-Blanquet approach. [In:] Whittaker R.H. (ed.): Classification of plant communities. Springer, Dordrecht, 287–399.
Woźniak G. & Sierka E. 2004: Importance of spontaneous succession in reclamation processes. Zeszyty Naukowe Uniwersytetu Zielonogórskiego, Inżynieria Środowiska, 131, 391–398.

1.4. Impact of Coal Mining Heaps on the Water Environment

Tadeusz Molenda
Institute of Earth Sciences,
Faculty of Natural Sciences, University of Silesia in Katowice;
60 Będzińska Str., 41-200 Sosnowiec, Poland;
tedimolenda@interia.pl; ORCID: 0000-0003-3000-8460

Joanna Kidawa
Institute of Earth Sciences,
Faculty of Natural Sciences, University of Silesia in Katowice;
60 Będzińska Str., 41-200 Sosnowiec, Poland;
joanna.kidawa@us.edu.pl; ORCID: 0000-0001-6479-3108

Damian Chmura
Institute of Environmental Protection and Engineering,
Faculty of Materials, Civil and Environmental Engineering, University of Bielsko-Biala;
2 Willowa Str., 43-309 Bielsko-Biała, Poland;
dchmura@ath.bielsko.pl; ORCID: 0000-0002-0242-8962

ABSTRACT: The influence of the coal mining industry, especially waste landfills, on the water environment was studied at three landfills, situated in the Upper Silesia. Changes that occurred in the surface hydrographic network under the influence of waste deposition were surveyed. Special attention was paid to leachate formation focusing on their hydrographic and hydrochemical characteristics. It has been shown that leachate outflows can be point-like (concentrated), and the efficiency of individual leachates can be several L/s. Leachate waters are characterized by a high degree of mineralization. The concentrations of individual ions, mainly sulphates, are very high and exceed several thousand mg/L.

KEYWORDS: Leachate, hard coal, water pollution, sulphates ions, chlorides ions, magnesium ions

INTRODUCTION

Deep mining of hard coal deposits is associated with the formation of a very large amounts of mineral byproducts, in which two main types are distinguished:

- mining waste which is a mixture of gangue, extracted from accessing and mining works, which are in most cases large rock crumbs;
- tailings (post-flotation wastes) consisting of gangue, mainly from bottom and coal seams, and from overgrowth excavated to the surface (Maciak 1999).

Waste is characterized by high variability of the petrographic composition with the usually dominant share of claystone, silt, sandstone and coal (Greszta & Morawski 1970). Gravel, limestones and tuffite occur sporadically. The share of mining waste in the total mass of waste is 25–40% and processing waste 60–70%, while the amount of mining and processing waste depends on both the geological conditions of deposit and the applied methods of exploitation technology. On average, 1,000 kg of extracted coal is from 400 kg of waste (Greszta & Morawski 1970). Currently, an average of 30 million tonnes of waste is generated annually in the mining industry in Poland.

On the territory of the Upper Silesian Coal Basin (Fig. 1), 219 dumps (post-colliery waste tips) are located within the administrative borders of Poland, and cover an area of over 4,000 ha (Gawor *et al.* 2014). The vast majority of waste dumps are located in the central part of the Upper Silesian Coal Basin. It is estimated that 480 million tonnes of waste have been deposited on them. Waste is usually deposited in the form of oversized heaps, called post-coal mine heaps. Three generations of waste dumps can be distinguished. The oldest (1st generation) were conical dumps with steep slopes and low material density. The relative height of these facilities could exceed 100 meters, e.g. the "Skalny" dump in Łaziska (Chmura & Molenda 2012) and "Szarlota" in Rydułtowy.

Another example of such heaps is the dumping ground in Czerwionka-Leszczyny. Second generation dumps are flat (table), and the waste is more dense. They occupy large areas, which in the case of the largest facilities reach 250 ha. The third generation of landfills, on the other hand, takes the form of "landscape buildings" with specifically shaped forms designed to blend in with the surrounding landscape. An example of such facilities can be a set of four heaps located in the north-eastern part of Bieruń. Waste placement can also be carried out by depositing it in former excavations after exploitation of building materials. Examples of this type of landfill are the "Maczki-Bór" and "Panewniki" landfills located in the excavations of former sand pits. Very often, after filling the excavation, waste storage is carried out and then levelling and above-level landfills are created. In the area of the Upper Silesian Coal Basin, other excavations of construction raw materials, e.g. clay, were used for waste deposition. There are also known cases of waste deposition in subsidence basins, sinkhole funnels and even breeding ponds. Currently, coal mine "Silesia" conducts waste deposition at the place of a closed reservoir, which was used for retention of salt mine waters. Nowadays some of the dumping grounds are treated as so-called anthropogenic deposits, and the material deposited on them can be used for commercial products (Gawor *et al.* 2014). This applies in particular to dumps that have burned out. In this type of landfills, the deposited material has undergone a thermal transformation (thermal metamorphism) as a result of fires occurring inside the landfill. Burned waste is characterized by high resistance to physical and chemical weathering, that is why they are a good material for the production of aggregates used in hydrotechnical and earth construction (Hanak & Porszke 2006). The abandoned waste heaps undergo spontaneous biological and ecological processes such as succession of vegetation (Rostański 2006; Woźniak 2010) in which micro-organisms, fungi, and soil fauna are, along with vascular plants, involved in habitat creation (Woźniak *et al.* 2015; Gucwa-Przepióra *et al.* 2016; Radosz *et al.* 2019). Only single papers are devoted to leachate waters from spoil heaps and their importance in influence on biodiversity (Chmura & Molenda 2007; Chmura *et al.* 2011). Thus, it is important to recognize the quality of these waters. The goal of the study was to characterize hydrochemical composition and determine ion type of leachate waters.

LOCALIZATION OF STUDIED OBJECTS

Three waste landfills after hard coal mining were selected for the study located in the Upper Silesian Coal Basin (Fig. 1). Special attention to research was concentrated on effluent from these landfills. The detailed location of the investigated leachate is described by geographical coordinates: "Smolnica" landfill (50°24'42.07" N, 18°52'45.71" E), "Knurów" waste tip (50°19'65.77" N, 18°67'18.48" E), and "Panewniki" waste tip (50°21'66.1" N, 18°88'30.83" E).

To assess changes in hydrographic conditions that have occurred in connection with the deposition of waste, hydrographic mapping was carried out in accordance with the methodology given by Gutry-Korycka & Werner-Więckowska (1996). The analysis also used cartographic materials, both archival and current, which were used to conduct detailed studies and spatial analyses as well as to present changes in the studied area. All cartographic studies were made by means of Arc-GIS and MapInfo computer software. Samples of leachate water for laboratory analyses were collected in polyethylene bottles with a capacity of 0.5 L. Samples were taken at monthly intervals for a period of one year (n = 12). The transport of water samples to the laboratory took place at a temperature of +4°C. Prior to analysis, the samples were filtered on a 0.45 μm filter (Millipore). Determination of the main cations and anions in water: Ca^{2+}, Mg^{2+}, Na^{+}, K^{+}, SO_4^{2-}, Cl^{-}, NO_3^{-} was performed on an ion chromatograph, Metromet 850 Professional IC. Bicarbonates (HCO_3^{-}) were determined by titration. The hydrochemical type of leachate waters was classified based on the Szczukariew-Prikłoński classification. The statistical analyses were done using the R language and environment (R Core Team 2019) and detailed comparisons of samples were done by means of Kruskal-Wallis tests followed by Conover tests for pairwise comparisons (de Mendiburu 2019) and visualization by the *lattice* package (Sarkar 2008).

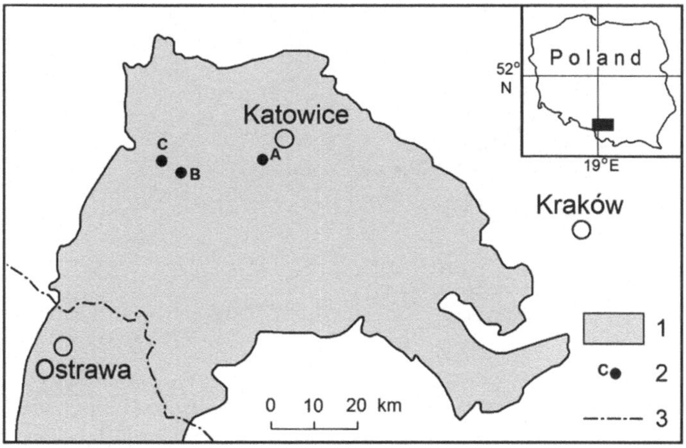

Figure 1. Location of the study area: 1 – Upper Silesian Coal Basin; 2 – Location of the studied objects: A – the mine waste tip "Panewniki", B – the mine waste tip "Knurów", C – the mine waste tip "Smolnica"; 3 – state border between Poland and the Czech Republic

RESULTS AND DISCUSSION

The observed direct changes are associated with the mechanical transformation of the surface hydrographic network, often backfilling the spring sections of river valleys (Fig. 2). There are also known cases of running watercourses in adits under landfills. An example is the coal mine "Borynia" landfill. In some cases, the deposition of waste carried out in the bottom of river valleys (without running adits carrying water) can lead to the accumulation of their waters, and the deposition of waste can also force the relocation of riverbeds and their regulations. Changes of this type are well illustrated by the example of the "Smolnica" landfill. The right-bank tributaries of the Bierawka River have been transferred and flow around the heap with an artificial channel (Fig. 3). Also, the Bierawka River was straightened (the meanders were cut off), and the newly formed riverbed was reinforced by fascine (Fig. 3). Along a section of several hundred meters, the boundary material was also changed, because it was originally made of sand, while now the shore is formed of coal waste.

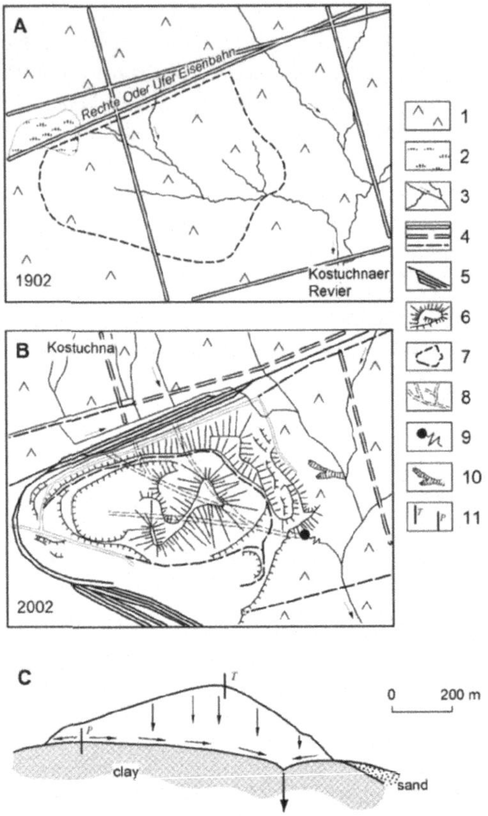

Figure 2. The vertical projection in 1902 (A) and in 2002 (B) and horizontal projection of scheme of water migration within the landfill "Kostuchna", and cross-section through the landfill (C) (according to Molenda (2011)): 1 – forest, 2 – meadows, 3 – water courses, 4 – roads, 5 – railways, 6 – margins of the landfill, 7 – borders of the contemporary landfill on the map from 1902, 8 – drainage system of buried river valleys, 9 – outflow of leachate waters, 10 – erosion cut, 11 – watersheds, T – topographic, P – underground

New linear hydrographic objects may arise around landfills, which are most often blindfolded ditches and canals, and their task is to drain surface runoff water outside the dumping area.

After the end of waste deposition, the negative impacts of heaps on the aquatic environment continues, because the impact of waste dumps on the surrounding environment is also associated with atmospheric precipitation on their surface. Rainwater on the surface of the dumping ground is divided into: surface runoff, subsurface runoff, infiltration, and evaporation.

The percentage share of individual forms of outflow depends mainly on two key factors, which are the type and degree of concentration (fragmentation) of waste. Other factors include the age of the landfill, the technique of waste storage and the slope gradient. The steeper the slope and the compacted material, the greater the proportion of rainfall which is transformed into surface runoff and the smaller the proportion which infiltrates into the landfill. Surface and subsurface runoff occurs episodically in landfills and is associated with rainfall or snowmelt. Most often surface runoff disappears shortly after rainfall.

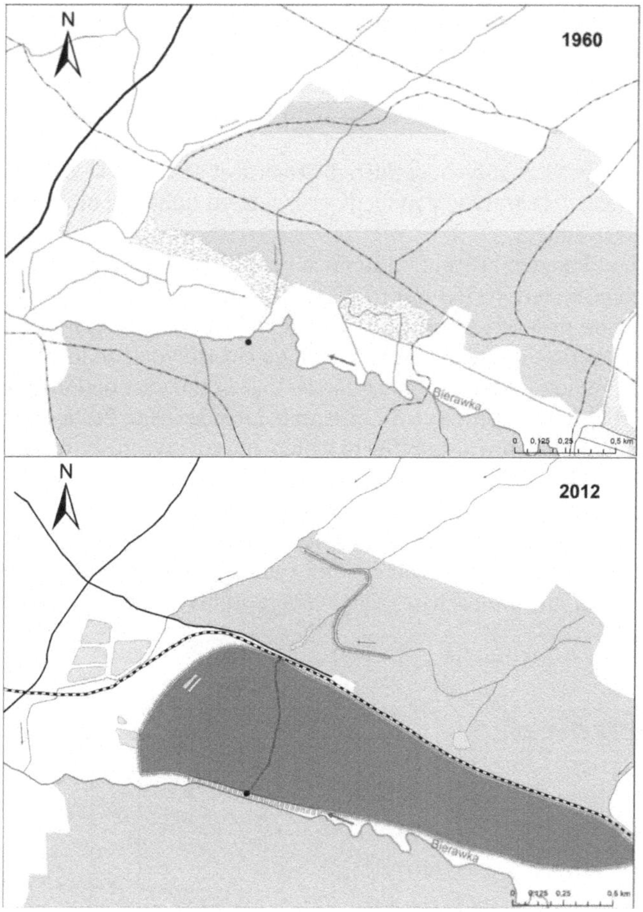

Figure 3. Surface changes of terrain and hydrographic network around the "Smolnica" landfill in 1960 and 2012: 1 – wetlands, 2 – meadows, 3 – landfill, 4 – forests, 5 – ponds, reservoirs, 6 – water courses, 7 – forest paths, 8 – drainage system of buried river valleys, 9 – roads, 10 – railways, 11 – edges morphological, 12 – outflow of leachate waters

Erosion processes are associated with surface runoff. The surface of slope heaps can be covered with numerous erosion gutters. The lack of plant cover favours erosion processes. In the foreground of the landfill, below outlets of the larger erosion gutters, inflow cones develop, which have a characteristic fan shape. In extreme cases, landslide processes can also develop on the surface of heaps. Concentrated surface runoff water can lead to mechanical transformation of riverbeds or create new erosive cuts. Transformations of river channels are mainly manifested by their deepening (deep erosion processes) and widening (lateral erosion processes). Surface runoff waters are heavily loaded with suspended matter, which is why there is deposition of rock debris taken from the landfill in the sections of troughs located in the foreground of the heap. Very often there is a transformation of the granulation of natural bottom sediments.

Rainfall infiltrates into the body of the landfill by gravitational displacement towards the heap base. Under certain morphological and geological conditions of the area where the heap has been built up, infiltration waters may flow out in the form of concentrated leachate (Molenda 2011). This situation occurs when the river valleys were filled during the deposition of waste (Fig. 2). In the case presented in the drawing, the landfill was built on clay soil, i.e. water-impermeable formations. Infiltration waters moving through the landfill heap after reaching the ground cannot further infiltrate the ground (due to an impermeable aluminium layer). They are captured by the drainage system of back-filled river valleys and discharged in the form of concentrated effluent outside the landfill (Fig. 2).

The infiltration water is highly mineralized because it leaches deposited waste as it moves through the landfill heap. Electrolytic conductivity is a good indicator of enrichment of infiltration waters with mineral substances.

In the case of the "Panewniki" and "Smolnica" landfills, the mean values of the electrolytic conductivity of leachate waters varied from 7.07 to 7.92 mS/cm (Fig. 4). Such values testify to the very high contamination of leachate waters with mineral substances. Compared to these landfills, the "Knurów" landfill definitely stands out; its average electrolytic conductivity of leachate waters was 12.73 mS/cm. A review of the literature shows that in the vast majority of landfills, leachate waters have an electrolytic conductivity <10 mS/cm (Molenda 2006; 2011; 2015).

However, objects are known whose effluent waters have an electrolytic conductivity exceeding 16 mS/cm (Molenda 2011; Chmura & Molenda 2012; Bauerek et al. 2017). Changes in electrolytic conductivity in the annual cycle show little variation, and the coefficient of variation (CV) does not exceed 8%. This demonstrates the stabilized geochemical processes occurring inside the landfill heap. The high value of electrolytic conductivity of waters is, among others, a consequence of a very high concentration of sulfates in leachate. The average sulfate concentration varied from 3,225 to 4,972 mg/dm^3 (Fig. 4).

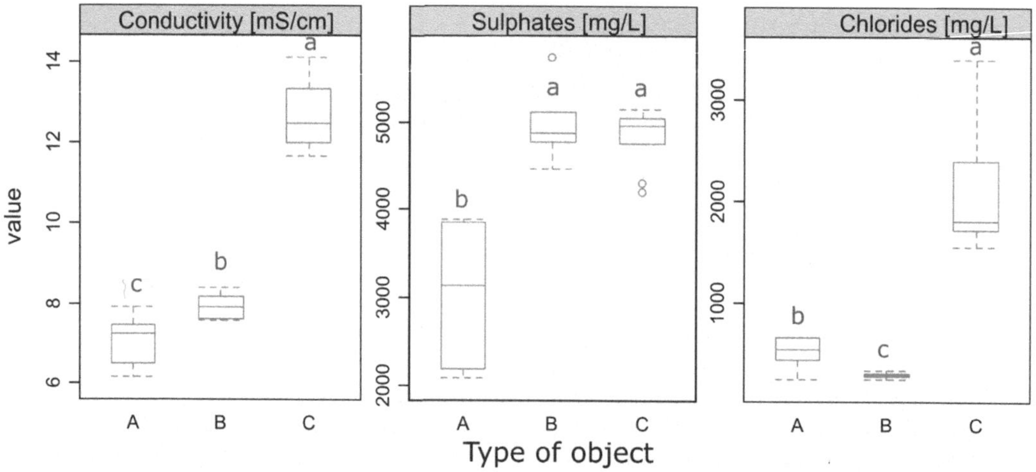

Figure 4. Comparison of mean electric conductivity, concentration of sulphates and chlorides (different letters denote significance among studied objects according to a Kruskal-Wallis test followed by a *post-hoc* Conover test). Abbreviations: A – "Smolnica", B – "Panewniki", C – "Knurów"

Compared to the concentration commonly found in surface and underground waters of this area, these are very high values, which are mainly a result of the weathering process of sulphides contained in the waste, because the S2 anion in a humid environment oxidizes to SO_4^{2-} (Dojlido 1995). The mean concentration of sulphates in leachate waters in the "Panewniki" and "Knurów" landfills was at a similar level and showed no statistically significant differences (Fig. 4). A lower average concentration was found for leachate waters of the "Smolnica" landfill (3,225 mg/dm^3). The sulfate concentrations in the leachate waters of the examined landfills is comparable to those found in leachate from other landfills located in the Upper Silesian Coal Basin (Twardowska *et al.* 1988; Molenda

2006; Bauerek *et al.* 2017). Chlorides also have a high concentration of anions, especially in the "Knurów" landfill (Fig. 4). The average chloride concentration in this landfill reached 2,096 mg/dm^3 and was statistically higher than the other two leachates, in which chloride concentrations were much lower. Chloride concentrations of several hundred mg/dm^3 is characteristic for most leachate waters (Molenda 2006). The source of chlorides are easily soluble chloride salts contained in the waste, and their content in the waste mainly depends on the salinity of the groundwater with which rocks are in contact. Considering the hydrochemical type of leachate waters of the "Panewniki" and "Smolnica" landfills, they represent the three-ion sulphate-sodium-magnesium type (Fig. 5). Leachate waters of the "Knurów" landfill are quaternary sulfate – chloride – sodium – magnesium. It is worth noting that the concentration of magnesium (meq/L) in all leachates was higher than the concentration of calcium, which gives the waters a specific hydrochemical type. This is a rare hydrochemical type because in the vast majority of cases magnesium concentration is lower than calcium concentration (Macioszyk & Dobrzyński 2002).

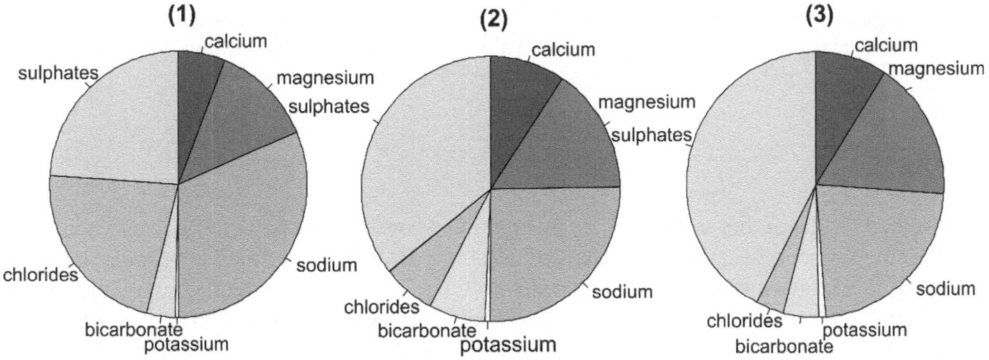

Figure 5. Chemical composition of leachate waters [meq/L].
Abbreviations: (1) – "Smolnica", (2) – "Panewniki", (3) – "Knurów"

Among the physical properties of leachate waters, temperature is an important parameter that should also be noted. Leachate waters that flow out in concentrated form have very stable thermal values (Fig. 6). In the studied outflows, the minimum temperature in winter does not fall below 12°C. It should be emphasized that this is warmer than temperatures found in natural outflows (sources) of this region in winter (Molenda 2014). In summer, the maximum temperature does not exceed 17°C. In the case of leachate from thermally active landfills, the temperature can reach 50°C and change in the annual cycle by only +/–1°C (Chmura & Molenda 2012). For the investigated leachate, the coefficient of variation (CV%) for individual outflows varied from 0.3 to 1.5%. The data presented indicate that stable thermal conditions occur inside the heap, and that the waters of the saturation zone are only slightly affected by atmospheric factors. Greater temperature variation is found in the case of natural sources fed with groundwater (Molenda 2011; 2014).

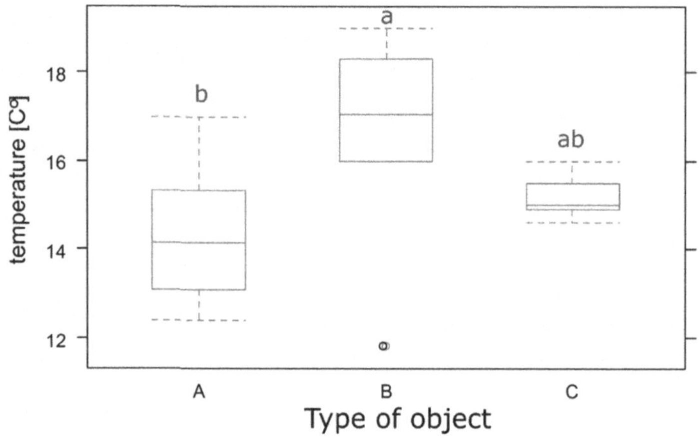

Figure 6. Comparison of mean temperature (different letters denote significance among studied objects according to a Kruskal-Wallis test followed by *post-hoc* Conover test). Abbreviations: A – "Smolnica", B – "Panewniki", C – "Knurów"

SUMMARY

The presented results of this study indicate that mining waste dumps have a very negative impact on the water environment, because it is caused by leaching of substances found in mining waste. Substances that to a large extent contribute to water pollution are mainly mineral compounds: sulfates, chlorides and toxic metals. The negative impact applies to both mechanical changes of the hydrographic surface network and the formation of highly contaminated leachate. Mining activity is one of the economic factors that causes the largest and most diverse and visible transformations of the environment in the Upper Silesian Coal Basin. Due to increasing interest in ecosystem services provided by industrialized sites including colliery waste tips (Woźniak *et al.* 2018) more attention should be paid to the problem of contamination of leachate.

REFERENCES

Bauerek A., Bebek M., Frączek R., Paw K. & Kasperkiewicz W. 2017: Zmienność składu chemicznego kwaśnych wód spływu powierzchniowego z czynnej hałdy odpadów górniczych reprezentujących osady krakowskiej serii piaskowcowej Górnośląskiego Zagłębia Węglowego. Przegląd Geologiczny, 65 (7), 450–458.

Chmura D. & Molenda T. 2007: Nowe stanowisko omiegu górskiego *Doronicum austriacum* Jacq. na Górnym Śląsku. Chrońmy Przyrodę Ojczystą, 63 (1), 20–24.

Chmura D. & Molenda T. 2012: Influence of thermally polluted water on the growth of helophytes in the vicinity of a colliery waste tip. Water, Air & Soil Pollution, 223 (9), 5877–5884.

Chmura D., Molenda T., Błońska A. & Woźniak G. 2011: Sites of leachate inflows on coalmine heaps as refuges of rare mountainous species. Polish Journal of Environmental Studies, 20 (3), 551–557.

de Mendiburu F. 2019: Agricolae: statistical procedures for agricultural research. R Package Version 1.3–1, https://CRAN.R-project.org/package=agricolae.

Dojlido J.R. 1995: Chemia wód powierzchniowych. Wydawnictwo Ekonomia i Środowisko, Białystok.

Gawor Ł., Warcholik W. & Dolnicki P. 2014: Możliwości eksploatacji złóż wtórnych (zwałowisk pogórniczych) jako przykład zmian w sektorze przemysłu wydobywczego. Prace Komisji Geografii Przemysłu Polskiego Towarzystwa Geograficznego, 27, 256–266.

Greszta J. & Morawski S. 1970: Zagospodarowanie nieużytków górnictwa węglowego. Liga Ochrony Przyrody, Warszawa.
Gucwa-Przepióra E., Nadgórska-Socha A., Fojcik B. & Chmura D. 2016: Enzymatic activities and arbuscular mycorrhizal colonization of *Plantago lanceolata* and *Plantago major* in a soil root zone under heavy metal stress. Environmental Science and Pollution Research, 23 (5), 4742–4755.
Gutry-Korycka M. & Werner-Więckowska H. 1996: Przewodnik do hydrograficznych badań terenowych. Państwowe Wydawnictwo Naukowe, Warszawa.
Hanak K. & Porszke A. 2006: Kierunki przeobrażeń minerałów w starych zwałowiskach odpadów powęglowych. Zeszyty Naukowe Politechniki Śląskiej, Górnictwo, 272, 51–60.
Maciak F. 1999: Ochrona i rekultywacja środowiska. Wydawnictwo SGGW, Warszawa.
Macioszyk A. & Dobrzyński D. 2002: Hydrogeochemia strefy aktywnej wymiany wód podziemnych. Wydawnictwo Naukowe PWN, Warszawa.
Molenda T. 2006: Charakterystyka hydrograficzno-hydrochemiczna wypływów wód odciekowych wybranych składowisk odpadów przemysłowych. Zeszyty Naukowe Politechniki Śląskiej, Górnictwo, 272, 95–103.
Molenda T. 2011: Naturalne i antropogeniczne uwarunkowania zmian właściwości fizyczno-chemicznych wód w pogórniczych środowiskach akwatycznych. Na przykładzie regionu górnośląskiego i obszarów ościennych. Uniwersytet Śląski, Wydawnictwo Gnome, Katowice.
Molenda T. 2014: Physical-chemical properties of coal mine waters of old adits and spring waters. Polish Journal of Environmental Studies, 23 (2), 393–399.
Molenda T. 2015: The role of wetlands in the removal of heavy metals from leachate (on the example of the Lipinka River Catchment, South of Poland). [In:] Tomaszek J. & Koszelnik P. (eds.): Progress in Environmental Engineering. Water, Wastewater Treatment and Environmental Protection Issues. CRC Press Taylor & Francis Group, 23–33.
R Core Team 2019: R: A language and environment for statistical computing. R Foundation for Statistical Computing. Vienna, Austria, https://www.R-project.org.
Radosz Ł., Ryś K., Chmura D., Hutniczak A. & Woźniak G. 2019: Rola fauny glebowej w zróżnicowaniu roślinności na zwałowisku karbońskiej skały płonnej. Inżynieria Ekologiczna, 20 (4), 21–28.
Rostański A. 2006: Spontaniczne kształtowanie się pokrywy roślinnej na zwałowiskach po górnictwie węgla kamiennego na Górnym Śląsku [Spontaneous plant cover on colliery spoil heaps in Upper Silesia (Southern Poland)]. University of Silesia, Katowice.
Sarkar D. 2008: Lattice: multivariate data visualization with R. Springer-Verlag, New York.
Twardowska I., Szczepańska J. & Witczak S. 1988: Wpływ odpadów górnictwa węgla kamiennego na środowisko wodne. Ocena zagrożenia, prognozowanie, zapobieganie. Prace i Studia IPIŚ PAN, 35.
Woźniak G. 2010: Zróżnicowanie roślinności na zwałach pogórniczych Górnego Śląska [Diversity of vegetation on coal-mine heaps of the Upper Silesia (Poland)]. W. Szafer Institute of Botany, Polish Academy of Sciences, Kraków.
Woźniak G., Markowicz A., Borymski S., Piotrowska-Seget Z., Chmura D. & Besenyei L. 2015: The relationship between successional vascular plant assemblages and associated microbial communities on coal mine spoil heaps. Community Ecology, 16 (1), 23–32.
Woźniak G., Sierka E. & Wheeler A. 2018: Urban and industrial habitats: how important they are for ecosystem services. [In:] Hufnagel L. (ed.): Ecosystem Services and Global Ecology. IntechOpen, 169–194.

1.5. Geological Basis of Changes in Water Conditions and Methods of Their Determination

Alicja Kicińska
Department of Environmental Protection,
Faculty of Geology, Geophysics and Environmental Protection,
AGH University of Science and Technology;
30 Mickiewicza Av., 30-059 Kraków, Poland;
kicinska@agh.edu.pl; ORCID: 0000-0003-2725-7319

Radosław Pomykała
Department of Environmental Engineering,
Faculty of Civil Engineering and Resource Management,
AGH University of Science and Technology;
30 Mickiewicza Av., 30-059 Kraków, Poland;
rpomyk@agh.edu.pl; ORCID: 0000-0003-2238-241X

ABSTRACT: The chapter describes changes in water relations (quantitative and qualitative) that may arise as a result of mining operations. The most important effects of the transformations were discussed, including: rock mass deformation and effects of this process for nature, destruction of the primary impermeable layers separating aquifers, discharge of mine water and post-mining sewage, mixing of various types of water and contamination of surface watercourses, landfilling of waste rocks and various post-mining waste on the surface. The study also characterized the impact of various types of mining: hard coal, lignite, metal ores, iron and chemical raw materials on the transformations of primary water conditions. The authors also presented methods of counteracting changes in water conditions, pointing to the importance of choosing appropriate exploitation systems, construction of watertight screens, carrying out gravity drainage of flooded areas, or filling and reclamation of post-mining areas.

KEYWORDS: Mining exploitation, water relations, deformations, eliminating the effects of transformations

INTRODUCTION

Mining exploitation transforms water relations on two very important levels: quantitative (changes in the regime of surface and groundwater, lowering the groundwater table, reduction or disappearance of underground and surface runoff, changes in the pressure gradient, deformation of watercourse beds, breaks of insulating layers, reduction of surface and groundwater resources, formation of wetlands and floodplains) and qualitative (changes in groundwater chemistry, contamination of groundwater and surface waters) changes. These changes apply to all types of water: surface, ground, underground, located in the immediate and further vicinities of the sites of exploitation (Posyłek *et al.* 2003; Bukowski 2009).

The transformation of water conditions in mining areas is closely related to the geological structure and hydrogeological conditions. It is caused by two main activities: underground mining and excavation drainage.

Underground mining exploitation causes deformation of the rock mass and deformation of the ground surface. In the first case, we observe them in the form of deflections – a subsidence trough appears in the area much several times larger than the area of the exploited deposit, as well as various interruptions of rock layers. This leads to breaks in the layers insulating aquifers, which in turn changes the permeability of rocks and the chemistry of groundwater (Gao *et al.* 2017). On the other hand, we observe deformations of the terrain surface in the form of various sinkholes (i.e. lowered fragments of the Earth's crust, the edges of which do not have to be parallel) and depressions (i.e. subsidence basins). In the case of surface watercourses, riverbeds may be damaged, the continuity of the water flow may be interrupted (flowing water collapsing below the surface), as well as an increase in infiltration and finally disturbance of the water balance. The results of these transformations are the floodplains, inundation and extensive wetlands that are formed in the areas adjacent to mining operations. These transformations most often also result in changes in groundwater chemistry and shifting the boundaries of surface catchments, as well as disturbing the water balance.

The second reason for the transformation of the water conditions in mining areas is the drainage of workings (the so-called rock drainage), which can leads to lowering the groundwater table (especially as a result of surface or shallow underground exploitation) and lowering the piezometric pressure. Lowering of the groundwater table and disappearance of the outflow causes the reduction or complete disappearance of the efficiency of springs, changes in the supply of surface watercourses leading to a reduction in flow and possibly their disappearance and/or change in the nature of watercourses from surface flow to infiltration and ground water flow. Lowering the piezometric pressure results in the compaction and settlement of the layers, as well as the reduction of pressure on the base of insulating layers (which results in additional subsidence processes). The direction and speed of water filtration change, and the processes of suffosion (mechanical washing of mineral particles from loose and weakly bonded rocks by water), clogging (the opposite of suffosion, consisting of fine mineral particles carried by water falling out of the filtration stream) and underground erosion (mechanical destruction of rocks by groundwater). Rock drainage also causes (as was the case with underground mining) the formation of sinkholes and disruption of the continuity of water flow in river beds, and thus transformation of the water balance of the catchment area and loss of water resources (Posyłek *et al.* 2003).

The greatest changes in water relations are observed in the first aquifer (groundwater occurring in the saturation zone, remaining in hydraulic communication, usually Quaternary). Drainage-related changes caused by mining up to the disappearance of water in the wells and damage to agricultural crops and the formation of permanent wasteland. In the case of deeper aquifers, it is assumed that the vertical range of the water-conducting fissures arising in results of caving system exploitation corresponds to 40 times the height of the workings (exploited layer), and the permeability of the post-cave fissure zone decreases exponentially as it moves away from the workings/exploitation zone.

Loss of water in the farmer wells hardly depends on the radius of the depression cone arised in result of dewatering (drainage). This difficulty is caused by the great variability of geological conditions and the extent to which it affects the drained rock mass. Factors influencing development of the depression cone are the geological structure of the deposit and its surroundings, as well as the hydrogeological conditions in its immediate and further neighbourhoods (Sui & Xu 2013). The size of the open pit and the depth of drainage as well as meteorological factors are also important. The size of the funnel is also influenced by the existing groundwater intakes and the modern hydrographic network. Generally, the extent of damage to agricultural crops is periodic. After the land drainage is stopped, the original state of soil moisture usually returns. Changes in the conditions of vegetation are not the most severe from the economic point of view. In the areas covered by post-exploitation ground surface depressions and changes in the position of the groundwater table, changes in the suitability of the site for construction purposes, changes in groundwater chemistry occur. There are also wetlands and floodplains whose positive role in the landscape and the environment is not assessed (Soni 2020). One of the methods of estimating changes in hydraulic drops that arise as a result of post-mining depressions is the vector method, consisting of summing up the vectors of the original slope of the terrain and the inclination of the edge of the depression basin.

An extremely important issue in the discussed subject matter is the storage of waste rocks in the form of heaps on the ground. Sub-level landfills affect the quality of groundwater. According to Posyłek *et al.* (2003), dumps contain over 40% of waste rocks mined together with the excavated material, e.g. coal. In the case of hard coal mining in Poland, these are mostly coarse-grained sandstone and clay waste. By infiltrating waste rock heaps, they dissolve minerals containing chlorides and sulphates (i.e. Zn, Fe, Cu, Pb, Mn, Hg, Ni, Cr, Co, etc.), and then carry them to surface, ground and underground waters (Kosa & Kicińska 2016; Kicińska *et al.* 2019). These substances, in excessive amounts, may affect living organisms in nearby habitats. Due to their size, dumps and heaps may also cause disturbances in surface runoff and the formation of floodplains and wetlands.

Mining works can be a direct cause of damage and devastation of water infrastructure (e.g. water pipelines), river beds and reservoir embankments, as a result of which significant amounts of water can enter the aquifers, raising the original groundwater levels (Sui & Xu 2013; Wang *et al.* 2015). Other types of transformations related to the aquatic environment are: increased bottom and lateral erosion (associated with increased flow velocity), increased sedimentation of sediments and shallowing of river beds (associated with reduced flow velocity), formation of landslides (due to the reduction of the internal friction angle of cohesive soils), settlement of structures (due to the reduction of the bearing capacity of the building substrate), and others. Therefore, an important condition determining the direction of changes in water conditions in mining areas is the existence of hydraulic insulation, eliminating or limiting the draining effect of mining excavations (Posyłek *et al.* 2003; Bukowski 2009).

GEOLOGICAL CONDITIONS OF CHANGES IN WATER CONDITIONS IN THE AREAS OF EXPLOITATION

Extensive and significant transformations of natural water conditions are observed in all mining areas: in Europe, the Upper Silesian Basin (PL), Ruhr and Saarland (D), Yorkshire and Northumberland (UK), Pechorskie (RU) andÆnieckie (UK); in Asia, the Datong, Kailuang and Fushun (CN), Damodar (IN), Kuznets (RU) and Karagandzkie and Ekibastuskie (KZ) basins; in North America, the Appalachian Basin, Inner, Wyoming and Colorado (USA); in Africa, around Johannesburg (South Africa); in Australia, the Newcastle basin. In Poland, hard coal is found in three basins: Dolnośląskie (DZW), Górnośląskie (GZW) and in Lubelskie (LZW). In the Upper Silesian Coal Basin (GZW), the strongest impact on the aquatic environment is observed due to the duration of operation (over 150 years) and the intensity of extraction. The GZW area is divided into two hydrogeological regions: uncovered (northern, central and eastern parts – the Carboniferous deposits are fed by infiltration of precipitation through the permeable layers of the Quaternary) and covered (southern and western parts – Carboniferous deposits are covered by clay and impermeable Tertiary layers). There are three aquifers of regional importance here: Quaternary, Triassic and Carboniferous. All three are subject to direct and/or indirect mining drainage (it is estimated that the area of the Upper Carboniferous outcrop covered by mining drainage in the GZW is approximately 1,750 km^2). The effect of the drainage is to lower the water table by about 100 m, and the transformations of the natural hydrogeological regime of the carbonate aquifer run much deeper, up to approx. 500 m below the Carboniferous top (Wilk 1990). The Triassic (in the area of Jaworzno) and the Quaternary (especially in the central and eastern parts) are also strongly drained. Within the youngest floor, the water table was lowered, the efficiency of numerous springs was lost and the phenomenon of intensive water infiltration into the ground was noted. The impact of fossil waste dumps and the discharge of mine water to surface watercourses were also significant in the area of GZW. As a result of post-mining surface subsidence, numerous permanent and periodic floodplains were created in this area.

In the Lower Silesian Coal Basin located within the Wałbrzyskie Mountains (Middle Sudetes), the influence of mining is based on the intensive drainage of Carboniferous rocks, complete disturbance of the natural water balance of the DZW, contamination of surface watercourses with mineralized mine water and infiltration contamination of groundwater by heaps and dumps (Wilk 1990).

The area of the third – Lublin Coal Basin (LZW) is located in the Lublin-Volyn artesian basin and the Łuków hydrogeological plinth. The transformations of the aquatic environment in the LZW are recorded deep below the surface (Jurassic and Carboniferous formations). The piezometric pressure in the Jurassic floor has decreased over a large area, and the remaining changes in water relations are small.

Lignite is mined in Poland in the following regions: Bełchatów, Wielkopolska (Konin, Turek) and Turoszów. The main impact of lignite mining on the aquatic environment is the drainage of aquifers: Quaternary, Neogene, Cretaceous and Jurassic. Surface waters also underwent a strong transformation. The waste rock dumps also had a significant impact, especially in terms of landscape transformation and the forms of use of the neighbouring areas (Grünewald 2001). In the Turek-Konin region, coal mining is associated with the drainage of the outlay formations and the coal complex, and with the discharge of mine water to surface watercourses.

In the lignite mining areas, there is a noticeable increase in the deficit of water resources due to the lack of mining water management and their direct discharges to surface watercourses, water transfers between catchments, which causes the drainage and stagnation of the depression cone to persist (Fagiewicz 2009). There are noticeable changes in unit runoff and changes in the physical parameters of water, a change (usually reduction) of retention is noted through the liquidation of small watercourses, drainage ditches, wetlands and other objects that can store significant amounts of water (Grünewald 2001).

The areas of exploitation of metal ores (Zn-Pb ores) occur mainly in the Bytom basin, located in the northern part of the GZW. The exploitation of these ores has been going on continuously since the 12th century, although initially only shallow (up to approx. 20 m) galena deposits were exploited. Currently, changes in water conditions are observed related to the Triassic drainage and the Jurassic-Quaternary aquifer. The maximum decrease in the level of the mirror in this area was recorded by approx. 150 m, and the area covered by drainage is over 110 km^2. A negative symptom of mining activity is the complete drying of surface watercourses (i.e. Baba or Biała) and the drying of several springs and many farm wells. In this area, the negative impact of the drainage on the nature of the Jurassic Landscape Park (approximately 15 km^2) was also noticed. Another negative effect is the discharge of mine water to surface watercourses (Kicińska *et al.* 2015) and the washing out of toxic substances (mainly sulphates and heavy metals) from flotation waste deposited on the surface of landfills. Numerous sinkholes are also noticeable, disturbing the outflow of ground and surface waters.

The exploitation of copper ores in Poland takes place in the Legnica-Głogów Copper District (LGOM). Geologically, this area is located in the central part of the Fore-Sudetic Monocline. In the deposit area, the upper (Cenozoic) and lower (Permian-Triassic) aquifer complexes are distinguished. As part of the transformation of the water environment of the LGOM, one should note the drainage of the rock mass with a regional range (approx. 600 km^2), the lowering of the piezometric table by over 850 m and changes in the circulation system of ground and surface waters. A significant environmental effect is also the discharge of mine water directly to the Odra River and the effluent of polluted water from post-flotation reservoirs.

The former exploitation of iron ore in the Częstochowa-Kłobuck region is currently only of historical importance to the aquatic environment. After the mines were closed and their dewatering ceased, the process of filling the previously prepared depression cone within the drained layers was started, which led to the disappearance of the effects of mining drainage. During the operation carried out in 15 small mines, in the 1960s, the water level dropped by about 100 m. In this area, there was an extensive depression crater and water infiltration from surface watercourses as well as changes in the chemical composition of groundwater. The most noticeable were the increases in the content of sulphate, chloride, iron, magnesium ions and lowering the pH of the water.

The areas of exploitation of chemical raw materials (the former Tarnobrzeg region of sulfur extraction and the Kujawy and Subcarpathian salt exploitation areas) are also of historical nature. The sulfur deposit in the Tarnobrzeg region was associated with a series of chemical Miocene sediments and was found within sulfur-bearing limestones. Sulfur was mined using the opencast ("Machów" mine) and borehole ("Grzybów" and "Jeziórko" mines) methods. One of the negative impacts of borehole mining was the contamination of the Quaternary layer of waters as a result of eruptions and uncontrolled flow of highly mineralized process waters and deposit waters. The discharge of water from drainage of the Quaternary aquifer directly into the Vistula River was a similar problem. Older waters, containing significant amounts of hydrogen sulphide, were discharged to a water purification plant where they underwent desulphurization processes. Substantial depressions of the terrain, resulting from the excavation of the deposit, resulted in the formation of extensive floodplains in this area, with a total area of approximately 30 ha.

In the case of salt deposits, the water hazard is associated with a special type of hazard during their exploitation. A necessary condition for the deposit to survive is its isolation from the flow and impact of groundwater (Changa *et al.* 2018). The very process of selecting salt deposits must take place in tight conditions in relation to the surrounding aquifers. In Poland, salt deposits are mined in the area of the Carpathian Foredeep (deposits of the Miocene Age) and in the Kujawy region (deposits of the Zechstein Age). In the Carpathian region, there are quite complex geologically and hydrogeologically conditions for the contact of salt deposits with groundwater, while in the Kujawy region the salt deposits have the character of a salt dome that breaks through to the surface. According to Wilk *et al.* (1990), under normal conditions, salt mines do not exert any drainage effect on the waters, and all kinds of hydraulic contacts of workings with groundwater (especially surface waters) may only be the result of errors in the art of mining or even failure. The very type of mining exploitation (underground with roof protection and borehole mining) minimizes the surface subsidence and the formation of flooding and flooding of the land surface. For obvious reasons, it is impossible to store the salt spoil on the ground surface, therefore no risks related to contamination of surface watercourses are noted. In the case of mining of this type of deposit, however, mining failures and disasters are dangerous, which can be very severe for the aquatic environment. Another dangerous thing may be the fact of the accumulation of surface subsidence above the leached salt chambers (which happened, for example, in the former "Łężkowice" mine) due to the slow loss of compressive strength of clay rocks and the properties of delayed rock salt deformation, which may result in the formation of sinkholes and, consequently, local salting of groundwater. In the case of borehole mining, groundwater may also be contaminated with chlorides as a result of penetration through unsealed epigenetic cap formations on the outcrops or directly along the columns of casing pipes. In the case of the Zechstein deposits, the drainage method is used, as a result of which extensive depression craters are created, which have a much more severe impact on the water relations in the mine decommissioning processes than during exploitation of the deposits.

Mining exploitation is also carried out due to the extraction of various building stones, crushed aggregates, carbonate binding materials, crumb raw materials and clay materials for building ceramics. In most of them, planned drainage is not carried out due to the fact that these are small and shallow plants, therefore their impact on the aquatic environment is small. However, it should be borne in mind that each mine, regardless of whether it exploits above or below the groundwater table, affects the water relations through changing conditions of surface runoff, field evaporation and infiltration (Wilk 1990; Changa *et al.* 2018; Ju *et al.* 2018). The scale of changes in water conditions significantly increases if we are dealing with exploitation below the groundwater table. The drainage necessary for the drainage of the excavation, causes not only the drying of the mining walls and the bottom of the excavation, but also ensures stability of the slopes, the possibility of removing the spoil and allows for safe blasting work in non-flooded blast holes. Drainage, in turn, is related to the phenomenon of mine water discharge to surface watercourses, which can significantly increase the content of nitrates or solid particles in the water.

WAYS TO PREVENT CHANGES IN WATER RELATIONS

Eliminating the effects of transformations, and even preventing their occurrence, can be planned and implemented almost at every stage of activity related to the extraction of minerals, from the planning stage, through the preparatory and operational stages, to the closure of the mine (Wang *et al.* 2015; Inkin 2018; Dong *et al.* 2021). As reported by Wilk (1990), in terms of changes caused by post-mining surface subsidence, one should use appropriate mining systems that limit the size of surface subsidence areas, coordinate exploitation in time and adapt the directions of exploitation fronts to the natural system of water conditions, correct the banks and the bottom of surface watercourses and surround troughs with embankments, build watertight screens to protect depressed areas against water infiltration from embanked rivers and canals, carry out gravity surface drainage of flooded areas, drain with vertical drains and lower water into lower mine workings or absorbent layers. According to Wilk (1990), there is also a possibility of limiting changes in the scope of changes caused by rock mass drainage, mainly through the use of mining systems that allow to maintain the tightness of aquifers. Other methods to protect water relations may include: limiting the simultaneous exploitation to specific tectonic blocks separated from each other by insulating fault zones, restoring the tightness of broken insulating screens through the waterproofing of vertical workings and liquidation of boreholes, selection of appropriate methods of crossing water-bearing faults and protecting them with pillars, waterproof lining of shafts and corridor workings, protection of surface waters against the influence of drainage by sealing river beds, clogging and silting of the bottom of reservoirs, construction of artificial insulating screens limiting the reach of the depression cone and the use of closed circulation of mine water, i.e. forcing pumped waters into the ground in the range zone of the depression crater (Wilk *et al.* 1990).

Hydrotechnical and utilization methods are used to protect surface waters against fossil waters (Gu *et al.* 2020; Dong *et al.* 2021). The first group includes: construction of dosing tanks in which saline waters are stored during periods of low water levels in rivers, and then in periods of high water levels they are discharged in a controlled manner into these rivers, the transfer of mine brines to lower sections of rivers with greater flow, and injection of brines into the substrate (Wilk *et al.* 1990). On the other hand, rendering devices use the phenomena of distillation, reverse osmosis and crystallization freezing, in order to pump out only a part of (the most saline) waters.

Securing the impact of waste rock landfills on the aquatic environment largely depends on the selection of the appropriate location and isolation of any leachate from groundwater (Gao *et al.* 2017; Soni 2020). For this purpose, vertical watertight barriers are made and the ground is sealed against possible leaks.

At the stage of liquidation of mines and mining excavations, cessation of drainage causes self-flooding with water from the natural inflow, and the water level depends on the geological structure, hydrogeological conditions and mining conditions of the mine being closed and the neighbouring mines (Gu *et al.* 2020). The inflow of flooded water to the workings is reduced in proportion to the decrease in the depression generated by the mine workings in the feeding aquifer, which in turn leads to a change in the chemical composition of the water and its mineralization in the caving zone of the liquidated mine and a change in the amount of water flowing to the neighbouring mine. There is also a change in water relations in the surface layers and on the ground surface, as well as a change in hydraulic connections between mine horizons draining different hydrochemical zones (Rogóż 2003; Gao *et al.* 2017).

REFERENCES

Bukowski P. 2009: Determining of water hazard zones for mining exploitation planned in the vicinity of reservoirs in abandoned mines. Gospodarka Surowcami Mineralnymi – Mineral Resources Management, 25 (3), 203–215.

Changa Q., Sun X., Zhoue H. & Shic X. 2018: A multivariate matrix model of analysing mine water bursting and its application. Desalination and Water Treatment, 123, 20–26.

Dong S., Wang H., Guo X. & Zhou Z. 2021: Characteristics of water hazards in China's coal mines: A review. Mine Water and the Environment, 40, 325–333.

Fagiewicz K. 2009: Górnictwo odkrywkowe jako czynnik kształtowania stosunków wodnych (na przykładzie kopalni węgla brunatnego „Adamów"). [In:] Jankowski A.T., Absalon D., Machowski R. & Ruman M. (eds.): Przeobrażenia stosunków wodnych w warunkach zmieniającego się środowiska. Wydział Nauk o Ziemi Uniwersytetu Śląskiego, Sosnowiec, 129–140.

Gao L., Bryan B.A., Liu J., Li W., Chen Y., Liu R. & Barrett D. 2017: Managing too little and too much water: Robust mine-water management strategies under variable climate and mine conditions. Journal of Cleaner Production, 162, 1009–1020.

Grünewald U. 2001: Water resources management in river catchments influenced by lignite mining. Ecological Engineering, 17 (2–3), 143–152.

Gu Q., Huang Z., Li S., Zeng W., Wu Y. & Zhao K. 2020: An approach for water-inrush risk assessment of deep coal seam mining: a case study in Xinlongzhuang coal mine. Environmental Science and Pollution Research, 27, 43163–43176.

Inkin O. 2018: Evaluation of water-bearing strata collective properties change at the storage of gaseous hydrocarbons. Visnyk of Taras Shevchenko National University of Kyiv: Geology, 3 (82), 80–86.

Ju F., Xiao M., Guo S. & Li B. 2018: Geological characteristics of the main aquifer coal seam in the Ordos Basin coalfield, China. Arabian Journal of Geosciences, 11, 646.

Kicińska A., Gruszecka-Kosowska A., Wardas-Lasoń M. & Wójcik R. 2015: Wpływ przemysłu wydobywczo-przeróbczego rud Zn-Pb na środowisko w rejonie olkuskim. Wydawnictwo AGH, Kraków.

Kicińska A., Smreczak B. & Jadczyszyn J. 2019: Soil bioavailability of cadmium, lead, and zinc in the areas of Zn-Pb ore mining and processing (Bukowno, Olkusz). Journal of Ecological Engineering, 20 (1), 84–92.

Kosa B. & Kicińska A. 2016: Coal from the waste disposal site of the Siersza mine (Trzebinia, Poland) and its properties as a possible alternative fuel. E3S Web of Conferences, 10, 00039.

Posyłek E., Rogoż M., Staszewski B. & Wilk Z. 2003: Wpływ działalności górnictwa węgla kamiennego na środowisko wodne. [In:] Wilk Z. (ed.): Hydrogeologia polskich złóż kopalin i problemy wodne górnictwa. Wydawnictwo AGH, Kraków, 290–338.

Rogóż M. 2003: Problemy wodne związane z likwidacją kopalń. [In:] Wilk Z. (ed.): Hydrogeologia polskich złóż kopalin i problemy wodne górnictwa. Wydawnictwo AGH, Kraków, 321–336.

Soni A.K. 2020: Estimation of mine water quantity: development of guidelines for Indian mines. Mine Water and the Environment, 39, 397–406.

Sui W.H. & Xu Z.M. 2013: Risk assessment for coal mining under sea area. [In:] Huang Y., Wu F., Shi Z. & Ye B. (eds.): New Frontiers in Engineering Geology and the Environment. Springer Geology. Springer, Berlin-Heidelberg, 199–202.

Wang H.K, Guo J. & Shi Y.C. 2015: Type of major water hazards and study of countermeasures in Shennan Mining Area. Journal of Groundwater Science and Engineering, 3 (1), 70–76.

Wilk Z. 1990: Przeobrażenia stosunków wodnych. [In:] Kozłowski S. (ed.): Zasady ochrony i kształtowania środowiska przyrodniczego na obszarach eksploatacji złóż kopalin. Wydawnictwo SGGW-AR, Warszawa, 123–133.

Wilk Z., Adamczyk A.F. & Nałęcki T. 1990: Wpływ działalności górnictwa na środowisko wodne w Polsce. Wydawnictwo SGGW-AR, Warszawa.

Part Two

Novel Ecosystems – the Image of Anthropocene Epoch Environmental Conditions

2.1. Post-mineral Excavation Sites as Novel Ecosystems and Examples of Socio-environmental Resilience

Gabriela Woźniak
Institute of Biology, Biotechnology and Environmental Protection,
Faculty of Natural Sciences, University of Silesia in Katowice;
28 Jagiellońska Str., 40-032 Katowice, Poland;
gabriela.wozniak@us.edu.pl; ORCID: 0000-0003-1936-2880

Andrzej M. Jagodziński
Department of Ecology,
Institute of Dendrology, Polish Academy of Sciences;
5 Parkowa Str., 62-035 Kórnik, Poland;
amj@man.poznan.pl; ORCID: 0000-0001-6899-0985

ABSTRACT: Humans have altered biotic and abiotic ecological components on a local, regional, and global scale. Landscape modification, ecosystem fragmentation, wetland loss, mass species extinction, invasive species expansion, ecosystem function destabilization, e.g., of worldwide nitrogen and carbon cycles, are among the leading causes of global climate change. Some of the listed significant examples of human influence on environmental change work as drivers of Novel Ecosystem establishment. The new ecological phenomena described as Novel Ecosystems requires many studies because they have no natural or semi-natural analogs and currently cover an expanding land area.

Novel Ecosystems are part of the socio-economic-environmental landscapes consisting of abiotic and biotic elements (including urban, suburban, industrial, and rural areas). Novel Ecosystems are expanding and cover more than three-quarters of the formerly wild Earth. The driver of novelty is human activity, which inevitably causes disturbance, the close association of the Novel Ecosystem components and adaptations in relation to the prevailing socio-ecological conditions.

On a local scale, the vegetation assemblages on old buildings or along field boundary stone walls in former agricultural and urban-industrial landscapes are examples of places where studies into Novel Ecosystem ecology are developing.

KEYWORDS: Species composition, spontaneous plant assemblages, post coal-mining sites, ecosystem services

INTRODUCTION

Human impacts on the environment include modifications, alterations, or changes to biochemical and biophysical conditions of habitats, which cause changes in ecosystem processes, biodiversity, and distribution of natural resources and species occurrence (Hawksworth & Bull 2008; Sahney *et al.* 2010; Wuebbles *et al.* 2017). Direct or indirect effects of human activities result in, e.g., global warming, biodiversity loss and biodiversity mass extinction. One of the readily observed impacts of human activity is the transformation of landscapes by removing vegetation cover and the resulting vast areas left as open abandoned land. In those situations, a critical question remains: what to do with the large surface area that is significantly altered, changed, or established *de novo* by human activity?

In the traditional approach in many countries, restoration ecologists try to reverse and undo the effects of changes caused by human activity and bring ecosystems back to their previous state. Restoration is never smooth and rarely successful (Tropek *et al.* 2012), similar to the Sisyphean task (Schoukens 2019). The widespread lack of success of restoration efforts has led some restoration ecologists, like Richard Hobbs, to conclude that some ecosystems have changed so much that it is impossible to restore them to their previous state. According to Hobbs, sites that are changed very profoundly, or those that emerged *de novo*, provide a set of unusual conditions for developing a Novel Ecosystem (Fig. 1; Hobbs *et al.* 2013).

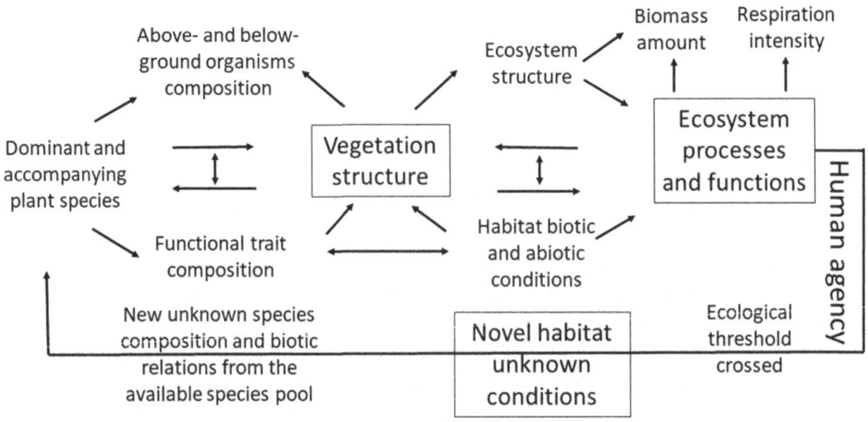

Figure 1. Simplified scheme of relationships among habitat conditions, vegetation structure, above- and below-ground species composition, and ecosystem functioning in natural and semi-natural habitats; modification of these relationships by intense human agency can lead to crossing a habitat threshold, resulting in a Novel Ecosystem

According to some researchers, Novel Ecosystems represent new ecological entities and they should be identified and enhanced according to their own independent biological, ecological, and environmental rules. Following the Novel Ecosystems concept and pathway of development will bring more benefits to environmental functioning (ecosystem services and human quality of life), than taking the risk of losing time and financial resources to restore them to the previous stage (Morse *et al.* 2014; Evers *et al.* 2018). According to one of the available descriptions, a site assigned as a Novel Ecosystem habitat has to fulfil five prerequisite conditions: I) The establishment of the habitat has to take place due to human agents; II) The plant and animal species composition is unknown in the existing ecosystems; III) The Novel Ecosystem has to grow in a new non-analogous manner; IV) It has to develop further as an autonomous, independent system after the human agent stops action; V) Finally, the Novel Ecosystem has to establish a self-maintaining ecosystem. Regardless of this description, there are still many misunderstandings.

Many discussions have focused on the definition of novelty. As a result, Morse *et al.* (2014) presented some concepts of the description. Novel Ecosystems are a unique assemblage of environmental conditions and biota set up due to human (human agency) alteration. The alteration must be sufficient to cross an ecological threshold that moves and maintains a new ecosystem trajectory and prevents its return to a historical, previous course even without additional human disturbance. Furthermore, the resulting Novel Ecosystem must be self-sustaining in ecosystem functioning, including best-adapted species composition, biogeochemistry, feedback relationships, structure, processes, ecosystem functioning and, consequently, ecosystem services.

Regardless of the continuous and very fruitful discussions on the meaning and definition of the novelty idea, some sites should undoubtedly be identified as Novel Ecosystems (Fig. 2). Among them are the mineral post-excavation sites like, e.g., opencast sandpits, post-coal-minė heaps, lead-zinc heaps, and limestone quarries.

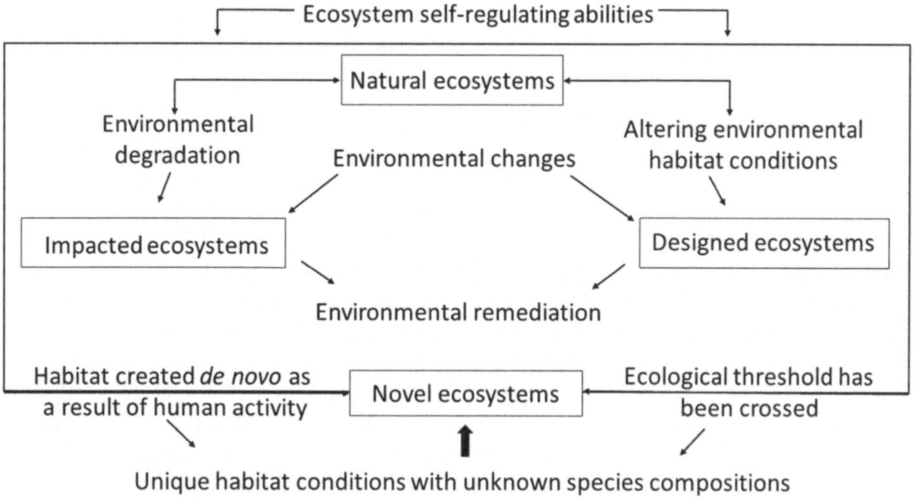

Figure 2. Scheme of the possible ecological changes and thresholds leading to Novel Ecosystems. The relations between ecosystems of different level of ecosystem disturbance undergoing different trajectories

The natural world and social interactions are interrelated and shaped by a range of values that come together to rule the processes of decision-making and influence the alterations of the wild, spontaneous ecosystem development trajectories (Woźniak 2010; Hallett *et al.* 2013). The social, cultural, and institutional contexts of the interrelated processes complicate the estimation and proper assessment of ecological thresholds and the best way to enhance Novel Ecosystems. The managers are overwhelmed by the idea of utility and financial benefits. They are not familiar with ecosystems developed while the processes acting in nature save enormous amounts of money only because they are possible and present in particular habitats. The floodplain forests along river embankments and mangrove forests along oceans are good examples of priceless, cost-free, and very efficient protection against floods and tsunamis provided by natural processes.

Another underestimated power of the natural process are the increasingly apparent plant trait adaptations caused by human-induced rapid environmental change (HIREC), which expands the spectra of unknown ecological processes and parameters that plants and other organisms respond to (Kueffer & Daehler 2009; Jackson 2013; Perring *et al.* 2013). On the one hand, the direct and indirect human influences can cause such profound changes in species composition and ecosystem structure that the limits of ecological resistance will be not enough to prevent crossing the thresholds of ecosystem self-regulation ability (Williams & Jackson 2007; Chen *et al.* 2011; Yamano *et al.* 2011; Grimm *et al.* 2013). On the other hand, the quick adaptation performed by many plants to the new, even very harsh environmental conditions is astonishing. The dynamic transformations of the living organisms to the HIREC factors provide additional proof that the establishment of Novel Ecosystems processes and its autonomous functioning can be a type of natural support for humankind facing global environmental changes.

The direct factors altering ecosystems cover increasing areas of Earth, e.g. wetlands and water bodies. Among the indirect factors, the decreasing abundances of wildlife areas, biodiversity loss, disturbed carbon and nitrogen cycles, and climate change resulting in changing temperatures and precipitation levels have to be listed (Morse *et al.* 2014; Evers *et al.* 2018). The rising influence of direct and indirect factors increases the probability that some ecosystems become pushed outside the borders of ecological resistance limits (thresholds) and become Novel Ecosystems (Morse *et al.* 2014).

THE CONCEPT OF NOVEL ECOSYSTEMS AS AN EXAMPLE OF SOCIO-ECOLOGICAL RESILIENCE

Novel Ecosystems are unintentional results of deliberate human actions (Perring *et al.* 2013). However, the emerged Novel Ecosystems can continue with self-sufficient development and are independent of further human activity (Hallett *et al.* 2013; Hobbs *et al.* 2013; Morse *et al.* 2014).

A growing body of proof reveals that the old restoration, rehabilitation, and restoration methods used for Novel Ecosystem habitats are not working. In addition, some reclamation work that ignores biological and environmental knowledge brings harm instead of improving ecosystem and environmental functioning (Hobbs *et al.* 2009; Belnap *et al.* 2012).

The lack of any existing equivalent ecosystem makes restoration of Novel Ecosystems to their previous state impossible (Hallett *et al.* 2013). The biological study of some of the Novel Ecosystems has revealed that these ecosystems can provide many services in the urban and industrialized landscape (Woźniak *et al.* 2018). Some environmentalists regard particular types of Novel Ecosystems as potential reservoirs for ecosystem services (Chapin *et al.* 2006; Perring *et al.* 2013). The drivers of novelty regard human activity, which causes the inevitable disturbances and close association of Novel Ecosystem components and adaptations with socio-ecological systems (Collier & Devitt 2016). Some environmentalists suggest that Novel Ecosystems represent examples of socio-ecological resilience (Collier & Devitt 2016) and can be as valuable for humans as the historical ecosystems that they replaced (Hallett *et al.* 2013). Novel Ecosystems can play an essential role in climate change mitigation (Zedler *et al.* 2012). Once their presence in the ecosystem mosaic is documented and proved, the main point of management action for Novel Ecosystems and habitats should enhance their critical ecosystem function (Zedler *et al.* 2012; Moyle 2014; Trueman *et al.* 2014). How can a system that we do not know and understand be managed?

Among the potential Novel Ecosystems, there are sites on which planned or unplanned human activities have led to establishment of ecosystems with natural values. The accidental or deliberate occurrence of new species composition is a sign of novelty (Hobbs *et al.* 2009; 2013). Furthermore, the new species assemblages are associated with unique above- and below-ground traits, interactions and processes (Kompała-Bąba *et al.* 2019; Milewska-Hendel *et al.* 2021).

THE NOVEL ECOSYSTEM IN RESEARCH

The Novel Ecosystem concept approach is present in a growing number of research results. Doley & Audet (2013) have been studying post-mining disturbance sites in altered landscapes. There were no natural reference points. They showed that the spontaneously emerged new ecosystems on these post-industrial landscapes present dynamic balances and functional development of ecosystems that can provide critical environmental goods for the local community. Such unique circumstances should be feasible for managers and satisfy the stakeholder's expectations. The Novel Ecosystems are always composed of new sets of species (Woźniak 2010). The new species composition of plants, as the primary producers, causes change in vegetation communities, ecosystem structure, and function (Martínez *et al.* 2010). Novel Ecosystems develop as independent ecological self-sustaining entities. Such a system can take over the ecosystem and environmental functions of the lost ecosystem and support native species establishment. The new environmental systems have not been previously quantified, and the potentially available new services are not identified and understood. Therefore, they are not deliberately enhanced (Fig. 2; Lin & Petersen 2013).

Novel Ecosystem theory requires a redefinition of the traditional interactions of humans with nature. Novel Ecosystem theory offers a new opportunity for transformational concepts. However, it still needs a cultural shift and social understanding of natural processes in any ecosystem. Rethinking how we deal with the management of transformed land, restoration actions, and conservation is necessary (Marris 2011; Yung *et al.* 2013). A comprehensive scientific approach and public dialogues about the heavily transformed, and in particular, the post-mineral excavation sites, should be performed. The interdisciplinary research conducted on an understanding of the adaptation processes, evolutionary, and ecological implications of how Novel Ecosystems function in the long term, would enable a better experience and more appropriate management considerations (Belnap *et al.* 2012).

Interdisciplinary environmental knowledge is fundamental for understanding the phenomenon of Novel Ecosystems because of the growing demand for the provision of particular ecosystem services in densely populated urban-industrial areas, which is often in contrast with social awareness and economic expectations (Seastedt *et al.* 2008; Stafford Smith *et al.* 2009; Belnap *et al.* 2012; Woźniak *et al.* 2018). Such an approach demands more appropriate and flexible management strategies if necessary (Harris *et al.* 2006; Hobbs *et al.* 2006; Seastedt *et al.* 2008; Bridgewater & Yung 2013; Hallett *et al.* 2013; Perring *et al.* 2013). Novel Ecosystem theory is forcing management plans and scenarios to introduce an interdisciplinary approach, focused on studies concerning the use of a range of possible practical and strategic opportunities for the enhancement of ecosystem and environmental functioning, for the environmental profit of social transformation (Bridgewater & Yung 2013; Hulvey *et al.* 2013).

The current research is providing evidence that the possibilities of taking a Novel Ecosystem approach instead of the non-feasible restoration to a historical reference is often the only rational solution for some sites in the urban-industrial landscape. There are places where entirely new environmental ecology quality has emerged, and it is pointless and not possible to drag the site back to its previous state (Tropek *et al.* 2012; Doley & Audet 2013). Novel Ecosystem elements in urban-industrial landscapes present new possibilities and potentially have high values for reorganization to create a unique mosaic of habitats and ecosystems with new environmental qualities that the designers and planners have to understand and learn to combine (Dooling 2015). The expanding available options provide more comprehensive possibilities for environmentally-friendly change in urban and industrial landscapes (Hobbs *et al.* 2013).

The recent arrival of Novel Ecosystem theory in scientific papers has spawned a debate on the environmental meaning of connecting the new biotic and abiotic interactions to benefit ecological functioning. This attitude consists of a challenge for the responsible authorities and managers (Bridge-water & Yung 2013; Collier & Devitt 2016). Due to the lack of practical examination of the long-term functioning of Novel Ecosystems, there are some challenges to the utility of Novel Ecosystem theory. It is becoming recognized that the existing environmental management frameworks are insufficient and sometimes harmful under the new circumstances where human activity leads to the creation of Novel Ecosystems more frequently (Morse *et al.* 2014; Evers *et al.* 2018).

Due to the spontaneous occurrence of nutrient-poor oligotrophic water, wetland, and terrestrial vegetation assemblages, the natural processes show insufficient conservation efforts through ecological reserves. The Novel Ecosystem approach can help ensure long-term protection of the oligotrophic habitats and ecosystems generated in the Anthropocene (Waltert *et al.* 2011). Furthermore, the current examples of the intensity and dynamics of ecosystem and environmental phenomena expand the current knowledge about the mechanisms involved (Evers *et al.* 2018). The poor oligotrophic mineral habitats with their moisture gradients (opencast sandpits, quarry excavations, e.g., base-rich limestone, Cambrian sandstone and post-coal mine heaps) provide refuge habitats for many organisms that had been diminished or extirpated along with the intense application of nutrients on agricultural land and subsequent eutrophication.

Studies to improve the understanding of Novel Ecosystems functioning and the enhancement of their recognized natural values, should be intensified to compensate for environmentally harmful human activities and related environmental transformations (Chapin *et al.* 2006; Hobbs *et al.* 2013; Perring *et al.* 2013; Graham *et al.* 2014; Murcia *et al.* 2014; Rotherham 2017).

NOVEL ECOSYSTEMS AND ENVIRONMENTAL FUNCTIONING

The available studies have revealed that the critical challenge for Novel Ecosystem theory to be applied in practice is successfully introducing the whole comprehensive Novel Ecosystem idea to managers and practitioners. The Novel Ecosystems approach is facing the same problems as the ecosystem services concept. Both Novel Ecosystem theory and the ecosystem services concept lack investigations appropriate for the applied policy implications (Fig. 2; Bridgewater *et al.* 2011; Graham *et al.* 2014; Perring *et al.* 2014).

Currently, there is a lack of attention to the biological, hydrological, and hydrogeological basis of environmental functioning in environmental policy. The environment is considered a technical mechanism that must be repaired and work for human economic benefit and usefulness (Robbins & Moore 2013; Morse *et al.* 2014; Murcia *et al.* 2014). Identifying the challenges connected with dynamic environmental alternations must be recognized and understood by the management systems and planning practitioners. According to Collier & Devitt (2016), there are three main groups of difficulties. The first group is connected with the complicated nature of bio-geo-chemical environmental processes themselves. Understanding the ecological complexities, such as assembly rules and colonizing stressors, and constant ecological change caused by pressure from human activity, which leads to the crossing of environmental thresholds, is critical. The second group of challenges underlined by Collier & Devitt (2016) is regarding the social values and perspectives, the social expectations from business, and economic compensation directed to the socio-environmental aspects of corporate social responsibility. The third group of issues is dealing with transfer of knowledge by practicing communication with stakeholders. Each decision and action in the environmental ecosystem mosaic must strengthen natural systems, natural capital, and resilience to mitigate further changes. There is no doubt that the ecological ecosystem mosaic will adapt to any changes, including global changes. The question remains if humankind will manage to adapt, too.

Novel Ecosystems can provide expanding and even augmented ecosystem and environmental functioning, which is the basis for ecosystem services. Novel Ecosystems offer opportunities to fulfil objectives focused on maintaining, e.g., biodiversity, water retention, and carbon sequestration. Biodiversity is the essential condition that integrates all ecosystem functions (ecosystem services), influencing and enhancing positive human-ecological interactions. For that to happen, socio-economic and administrative instruments that can recognize the potential environmental value of Novel Ecological assemblages have to be introduced (Tassin & Kull 2015). The fundamental prerequisite is fluent communication between scientists and policymakers, followed by extending cooperation between public society and policymakers and dialogue between stakeholders and practitioners.

These are the main opportunities for environmentally-focused transformation, especially in changing socio-economic-environmental systems (Morse *et al.* 2014).

Many sites fulfil all the conditions listed by Morse *et al.* (2014). Many such places are established as a side product of mineral excavation, including post-hard coal mining and quarrying of various mineral resources. The side effect of mineral excavation develops in places with many watercourses, wetlands, and terrestrial mineral oligotrophic habitats. Many of them provide unusual (sometimes extreme) abiotic conditions. As a result, the set of organisms colonizing such sites present unique compositions associated with outstanding features and biotic opportunities (Hobbs *et al.* 2013; Kueffer 2013; Robbins & Moore 2013; Woźniak *et al.* 2015). Sufficient adoption of the Novel Ecosystem concept for such sites and landscapes should not be prevented and should be introduced without delay. Novel Ecosystems, for which there is no doubt about ecological novelty, should be divided from those which need more work to assign clear criteria for assessing their environmental status. Such division would provide an adequate response to the uncertainties and challenges for a more open, dynamic approach to conservation effectiveness for the ecosystem, environmental system and enhancement of biodiversity and other ecosystem elements (Zedler *et al.* 2012; Bridgewater & Yung 2013). The lack of clear criteria for ecological novelty hinders managers and environmental legal frameworks from improving the adaptive capacity of the human-induced transformation. Applying some aspects of the ecological wonder of the Novel Ecosystems mosaic would allow for the enhancement of ecological functioning and human living conditions closer to wildlife. The solutions proposed above will significantly improve environmental conditions (Bridgewater & Yung 2013).

The requirement of greater collaboration between scientists and policy decision-makers, managers, and practitioners should focus on understanding how the processes and functional parameters of Novel Ecosystems work. This knowledge will provide the basis for how Novel Ecosystems should be supported (Fig. 3). In principle, further research has to be conducted to find out how to protect specific vegetation species composition and the associated ecosystem relationships, such as the unique biodiversity of Novel Ecosystems, which should be maintained (Hulvey *et al.* 2013; Morse *et al.* 2014). The dialogue has to be conducted from the local to the macro-regional level to understand and acknowledge the occurrence, extent, and importance of ecological novelty for the limitation and mitigation of global change.

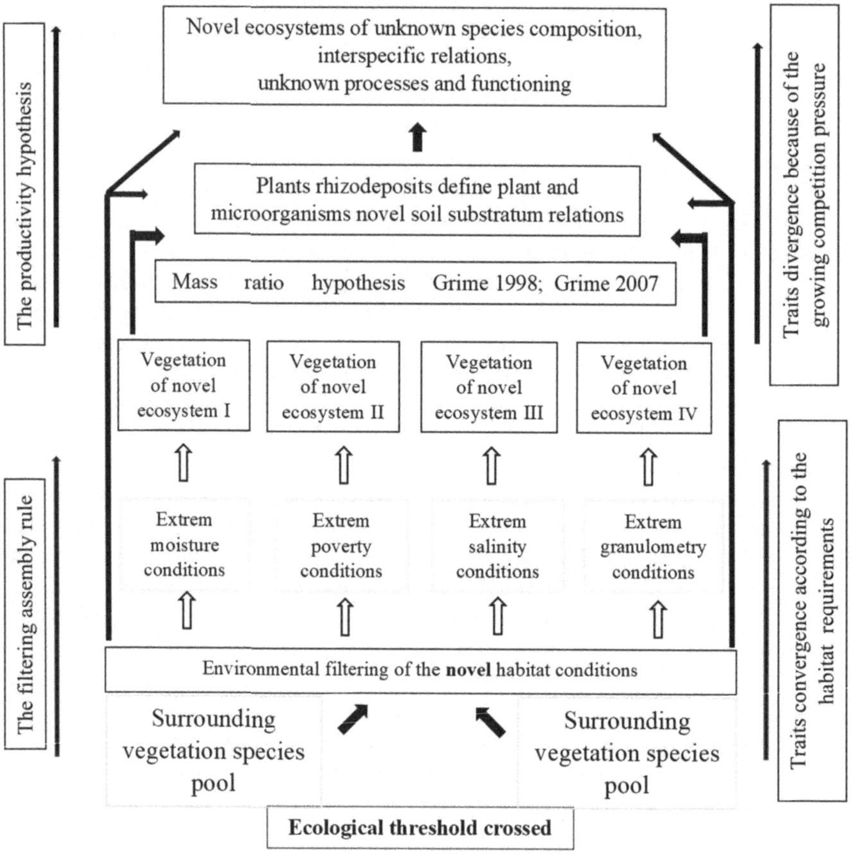

Figure 3. The possible research questions to study, in order to understand the processes and functioning of Novel Ecosystems

REFERENCES

Belnap J., Ludwig J.A., Wilcox B.P., Betancourt J.L., Dean W.R.J., Hoffmann B.D. & Milton S.J. 2012: Introduced and invasive species in novel rangeland ecosystems: friends or foes? Rangeland Ecology & Management, 65 (6), 569–578.

Bridgewater P. & Yung L. 2013: The policy context: Building laws and rules that embrace novelty. [In:] Hobbs R.J., Higgs E.S. & Hall C.M. (eds.): Novel Ecosystems: Intervening in the New Ecological World Order. John Wiley & Sons, Chichester, 272–283.

Bridgewater P., Higgs E.S., Hobbs R.J. & Jackson S.T. 2011: Engaging with novel ecosystems. Frontiers in Ecology and the Environment, 9 (8), 423.

Chapin III F.S., Lovecraft A.L., Zavaleta E.S., Nelson J., Robards M.D., Kofinas G.P., Trainor S.F., Peterson G.D., Huntington H.P. & Naylor R.L. 2006: Policy strategies to address sustainability of Alaskan boreal forests in response to a directionally changing climate. Proceedings of the National Academy of Sciences of the United States of America, 103 (45), 16637–16643.

Chen I.-C., Hill J.K., Ohlemüller R., Roy D.B. & Thomas C.D. 2011: Rapid range shifts of species associated with high levels of climate warming. Science, 333 (6045), 1024–1026.

Collier M.J. & Devitt C. 2016: Novel ecosystems: Challenges and opportunities for the Anthropocene. The Anthropocene Review, 3 (3), 231–242.
Doley D. & Audet P. 2013: Adopting novel ecosystems as suitable rehabilitation alternatives for former mine sites. Ecological Processes, 2, 22.
Dooling S.E. 2015: Novel landscapes: Challenges and opportunities for educating future ecological designers and restoration practitioners. Ecological Restoration, 33 (1), 96–110.
Evers C.R., Wardropper C.B., Branoff B., Granek E.F., Hirsch S.L., Link T.E., Olivero-Lora S. & Wilson C. 2018: The ecosystem services and biodiversity of novel ecosystems: A literature review. Global Ecology and Conservation, 13, e00362.
Graham N.A.J., Cinner J.E., Norström A.V. & Nyström M. 2014: Coral reefs as novel ecosystems: embracing new futures. Current Opinion in Environmental Sustainability, 7, 9–14.
Grime J.P, Hodgson J.G. & Hunt R. 2007: Comparative plant ecology. A functional approach to common British species. 2nd ed. Castlepoint Press.
Grime J.P. 1998: Benefits of plant diversity to ecosystems: immediate, filter and founder effects. Journal of Ecology, 86, 902–910.
Grimm N.B., Chapin III F.S., Bierwagen B., Gonzalez P., Groffman P.M., Luo Y., Melton F., Nadelhoffer K., Pairis A., Raymond P.A., Schimel J. & Williamson C.E. 2013: The impacts of climate change on ecosystem structure and function. Frontiers in Ecology and the Environment, 11 (9), 474–482.
Hallett L.M., Standish R.J., Hulvey K.B., Gardener M.R., Suding K.N., Starzomski B.M., Murphy S.D. & Harris J.A. 2013: Towards a conceptual framework for novel ecosystems. [In:] Hobbs R.J., Higgs E.S. & Hall C.M. (eds.): Novel Ecosystems: Intervening in the New Ecological World Order. John Wiley & Sons, Chichester, 16–28.
Harris J.A., Hobbs R.J., Higgs E. & Aronson J. 2006: Ecological restoration and global climate change. Restoration Ecology, 14 (2), 170–176.
Hawksworth D.L. & Bull A.T. (eds.) 2008: Biodiversity and Conservation in Europe. Springer, Dordrecht.
Hobbs R.J., Arico S., Aronson J., Baron J.S., Bridgewater P., Cramer V.A., Epstein P.R., Ewel J.J., Klink C.A., Lugo A.E., Norton D., Ojima D., Richardson D.M., Sanderson E.W., Valladares F., Vilà M., Zamora R. & Zobel M. 2006: Novel ecosystems: theoretical and management aspects of the new ecological world order. Global Ecology and Biogeography, 15 (1), 1–7.
Hobbs R.J., Higgs E. & Harris J.A. 2009: Novel ecosystems: implications for conservation and restoration. Trends in Ecology & Evolution, 24 (11), 599–605.
Hobbs R.J., Higgs E.S. & Hall C.M. (eds.) 2013: Novel Ecosystems: Intervening in the New Ecological World Order. John Wiley & Sons, Chichester.
Hulvey K.B., Standish R.J., Hallett L.M., Starzomski B.M., Murphy S.D., Nelson C.R., Gardener M.R., Kennedy P.L., Seastedt T.R. & Suding K.N. 2013: Incorporating novel ecosystems into management frameworks. [In:] Hobbs R.J., Higgs E.S. & Hall C.M. (eds.): Novel Ecosystems: Intervening in the New Ecological World Order. John Wiley & Sons, Chichester, 157–171.
Jackson S.T. 2013: Perspective: Ecological novelty is not new. [In:] Hobbs R.J., Higgs E.S. & Hall C.M. (eds.): Novel Ecosystems: Intervening in the New Ecological World Order. John Wiley & Sons, Chichester, 61–65.
Kompała-Bąba A., Bierza W., Błońska A., Sierka E., Magurno F., Chmura D., Besenyei L., Radosz Ł. & Woźniak G. 2019: Vegetation diversity on coal mine spoil heaps – how important is the texture of the soil substrate? Biologia, 74, 419–436.
Kueffer C. & Daehler C.C. 2009: A habitat-classification framework and typology for understanding, valuing and managing invasive species impacts. [In:] Inderjit (ed.): Management of Invasive Weeds. Invading Nature: Springer Series in Invasion Ecology. Volume 5. Dordrecht, Springer, 77–101.
Kueffer C. 2013: Integrating natural and social sciences for understanding and managing plant invasions. [In:] Larrue S. (ed.): Biodiversity and Society in the Pacific Islands. Presses Universitaires de Provence, Marseille, France & ANU ePress, Canberra, Australia, 71–95.
Lin B.B. & Petersen B. 2013: Resilience, regime shifts and guided transition under climate change: Examining the practical difficulties of managing continually changing systems. Ecology and Society, 18 (1), 28.
Marris E. 2011: Rambunctious Garden: Saving Nature in a Post-Wild World. Bloomsbury, New York.
Martínez O.J.A., Rodríguez M.A., Rosario I., Soto N., López A. & Lugo A.E. 2010: Structure and species composition of novel forests dominated by an introduced species in northcentral Puerto Rico. New Forests, 39 (1), 1–18.

Milewska-Hendel A., Chmura D., Wyrwał K., Kurczyńska E.U., Kompała-Bąba A., Jagodziński A.M. & Woźniak G. 2021: Cell wall epitopes in grasses of different novel ecosystem habitats on post-industrial sites. Land Degradation & Development, 32 (4), 1680–1694.
Morse N.B., Pellissier P.A., Cianciola E.N., Brereton R.L., Sullivan M.M., Shonka N.K., Wheeler T.B. & McDowell W.H. 2014: Novel ecosystems in the Anthropocene: a revision of the novel ecosystem concept for pragmatic applications. Ecology and Society, 19 (2), 12.
Moyle P.B. 2014: Novel aquatic ecosystems: The new reality for streams in California and other Mediterranean climate regions. River Research and Applications, 30 (10), 1335–1344.
Murcia C., Aronson J., Kattan G.H., Moreno-Mateos D., Dixon K. & Simberloff D. 2014: A critique of the 'novel ecosystem' concept. Trends in Ecology & Evolution, 29 (10), 548–553.
Perring M.P., Audet P. & Lamb D. 2014: Novel ecosystems in ecological restoration and rehabilitation: Innovative planning or lowering the bar? Ecological Processes, 3, 8.
Perring M.P., Standish R.J. & Hobbs R.J. 2013: Incorporating novelty and novel ecosystems into restoration planning and practice in the 21st century. Ecological Processes, 2, 18.
Robbins P. & Moore S.A. 2013: Ecological anxiety disorder: diagnosing the politics of the Anthropocene. Cultural Geographies, 20 (1), 3–19.
Rotherham I.D. 2017: Recombinant Ecology – A Hybrid Future? Springer International Publishing.
Sahney S., Benton M.J. & Ferry P.A. 2010: Links between global taxonomic diversity, ecological diversity and the expansion of vertebrates on land. Biology Letters, 6 (4), 544–547.
Schoukens H. 2019: Legal considerations in operationalizing eco-restoration in the European Union : a Sisyphean task or unlocking existing potential? [In:] Akthar-Khavari A. & Richardson B.J. (eds.): Ecological Restoration Law: concepts and case studies. Routledge, Abington-New York, 167–192.
Seastedt T.R., Hobbs R.J. & Suding K.N. 2008: Management of novel ecosystems: are novel approaches required? Frontiers in Ecology and the Environment, 6 (10), 547–553.
Stafford Smith D.M., Abel N., Walker B. & Chapin III F.S. 2009: Drylands: Coping with uncertainty, thresholds and changes in state. [In:] Chapin III F.S., Kofinas G.P. & Folke C. (eds.): Principles of Ecosystem Stewardship. Resilience-Based Natural Resource Management in a Changing World. Springer, New York, 171–195.
Tassin J. & Kull C.A. 2015: Facing the broader dimensions of biological invasions. Land Use Policy, 42, 165–169.
Tropek R., Kadlec T., Hejda M., Kocarek P., Skuhrovec J., Malenovsky I., Vodka S., Spitzer L., Banar P. & Konvicka M. 2012: Technical reclamations are wasting the conservation potential of post-mining sites. A case study of black coal spoil dumps. Ecological Engineering, 43, 13–18.
Trueman M., Standish R.J. & Hobbs R.J. 2014: Identifying management options for modified vegetation: Application of the novel ecosystems framework to a case study in the Galapagos Islands. Biological Conservation, 172, 37–48.
Waltert M., Bobo K.S., Kaupa S., Montoya M.L., Nsanyi M.S. & Fermon H. 2011: Assessing conservation values: Biodiversity and endemicity in tropical land use systems. PLoS ONE, 6 (1), e16238.
Williams J.W. & Jackson S.T. 2007: Novel climates, no-analog communities, and ecological surprises. Frontiers in Ecology and the Environment, 5 (9), 475–482.
Woźniak G. 2010: Zróżnicowanie roślinności na zwałach pogórniczych Górnego Śląska [Diversity of vegetation on coal-mine heaps of the Upper Silesia (Poland)]. W. Szafer Institute of Botany, Polish Academy of Sciences, Kraków.
Woźniak G., Markowicz A., Borymski S., Piotrowska-Seget Z., Chmura D. & Besenyei L. 2015: The relationship between successional vascular plant assemblages and associated microbial communities on coal mine spoil heaps. Community Ecology, 16 (1), 23–32.
Woźniak G., Sierka E. & Wheeler A. 2018: Urban and industrial habitats: How important they are for ecosystem services. [In:] Hufnagel L. (ed.): Ecosystem Services and Global Ecology. IntechOpen, 169–194.
Wuebbles D.J., Fahey D.W., Hibbard K.A., DeAngelo B., Doherty S., Hayhoe K., Horton R., Kossin J.P., Taylor P.C., Waple A.M. & Weaver C.P. 2017: Executive summary. [In:] Wuebbles D.J., Fahey D.W., Hibbard K.A., Dokken D.J., Stewart B.C. & Maycock T.K. (eds.): Climate Science Special Report: Fourth National Climate Assessment. Volume I. U.S. Global Change Research Program, Washington DC, USA, 12–34.
Yamano H., Sugihara K. & Nomura K. 2011: Rapid poleward range expansion of tropical reef corals in response to rising sea surface temperatures. Geophysical Research Letters, 38 (4), L04601.

Yung L., Schwarze S., Carr W., Chapin III F.S. & Marris E. 2013: Engaging the public in novel ecosystems. [In:] Hobbs R.J., Higgs E.S. & Hall C.M. (eds.): Novel Ecosystems: Intervening in the New Ecological World Order. John Wiley & Sons, Chichester, 247–256.
Zedler J.B., Doherty J.M. & Miller N.A. 2012: Shifting restoration policy to address landscape change, novel ecosystems and monitoring. Ecology and Society, 17 (4), 36.

Green Scenarios..., Dyczko, Jagodziński & Woźniak (eds.)

2.2. Returning Collieries Back to Nature in England, UK

Lynn Besenyei
Faculty of Science & Engineering,
University of Wolverhampton;
Wulfruna Str., Wolverhampton WV1 1LY, West Midlands, England, UK;
L.Besenyei@wlv.ac.uk; ORCID: 0000-0003-2975-1740

ABSTRACT: The UK coal mining industry has been in sharp decline since the 1970s with only a small number of active mines still remaining. Inactive coal mine sites are large areas of dereliction and devastation. Most of the early mine closures were back-filled, made safe and then left to nature to undergo successional processes. Three case studies of closed collieries in the West Midlands provide evidence that post-industrial sites left to naturally regenerate, with little human intervention, have provided opportunities for colonisation by local biodiversity, often providing the specific conditions to be colonised by rare species, and are now designated nature reserves. A fourth case study shows how these sites can provide opportunities for facilitated restoration and enhanced habitat creation including creating grasslands using strewn green hay from a local meadow. Some features of the buildings associated with these mines have been left intact to provide features of industrial heritage which have been explained by strategic environmental interpretation and now provide additional interest for the leisure activities of local residents. With careful planning and strategic restoration building on previous knowledge and paying attention to local topography and soil conditions these sites have the potential to have high biodiversity value and thus be effectively returned to nature and reintegrated into the landscape.

KEYWORDS: Biodiversity, industrial heritage, restoration, succession processes

INTRODUCTION

Many post-industrial sites present the opportunity to provide open spaces for people and for biodiversity by provisioning different types of habitats depending on the environmental conditions left as a consequence of their previous industrial use. The UK has a long history of coal mining but many of these coal mines have closed since nationalisation in 1947. Further contraction took place in the mid-1980s with the closure of unproductive mines and the demise of the UK coal mining industry. Mines which were closed in the 1960s and 70s were back filled and made safe with spoil and pit mounds being regraded. Planning legislation was introduced in 1948 and progressively amended and consolidated into statutes, the most recent of which is the National Planning Policy Framework (2019). Traditional restorations aimed to establish visually acceptable vegetation as quickly as possible (Bradshaw & Chadwick 1980) and often utilised agricultural or "do-nothing" approaches aiming to utilise the land for agriculture, or forestry by tree planting. By the 1990s priorities had changed and nature conservation amenity use became valid end-points of restoration (Anonymous 1994a; 1994b). More recently the ultimate goal of restoration is to create self-supporting ecosystems that are resilient to perturbation without further assistance (Society for Ecological Restoration 2004). Now the needs of restoration, to reclaim spoil, to eliminate toxicities and to return to a functioning ecosystem have to be considered before planning approval is granted (HMSO 2021). To achieve effective restoration planning it is necessary to prioritize areas for restoration, establish long-term monitoring, develop realistic goals, avoid the use of invasive species and consider the human population and their needs (Rocha-Nicoleite *et al.* 2017).

Natural Regeneration

Once collieries closed and the movement of substrate ceased these sites were initially invaded by lichens, mosses, and liverworts, later by grasses, sedges and rushes and ultimately by herbaceous species, which can be slow to colonise, particularly in areas of high toxicity or poor substrates. Many of the early flowering colonists were small ephemeral species, with short life cycles, quickly exploiting available resources and taller ruderal species. Plants will exploit localised resources as these become available, and some species will even tolerate the relatively high toxicities and extremes of high, or low, pH levels associated with post-industrial sites. Such sites can be very variable in their water-holding capacities, with localised water accumulation where soils are heavily compacted, or very well drained in locations where substrate particle sizes are large and water can percolate through rapidly. Interesting assemblages of aquatic plant species can quickly colonise open areas of ponds with associated colonisation over time by amphibians and aquatic invertebrates. Reptiles are known to bask on the open sun-facing slopes of pit mounds but are usually sufficiently uncommon to increase the value of sites. Where plants have naturally colonised and become established it is important to ensure that the implementation of any restoration scheme is mindful of safeguarding these (Lunn 2000).

Urban Post-industrial Sites

Many urban post-industrial (often known as brownfield) sites are known to be ecologically interesting for their assemblages of plant species that have arisen spontaneously. The slow initial colonisation often results in a variety of vegetation assemblages at these sites. This type of open mosaic habitat, on previously developed land (Rodwell 2000), is a priority for conservation and is protected by the United Kingdom's Biodiversity Action Plan (BAP) and as such is a legally protected habitat. The halt of species decline in recent years in urban areas is often due to the colonisation of such sites and their resultant classification as nature reserves for specific species conservation. Preserving this kind of vegetation on existing brownfield sites is important for town and city planners as it helps with international biodiversity commitments (Buglife 2009). Where spontaneous open habitats have developed, these can be further enhanced by expedient habitat creation (e.g. meadows), if the vegetation is less developed in some places, and where appropriate depending on the localised conditions and the size of the site. On-site substrate conditions may facilitate particular opportunities depending on previous history and future development plans.

Bees and hoverflies require support across towns and cities as well as on agricultural areas and this is being provisioned under the National Pollinator Strategy (Defra 2014). Increased plant species, and therefore floral diversity, in the open habitats of post-industrial sites, can help to promote habitat for pollinators and key invertebrate species, and these sites can be refuges in urban areas (Buglife 2009; 2019). Around 30 species of butterfly, and many species of moths, are associated with brownfield sites and respond rapidly to any changes in condition of these sites (Butterfly Conservation 2019). Butterflies in particular have a preference for the open, varied habitats of post-industrial sites, and can naturally colonise sparsely scattered patches of vegetation. Post-industrial sites should also be managed for pollinators as these post-industrial sites provide safe havens from pesticides. This can be done after a site evaluation for its value for wildlife (Defra 2014). Land managers need to provide essential resources and habitats for pollinators, which require pollen and nectar as food sources between February and October, so plants need to be allowed to flower continuously throughout the season, and places to nest and shelter should be left undisturbed during the hibernation period. To prevent sparse vegetation for butterflies from disappearing from these sites, they need occasional disturbance and scrub clearance (Butterfly Conservation 2019). The halt of species decline in recent years in urban areas is often due to the colonisation of these brownfield sites and their resultant classification as nature reserves for specific species conservation.

Green Infrastructure in Towns and Cities

The presence of biodiversity has been seen as an important success criterion of post-industrial greening projects and the UK government has seen this as a prime tool for regional economic regeneration, neighbourhood renewal, and achieving its international biodiversity commitments (Doick *et al.* 2009; Pediaditi *et al.* 2010). However, post-industrial sites which have been neglected, and left undisturbed, are often peaceful sanctuaries for biodiversity. Post-industrial sites can be used as a component of green infrastructure development for sustainable cities. The promotion of sustainable land use by developing these kinds of sites is preferable to the loss of greenfield sites from the countryside and adding to existing urban sprawl. Many brownfield sites form mosaics of different land uses with their diversity of soil substrates leading to different plant communities, also at different successional age stages, along with patches of native habitat within urban landscapes containing different species of plants and animals (Bonthoux *et al.* 2014). Landscape connectivity and recolonization dynamics are key factors in continuing species population persistence at these post-industrial sites. In addition, these post-industrial greenspaces can provide opportunities for local people to widen their access to nature, in order for them to enjoy being in the natural world (Millward & Mostyn 1989; Icarus 2014) and to increase their outdoor recreation through dog walking, walking in green gyms, children playing in wild play areas, bird watching and heritage appreciation.

CONSERVING NATURE AND HERITAGE

Creating and Designating Nature Reserves

Traditionally nature conservation in Britain was focused on pristine sites with significant biodiversity interest (Ratcliffe 1976). Sites with considerable interest were awarded statutory protection under the Wildlife & Countryside Act (1981) as Sites of Special Scientific Interest (SSSIs) and National Nature Reserves (NNRs) under rigorously defined criteria (Nature Conservancy Council 1989). Also under EU legislation sites such as Special Protection Areas (SPAs) under the EC Directive (79/109) for the Conservation of Wild Birds and Special Areas of Conservation (SAC) under the Habitats Directive (1992) were designated. These sites are all natural, or semi-natural. Initially when sites were being designated there was little interest in so called second-tier sites which had nature conservation interest but did not meet the exacting criteria for designation as SSSIs.

Nature Conservation Value of Post-industrial Sites

A Local Nature Reserve (LNR) is an area of land designated by a local authority because of its special local natural interest and, where possible, educational and community value (Natural England 2010; Gov.UK 2019). More and more of these are post-industrial and inevitably occur in urban areas. The land is managed for conservation purposes, and also for recreational purposes (if this doesn't compromise its conservation purpose) (Natural Environment and Rural Communities Act 2006). The site should be publicly accessible where visitors would not damage or destroy the wildlife. These sites have the lowest level of statutory protection. Local Nature Reserves can contribute to Natural England's targets and the ANGSt standard of a minimum of 1 ha of LNR per 1000 head of population (Natural England 2013). Thus, they are mainly managed by local authorities for their amenity value.

Country Parks have no statutory protection and were mainly (but not solely) designated in the 1970s, under the Countryside Act 1968 (Gov.UK 2014). Their purpose is to provide areas in towns for local people to visit nature and enjoy recreation, but who may not wish to venture into the wilder countryside. They are mainly managed by local authorities, but can be run by private individuals and other organisations. They are mainly areas of public open space with an informal atmosphere close to the edge of urban conurbations and have some formal facilities such as toilets, cafe and car park. They were often post-industrial sites, hence their proximity to urban areas.

There is a level of designation below the category of Local Nature Reserve and Country Park. Many of these were also post-industrial sites, particularly old collieries. These non-statutory sites have no legal protection but have been recognised by the local authority responsible for the land on which they are found as being of interest for nature conservation. These are called slightly different names in different parts of the country. In the West Midlands they are known as Sites of Interest for Nature Conservation (SINC). There is also a level below this known as Sites of Local Interest for Nature Conservation (SLINC). In Staffordshire, Cheshire and Greater Manchester similar types of site are known as Sites of Biological Interest (SBI) and the lower designation is Biological Action Sites (BAS). In Shropshire similar Sites of Interest are designated by Shropshire Wildlife Trust (McKelvey 2014) to sites which are of county significance by virtue of the range of diversity, rarity, scientific, research or educational value of their fauna, flora or geology. Criteria for assessment of nature conservation value are kept under regular review through the Local Sites Partnership (Defra 2020). There are similar designations for geological sites which acknowledge the need for the wider protection of geological features for educational purposes (Box & Cutler 1988). Such sites also provide a link with past industrial activity.

In the early days of designation (1970s and 1980s) a number of post-industrial sites in the West Midlands were designated areas for nature conservation in a revision of the selection criteria for Sites of Importance for Nature Conservation (SINCs) which recognised the value of post-industrial sites across the Black Country and wider West Midlands (Box *et al.* 1994). Box & Cossons (1988) had earlier argued that post-industrial sites had an intrinsic value for rare species, such as clubmosses (*Lycopodiaceae*), which they found growing in moss and lichen dominated communities on the acidic clay spoil of an abandoned opencast coal mine in Telford. Lunn (2000) evaluated colonising plant species, on derelict sites from collieries in the Yorkshire coal field, according to how rare they were by using an assessment category which utilised red data book species, nationally scarce species lists and local scarcity from published county floras.

Putting nature back into towns and cities involves community involvement and communities are directly involved with the economic redevelopment of urban areas. Often such sites require local communities to take ownership of them and to form groups that will manage the site and be responsible for it. "Friends Groups" that take an active role in site management gain an ownership of the site which helps to keep down management costs from employing on site staff. Ultimately such sites can also promote aspects of the healthy living agendas for the local people who live nearby, providing amenity areas for them.

Conserving Nature Through a Landscape Scale Approach

Sir John Lawton made an appeal to make space for nature in his review of England's wildlife sites (Lawton *et al.* 2010). The report suggested a targeted approach to conservation efforts to improve existing natural sites in conjunction with linking to adjacent sites in surrounding areas to strengthen and expand areas available to wildlife. It advocated a landscape scale approach to targeted nature conservation across the country. Consequently, the current National Planning Policy Framework (NPPF Section 15) (HMSO 2021) requires that councils ensure the effective use of land and sets out the policy measures necessary to ensure the sustainable use of previously developed land. It acknowledges that available land should be set aside for its most appropriate use depending on the site's characteristics and urges a better use of these finite resources. This use could be for nature and amenity purposes and would provide an opportunity for nature conservation by allowing land that has naturally re-colonised to persist as a nature reserve. Thus, land from old collieries offer opportunities to provide additional land for nature which is unpopulated by humans. Facilitating its ecological restoration through re-colonisation by nature and providing a helping hand through strategic habitat creation will extend the amount of land available for wildlife and strengthen ecological networks, especially if this is near to existing sites of interest.

Conserving the Industrial Heritage of a Site

In England in 1913, at the peak of the coal industry, 2600 deep mines employed more than a million miners. At Nationalisation, in 1947, a thousand collieries came into public ownership, but now only a few are still operating (Falconer & Gould 2011). The Department of the Environment recognised the importance of industrial heritage in the early 1960s and implemented a programme of designation of the most important sites and decided that the state would not become the guardian of these industrial sites but they would be left to the voluntary sector to look after (Cussons 2011). The approach to preserving the legacy of England's industrial heritage over the last 50 years has not been strategic but random. Some parts have been partly recorded over the last 50 years, some parts have been preserved and some have not. Industrial museums have been created and six World Heritage Sites have been inscribed; one of these recognises the site of the birthplace of industry in Ironbridge, near Telford, Shropshire (Falconer & Gould 2011). What is being achieved is being done by official agencies, commercial partners and the voluntary sector. However, it is inadequately resourced, lacks vision and strategic thinking (Falconer & Gould 2011).

A good example of the state of England's industrial heritage is Chatterley Whitfield, near Stoke-on-Trent, Staffordshire, which was Britain's first one million ton colliery. It closed in 1977 and was designated a scheduled monument by Historic England as the most complete example of a large colliery in England. It temporarily became a museum but had to close. It is now run as a registered charity by an active "Friends group" who meet regularly to preserve the history of the site by digitising documents, recording personal histories, organising open days for the public and running the Heritage Centre (Chatterley Whitfield 2019).

CASE STUDIES OF COLLIERIES RETURNED TO NATURE

Case Study – Baggeridge Colliery, Now Baggeridge Country Park, Gospel End, Staffordshire

Baggeridge Colliery was owned by the Earl of Dudley who sank the first shaft in 1902. Full production started in 1912 and 550,000 tons were produced annually by 1933. In its heyday it employed 3,000 men producing 12,000 tons of coal per week and at Nationalisation (when it was taken into public ownership) 1,500 people were employed there (Morris 2019). It closed in 1968 and today little remains to suggest that this was once the largest coal mine in the world. The sole remaining items at the site are the concrete blocks that once supported the head gear and a few short lengths of rails with a cart (AditNow 2019a). There is an information board on-site about the old colliery which includes some old photographs.

The derelict land was bought in 1969 by Seisdon Rural District Council (now South Staffordshire Council) and initial reclamation works were carried out. The site was designated a Country Park in 1970. Government grant aid enabled improvement works to be carried out in 1971. Some adjacent land was purchased in 1980 which enabled a new entrance to be built to replace the old colliery access and an area behind the pit mounds (Mailding 1980a) was also purchased which enabled the spoil to be redistributed, reduced in height and re-contoured (Mailding 1980b). Full reclamation started in 1981 and involved extensive earth movement (375,000 m^3) and the reconstruction of vegetation (Land Use Consultants (no date)). Coal slurry (80,000 tonnes) was removed off site and also dredged from pools (40,000 tonnes) and removed. Topsoil and subsoil were stripped and spread thinly across a wider area. All non-wooded areas were covered with a standard grass mix (to become amenity grassland) and a herb rich seed mixture (3,250 kg) was scattered in some areas. On the pit mound (now known locally as the toposcope facilitating views into Wales) the slopes were hydroseeded with a slurry mix to make the seeds stick and to provide initial nutrients to become established. 19,500 trees were planted which were mostly birch (*Betula pendula*), oak (*Quercus robur*) and hawthorn (*Crataegus monogyna*).

The site now has 15 different habitat types (Trueman 1991) of which eight are included in the county Biodiversity Action Plan (Staffordshire Biodiversity Action Plan 2019). These are: ancient woodland (undamaged by industrial activity, Potter 1991) including calcareous woodland with *Fraxinus excelsior* and *Allium ursinum*, neutral woodland with *Quercus robur* and wet woodland (mainly *Alnus glutinosa*) (35 ha total), lowland heathland (4 ha), grassland (mostly amenity and some species-rich, 10 ha), reed bed, lakes, streams and ponds. Naturally regenerated secondary woodland is dominated by sycamore (*Acer pseudoplatanus*). The site is managed with conservation work and maintenance, by a small team of on-site wardens employed by South Staffordshire Council with help from some local volunteers. The ongoing management aims to enhance the biodiversity where possible and to keep invasive species at bay (Gallis *et al.* 2015). Plant diversity is enhanced by the wide range of soil types in terms of their pH and nutrient content which support both calcifuges (*Calluna vulgaris* and *Ulex europaeus*) and calcicoles (*Ophrys apifera* and *Blackstonia perfoliata*) in close proximities. Tufa springs and calcareous seepages have recently been identified on site and may hold unusual invertebrate species. The whole site has revegetated well and recent visitors find it difficult to imagine a large coal mine was ever at this site. The pit mounds are especially diverse due to their nutrient poor soils.

The biodiversity at the site is considered to be good with a good diversity of species mainly as a consequence of the range of habitats on site, with some rarer than others (Fig. 1). The site was designated a Local Nature Reserve (LNR) in 2005 and is also a Site of Biological Importance (SBI) grade1 (Staffordshire Biodiversity Action Plan 2019). Over 350 species of plants have been recorded here, 378 species of moths, 29 species of butterflies, 15 species of dragonflies and 127 species of birds. Species of note for the region, and nationally, are white clawed crayfish (*Austropotamobius pallipes*) (Decapoda, nationally scarce), great crested newt (*Triturus cristatus*) (Amphibia, Caudata, European protected species), six-belted clearwing (*Bembecia ichneumoniformis*) (Lepidoptera, nationally scarce), currant clearwing (*Synanthedon tipuliformis*) (Lepidoptera, nationally scarce), hornet moth (*Sesia apiformis*) (Lepidoptera, nationally scarce), white letter hairstreak (*Satyrium w-album*) (Lepidoptera, species of principal importance in England and Wales) and dingy skipper (*Erynnis tages*) (Lepidoptera, species of principal importance in England, Wales and Northern Ireland). In terms of plant species, the increase seen in orchids over recent years is of note. In the case of southern marsh orchids (*Dactylorhiza praetermissa*) four spikes were initially found on site and these spread over the years to 13600 and keep spreading. Some species have been lost recently, such as lesser spotted woodpecker (*Dryobates minor*) in line with the national trend for this species (Gallis *et al.* 2015).

Despite being located on the western edge of the West Midlands conurbation, Baggeridge Country Park is an intrinsic part of a much larger ecologically significant landscape, which contains a green wedge of countryside, including ancient woodland, heathland, parkland, grassland and wetland, between the city of Wolverhampton (to the north) and the Metropolitan Borough of Dudley (south east) and a buffer between the large village of Wombourne (west) and the conurbation. The overall nature conservation value of the wetlands in the neighbouring area is enhanced by the presence of the Park's open water, pools and wetlands (Gallis *et al.* 2015). There are links to the wider countryside of Shropshire to the west through adjacent farmland and connectivity is enhanced by the old railway network with the old branch line from the colliery leading onto the South Staffordshire Railway Walk and to Smestow Valley Local Nature Reserve (to the north) and to Pensnett Railway walk (to the south) which in turn link to, and cross, the canal network at several points in the local area.

The site is now 60 hectares of country park comprising children's play area, high ropes adventure area, large car park, café (which also hosts old photographs of the colliery), miniature railway (run by local enthusiasts in the summer months) and offers a range of outdoor pursuits (South Staffordshire Council 2019). The site has been conferred the Green Flag award for the last 21 years, mainly for its amenity value, but the judging criteria also take into account the ecological and environmental management of the site (Green Flag Award 2019). A parcel of land (2 ha) from the demolition of the adjacent brickworks which recently became a housing estate, has been seeded with a wildflower mix and planted with trees and scrub and also includes a sustainable urban drainage pond (SUDS), and has been incorporated into the park for management as part of the planning proposal (Town & Country Planning Act Section 106) mitigation.

Figure 1. View of Baggeridge Country Park from the top of an old spoil heap
(Photo: Kane Wilkinson-Knowles)

Case Study – Fair Lady Coppice Colliery, Heath Hayes, Staffordshire

Fair Lady Coppice Colliery opened in 1893 and closed in April 1964. Whilst in operation spoil was dumped on the heathland and grassland in the area surrounding the pit head (Cannock Chase Council 2016). Production ranged between 150,000 and 180,000 tons per annum in the first fifty years and up to 950 men were employed there (Cannock Chase Mining Historical Society, no date). After Nationalisation new and better machinery was installed allowing up to 420,000 tons to be extracted per year and less than 400 men were employed. Nothing now remains of the former colliery buildings. After closure the site was made safe and backfilled (Barnes 2019) and natural regeneration was allowed to take its course. The coal spoil heaps were extensively reshaped in 1990 when opencast mining commenced to the east of the site (Cannock Chase Council 2017) and this recontouring produced two flat plateaus on the top of the spoil, the southern one slightly higher than the northern one. A sandy material, with a high silt content, was thinly spread across the surface and lime was added in some places to increase the pH of the substrate and in other places topsoil was added. Further reclamation took place in 2001 when a 360° digger was used to break up the surface material on the steep slopes (south, east and west) of the pit mound and granular sewage sludge was added in order to facilitate tree planting on the northern slope by a carbon neutral company that had entered into an agreement with the landowners, Cannock Chase Council. Silver birch (*Betula pendula*), oak (*Quercus robur*), hawthorn (*Crataegus monogyna*), field maple (*Acer campestre*) and common alder (*Alnus glutinosa*) were planted, but no ground flora were seeded and few woodland ground flora species have colonised other than nettles (*Urtica dioica*) and bramble (*Rubus fruticosus*). Two ponds were created on top of the plateaus at this time. The very steep western slope recolonised naturally with silver birch, oak and hawthorn. The poor soil conditions resulted in an open mosaic habitat consisting of early successional stage species across most of the site (Fig. 2). There are also scattered patches of heathland with naturally seeded heather (*Calluna vulgaris*), particularly on the southern slope, where the substrate is more acidic, along with purple moor-grass (*Molinia caerulea*) and wavy hair grass (*Deschampsia flexuosa*). In some areas, grassland is widespread and dominated by wild carrot (*Daucus carota*), bird's-foot trefoil (*Lotus corniculatus*), ox-eye daisy (*Leucanthemum vulgare*), ladies' bedstraw (*Galium verum*) and common century (*Centaurium erythraea*). Southern marsh orchids (*Dactyloriza praetermissa*) are locally common. Penny royal (*Mentha pulegium*) and water purslain (*Lythrum portula*) are widespread on the less well drained areas on top of the plateau. Lesser reed mace (*Typha angustifolia*) is common in one of the ponds and bog pondweed (*Potamogeton polygonifolius*) is present in another, both of which are considered rare in Staffordshire (Cannock Chase Council 2015). There are a few non-native species at the site which have either been introduced by local people or have naturally colonised.

The former colliery site is located 3.4 km east of Cannock and is now occupied by two football pitches and a nature reserve. The reserve is owned and managed by Cannock Chase Council, covers 10.1 hectares and mainly consists of the former colliery spoil mound. A large part of the site, the two plateaus on the top of the pit mound, is designated as a wildlife sanctuary and is not accessible to the public. There are two information boards located at the main entrance to the site. One conveys information about the history of the colliery and contains some old photographs (Cannock Chase Mining Historical Society, no date) and the other contains information about the site's current nature conservation interest (Cannock Chase Council 2017). Other than the pit mound there are no remaining features from the colliery.

The site is of value to butterflies. Dingy skipper (*Erynnis tages*) and common blue (*Polyommatus icarus*) are doing well at the site. Small heath (*Coenonympha pamphilus*), large skipper (*Ochlodes sylvanus*), small skipper (*Thymelius sylvestris*), Essex skipper (*Thymelicus lineola*), meadow brown (*Maniola jurtina*) and ringlet (*Aphantopus hyperantus*) are all common. Skylark (*Alauda arvensis*) and meadow pipit (*Anthus pratensis*) are known to breed at the site. The wetland areas on the southern plateau supported breeding populations of little ringed plover (*Charadrius dubius*) and lapwing (*Vanellus vanellus*) until 1993 and 2013, respectively. Due to successional vegetation changes wading birds no longer breed here. The site has recently supported winter populations of snipe (*Gallinago gallinago*) and migrant jack snipe (*Lymnocryptes minimus*) in large numbers. Six species of birds of prey have been regularly recorded at the site. Common lizards (*Zootoca vivipara*) have been recorded on the southern slope and common toad (*Bufo bufo*), common frog (*Rana temporaria*) and smooth newt breed in the ponds. Red (*Cervus elaphus*) and muntjac (*Muntiacus reevesi*) deer regularly use the site, and badgers (*Meles meles*) and rabbits (*Oryctolagus cuniculus*) breed on site.

The site is currently (from 2011) in a ten year Higher Level Stewardship agreement with Natural England for its management (Natural England 2012). The two main objectives for the site are heathland creation and the provision of habitat for wading birds. The tree plantations are to be managed for one hundred years by the company that planted them in accordance with good silvicultural practice (Cannock Chase Council 2015). The bare areas are becoming reduced in number and in extent.

However, additional management of scalloping woodland edges, grazing with Dexter cattle, repairing fences and keeping paths open are undertaken annually. Periodically, when tall vegetation becomes established the plateaus are scraped with a small digger to retain short plant species and shallow water in the winter for birds and amphibians.

The site lies between Cannock Chase and Brownhills Common, both large areas of remaining semi-natural heathland, and close to the Chasewater and Southern Staffordshire Coalfield Heaths Site of Special Scientific Interest, thus, within a landscape context increasing the size of local patches of heathland within this corridor are important. The site is used by local dog walkers and children who use the site for wild/informal play. Two football pitches have been on-site since the colliery closed.

Figure 2. View of Fair Lady Coppice Colliery from the southern pit mound (Photo: Lynn Besenyei)

Case Study – Granville Colliery Telford, Shropshire Now Granville Country Park

Granville Colliery was owned by Earl Gower and opened in 1860. It survived Nationalisation in 1947 but closed in June 1979 (AditNow 2019b) when it employed 560 men and was the last working coal mine in Shropshire (Friends of Granville, no date). At its peak in the 1960s it produced 600,000 tons of coal per annum (Shropshire History 2016). It was located in a semi-rural area with agriculture to the north and east and the industrial villages that now make up the new town of Telford to the south and west.

Extensive natural regeneration has taken place on the land that was once Granville Colliery over the intervening years since its closure (Latham 2019). Telford Nature Conservation Project (1985) provided an ecological report for Telford Development Council from surveys undertaken in 1984 and when there were still remains of mining buildings and associated coal yards where the colliery had stood. The immediate surroundings of the site suggested they had been modified by mining activities with the result that a patchwork of heavily grazed small meadows, hedgerows, scrub (dominated by hawthorn *Crataegus monogyna*) and areas of impeded drainage had resulted. At that time four main habitats were present that had probably developed since closure, these were heathland, grassland, scrub and wetland with small pools (with *Juncus* spp. and *Typha latifolia*). There are no records of tree planting taking place at the site in the report but the large area of coal shale deposits in the centre of the site was rapidly being colonised by birch (*Betula pendula* and *B. pubescens*) (suggesting acidic soil) with occasional Scots pine (*Pinus sylvestris*) and sallow (*Salix cinerea*). The flora here at that time included white melilot (*Melilotus alba*) and ribbed melilot (*M.*

officinalis) which are rare in Shropshire, and blue fleabane (*Erigeron acer*) (suggesting alkaline soils) which is uncommon in Shropshire. The pit mound was being colonised by *B. pendula* and some ruderal species, dyer's weed (*Reseda luteola)* and docks (*Rumex* spp.). Over the intervening years the colliery site has naturally regenerated further and has become incorporated into the now revegetated post-industrial areas in the vicinity (Telford & Wrekin Council 2019). This wider area has been incorporated into Telford's development plan for the northeast of the town.

Granville Country Park was opened to the public in 1986 and was designated a Local Nature Reserve (LNR) of 22.3 ha in 1989 which was later increased in size to 61.4 ha in 2016 by incorporating some adjacent areas resulting in the consequent re-designation (Latham 2019). It is now one of the largest green spaces within the boundary of Telford & Wrekin District Council and one of the richest in wildlife (Telford & Wrekin Council 2019). It is owned and managed by the council in conjunction with Shropshire Wildlife Trust and the Friends of Granville volunteer group (Friends of Granville Country Park, no date; Shropshire Wildlife Trust 2008). There are four information and two interpretation boards at key entrance points. This site, along with others within this area of Telford, forms an ecologically important area of post-industrial sites interconnected with older habitats by an old canal and a now disused railway line, both of which served to transport goods from coal mines and factories in the area (Fig. 3).

The site is situated two miles northeast of Telford town centre and forms a part of the wider Telford Green Network which includes a range of Local Wildlife Sites, LNRs and Green Guarantee sites (with a council commitment not to build on them) in the locality and these provide a very high level of connectivity within the wider region, which means that the whole area is now one of the greenest towns in England (Telford & Wrekin Council 2019). The Telford Green Network was established to provide protection to 2300 ha of open land in Telford with prominent landscape features (Landscape Architects 2002). Telford & Wrekin District Council worked with parish councils, "friends groups" and local environmental organisations to make a commitment to increase the provision of easily accessible green spaces across the borough (Latham 2019). They identified and protected green spaces within the borough boundary. Many of these are naturally regenerated pit mounds, or other features of the area's post-industrial heritage as the birthplace of industry, and together these form important areas within the town and wider district for nature conservation and informal recreation. Individual communities have expanded within these wedges of green land yet still retain their sense of identity. This was in order to promote an image of Telford of being a place to appeal to investment and to be an attractive place to work and live, and this was achieved by linking its diverse natural habitats, open spaces and post-industrial sites by footpaths, cycle ways and ecological linkages joining different parts of the town. Since statutory designation of the Green Network, actual area of habitats lost has stabilised to approximately 1% per year. However, the main concern now is loss of quality, which is being addressed by the local Wildlife Trust (Landscape Architects 2002).

The site now supports twelve different plant communities within four different habitat types, woodland, grassland, heathland and wetland (S7 *Carex acutiformis* swamp 2 ha). The largest areas are covered by existing ancient and naturally regenerated woodland of a range of phytosociological communities, mainly W10 *Quercus robur-Pteridium aquilinum-Rubus fruticosus* woodland (10 ha), W21 *Crataegus monogyna-Hedera helix* scrub (17 ha), W23 *Ulex europaeus-Rubus fruticosus* scrub (2 ha) and W8 *Fraxinus excelsior-Acer campestre-Mercurialis perennis* woodland (1 ha). The heathland area of H9 *Calluna vulgaris-Deschampsia flexuosa* heath (1 ha) is ecologically significant for this area. The pit mound on the site supports a population of dingy skipper butterfly (*Erynnis tages*) and green hairstreak (*Callophrys rubi*) is also found at the site. Both are Regionally Medium Priority BAP species. Wood White (*Leptidea sinapis*), a Shropshire priority BAP species, has been recorded here. Granville's scrubby habitats support good populations of warblers and other bird species. The fast-declining water vole (*Arvicola terrestris*), a UK and Shropshire priority BAP species, has also been recorded here.

Figure 3. Aerial View over Granville Country Park (Photo: Sam Bagnall Photography)

FACILITATING RESTORATION

Meadow Creation on Post-industrial Sites

Post-industrial sites can provide opportunities for meadow creation on different types of land including coal mines. Creation can be undertaken by sowing previously harvested seeds in proportion to their species constituents in natural phytosociological communities, or by introducing the species in other ways (Walker *et al.* 2004), such as the use of turf, or plug plants, or more naturally, by using species-rich green hay harvested from a semi-natural hay meadow and spread across the land (Trueman & Millett 2003). Grassland habitat creation increases plant diversity colonising these sites and facilitates animals with new habitats. It has been possible to create almost exact replicas of the phytosociological MG5 Crested dog's-tail – Black knapweed *Cynosurus cristatus – Centaurea nigra* hay meadow community (Rodwell 1992) (European Cynosurion) in the Midlands by using very fresh green hay (Trueman & Millett 2003) on various types of post-industrial sites.

Post-industrial sites can be very harsh environments due to a number of reasons: previous toxicity, high daytime temperatures (in sheltered locations in direct sunlight), poor or non-existent soils, poor or excessive drainage as a consequence of compaction from heavy machinery, and uneven substrates with large pore sizes in the sediments. These soils are often composed of various types of substrates which have arisen from the component materials from the site's previous use. Soils on such sites can take many years to form and result from the decomposition of invading plants, which leads to a slow accumulation of organic material over time. Such sites can be considered a blank canvas because they lack a seed bank and are dependent on plant species coming in from elsewhere, or being brought there by animals.

To establish a species-rich meadow on post-industrial sites any initial colonising vegetation should be killed or cut short. Killing is recommended where vigorous weeds are present, but it is only necessary to cut short the existing stand of vegetation where this contains desirable species (Grime *et al.* 1988). Or sites could be strewn with a thin layer of nutrient-poor topsoil which is raked into a fine tilth to prepare a seed bed. Large stones should be removed from the site to prevent later damage to cutting machinery. Species-rich green hay should be harvested in early morning during mid-July, baled for ease of transport and spread at the site immediately after cutting (Trueman & Millett 2003). Depending on the size of the site, it may be necessary to use a tedder (Gamble

2015) or a muck spreader, to distribute the hay evenly around the site, or it can be spread by hand with the help of volunteers. The hay should be left to shed its seed and will also act as a mulch for germinating seedlings; germination times of seeds will vary depending on their vernalisation requirements. A study by Jones *et al.* (1995) found that good ground preparation and spreading green hay, as well as the use of appropriate management techniques, led to species diversity as much as 90% similar to the source meadow. However, a number of species have proved too difficult to transfer for various reasons, including those which set seed outside the normal harvesting time. Vegetation rich in flowering dicotyledonous species can be obtained within two years of hay being spread on an area equivalent to a third the size of the source site using hay from a suitable flower-rich MG5 Crested dog's-tail – Black knapweed *Cynosurus cristatus–Centaurea nigra* hay meadow (Besenyei 2000). Good results can be achieved on post-industrial sites within five years by spreading the green hay at a ratio of between 4:1 and 1:4 depending on site fertility and provided that good meadow management can be maintained.

High and low soil fertility can be a major constraint in the creation of grasslands. This has led to investigations into the relationships between soil characteristics and site species-richness. In an attempt to identify optimal nutrient ranges, when evaluating soils for meadow creation, McCrea *et al.* (2001a) studied soil nutrient content in two created meadows in two post-industrial sites in Wolverhampton, England. From the range of soil parameters studied, they found that phosphorus and potassium concentrations separately accounted for the most variation in species composition. High levels of phosphorus were found to inhibit species richness, but intermediate levels of potassium were found to be beneficial for diversity. They recommend that soils containing less than 7 mg of extractable phosphorus per 100 grams, and between 10 and 30 mg of extractable potassium per 100 grams, would yield optimal plant species diversity (results were based on Truog's method of analysis for soil phosphorus and on ammonium acetate extraction for soil potassium). Thus, on sites where soil nutrients are found to be higher, or lower, than these levels, meadow creation is not recommended. Soils should be tested before any detailed planning starts.

Further work by McCrea *et al.* (2004) investigated the relationship between soil fertility and species diversity in eight meadows created on urban soils and 28 semi-natural meadows. The soils had pH ranging from 4.5 to 6.5 and thus classified as mesotrophic grasslands (Rodwell 1992). The main differences between the soils of the created and semi-natural meadows were that the created ones contained more extractable phosphorus and less mineralisable nitrogen and organic matter. High phosphorus concentrations were thought to be the main reason few species colonise, or survive, in grasslands on many urban soils. Thus, they proposed that phosphorus was the main edaphic factor controlling the establishment of created meadows in many urban areas.

When nutrient levels in soils are high this will result in the vigorous growth of undesirable species and it might be worth considering depleting nutrient levels by growing crops for a season or two, in order to deplete nutrient levels to the desired range (McCrea *et al.* 2001b). Where the topsoil is high in nutrients, another option might be to remove it off site and create the meadow in the subsoil. In such cases, seed germination will still take place, but, overall, plant height will be appreciably less than normal, with vegetation development taking longer, until organic matter builds up sufficiently to retain nutrients.

One of the main problems of habitat creation is the necessity to manage them once created, and many schemes fail as a consequence of not implementing the necessary management regime (Parker 1995). Meadow creation is more successful if the traditional management of semi-natural meadows is carried out. The very minimum requirement is the need to cut and remove the hay, ideally removed in late summer. If this cannot be undertaken the diversity of the vegetation can deteriorate within two seasons (Besenyei 2000). Therefore, it is necessary to provide access to sites for a tractor-mounted flail mower to cut the vegetation. Very small sites can be cut by hand with a strimmer, but this is hard work at larger sites and makes the hay difficult to collect and remove once cut. Where possible, after mowing, the site should be grazed until the vegetation stops growing when the temperature drops in autumn. This may not be possible in urban areas due to security issues with livestock. It is possible to imitate the effect of grazing by cutting and then damaging the vegetation in some way, e.g. for example, by scarifying, to expose the soil in places to facilitate regeneration gaps in which seeds can germinate. Failure to remove the vegetation results in taller-growing undesirable species out-competing the shorter species and results in the build-up of dead vegetation which stops the seeds from reaching the soil and germinating, quickly resulting in the loss of annuals from the grassland.

Deciding when to implement the first cut after creation can be difficult, however, it is not recommended to do so after the first season's growth. Much of the initial seed germination can be weed species (not members of the desired vegetation community). If possible these should be eliminated by spot spraying, or weed-wiping with herbicide in the first season before they set seed. At the end of the second season, depending on the growth of the required species, the sward should be cut and removed from the site. If there has been little growth cutting should be delayed until after the third season. On post-industrial sites, vegetation growth might be slow because of stress due to lack of nutrients in the soil, or lack of topsoil, or localised toxicity. Germination and growth of species in the first season can be disappointing, perhaps because of disturbance, which can result in the germination of weed species which grow more vigorously than the desired species and because many species do not flower for a couple of seasons after establishment. However, common spotted orchid (*Dactylorrhiza fuchsii*) and green winged orchid (*Anacamptis morio*) have been found flowering on post-industrial sites in the second season after seeding.

Case Study of Meadow Creation on an Old Colliery – Ettingshall Park Meadow, Wolverhampton

Ettingshall Park is a public open space close to the southern boundary of the city and located in a residential area. It is owned and managed by the local authority, Wolverhampton City Council. The site has had a mixed history due to mineral extraction. It is now used primarily by local residents for dog walking and informal recreation. The site was historically occupied by small collieries with resultant spoil accumulation. Clay was also extracted from a small on-site clay pit and was used at a brickworks (now gone). The derelict land was reclaimed in two phases, one in the 1970s when the land to the north and north-west was regraded and turned into a public open space with tree planting and sports pitches (and formed the larger portion of the open space). The remaining area of land was scheduled for reclamation in 1985–1986, but by that time a large population of southern marsh orchids (*Dactylorhiza praetermissa*) had established within the clay pit. The reclamation scheme was thus limited to path construction, rubbish clearing and some tree planting, and was left as a nature conservation area. As a result of an error which occurred at that time during reclamation, a species-rich grassland which had naturally colonised since industrial activity had ceased, was completely scraped off revealing silty subsoil. This meant it was necessary to consider a grassland creation scheme to revegetate the area. Creation involved using a seed mixture, which was devised by Professor Ian C. Trueman, to establish a meadow (Fig. 4.) that would only require summer management (Trueman 1993) (pre-dating the publication of the British grassland community descriptions (Rodwell 1992)). In preparation for seeding the subsoil was harrowed several times. As there was no topsoil it could not be tested for nutrient levels.

The seed mixture was sown in spring 1986 using a fertiliser spreader. The vegetation which resulted was very thin and open, suggesting infertility (associated with the lack of topsoil) and the subsequent vegetation developed very slowly. It was not necessary to cut the meadow until 1991, once the vegetation had become continuous with a diverse collection of perennial plant species, when a late summer cut was implemented and the vegetation was removed. Growth of the vegetation has been very open and short over most of the meadow area since its inception. In the summer of 1987, there was a good display of cornflowers (*Centaurea cyanus*). However, by this time the meadow area had been divided in two by tree planting, along with other tree plantings around the periphery as part of landscaping work. At a fairly early stage in the meadow development yellow rattle (*Rhinanthus minor*) was imported as seed from a local semi-natural source and scattered evenly onto the meadow, as this species had not established from the original seed mixture. Subsequent research (in 1996) at the site found that extractable potassium was between 11.92 and 28.94 mg 100 g^{-1} and extractable phosphorus was between 2.46 and 6.19 mg 100 g^{-1} in areas that were very plant diverse (McCrea *et al.* 2001a). The sports fields in the north are mown every two weeks to keep the grass sward short and the created meadows have been managed by cutting and removal of clippings in late summer. This has not always been consistent, but some degree of annual management has been performed by the local authority agents. More recently the meadows have been cut but the vegetation has not been removed, although a reasonable number of species continue to persist. Over the years there has been more and more tree planting and the orchids have died out in the clay pit along with some diverse acid coal spoil vegetation. For educational purposes, there has been some clearance of vegetation which covered important geological exposures.

Figure 4. Created meadow vegetation at Ettingshall Park (Photo: Ian C. Trueman)

CONCLUSIONS

The restoration of old collieries takes time and the site may never be as biodiverse as it was originally. However, it is possible that these post-industrial sites can be colonised over time and ultimately become nature reserves. Much depends on the chosen end point of the restoration and the amount of time and money that are invested. The case studies have shown what can be achieved with very little investment other than being left alone to naturally colonise. Where nature has been given a helping hand by introducing species, this has enabled coverage of spoil by vegetation in a fairly short time frame. Here management will ultimately become necessary, especially in order to retain a mosaic of early successional and mature vegetation stages. Where it is desirable to retain rare species and early successional species this management will, however, be more costly, especially if shallow water bodies are to remain open.

It is possible to retain some on-site features as relics of industrial heritage and on-site information boards can alert visitors to what happened at the site in the past. This may increase the amenity value of the site. Pit mounds form notable features within the landscape that people have an association with, children play there and walkers enjoy the views from the top. The colonisation of old colliery buildings by plants or securing them and leaving them in place can create a special atmosphere for a location.

Coal mines in the West Midlands (and especially Telford) have been effectively converted into sports pitches and golf courses (and other amenities) and have been linked with areas of existing nature conservation interest by other old transport links (old railway lines and canals) which themselves are seeing a renaissance due to tourism and providing useful access to the wider countryside. It is possible to plan landscape links to facilitate greater connectivity of sites for biodiversity. Careful planning by facilitating nature through the establishment of green corridors and strategic green planting should help to facilitate this. Footpaths will allow people access to the wider countryside from these post-industrial sites. Sites of old coal mines should be saved for their amenity and heritage value, as features in the landscape and most of all for their nature conservation potential.

REFERENCES

AditNow 2019a: Mine Exploration and Mining History – Baggeridge Colliery. https://www.aditnow.co.uk/Photo/Baggeridge-Colliery_83481/, access: 13.09.2019.

AditNow 2019b: Mine Exploration and Mining History – Granville Colliery. https://www.aditnow.co.uk/Mines/Granville-Coal-Mine_4428/, access: 29.09.2019.

Anonymous 1994a: Minerals Planning Guidance: Coal Mining and Colliery Spoil Disposal. MPG3, Department of the Environment. HMSO, London.

Anonymous 1994b: Amenity Reclamation of Mineral Sites. Department of the Environment. HMSO, London.

Barnes S. 2019: Cannock Chase Council – on Site Conversation with Author. 13.09.2019.

Besenyei L. 2000: The management of artificially created species-rich meadows in urban landscaping schemes. Unpublished PhD thesis, University of Wolverhampton.

Bonthoux S., Brun M., Di Pietro F., Greulich S. & Bouché-Pillon S. 2014: How can wastelands promote biodiversity in cities? A review. Landscape and Urban Planning, 132, 79–88.

Box J. & Cutler A. 1988: Earth Science Conservation. Nature Conservancy Council Regional Focus, 25, 29–35.

Box J., Douse A. & Kohler T. 1994: Non-statutory sites of importance for nature conservation in the West Midlands. Journal of Environmental Planning and Management, 37 (3), 361–367.

Box J.D. & Cossons V. 1988: Three species of clubmoss (Lycopodiaceae) at a lowland station in Shropshire. Watsonia, 17, 69–71.

Bradshaw A.D. & Chadwick M.J. 1980: The Restoration of Land. Blackwell, Oxford.

Buglife 2009: Planning for brownfield biodiversity: A best practice guide. Buglife – The Invertebrate Conservation Trust, Peterborough.

Buglife 2019: Brownfield sites under threat. https://www.buglife.org.uk/resources/habitat-hub/brownfield-hub/brownfields-sites-under-threat/, access: 11.10.2019.

Butterfly Conservation 2019: Brownfields for Butterflies. https://butterfly-con-servation.org/sites/default/files/habitat-brownfield-leaflet.pdf, access: 11.10.2019.

Cannock Chase Council 2015: Fair Lady Coppice Nature Reserve Management Plan. Unpublished.

Cannock Chase Council 2016: Parks and Open Spaces – Fair Lady Coppice. https://www.cannockchasedc.gov.uk/residents/leisure/parks-open-spaces/nature-reserves/fair-lady-coppice, access: 25.09.2019.

Cannock Chase Council 2017: Fair Lady Coppice Local Nature Reserve On-site Information Board.

Cannock Chase Mining Historical Society (no date): Coppice (Fair Lady) Colliery – On-site Information Board, access: 13.09.2019.

Chatterley Whitfield 2019: Chatterley Whitfield Friends. https://chatterleywhitfieldfriends.org.uk/, access: 30.09.2019.

Cossons N. 2011: Saving the Age of Industry. Conservation Bulletin, 67, Autumn, 3–6.

Defra 2014: The National Pollinator Strategy: for bees and other pollinators in England. November 2014. Department for Environment, Food and Rural Affairs.

https://assets.publishing.service.gov.uk/government/uploads/system/uploads/attachment_data/file/794706/national-pollinator-strategy.pdf, access: 27.11.2021.
Defra 2020: Nature conservation: Local Sites in positive conservation management in England, 2008/09 to 2018/19. Department for Environment, Food and Rural Affairs. https://assets.publishing.service.gov.uk/government/uploads/system/uploads/attachment_data/file/857437/Local_Sites_in_positive_conservation_management_England_2008-09-2018-19_FINAL.pdf, access: 27.11.2021.
Doick K.J., Sellers G., Castan-Broto V. & Silverthorne T. 2009: Understanding success in the context of brownfield greening projects: The requirement for outcome evaluation of success assessment. Urban Forestry & Urban Greening, 8 (3), 163–178.
Falconer K. & Gould S. 2011: Stewards of a Global Legacy. Conservation Bulletin, 67, Autumn, 2.
Friends of Granville Country Park (no date): https://friendsofgranvillecountrypark.com/about/, access: 29.09.2019.
Gallis S., Stanford S. & Westwood D. 2015: Baggeridge Country Park Management Plan (2015–2020). South Staffordshire Council.
Gamble D. 2015: Hay Time meadow restoration. Field Studies. https://fsj.field-studies-council.org/media/2123727/fs2015_gamblelr.pdf, access: 27.11.2021.
Gov.UK 2014: Get accreditation for a country park you manage in England. https://www.gov.uk/guidance/get-accreditation-for-your-country-park, access: 27.11.2021.
Gov.UK 2019: Local Nature Reserves, England. https://data.gov.uk/dataset/acdf4a9e-a115-41fb-bbe9-603c819aa7f7/local-nature-reserves-england, access: 27.11.2021.
Green Flag Award 2019: Judging Criteria. http://www.greenflagaward.org.uk/how-it-works/judging-criteria/, access: 27.11.2021.
Grime J.P., Hodgson J.G. & Hunt R. 1988: Comparative Plant Ecology: a functional approach to common British species. Unwin Hyman, London.
HMSO 2021: National Planning Policy Framework. Ministry for Housing, Communities & Local Government, London. https://assets.publishing.service.gov.uk/government/uploads/system/uploads/attachment_data/file /1005759/NPPF_July_2021.pdf, access: 27.11.2021.
Icarus 2014: Access to Nature: inspiring people to engage with their natural environment. Final Evaluation Report, Part 2, Natural England.
Jones G.H., Trueman I.C. & Millett P. 1995: The use of hay strewing to create species-rich grasslands. (I) General principles and hay strewing versus seed mixes. Land Contamination and Reclamation, 3, 104–107.
Land Use Consultants (no date): Baggeridge Development Plan. Report to Seisdon Rural District Council.
Landscape Architects 2002: Green Network Guidance. Unpublished report to Wrekin Council.
Latham M. 2019: Ecology & Green Infrastructure Specialist Telford & Wrekin Council, email to author 26.09.2019.
Lawton J.H., Brotherton P.N.M., Brown V.K., Elphick C., Fitter A.H., Forshaw J., Haddow R.W., Hilborne S., Leafe R.N., Mace G.M., Southgate M.P., Sutherland W.J., Tew T.E., Varley J. & Wynne G.R. 2010: Making Space for Nature: a review of England's wildlife sites and ecological network. Report to Department for Environment, Food and Rural Affairs (DEFRA).
Lunn J. 2000: Ecological and nature conservation aspects of land affected by mining in the Yorkshire Coalfield. Unpublished Master's thesis, Sheffield Hallam University.
Mailding 1980a: Tree Planting. Sheet number 179/01/86, South Staffordshire District Council.
Mailding 1980b: Soil Reconstruction. Sheet number 179/01/127, South Staffordshire District Council.
McCrea A.R., Trueman I.C. & Fullen M.A. 2001b: A comparison of the effects of four arable crops on the fertility depletion of a sandy silt loam destined for grassland habitat creation. Biological Conservation, 97, 181–187.
McCrea A.R., Trueman I.C. & Fullen M.A. 2004: Factors relating to soil fertility and species diversity in both semi-natural and created meadows in the West Midlands of England. European Journal of Soil Science, 55, 335–348.
McCrea A.R., Trueman I.C., Fullen M.A., Atkinson M.D. & Besenyei L. 2001a: Relationships between soil characteristics and species richness in two botanically heterogeneous created meadows in the Urban English West Midlands. Biological Conservation, 97, 171–180.
McKelvey J. 2014: Land Acquisition & Disposal Policy. Unpublished document Shropshire Wildlife Trust, Shrewsbury.
Millward A. & Mostyn B. 1989: People and nature in cities: the social aspects of planning and managing natural parks in urban areas. Urban Wildlife Now, 2, 1–161. Nature Conservancy Council, Peterborough.

Morris C.D. 2019: Lower Gornal. http://www.lowergornal.co.uk/i_mining_baggeridge.htm, access: 23.09.2019.
Natural England 2010: Local Nature Reserves in England: A guide to their selection and declaration. www.naturalengland.org.uk/publications, access: 27.11.2021.
Natural England 2012: Higher Level Stewardship: Environmental Stewardship Handbook. 4th ed. http://publications.naturalengland.org.uk/publication/2827091, access: 27.11.2021.
Natural England 2013: Monitor of Engagement with the Natural Environment: The national survey on people and the natural environment. Wellbeing and the natural environment. http://publications.naturalengland.org.uk/publication/6710511932538880, access: 27.11.2021.
Nature Conservancy Council 1989: Guidelines for Selection of Biological SSSIs. NCC, Peterborough.
Parker D.M. 1995: Habitat creation – a critical guide. English Nature, Peterborough.
Pediaditi K., Doick K.J. & Moffat A.J. 2010: Monitoring and evaluation practice for brownfield, regeneration to greenspace initiatives. A meta-evaluation of assessment and monitoring tools. Landscape and Urban Planning, 97 (1), 22–36.
Potter S. 1991: A Silvicultural Management Plan for Whites Wood. Baggeridge Country Park, Staffordshire County Council.
Ratcliffe D.A. 1976: Thoughts towards a policy of nature conservation. Biological Conservation, 9 (1), 45–53.
Rocha-Nicoleite E., Overbeck G.E. & Müller S.C. 2017: Degradation by coal mining should be priority in restoration planning. Perspectives in Ecology and Conservation, 15, 202–205.
Rodwell J.S. (ed.) 1992: British Plant Communities. Volume 3. Grassland and montane communities. Cambridge University Press, Cambridge.
Rodwell J.S. (ed.) 2000: British Plant Communities. Volume 5. Maritime communities and vegetation of open habitats. Cambridge University Press, Cambridge.
Shropshire History 2016: Granville Colliery. http://shropshirehistory.com/mining/mines/granville.htm, access: 24.09.2019.
Shropshire Wildlife Trust 2008: Management Brief Granville Nature Reserve 2008–2013. Unpublished.
Society for Ecological Restoration 2004: The SER International Primer on Ecological Restoration. Society for Ecological Restoration International, Tucson, Arizona. www.ser.org, access: 11.10.2019.
South Staffordshire Council 2019: Baggeridge Country Park. https://www.sstaffs.gov.uk/leisure/baggeridge-country-park.cfm, access: 23.09.2019.
Staffordshire Biodiversity Action Plan 2019: Staffordshire Site Designations. http://www.sbap.org.uk/staffordshire/index.php?page=sitedesignations, access: 23.09.2019.
Telford & Wrekin Council 2019: Granville Country Park. https://www.telford.gov.uk/info/20629/local_nature_reserves/3866/granville_country_park, access: 26.09.2019.
Telford Nature Conservation Project 1985: North-east Telford an Ecological Report for Telford Development Council. Unpublished report.
Trueman I.C. & Millett P. 2003: Creating wild-flower meadows by strewing green hay. British Wildlife, 15 (1), 37–44.
Trueman I.C. 1991: Baggeridge Country Park Management Plan. Wolverhampton Polytechnic Environmental Consultancy.
Trueman I.C. 1993: Ettingshall Park Management Plan. University of Wolverhampton Environmental Consultancy.
Walker K.J., Stevens P.A., Stevens D.P., Mountford J.O., Manchester S.J. & Pywell R.F. 2004: The restoration and recreation of species-rich lowland grassland on land formerly managed for intensive agriculture in the UK. Biological Conservation, 119, 1–18.

2.3. Brownfield Sites as Hot Spots of Plant Diversity

Ian C. Trueman
*Emeritus Professor of the University of Wolverhampton,
Wulfruna Str., Wolverhampton WV1 1LY, West Midlands, England, UK;
ianctrueman@gmail.com; ORCID: 0000-0003-0129-877X*

Barbara Tokarska-Guzik
*Research Team of Botany and Nature Protection,
Institute of Biology, Biotechnology and Environmental Protection,
Faculty of Natural Sciences, University of Silesia in Katowice;
28 Jagiellońska Str.,40-032 Katowice, Poland;
barbara.tokarska-guzik@us.edu.pl; ORCID: 0000-0002-4058-1220*

Eleanor V.J. Cohn
*Retired member of staff of the University of Wolverhampton;
Wulfruna Str., Wolverhampton WV1 1LY, West Midlands, England, UK;
evjcohn@googlemail.com; ORCID: 0000-0003-4456-3406*

ABSTRACT: The present study focuses on those brownfield lands in which the entire soil and subsoil has been replaced by the waste products of industry. The ability of plants to occupy such material is limited and the rate of colonisation may be very slow or patchy. It is one of the tasks of the mining engineer to deal with these soil and topographical problems to generate a stable vegetation where instability would be dangerous or, sometimes, just for aesthetic reasons. There are, however, many situations in which such spoil has been left, sometimes for long periods of time, to develop vegetation spontaneously. On occasion, and for different kinds of reasons, known and unknown, types of vegetation of nature conservation interest can develop in such situations. This paper reviews the types of vegetation developing on the waste materials of industry with particular regard to examples of plant diversity hotspots in the UK and in Poland. We also try to address the important question of how brownfield sites could be managed to conserve or even develop biodiversity hotspots.

KEYWORDS: Post-industrial sites, biodiversity hotspots, post-industrial plant communities, Calaminarian Grasslands, Open Mosaic Habitats on Previously Developed Land

INTRODUCTION

In UK urban planning, brownfield sites are defined as any previously developed land not currently in use (Grimski & Ferber 2001; Alker *et al.* 2010)[1]. This definition includes a vast array of sites

[1]In Poland, according to the Government Programme for post-industrial areas, these are "degraded, unused or not fully used areas originally intended for economic activity which has been completed" (Ministerstwo Środowiska 2004). Domański (2000) adds to this definition the issue of areas degraded by industrial activities, including landfills, water and soil pollution, i.e. the immediate surroundings of industrial plants.

which have previously been covered with factory buildings and demolished housing which are not the principal focus of the present study. Gilbert (1989) described the typical ecological succession on such sites. The initial colonists tend to be annual herbs, and this stage offers opportunities for many species associated, as weeds, with agriculture which are becoming rare in the countryside due to changes in agricultural practices. A continuous succession of flowers is produced through the growing season which is important for bees and other pollinating insects. For both these reasons, this initial stage after land clearance has nature conservation significance and should not be subjected to strenuous "greening-up" procedures. However, after a few years, populations of slugs and other small grazing invertebrates build up and seedlings of annual species are selected against. The vegetation of perennials which succeeds the annuals is usually of less ecological interest and common wind-pollinated grasses usually eventually predominate (Angold *et al.* 2006). Colonisation by trees may eventually produce unusual "urban woodland" which may be of increased interest but generally, by this time, the land will have come back into human use. A good example of such woodlands are in the railway area around the traffic junction at Gleisdreieck in Kreuzberg, Berlin, which became a wasteland at the end of the Second World War, and developed a rich diversity of vegetation. The area was developed into a park, completed in 2013, which has attempted to preserve the ecosystems which had developed (Kowarik 2015).

The focus of the present study is the brownfield lands in which the entire soil and subsoil has been replaced by the products of industry. Such sites may well include patches where equipment was placed which resemble those described above and also fragments of pre-industrial landscapes which may bear remnants of pre-existing semi-natural vegetation, but predominantly the substrate is the spoil from mining or from some other industrial process. These materials tend to share a number of characteristics inimical to rapid colonisation by vegetation. Often the material is sorted by particle size, so that the finer material lacks the air spaces necessary for plant roots and the coarser material holds little water and becomes liable to drought (Gilbert 1989; Jędrzejczyk-Korycińska & Rostański 2015a; Rich *et al.* 2015). Passages for water and air are disrupted and have to develop anew. The material is almost always devoid of organic matter and of forms of nitrogen available to plants, and usually also lacks appropriate forms and concentrations of other major plant nutrients, notably potassium and phosphorus. The pH of the material may be too low or high for most plants and the processes of weathering may release chemicals which make the pH even more extreme (Wiggering 1993; Gehrke *et al.* 1998). The material may well have an unusual balance of elements compared with normal soils or be contaminated with elements such as lead or zinc which are toxic to plants (Jędrzejczyk-Korycińska & Rostański 2015a; Pohrebennyk *et al.* 2019). At least initially it will be devoid of biota: micro-organisms (of which mycorrhizae are probably the most important), plant propagules, earthworms and invertebrates generally.

As a result of problems such as those described above, the ability of plants to occupy such material is limited and the rate of colonisation may be very slow or patchy. It is one of the tasks of the mining engineer to deal with these soil and topographical problems to generate substrate stability by means of vegetation where instability would be dangerous or, sometimes, just for aesthetic reasons. There are however, many situations in which such spoil has been left, sometimes for long periods of time, to develop vegetation spontaneously. On occasion, and for different kinds of reasons, known and unknown, types of vegetation of nature conservation interest can develop in such situations (Tokarska-Guzik *et al.* 1991; Tokarska-Guzik 1996; Gibson & Lyons 1998; Tokarska-Guzik & Rostański 2001; Jędrzejczyk-Korycińska 2009; Ishimatsu & Ito 2013; Jędrzejczyk-Korycińska *et al.* 2015b; Rostański *et al.* 2015a). It is of interest here to discuss why this should be the case.

Plants have been divided, very broadly and with intermediates, into competitors, ruderals and stress-tolerators (Grime 1979). Competitor species are typically vigorous and strong-growing. In order to achieve this they require low levels of physical disturbance (both of the substrate and the existing plant growth) and stress (water stress, the stress of heavy shade and particularly stress associated with low soil fertility or high toxicity). Under conditions of disturbance and/or stress they grow very poorly, if at all. Competitors are so-called because under conditions of low stress and low disturbance they will out-compete other plants and tend to produce vegetation dominated by a small number of such species. Such vegetation is unlikely to be rich in plant species or to provide biodiversity hotspots.

The ruderals are the plant "opportunists". They take advantage of brief periods of disturbance (particularly of the substrate), resulting in the loss of existing vegetation, to appear and rapidly go through their life cycle (typically very short) and produce quantities of seed or other propagules which can be widely dispersed and lie in the soil to await the next phase of disturbance. Such plants will require low levels of stress during their growth phase since it will necessarily be short. The vegetation can be species-rich but is of little long-term interest in nature conservation terms since it exists for such a short period of time. This is the kind of vegetation described in the first paragraph of this study. Under some circumstances such species can contribute to spoil communities, for example areas which consistently experience high levels of drought in summer can sometimes bear unusual and species-rich communities of annual species consistently, year after year.

The third group, the stress-tolerators, includes many types of xerophytes, hydrophytes and others adapted to particular kinds of stress, even heavy-metal toxicity, as described above. Typically, they have a fairly narrow range of adaptation and are therefore specific to a particular set of circumstances. Because of this, they tend to be relatively slow-growing and not strongly competitive. Stress-tolerators are not usually tolerant of physical disturbance, since they are dependent on the efficient use of limited environmental resources, nor can they compete with competitor species under ideal conditions for plant growth. They will generally coexist with other stress-tolerators, each plant exploiting the particular niche to which it is adapted. Such communities tend to be relatively species-rich and since the adaptations of the species are usually very specific, the species which make them up tend to be uncommon and predominantly native. For these reasons, stress-tolerators tend to feature prominently in biodiversity hotspots, and the relatively extreme conditions set by brownfield sites made up largely of the waste material of industries have the possibility for the development of such biodiversity hotspots. Furthermore, habitat conditions similar to more extreme and uncommon natural habitats can sometimes be provided, and these can help maintain populations of some of their characteristic rare species (Rostański *et al.* 2015b).

TYPES OF VEGETATION DEVELOPING ON THE WASTE MATERIALS OF INDUSTRY

Some old post-industrial sites have long been celebrated for their floristic richness (Davis 1976; Greenwood & Gemmell 1978) and a few are managed as nature reserves, but all conurbations include both large and small examples (Tokarska-Guzik & Rostański 2001; Trueman *et al.* 2001; Bielczyk *et al.* 2009; Rich *et al.* 2015; Rostański *et al.* 2015a).

Biodiversity hotspots associated with natural and semi-natural vegetation have detailed and long-established systems of description and comparison, and hence relatively straightforward processes for evaluation for conservation. In the UK, under the Wildlife and Countryside Act 1981 (amended 1985), the government has a duty to notify as a Site of Special Scientific Interest (SSSI) any land which in its opinion is of special interest by reason of any of its flora, fauna, geological or physiographical features. An SSSI has legal protection against potentially damaging operations including development. Despite the typical predominance of native species, the biodiversity hotspots which develop on the waste products of industry may not closely resemble those seen in the countryside. New combinations of native and alien species often come together in these sites, particularly in urban areas and, at least in theory, have no parallel in natural and semi-natural plant communities. The term "ecological recombination" (Barker 2000; Rotherham 2017) has been coined to encompass these *de novo* communities.

As a result of these characteristics, the evaluation of post-industrial plant communities has rarely been straightforward and has usually been based directly on individual examples and the conservation status of the particular species which they support. The UK Biodiversity Action Plan attempted to deal with these problems by including two relevant habitats which were given nature conservation priority: "Calaminarian Grasslands" (Defra 2008), characterised by high levels of heavy metals and "Open Mosaic Habitats on Previously Developed Land" (Defra 2010), characterised by mosaics of bare substrate and early successional communities made up of stress-tolerant species. The former have some semi-natural examples on natural outcrops of serpentine and related rock types, but also occur in urban and post-industrial areas on mine workings and spoil heaps. The Calaminarian Grasslands type is synonymous with the wider European H6130 Calaminarian Grasslands of the *Violetalia calaminariae*. In southern Poland, mainly in the Silesian-Kraków Upland, there are areas with high concentrations of heavy metals (zinc and lead, also silver) in the ground. Exploitation carried out since the 12th Century in this area contributed to various local modifications of the area, while arousing great interest from naturalists/scientists concerning the development of specific plant cover in these areas (Jędrzejczyk-Korycińska 2006; Nowak *et al.* 2011; Godzik 2015; Woch *et al.* 2017). Open Mosaic Habitats are entirely characteristic of post-industrial sites with the soil severely modified or replaced, usually by industrial spoil and where the normal processes of succession are curtailed or even prevented by the nature of the substrate.

EXAMPLES OF PLANT DIVERSITY HOTSPOTS ON THE WASTE MATERIALS OF INDUSTRY

Coal spoil is a widespread and abundant kind of industrial waste. Where it contains sulphides it can oxidise on exposure to the air to produce extremely acidic substrates which are only colonised with difficulty by vascular plants. Such sites can develop diverse cryptogamic floras (Middleton & Lunn 2001; Bielczyk *et al.* 2009; Rola *et al.* 2014), in particular unusual lichen heaths in which the dominant species are fruticose *Cladonia* lichens. Where the acidification is less extreme, heathland communities with *Calluna vulgaris*-dominated vegetation can develop similar to that developing on natural sands and gravels. However, coal spoil tips can offer a wide range of possibilities for the development of biodiversity. Coal mining took place in South Wales over a long period and has left extensive spoil tips. These are complex areas of varied topography, aspect, substrate, composition, hydrology, pH (from acid to calcareous) and levels of disturbance (Olds & Wistow 2018), with calcifuges such as *Erica tetralix* and *Galium saxatile* growing side by side with calcicoles such as *Carlina vulgaris* and *Linum catharticum*. Waterlogged areas on flatter ground can develop communities comparable with those of dune slacks, with Marsh-orchids *Dactylorhiza* spp. and nationally scarce Wintergreens *Pyrola* spp. Lichen heaths are a particular feature, and Humphries (2012) gives *Cladonia portentosa*, *Cladonia cervicornis* ssp. *verticillata*, *Cladonia floerkeana, Dibaeis baeomyces* and *Peltigera hymenina* as characteristic species. In a bryophyte survey of Varteg Colliery in Gwent, Bosanquet (2005) recorded 83 moss and 18 liverwort taxa, of which ten were of national or local interest. The Yorkshire coalfields have also developed similar sites of nature conservation significance (Middleton 2000; Lunn 2001).

The "Black Country" of the English West Midlands also has a long coal-mining history and it was possible to describe large tracts in 1950 as "covered with coal spoil" (Rees & Skelding 1950) but this was much less acidic than the coal spoil of South Wales or Yorkshire coalfields and has since largely disappeared under redevelopment and ecological succession. In one interesting exception at Clayhanger in Walsall, spoil tips and mining subsidence have combined with remnants of unimproved pastures to give a rich blend of open water, reed swamp, moderately acidic spoil vegetation and species-rich grassland which was scheduled as a Site of Special Scientific Interest in the 1980s (Insert 1).

Insert 1. Clayhanger Site of Special Scientific Interest, Walsall, UK West Midlands (Photo: Ian C. Trueman)

Spoil tips and mining subsidence have combined with unimprovable remnants of pastures to give a rich blend of open water, reed swamp, moderately acidic spoil vegetation and species-rich grassland.

Common Restharrow
Ononis repens

Tubular Water-dropwort
Oenanthe fistulosa

Inland Club-rush
Bolboschoenus laticarpus

General view of the area

Silver Hair-grass
Aira caryophyllea

Meadow Thistle
Cirsium dissectum

This ability of post-industrial sites to "trap" remnants of pre-industrial landscapes is a further factor in the development of brownfield biodiversity hotspots. An opencast coalmine at Stoneyhill in Telford, Shropshire, abandoned without remediation in the 1950s, developed by 1983 a vegetation rich in *Dactylorhiza fuchsii* in clay areas and a much more open vegetation with three clubmosses: *Lycopodium clavatum*, *Huperzia selago* and *Diphasiastrum alpinum* on stony clay. This was the first record for *D. alpinum* in Shropshire since 1723 (Insert 2).

The UK Black Country was also the site of much of the early development of iron and steel smelting in Britain in the 17th and 18th Centuries. This resulted in many tips of furnace slag. This is a very different material from coal spoil, consisting largely of silica and calcium oxide, with smaller proportions of aluminum, iron, manganese, magnesium oxides, sulphates and phosphates. There may be large glass-like blocks and finely-divided material. The tips often included quite high proportions of moulding sand. The material has uses in building and these tips have largely disappeared from the landscape, but a few remain and offer similar difficulties for plant colonisation to those posed by coal spoil. One at Burnt Tree in Sandwell in the UK Black Country overlies an astonishing complex of old shafts, collieries, furnaces, tramways, even a lime works. There are areas in which neophytes and native species coexist in recombinant communities and drought-stressed areas with lichen heaths and communities of winter annuals which appear every year (Insert 3).

A more recently tipped material is Pulverised Fuel Ash (PFA), the fine ash produced during the combustion process used at coal-fired power stations. Much of this has been disposed of as waste, mostly in landfill or as mounds and covered with topsoil, but Shaw (1992) described the vegetational changes and soil development where PFA lagoons in the Lee Valley in southern England had been allowed to develop without remediation. These resemble those taking place on coal spoil, with open birch-willow scrub rich in *Dactylorhiza* orchids developing within 24 years.

Spoil from heavy metal extraction and smelting is in a different category, because the material may well be toxic to plants as well as being poorly endowed for plant growth (Insert 4). In extreme examples, typical races of all plant species are completely unable to grow on the substrate. This sets up the conditions for rapid evolution of races tolerant to specific heavy metals and this has been widely shown to occur (Antonovics *et al.* 1971; Wu *et al.* 1975; Rostański *et al.* 2015b). These races do not develop in all species and are usually poorly adapted to conditions where the heavy metal is absent. Such races, however, clearly have value in the revegetation of spoil and some concentrate heavy metals to the extent that they have been used to remove heavy metals from contaminated soils (Saxena *et al.* 1999).

The vegetation which develops naturally on such sites can sometimes have biodiversity value, although of course there must be evaluation of the risk to humans and animals in the general locality from the contamination of water and the dispersal of contaminated dust. The vegetation develops particularly slowly, and where heavy metal levels are moderate, highly species-rich vegetation can eventually develop. Individual species can become so characteristic of such sites that they can be used as indicators of heavy metal contamination. The lead mine spoil tips of Wales and Derbyshire in England are characterised by the presence of *Minuartia verna*, a plant otherwise found in upland limestone and base-rich volcanic rock and *Noccaea caerulescens*, a plant from high limestone mainly in Scotland away from the spoil tips. Similarly, *Biscutella laevigata* is characteristic of the lead mine spoil tips of Olkusz, between Katowice and Kraków in Poland (Insert 5). *Biscutella laevigata* is a species limited in Poland to only two areas – lowland in the Olkusz post-mining area (Kowolik *et al.* 2010) and mountain in the limestone Western Tatras (Wierzbicka & Pielichowska 2002). The lowland site is considered a glacial relic (Kowolik *et al.* 2010 after Szafer 1959). An abundant, new locality of this species is also known from rocky grasslands in the surroundings of Kozubów in Nida Basin (Rostański 2014 after Przemyski & Piwowarczyk 2012). The abandoned metalliferous mining sites of Cornwall, where copper, tin, arsenic and lead were extracted, have also developed nature conservation value and *Armeria maritima* is a local metallophyte (Spalding 2005). Mercury and copper spoil tips in central Slovakia conserve some rare and valuable bryophyte species (Širka *et al.* 2018).

Insert 2. An opencast coalmine at Stoneyhill in Telford, Shropshire, UK (Photo: Ian C. Trueman)

Abandoned without remediation in the 1950s, it had developed by 1983 a vegetation rich in *Dactylorhiza fuchsii* in fine clay areas and a much more open vegetation with three clubmosses – *Lycopodium clavatum*, *Huperzia selago* and *Diphasiastrum alpinum* – on stony clay. This was the first record for *D. alpinum* in Shropshire since 1723.

Alpine Clubmoss
Diphasiastrum alpinum

Fir Clubmoss
Huperzia selago

General view of the site

Stag's-horn Clubmoss
Lycopodium clavatum

Common Spotted-orchid
Dactylorhiza fuchsii

Insert 3. Flora of a furnace spoil tip at Burnt Tree, Black Country, UK
(Photos: Ian C. Trueman, Mike Poulton, EcoRecord)

The tip overlies an astonishing complex of old shafts, collieries, furnaces, tramways, even a lime works. There are areas in which neophytes and native species coexist in recombinant communities and drought-stressed areas with lichen heaths and communities of winter annuals which appear every year.

Blue Fleabane
Erigeron acris

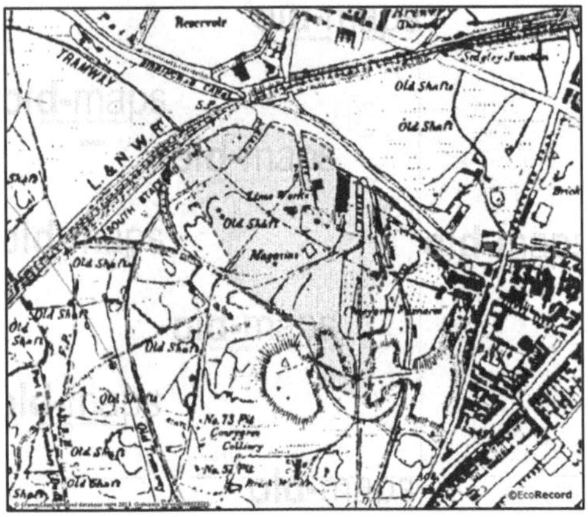

Site overlain (red area) on 1880s Ordnance Survey Map

Bee Orchid
Ophrys apifera

Common Cudweed
Filago vulgaris

Little Mouse-ear
Cerastium semidecandrum

General views of the area

Insert 4. Lead/zinc spoil in Poland (Photo: Barbara Tokarska-Guzik)

The zinc smelter spoil heap in Katowice-Wełnowiec without (left) and with vegetation cover (right)

The population of *Cardaminopsis hallerii* is the most abundant here

Degraded areas around the former non-ferrous (high thallium and zinc contamination) Metals Plant in Katowice-Szopienice (1834–2008)

Species composition is poor – mainly Sheep's-fescue *Festuca ovina*, Sand Rock-cress *Arabis arenosa* and Bladder Campion *Silene vulgaris* occur here

Insert 5. Flora of lead/zinc contaminated sites (Photo: Ian C. Trueman & Barbara Tokarska-Guzik)

Phytotoxicity greatly limits the possibility of colonisation of these sites, and those contaminated with other heavy metals. Nevertheless, extreme selective pressure can result in the evolution of heavy metal tolerant races of some species, which become valuable resources for the revegetation of contaminated sites and in some instances are even capable of removing the contaminants from the substrate by concentrating them in plants of the tolerant races. Certain species are also characteristic of contaminated sites, to the extent of being indicators for contamination. These are often species scarce in the typical lowland situations in which these sites occur and are much more characteristic of higher altitude sites where there is no obvious contamination.

Alpine Penny-cress
Noccaea caerulescens
Derbyshire, UK

Buckler-mustard
Biscutella laevigata
Olkusz, Poland

Olkusz site, Poland

Spring Sandwort
Minuartia verna
Wales, UK

Thrift
Armeria maritima
Olkusz, Poland

Many industrial processes generate huge quantities of calcium-rich spoil. The high pH and the abundance of calcium interact with the poverty of nutrient and organic matter to produce some astoundingly rich and diverse plant communities. One British example, early recognised as a biodiversity hotspot, is the LeBlanc process tip at Nob End, Bolton in Lancashire, created in the 19th Century. Originally it would have been quite toxic to most plants (and humans!), consisting as it did of calcium sulphide, calcium oxide and limestone, but it has weathered to allow the development of open vegetation including many calcicolous species, many otherwise very uncommon or even absent in Lancashire away from the coastal dunes (Insert 6). It was declared a Site of Special Scientific Interest in 1988 (Shaw & Halton 1998).

The Solvay process produced much less toxic calcareous waste and has generated extremely biodiverse plant communities, for example in Cheshire in the UK (Lee & Greenwood 1976) and in Jaworzno (Upper Silesia) in Poland (Cohn *et al.* 2001 and Insert 7). The flora of the "soda heap" in Jaworzno, small in terms of surface area, has a population of nearly 140 species of vascular plants. Despite the anthropogenic origin of the area, native plants dominate the flora. Particularly noteworthy is the group of several dozen species of plants associated with thermophilic (xerothermic) grassland communities and over a dozen calcifuge species of peat bogs. The species composition of the flora includes many protected, locally and regionally rare species, including 7 orchid species. The population of *Epipactis palustris*, a rare species on the Polish scale, is particularly numerous. It is also the most abundant location of this species within the city limits (Tokarska-Guzik 1996; Cohn *et al.* 2001).

HOW CAN BROWNFIELD SITES BE MANAGED TO CONSERVE (OR EVEN DEVELOP?) BIODIVERSITY HOTSPOTS?

As has been discussed above, there are problems with evaluating post-industrial sites and as a result, many are lost to "reclamation" or development without sufficient consideration (Harrison & Davies 2002). This danger is perhaps particularly acute in Britain, where greater and greater pressure is placed on urban local authorities to develop their brownfield sites to defend "Green Belts" around conurbations, the extremely successful idea to prevent encroachment and development in the countryside. Even without development, brownfield sites are often threatened by misuse. Where there is public access it is probable that, without publicity, the ecological value of important brownfield sites will not be widely appreciated, and such sites are prone to the dumping of miscellaneous rubbish and inappropriate use such as motorcycle racing. The owners of the sites may themselves destroy the botanical interest through misuse.

The first principle in conserving biodiversity hotspots on brownfield sites is to be aware that they occur and to encourage botanical surveys to identify and evaluate them. It should also be pointed out, that these sites may well have wider interest for their birds, invertebrates etc., and therefore survey should extend beyond the botany. Where there is a case for conservation, it is not enough just to protect against public misuse and, wherever possible, to encourage public appreciation and enjoyment. It is also usually necessary to understand that the processes of ecological succession are likely to be occurring, however slowly, and there may be a need to instigate programmes of management to limit the degradation of areas at particularly biodiverse stages. As more normal soils start to develop, it is possible that invasive alien or native species may appear, and may require management to limit their encroachment where they threaten the existing diversity.

Because of their initial characteristics such as stressful conditions and a lack of a bank of biota, brownfield sites may also, after evaluation, be suitable places for the creation of new botanically-rich habitats. Where the process of colonisation is well-understood it might be possible to propagate existing examples of high biodiversity onto new sites.

Insert 6. Nob End LeBlanc Process Tip, Bolton, UK (Photo: Ian C. Trueman)

This site originated as LeBlanc process spoil – Calcium sulphide plus Calcium carbonate – in the 19th Century. Declared a Site of Special Scientific Interest in 1988, the vegetation is unusual but resembles various calcicolous types of grassland and supports many notable species.

Early Marsh-orchid
Dactylorhiza incarnata

General view of the site

Fragrant-orchid
Gymnadenia conopsea

Northern Marsh-orchid
Dactylorhiza purpurella

Marsh Helleborine
Epipactis palustris

Southern Marsh-orchid
Dactylorhiza praetermissa

Insert 7. The old solvay process spoil tip in Upper Silesia, Poland (Photo: Barbara Tokarska-Guzik)

This soda heap, so-called in Polish "Wapniówka", the calcium storage site of the Austrian Ammonia Soda Factory, was closed in 1909. The habitat conditions here are specific: alkaline soil which limits decay and biological activity generally, creating extreme conditions for the flora (Tokarska-Guzik 1996; Cohn *et al.* 2001).

It is a table-type heap with a flat surface, with characteristic curved edges (specificity related with the method of waste storage). The types of vegetation developed here, together with their special, rich floristic and fauna composition, are extremely rare in the country

The population of Marsh Helleborine – *Epipactis palustris* is the most abundant orchid here

Broad-leaved Marsh-orchid
Dactylorhiza majalis

Fragrant-orchid
Gymnadenia conopsea

Dark-red Helleborine
Epipactis atrorubens

Marsh Helleborine
Epipactis palustris

CONCLUSIONS: THE RESEARCH SIGNIFICANCE OF BROWNFIELD BIODIVERSITY HOTSPOTS

Mining and other industrial waste tips are extremely unusual in ecological terms since they are, or can be, biologically sterile and can offer the possibility of studying the processes of primary colonisation both at the surface and in the substrate. The difficult conditions which they present mean that they have been and continue to be sources of information about micro-evolution in response to extreme selective pressures, for example in spoil contaminated with heavy metals. All this work has already informed revegetation techniques and will continue to do so (Prach & Pyšek 2001; Prach & Hobbs 2008). Apart from the information these communities give us about the requirements of individual species, they offer the possibility of investigating and understanding biodiversity and the nature of biological communities and developing schemes for their management. For these reasons there is a case to be made for the conservation of brownfield biodiversity hotspots and for the ecology of brownfield sites to be carefully and comprehensively examined when they are being considered for development.

REFERENCES

Alker S., Joy V., Roberts P. & Smith N. 2010: The definition of brownfield. Journal of Environmental Planning and Management, 43 (1), 49–69.

Angold P.G., Sadler J.P., Hill M.O., Pullin A., Rushton S., Austin K., Small E., Wood B., Wadsworth R., Sanderson R. & Thompson K. 2006: Biodiversity in urban habitat patches. Science of the Total Environment, 360, 196–204.

Antonovics J., Bradshaw A.D. & Turner R.G. 1971: Heavy metal tolerance in plants. Advances in Ecological Research, 7, 1–85.

Barker G. (ed.) 2000: Ecological Recombination in Urban Areas: Implications for Nature Conservation. A workshop held at CEH (Monks Wood).

Bielczyk U., Jędrzejczyk-Korycińska M. & Kiszka J. 2009: Lichens of abandoned zinc-lead mines. Acta Mycologica, 44 (2), 139–149.

Bosanquet S.D.S. 2005: Varteg colliery site bryophyte survey. Report for Torfaen County Borough Council.

Cohn E.V.J., Rostański A., Tokarska-Guzik B., Trueman I.C. & Woźniak G. 2001: The flora and vegetation of an old solvay process tip in Jaworzno (Upper Silesia, Poland). Acta Societatis Botanicorum Poloniae, 70 (1), 47–60.

Davis B.N.K. 1976: Wildlife, urbanization and industry. Biological Conservation, 10 (4), 249–291.

Defra 2008: UK Biodiversity Action Plan: Priority Habitat Descriptions. Calaminarian Grasslands. http://jncc.defra.gov.uk/page-5706.

Defra 2010: UK Biodiversity Action Plan: Priority Habitat Descriptions. Open Mosaic Habitats on Previously Developed Land. http://jncc.defra.gov.uk/page-5706.

Domański B. 2000: Restrukturyzacja terenów poprzemysłowych w miastach. [In:] Ziobrowski Z., Ptaszycka-Jackowska D., Rębowska A. & Geissler A. (eds.): Rewitalizacja, rehabilitacja i restrukturyzacja – odnowa miast. Instytut Gospodarki Przestrzennej i Komunalnej Oddział w Krakowie, Kraków, 107–142.

Gehrke T., Telegdi J., Thierry D. & Sand W. 1998: Importance of extracellular polymeric substances from *Thiobacillus ferrooxidans* for bioleaching. Applied and Environmental Microbiology, 64 (7), 2743–2747.

Gibson N. & Lyons M.N. 1998: Flora and Vegetation of the Eastern Goldfields Ranges: Part 3. Parker Range. Journal of the Royal Society of Western Australia, 81, 119–129.

Gilbert O.L. 1989: The Ecology of Urban Habitats. Chapman & Hall, London-New York.

Godzik B. (ed.) 2015: Natural and historical values of the Olkusz Ore-bearing Region. W. Szafer Institute of Botany, Polish Academy of Sciences, Kraków.

Greenwood E.F. & Gemmell R.P. 1978: Derelict industrial land as a habitat for rare plants in S. Lancs. (v.c. 59) and W. Lancs. (v.c. 60). Watsonia, 12, 33–40.

Grime J.P. 1979: Plant Strategies and Vegetation Processes. John Wiley & Sons, Chichester-New York-Brisbane-Toronto.

Grimski D. & Ferber U. 2001: Urban brownfields in Europe. Land Contamination & Reclamation, 9 (1), 143–148.
Harrison C. & Davies G. 2002: Conserving biodiversity that matters: practitioners' perspectives on brownfield development and urban nature conservation in London. Journal of Environmental Management, 65 (1), 95–108.
Humphries R.N. 2012: Case study: The contribution of active surface mines in the conservation of lichen communities in the South Wales coalfield, United Kingdom. Journal of The American Society of Mining and Reclamation, 2 (1), 80–98.
Ishimatsu K. & Ito K. 2013: Brown/biodiverse roofs: a conservation action for threatened brownfields to support urban biodiversity. Landscape Ecology and Engineering, 9, 299–304.
Jędrzejczyk-Korycińska M. & Rostański A. 2015a: Tereny o wysokiej zawartości metali ciężkich w podłożu na Górnym Śląsku. [In:] Wierzbicka M. (ed.): Ekotoksykologia: rośliny, gleby, metale. Wydawnictwo Uniwersytetu Warszawskiego, Warszawa, 175–188.
Jędrzejczyk-Korycińska M. 2006: Floristic diversity in calamine areas of the Silesia-Cracow monocline. Biodiversity: Research and Conservation, 3–4, 340–343.
Jędrzejczyk-Korycińska M. 2009: Obszary dawnej eksploatacji złóż cynkowo-ołowiowych – ich bogactwo florystyczne a możliwości ochrony. Problemy Ekologii Krajobrazu, 24, 71–80.
Jędrzejczyk-Korycińska M., Zagórna M. & Godzik B. 2015b: Protected and protection-worthy ecological areas or features in the Olkusz Ore-bearing Region. [In:] Godzik B. (ed.): Natural and historical values of the Olkusz Ore-bearing Region. W. Szafer Institute of Botany, Polish Academy of Sciences, Kraków, 315–327.
Kowarik I. 2015: Gleisdreieck: how urban wilderness became possible in the new park. [In:] Lichtenstein A. & Mameli F.A. (eds.): (Gleisdreieck/Park Life Berlin), Transcript, Verlag, 216–221.
Kowolik M., Szarek-Łukaszewska G. & Jędrzejczyk-Korycińska M. 2010: Użytek ekologiczny „Pleszczotka górska" w cynkowo-ołowiowym terenie pogórniczym – potrzeba aktywnej ochrony. Chrońmy Przyrodę Ojczystą, 66 (1), 35–38.
Lee J.A. & Greenwood B. 1976: The colonisation by plants of calcareous wastes from the salt and alkali industry in Cheshire, England. Biological Conservation, 10, 131–149.
Lunn J. 2001: Wildlife and mining in the Yorkshire coalfield. British Wildlife, 12 (5), 318–326.
Middleton P. & Lunn J. 2001: Bryophytes and Lichens of Colliery Spoil Heaps in Yorkshire. The Naturalist, 126 (1038), 151–156.
Middleton P. 2000: The wildlife significance of a former colliery site in Yorkshire. British Wildlife, 11 (5), 333–339.
Ministerstwo Środowiska 2004: Program rządowy dla terenów poprzemysłowych przyjęty przez Radę Ministrów w dniu 27 kwietnia 2004 roku. Warszawa.
Nowak T., Kapusta P., Jędrzejczyk-Korycińska M., Szarek-Łukaszewska G. & Godzik B. 2011: The vascular plants of the Olkusz Ore-bearing Region [Rośliny naczyniowe Olkuskiego Okręgu Rudnego]. W. Szafer Institute of Botany, Polish Academy of Sciences, Kraków.
Olds L. & Wistow R. 2018: Colliery-spoil biodiversity of the South Wales Valleys. British Wildlife, 30 (2), 108–115.
Pohrebennyk V., Koszelnik P., Mitryasova O., Dzhumelia E. & Zdeb M. 2019: Environmental monitoring of soils of post-industrial mining areas. Journal of Ecological Engineering, 20 (9), 53–61.
Prach K. & Hobbs R.J. 2008: Spontaneous succession versus technical reclamation in the restoration of disturbed sites. Restoration Ecology, 16, 363–366.
Prach K. & Pyšek P. 2001: Using spontaneous succession for restoration of human-disturbed habitats: Experience from Central Europe. Ecological Engineering, 17 (1), 55–62.
Przemyski A. & Piwowarczyk R. 2012: *Biscutella laevigata* L. in the Polish Uplands – new data on its distribution from Nida Basin. Botanika – Steciana, 16, 21–29.
Rees W.J. & Skelding A.D. 1950: Vegetation. [In:] Kinvig R.H., Smith J.G. & Wise M.J. (eds.): Birmingham and its Regional Setting. British Association, Birmingham, 65–76.
Rich K.J., Ridealgh M., West S.E., Cinderby S. & Ashmore M. 2015: Exploring the links between post--industrial landscape history and ecology through participatory methods. PLoS ONE, 10 (8), e0136522.
Rola K., Osyczka P. & Nobis M. 2014: Cryptogamic communities dominated by the lichen *Cladonia rei* – a case study of Polish post-smelting dumps in a worldwide context. Herzogia, 27 (1), 121–135.

Rostański A. 2014: Intentional introduction of *Biscutella laevigata* L. on the post-flotation lead-zinc ore spoil heap in Piekary Śląskie (Upper Silesia, Southern Poland). Environmental & Socio-economic Studies, 2 (3), 24–32.

Rostański A., Jędrzejczyk-Korycińska M. & Nowak T. 2015a: Ochrona przyrody obszarów metalonośnych. [In:] Wierzbicka M. (ed.): Ekotoksykologia: rośliny, gleby, metale. Wydawnictwo Uniwersytetu Warszawskiego, Warszawa, 507–521.

Rostański A., Nowak T. & Jędrzejczyk-Korycińska M. 2015b: Metalolubne gatunki roślin we florze Polski. [In:] Wierzbicka M. (ed.) Ekotoksykologia: rośliny, gleby, metale. Wydawnictwo Uniwersytetu Warszawskiego, Warszawa, 299–322.

Rotherham I.D. 2017: Recombinant Ecology – A Hybrid Future? Springer International Publishing.

Saxena P.K., KrishnaRaj S., Dan T., Perras M.R. & Vettakkorumakankav N.N. 1999: Phytoremediation of heavy metal contaminated and polluted soils. [In:] Prasad M.N.V., Hagemeyer J. (ed.): Heavy Metal Stress in Plants. From Molecules to Ecosystems. Springer-Verlag, Berlin-Heidelberg, 305–329.

Shaw P.J.A & Halton W. 1998: Classic Sites. Nob End, Bolton. British Wildlife, 10 (1), 13–17.

Shaw P.J.A. 1992: A preliminary study of successional changes in vegetation and soil development on unamended fly ash (PFA) in southern England. Journal of Applied Ecology, 29 (3), 726–736.

Širka P., Kubešowá S. & Mišiková K. 2018: Bryophytes of spoil heaps rich in toxic metals in Central Slovakia. Thaiszia – Journal of Botany, Košice, 28 (1), 59–77.

Spalding A. 2005: The nature-conservation value of abandoned metalliferous mine sites in Cornwall. British Wildlife, 16 (3), 175–183.

Szafer W. 1959. Szata roślinna Polski. Tom I i II. Państwowe Wydawnictwo Naukowe, Warszawa.

Tokarska-Guzik B. & Rostański A. 2001: Możliwości i ograniczenia przyrodniczego zagospodarowania terenów poprzemysłowych. [In:] Rostański K. (ed.): Hałdy poprzemysłowe – obiekty badań botanicznych. Natura Silesiae Superioris, Suplement, 5–17.

Tokarska-Guzik B. 1996: Rola hałd zasadowych w utrzymaniu lokalnej bioróżnorodności. Przegląd Przyrodniczy, 7 (3–4), 261–266.

Tokarska-Guzik B., Rostański A. & Klotz S. 1991: Roślinność hałdy pocynkowej w Katowicach-Wełnowcu [Vegetation of the zinc smelter spoil heap in Katowice-Wełnowiec]. Acta Biologica Silesiaca, 19 (26), 94–102.

Trueman I.C., Cohn E.V., Tokarska-Guzik B., Rostański A. & Woźniak G. 2001: Calcarous waste slurry as wildlife habitat in England and Poland. [In:] Sarsby R.W. & Meggyess T. (eds.): "GREEN 3 – The Exploitation of Natural Resources and the Consequences". Thomas Telford, London, 527–534.

Wierzbicka M. & Pielichowska M. 2002: *Biscutella laevigata* – roślina Tatr i hałd przemysłowych. [In:] Borowiec W., Kotarba A., Kownacki A., Krzan Z., & Mirek Z. (eds.): Przemiany środowiska przyrodniczego Tatr. TPN-PTPNoZ, Kraków-Zakopane, 191–195.

Wiggering H. 1993: Sulfide oxidation – an environmental problem within colliery spoil dumps. Environmental Geology, 22 (2), 99–105.

Woch M.W., Stefanowicz A.M. & Stanek M. 2017: Waste heaps left by historical Zn-Pb ore mining are hotspots of species diversity of beech forest understory vegetation. Science of the Total Environment, 599–600, 32–41.

Wu L., Bradshaw A.D. & Thurman D.A. 1975: The potential for evolution of heavy metal tolerance in plants. III. The rapid evolution of copper tolerance in *Agrostis stolonifera*. Heredity, 34 (2), 165–187.

Green Scenarios..., Dyczko, Jagodziński & Woźniak (eds.) © 2022 Taylor & Francis Group, London, UK

2.4. Returning Coal Mine Sedimentation Pools to Nature: Technical Versus Natural Processes in Population, Vegetation and Landscape Contexts

Agnieszka Kompała-Bąba
Institute of Biology, Biotechnology and Environmental Protection,
Faculty of Natural Sciences, University of Silesia in Katowice;
28 Jagiellońska Str., 40-032 Katowice, Poland;
agnieszka.kompala-baba@us.edu.pl; ORCID: 0000-0002-4736-4704

Gabriela Woźniak
Institute of Biology, Biotechnology and Environmental Protection,
Faculty of Natural Sciences, University of Silesia in Katowice;
28 Jagiellońska Str., 40-032 Katowice, Poland;
gabriela.wozniak@us.edu.pl; ORCID: 0000-0003-1936-2880

ABSTRACT: The paper presents the results of research on the processes of spontaneous vegetation succession on coal-mine sedimentation pools located in Upper Silesia (Southern Poland). Frequently, saline mine waters are discharged to them before they are introduced into rivers. As a result of sedimentation, coal dust settles and treated waters are discharged into rivers. Different stages of vegetation succession from the initial ones through intermediate to the later stages of succession are presented. The course of succession was considered on different levels of organization: population, community and landscape. Moreover, abiotic and biotic factors that may influence the process of succession on the sedimentation pools were taken into account. Due to the diversity of habitats occurring on the sedimentation pools, presence or lack of salinity, various plant communities (aquatic, rushes, halophilous, ruderal, meadow, grassland) may develop.

KEYWORDS: Spontaneous succession, biodiversity, salinity, dynamics of populations, underground mine waters

INTRODUCTION

Mining significantly contributes to the production of large amounts of solid and waters (brines, saline waters) mine side products, which are produced both at the stage of mineral extraction and processing of mineral resources. In Poland, until 1997, the amount of side products produced by the hard coal mining industry prevailed (over 50%) in comparison with the extraction of other minerals (zinc and lead ores, natural aggregates, copper ores). As a result of changes in the mining industry, it started to decrease systematically and in 2004 it was only slightly above 30% (Przeniosło *et al.* 2006).

The processes of mining of hard coal are accompanied by the extrusion of huge amounts of mine water to the surface, as a result of natural phenomena such as horizontal and vertical hydrogeochemical zoning within a given coal basin. The amount of underground waters pumped out annually from all mines in Poland is estimated at 118 million m^3 (Strozik 2017). These waters contain chlorides and sulphates in high concentrations since they have been extracted from the increasing depth because of the necessity to protect mines from overflooding (Paczyński & Sadurski 2007; Policht-Latawiec 2014). Surface waters of Poland receive about 3.5 million m^3/day of mine discharge waters. Surface water containing saline mine waters has to be monitored. In 2015, 119 million m^3 of saline water was introduced into the environment, which contained 17.9 kg/Mg salt (Ochrona środowiska 2016). Saline water harms the state of surface and ground water bodies (Jankowski 1986; Różkowski & Różkowski 1994) as well as plant and animal communities that occur in their vicinity (Molenda 2014; Zgórska *et al.* 2016). The management of saline mine waters is very diverse, depending on the size, location, and technical condition of the mine. This determines the quantity and quality of mine waters which are discharged from mine workings to surface receivers and the quantity and quality of technological and living waste-water. The salinity of underground waters also depends on the depth of exploitation. With the increase in the depth of exploitation, an increase in water mineralization and a decrease in the amount of mine water is observed (Lebecka & Chałupnik 1995; Lebecka 1996). Mine waters pumped to the surface are a mixture of water from liquid backfilling, water from fire-fighting pipelines, and natural tributaries. The amount of natural water inflows to the mine workings decreases with the depth of exploitation. However, this is accompanied by a significant increase in their mineralization.

Regardless of the harsh habitat conditions, the human-created sites of coal-mine water sedimentation pools are colonized by living organisms. A large variety of life forms (lichens, liverworts, mosses, ferns, vascular plants) was observed on these objects. The natural processes going on in post-industrial areas allows the formation of plant communities adapted to the abiotic conditions prevailing in the area (Woźniak 2003; 2005; 2010; Woźniak & Kompała 2000a; 2001a; 2001b; Woźniak & Rostański 2000; Woźniak & Sierka 2004). Since most of these areas provide habitats poor in nutrients, the physicochemical parameters of the sedimentation pools are significantly different from those occurring in natural and semi-natural areas. These differences cause affected sites to be inhabited by species that have specific habitat preferences (e.g. halophytes, metallophytes), or species under protection in Poland. Some of these species are included in the country or regional "Red List of Species" (Woźniak & Kompała 2000a; 2000b).

THE DIVERSITY OF VEGETATION RECORDED ON COAL MINE SEDIMENTATION POOLS

Abiotic factors that determine the direction and rate of succession as well as species composition of vegetation that develops on coal mine sedimentation pools

In spite of harsh conditions, the plant species composition of the coal mine sedimentation pools is characterized by high diversity. Due to the large variety and number of available habitats (flat surfaces, hollows filled with water, slopes), as well as the time that has elapsed since the establishment of the sedimentation pools and time since the beginning of their colonization, communities of aquatic, rushes, trampled places, ruderal, grasslands, meadows, floodplains, and in saline places, communities of halophytes are formed there. The factors that influence the diversity of communities include: moisture (habitats from dry through fresh to waterlogged), substrate structure (fine-grained, drought prone during summer months and breaking into smaller pieces) and its granulometric composition (coal dust, hard dust), as well as lack or presence of salinity (Woźniak 1992; Fig. 1, 2; Tab. 1).

Characteristics of vegetation that develop spontaneously during succession

The resident plant communities differ in reference to frequency of occurrences, community structure, area they occupy, as well as the role they play in plant cover transformation. Considering the latter criterion, migratory and invasive communities, phases of increased competition, and communities with stabilizing structure can be distinguished (Woźniak 1998a; 1998b; 1998c). Depending on the stage of successional advancement, they also have different structures. The structure of initial communities is formed by 1 to 3 dominants, with a small number of species and a small coverage; these progress to single- (two- or three-layer) communities, species-rich communities, as well as multispecies and multi-layer communities (Tab. 2).

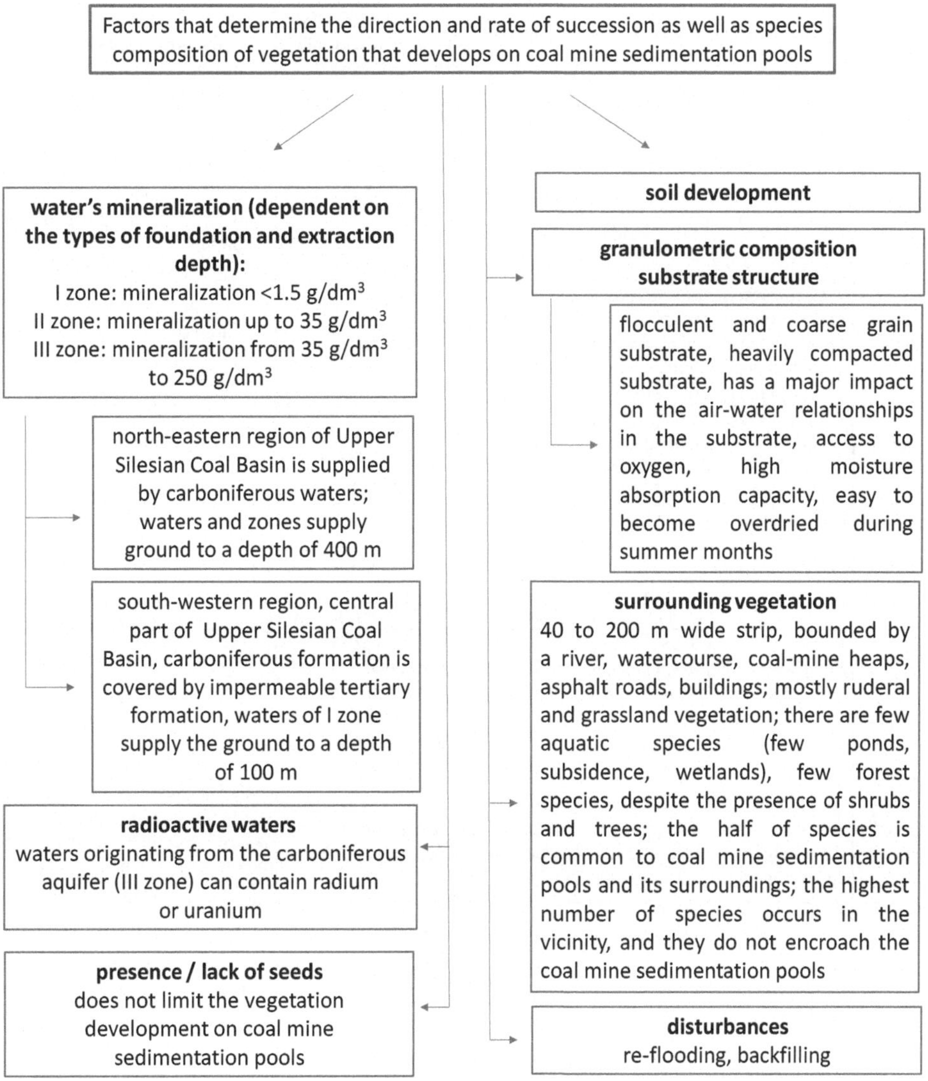

Figure 1. Factors that determine the direction and rate of succession on sedimentation pools.
Source: Woźniak (1992; 2001a; 2001b); Lebecka (1996); Pluta (2005); Smoliński (2006); Olkuski & Stala-Szlugaj (2009); Pluta *et al.* (2010)

Table 1. Characteristics of vegetation that develop on coal mine sedimentation pools (Fig. 2, Woźniak (1998a))

COMMUNITY	SPECIES RICHNESS FLORISTIC COMPOSITION	HABITAT CONDITIONS	FUNCTIONAL TRAITS
	Initial and migration communities		
grasslands with *Puccinellia distans*	1–7 species, *Chenopodium album*, *Ch. glaucum*, *Atriplex prostrata* ssp. *prostrata*	salty, damp, clayey, silty places	tufts up to 0.5–0.7 m, hemicryptophytes
Atriplex prostrata ssp. *prostrata* com.	4–11 species, *Atriplex prostrata* ssp. *prostrata*, *Tussilago farfara*, *Polygonum aviculare*	salted areas, large patches (50–200 m^2), hillsides, slimestones	therophytes, fleshy, 35-40 cm diameter clusters of overlapping shoots, compact vegetation
Plantago intermedia com.	2–9 species, *Plantago intermedia*, *Puccinellia distans*, *Atriplex prostrata* ssp. *prostrata*	large areas (100–300 m^2), clayey-dusty substrate with variable moisture	cover of vegetation 60–90%, low turf (up to 20 cm in height)
	Communities in the phase of intensified competition		
Puccinellio-Chenopodietum glauci	3–12 species, *Puccinellia distans*, *Chenopodium glaucum*, *Tussilago farfara*, *Atriplex prostrata* spp. *prostrata*, *Chenopodium rubrum*	salty substrate	patches of communities can develop into tall communities *Calamagrostis epigejos*, *Solidago canadensis*, tall herb fringe communities or meadow communities; patches 50–900 m^2, cover 40–90%
Chenopodietum glauco-rubri	4–19 species, *Atriplex prostrata* spp. *prostrata*, *Festuca rubra*, patches with share of xerothermic grassland species (low stature species connected with dry and warm habitats) and ruderal habitats	salty substrate	large complexes (50–300 m^2), cover 20–100%
Medicago lupulina com.	14–17 species, *Medicago lupulina*, *Puccinellia distans*, *Taraxacum officinale*, *Daucus carota*, *Epilobium hirsutum*, *Lepidium ruderale*, *Sonchus arvensis*	in succession after pioneer communities (halophytes)	small patches (5–9 m^2) (sporadically up to 300 m^2)
Tussilago farfara com.	4–13 species, low share of meadow and ruderal species (*Daucus carota*, *Artemisia vulgaris*, *Melilotus alba*, *Calamagrostis epigejos*)	low concentrations of mineral compounds	small patches (5–30 m^2), abundance 40–70%,
	Group of communities with relatively stabilized structure		
Polygono-Matricarietum	*Polygonum aviculare*, *Chamomilla suaveolens*, *Puccinellia distans*, *Taraxacum officinale*, *Cirsium vulgare*, *Epilobium hirsutum*	with time, expansive species appear in the patches *Calamagrostis epigejos*, *Tussilago farfara*, *Tanacetum vulgare*, *Cirsium vulgare*	occur often but do not form a large patches
Melilotus alba com.	8–11 species, *Melilotus alba* dominates, other species: *Tussilago farfara*, *Tanacetum vulgare*, *Dactylis glomerata*, *Hieracium sabaudum*, *Chamomilla suaveolens*	the driest parts of sedimentation pools	significantly homogeneous patches particularly in blooming phase, two-layered structured (higher layer 1–1.5 m, lower 0.5 m); patches can cover the area of 4–20 m^2
Daucus carota com.	7–18 species, mainly meadow species *Achillea millefolium*, *Dactylis glomerata*, *Pastinaca sativa*, *Poa pratensis*, rarely ruderal species *Artemisia vulgaris*, *Tussilago farfara*	the less fine soil substrate material	small patches (8–20 m^2) scattered on the sedimentation pool that cover the area 300–700 m^2 in a whole on those sedimentation pools where pioneer communities were dominated by *Tussilago farfara*, *Cirsium arvense*
Tanaceto-Artemisietum	6–31 species, *Tanacetum vulgare*, *Artemisia vulgaris*, *Melilotus alba*, *Echium vulgare*, *Dactylis glomerata*, *Achillea millefolium*, juvenile individuals of *Populus tremula* and *Salix caprea*	slightly elevated parts of the sedimentation pools	the participation of *Tanacetum vulgare* is extending the other species in those vegetation patches

COMMUNITY	SPECIES RICHNESS FLORISTIC COMPOSITION	HABITAT CONDITIONS	FUNCTIONAL TRAITS
Echio-Melilotetum	7–12 species, *Echium vulgare*, *Melilotus alba*, *Artemisia vulgaris*, *Oenothera biennis*, *Tanacetum vulgare*, *Arctium lappa*	large patches in unsalted sedimentation pools	cover of 40–90%
Artemisio-Tanacetetum	6–35 species, presence of meadow species (*Dactylis glomerata*, *Achillea millefolium*) and juvenile individuals (*Populus tremula*, *Salix caprea*), *Artemisia vulgaris*, *Tussilago farfara*, *Cirsium arvense*, *Echium vulgare*	10–30 m^2 patches, cover of herb layer 60–90%	intermediate stage with a relatively stable structure, abundance 60–90%
Communities preceding the tree entry phase			
Calamagrostis epigejos com.	3–25 species, *Calamagrostis epigejos*, *Tussilago farfara, Solidago canadensis*, low share of tree seedlings (*Betula pendula*, *Quercus rubra*, *Populus tremula*, *Salix cinerea*)	peripheral parts of sedimentation pools, large patches (100–500 m^2)	cover 50–80%, tall herbs
Solidago canadensis com.	6–17 species, *Tanacetum vulgare*, *Artemisia vulgaris*, *Melilotus alba*, *Echium vulgare*, *Dactylis vulgaris*, *Achillea millefolium*, juvenile individuals of poplar aspen and willows	large patches (500–1000 m^2) near escarpments	cover 40–100%, large patches 600–1000 m^2, tall species with high competitive ability

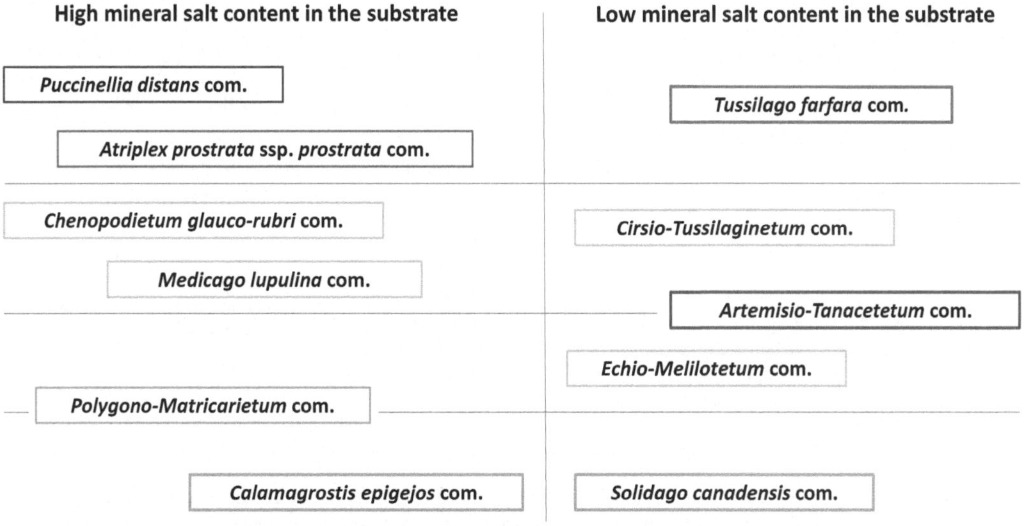

Figure 2. The possible pathways of succession on substrates with high or low salt contents

PATTERNS OF COLONIZATION OF THE SURFACE OF SEDIMENTATION POOLS ALONG TRANSECTS

The transect method is commonly used in studies on vegetation dynamics. It allows detection of directional changes of factors as well as some spatial phenomena (Falińska 1974; Wilkoń-Michalska 1986). The transects are set up deliberately across areas (objects) where rapid changes in vegetation and steep environmental gradients exist (Kent & Coker 1992). The colonization of coal mine sedimentation pools took place from their banks (escarpments) to the centre. In places where depressions or troughs appeared on the surface of the pool where water had accumulated, aquatic or rush communities developed. The rate of colonization of distinct sedimentation pools differed considerably. In places where halophytes and rushes appeared as pioneer species, the migration process was faster than on other sedimentation pools. Over time, the percentage share of individual species in particular squares increased. This fact has influence on the increase of intra- and interspecific competition. Increase in cover was influenced by the increase in the number of individuals per unit area, as a result of the multiplication of mainly young individuals (Fig. 3) (Woźniak 1998a).

Table 2. Transect distinguished on the basis of changes in species composition along environmental gradients and in time

GROUP	NUMBER OF SQUARES	COMPOSITION OF THE FLORA	CHANGES DURING SUCCESSION	TIME
I	low number of squares, pioneer squares 0–3 number of all squares 1–11	from 2 to 7 species ruderal species dominate	significant quantitative changes in plant cover with slight qualitative changes	unification of floristic composition, pioneer species are dominants in the later stages of succession *Tussilago farfara*, *Cirsium arvense*, *Calamagrostis epigejos*
II	number of pioneer squares – 1–17 number of all squares 6–29	halophilous and ruderal groups; from 1 to 3 pioneer species (facultative halophytes)	cover of pioneer species decreases and cover of ruderal species increases; salinity plays an important role (gradient of salinity)	pioneer species with low abundance *Puccinellia distans*, *Atriplex prostrata* ssp. *prostrata*, the floristic composition of the youngest quadrats is significantly different from older ones in which *Calamagrostis epigejos* dominates
III	number of pioneer squares – 2–144 number of all squares 4–147	from 2 to 9 species; halophytes are core species, *Puccinellia distans*, *Atriplex prostrata* spp. *prostrata*	pioneer species are also dominants in older squares, quantitative changes, small changes in mineral compounds during succession	uniform floristic composition, more species in older squares
IV	number of pioneer squares – 1–13, total number of squares 5–24	rush species (*Phragmites australis*, *Bulboschoenus maritimus*, *Typha latifolia*)	physicochemical factors: chemical composition of water, type of soil substrate, access of diasporas	flooding of transects with aquatic, rush communities associated with fresh waters inland salt waters
V	number of pioneer squares – 1	psammophilous species *Corynephorus canescens*	sand from sandy escarpments continuously covers the surface of sedimentation pools	*Corynephorus canescens*, pioneer species and dominant of later stages of succession

Explanations: I group – corresponds to stages of pioneer communities, that develop on sedimentation pools without salinity, II – belt strips running perpendicularly to the zonal vegetation system belt (strong gradient of salinity), III – correspond to stage of pioneer communities that develop on long-lasting substrate, IV, V – transformations of zonal communities within hydro- and psammospheres

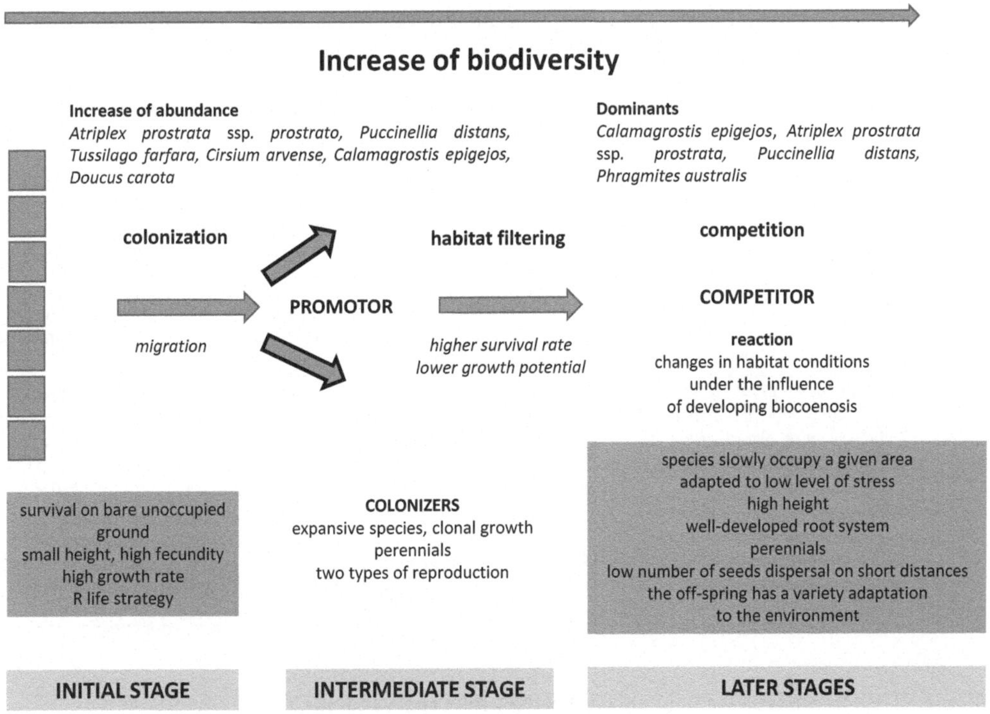

Figure 3. The functional traits of species in subsequent stages of succession in coal mine sedimentation pools (source: Woźniak (1998a) – changed)

DYNAMICS OF VASCULAR PLANTS POPULATIONS THAT COLONIZE COAL MINE SEDIMENTATION POOLS

Studies on the dynamics of populations of particular species (Falińska 1991; 2002), that colonize the surface of the coal mine sedimentation pools, play an important role in understanding successional processes. The studies have been carried out for several years on permanent study plots founded on sedimentation pools since the appearance of the first individuals of plants. The spatial structure of the population is influenced by parent organisms, habitat preferences and interactions between individuals and the environment. They include changes in species composition and abundance (cover) of individuals of particular species over time.

The research also allows us to learn about abiotic (structure and granulometric composition of the substrate, content of nutrients in the substrate, salinity, water relations), and biotic factors (seeds availability, type of community in the vicinity) that may influence the colonization process of species on coal mine sedimentation pools.

The research showed that the diversity of land microrelief (hollows, troughs, flooding, periodic drying of the settling area), affects the distribution of individuals of some species, e.g., *Puccinellia distans*, *Atriplex prostrata* ssp. *prostrata*, *Chenopodium glaucum* (Tab. 3). The research revealed that at that time the number of seedlings of the species was not large, while the number of already existing specimens was continuously growing and often combined into one lush clump. On the other hand, there was little evidence of successful recruitment. The growth form and vegetative propagation affects the distribution in the microscale. Spatial distribution of clumps depends on the mosaic of habitats, which shapes the patterns of population structure when interspecies interactions are superimposed on the social pattern during the initial colonization stage.

Table 3. Dynamics of chosen vascular plant species spontaneously colonizing sedimentation pools and representing different ecological groups of species

GROUP	SPECIES	ECOLOGICAL GROUP OF SPECIES	FUNCTIONAL TRAITS OF DOMINANT SPECIES
I	*Chamaenerion palustre, Chenopodium botrys, Diplotaxis muralis, Epipactis atrorubens, Myricaria germianica*	species of riverine stones and gravels, gravel pits, railway platforms and intersections	seeds equipped with a volatile apparatus, new specimens of the species appeared near the mother plant (up to 4 m), monospecific systems with a coverage of 20–30%
II	*Calamagrostis epigejos, Chenopodium glaucum, Ch. rubrum, Cirsium arvense, Echinochloa crus-galli, Equisetum arvense, Tussilago farfara*	species of ruderal habitats, psammophilous grasslands, segetal plants, species of fallow lands	the population remained stable, juveniles appeared around the parent plants, development of the population is characterized by the multiplication of seedlings and, to a small extent, by the overgrowth of the already existing clumps, the adult plants have a lower vitality
III	*Atriplex prostrata* spp. *prostrata, Plantago intermedia, Puccinellia distans, Spergularia salina*	species of salt flats	the species spread from the edge of the sedimentation pools to its interior, populations with an almost uniform density are created, seeds can be produced several times a year (spring and autumn), juveniles differ in morphology from adults

INFLUENCE OF VEGETATION OCCURRING IN THE VICINITY ON THE COMPOSITION OF VASCULAR FLORA OF SEDIMENTATION POOLS

Sedimentation pools, in which the coal-mine underground waters were located, occurred in the vicinity of, e.g., other sedimentation pools, urban-industrial wastelands, fallow land, orchards and allotment gardens, ditches, houses or forest thickets. Studies on the influence of surrounding vegetation on the vascular flora of the sedimentation pools showed that half of the plant species occurred both on the surface and in surroundings (Woźniak 1998a). The most numerous were ruderal, meadow and grassland plants, followed by brush and forest species, and rushes in water habitats (Woźniak 2001a; 2001b). The study showed that most species that occurred in the vicinity of the studied objects did not appear on the surface of the sedimentation pools. Thus, it did not influence the formation of spontaneous vegetation (Fig. 3, 4).

OVERVIEW OF METHODS FOR UTILIZATION OF SALT WATERS COMING FROM UNDERGROUND MINING

As an engineering approach the desalination with reverse osmosis and subsequent placement of brines in mine goafs seems to be an optimal way to solve the problem of the extreme (about 20,000 m^3/day) inflow of salt waters (Gunther & Mey 2006). This method has some positive features: improvement of the conditions for mining works, reduction of investment costs and environmental fees. Excess water is discharged to surface waters in a controlled way (hydrological method), which does not allow for exceeding the concentrations of salt permitted by environmental protection regulations (Strozik 2017).

Brine is a by-product of all desalination methods. The undertaken salt crystallization procedures (multistage equilibrium distillation) involve significant consumption of electricity (20–25 kWh/m^3) and generate a carbon footprint, which reaches 19–23 kg CO_2 per 1 m^3 of evaporated brine (Tong & Elimelech 2016). Methods of limiting the concentration and management of underground waters from the mines of Silesia include: desalination of underground waters and salt production (used by the Dębieńsko Desalination Plant), use of mine waters in underground mining technologies, in processing, sprinkling, production of sodium hydroxide and carbonated chlorine.

However, these methods have some limitations such as high costs of production (Bobik & Labus 2014; Gromiec *et al.* 2014; Gruszczyński *et al.* 2014). Moreover, there is currently a lack of demand for salt and carbonated chlorine due to their provision by the mines (Gruszczyński *et al.* 2014).

Apart from the technical approach the potential of natural processes should be taken into consideration. Inland halophitic habitats would provide a considerable biodiversity increase by developing halophyte ecosystems.

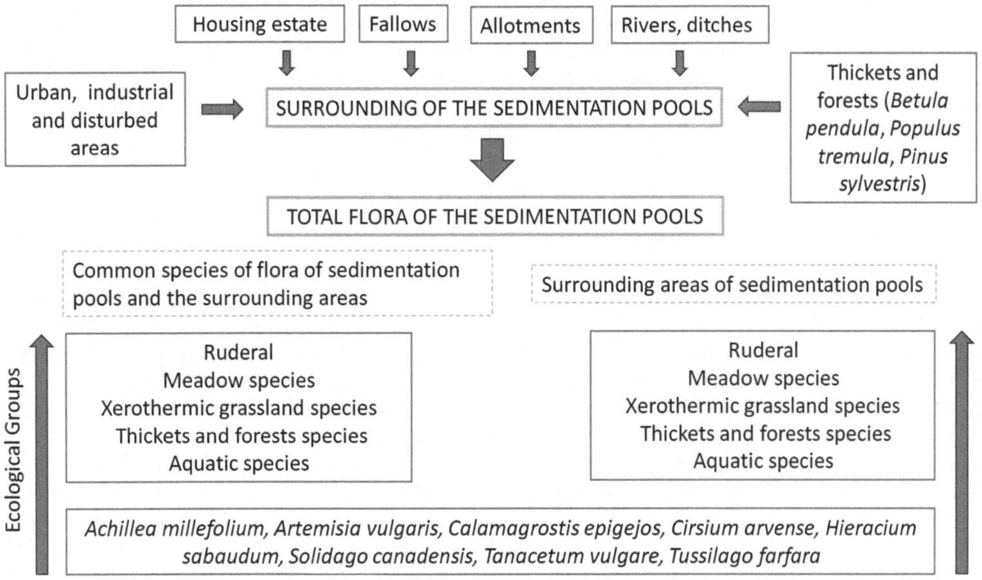

Figure 4. The origin of species and ecological groups that form vascular flora in coal mine sedimentation pools (source: Woźniak (1998a) – changed)

SUMMARY

The naturally developed vegetation on coal-mine sedimentation pools is characterized by high diversity in spite of the harsh habitat conditions. Species colonizing them face a variety of stresses such as water, temperature, and salinity extremes. The vegetation diversity is a consequence of (among other factors) the availability of varied habitats such as flat surfaces, hollows filled with water, and slopes that facilitate development of aquatic, rush, trampled places, ruderal, grassland, meadow, floodplain, and in saline places, halophyte communities.

The plant communities differ in reference to frequency of occurrence, structure, and occupied area, as well as the role they play in plant cover transformation. The structure of initial communities is formed by 1 to 3 dominants, with a small number of species and small cover. Further in succession single- (two- or three-layer) communities, species-rich communities, as well as communities built by many species and having multi-layer structure develop.

The research showed that the diversity of land microrelief (hollows, troughs, flooded areas, periodic drying of the settling area), affected the distribution of individuals of some species, e.g. *Puccinellia distans*, *Atriplex prostrata* spp. *prostrata*, *Plantago intermedia*, *Spergularia salina*. The research revealed that the number of seedlings of the species was not large, while the number of already existing specimens was continuously growing and often combined into one lush clump. On the other hand, there was little evidence of successful recruitment. The growth form and vegetative propagation of plant species affects the distribution at the microscale. Spatial distribution of clumps depends on the mosaic of habitats, which shapes the patterns of population structure. When interspecies interactions are superimposed on this, then we observe colonization of the area according to the social pattern.

REFERENCES

Bobik M. & Labus K. 2014: Metody odsalania wód kopalnianych w praktyce przemysłowej – stan obecny technologii i nowe wyzwania. Przegląd Górniczy, 4, 99–105.

Falińska K. 1974: Reakcja populacji roślinnych na zróżnicowanie ekosystemów oraz układów ekotonowych [Reaction of plant populations to differences in ecosystems and ecotone systems]. Wiadomości Ekologiczne, 20, 356–376.

Falińska K. 1991: Plant Demography and Vegetation Succession. Kluwer, Academic Publishers, Dordrecht-Boston-London.

Falińska K. 2002: Przewodnik do badań biologii populacji roślin. Vademecum Geobotanicum. Wydawnictwo Naukowe PWN, Warszawa.

Gromiec M., Sadurski A., Zalewski M. & Rowiński P. 2014: Zagrożenia związane z jakością wody. Nauka, 1, 99–122.

Gruszczyński S., Motyka J., Mikołajczyk J. & Kasprzak A. 2014: Potrzeba wdrożenia zintegrowanego systemu monitorowania i dozowania wód kopalnianych do rzeki Wisły. Przegląd Górniczy, 8, 142–149.

Gunther P. & Mey W. 2006: Selection of Mine Water Treatment Technologies for the Emalahleni (Witbank) Water Reclamation Project. Anglo Coal, Biennial WISA Paper.

Jankowski A.T. 1986: Antropogeniczne zmiany stosunków wodnych na obszarze uprzemysłowionym i urbanizowanym (na przykładzie Rybnickiego Okręgu Węglowego). Prace Naukowe Uniwersytetu Śląskiego, 868, Wydawnictwo Uniwersytetu Śląskiego, Katowice.

Kent M. & Coker P. 1992: Vegetation Description and Analysis. John Wiley and Sons, Belhaven Press, London.

Lebecka J. & Chałupnik S. 1995: Oczyszczanie słonych wód kopalnianych z radu. Przegląd Górniczy, 4, 38–42.

Lebecka J. 1996: Oczyszczanie wód kopalnianych z radu. Bezpieczeństwo Pracy i Ochrona Środowiska w Górnictwie, 5, 38–42.

Molenda T. 2014: Impact of saline mine water: development of a meromictic reservoir in Poland. Mine Water and the Environment, 33, 327–334.

Ochrona środowiska 2016: Główny Urząd Statystyczny, Warszawa,163–164.

Olkuski T. & Stala-Szlugaj K. 2009: Występowanie pierwiastków promieniotwórczych w węglach kamiennych pochodzących z GZW w skałach przywęglowych, w wodach kopalnianych oraz w odpadach. Gospodarka Surowcami Mineralnymi, 25 (1), 5–17.

Paczyński B. & Sadurski A. 2007: Hydrogeologia regionalna Polski. Tom II – Wody mineralne, lecznicze i termalne oraz kopalniane. Państwowy Instytut Geologiczny, Warszawa.

Pluta I. 2005: Wody kopalń Górnośląskiego Zagłębia Węglowego – geneza, zanieczyszczenia i metody oczyszczania. Prace Naukowe Głównego Instytutu Górnictwa: Studia, Rozprawy, Monografie, 865, GIG.

Pluta I., Gruszka Z. & Musiolik A. 2010: Monitoring procesu oczyszczania z baru wód kopalni „Krupiński". Wiadomości Górnicze, 6, 349–353.

Policht-Latawiec A. 2014: The effect of saline mine waters discharge from hard coal mine on the ecological state of the Vistula River. Acta Horticulturae et Regiotecturae, 2, 43–47.

Przeniosło S., Malon A. & Tymiński M. 2006: Analiza zmian ilościowych odpadów pogórniczych i przeróbczych oraz solanek i wód zasolonych w Polsce. Przegląd Geologiczny, 54 (8), 667–670.

Różkowski A. & Różkowski J. 1994: Impact of mine waters on river water quality in the Upper Silesian Coal Basin. [In:] 5th International Mine Water Congress, Nottingham (UK), September 1994, 811–821.

Smoliński A. 2006: Gospodarka zasolonymi wodami kopalnianymi. Prace Naukowe GIG, Górnictwo i Środowisko, 1, 1–15.

Strozik G. 2017: Bilans i kierunki utylizacji słonych wód kopalnianych z czynnych i zlikwidowanych kopalń węgla kamiennego z uwzględnieniem ich zagospodarowania w podziemnych technologiach górniczych. Zeszyty Naukowe Instytutu Gospodarki Surowcami Mineralnymi i Energią PAN, 98, 199–210.

Tong T. & Elimelech M. 2016: The global rise of zero liquid discharge for wastewater management: drivers, technologies, and future directions. Environmental Science & Technology, 50, 6846–6855.

Wilkoń-Michalska J. 1986: Tendencje rozwojowe i ochrona halofitów w Polsce. Acta Universitatis Lodziensis, Folia Sozologica, 3, 123–129.

Woźniak G. & Kompała A. 2000a: Rola procesów naturalnych w rekultywacji nieużytków poprzemysłowych [The role of natural processes in the reclamation of postindustrial waste sites]. Inżynieria Ekologiczna, 1, Ochrona i rekultywacja gruntów. Wyd. Ekoinżynieria, Lublin, 87–94.

Woźniak G. & Kompała A. 2000b: Gatunki chronione i rzadkie na nieużytkach poprzemysłowych. Centrum Studiów nad Człowiekiem i Środowiskiem Uniwersytetu Śląskiego, Problemy środowiska i jego ochrony, 8, 101–109.

Woźniak G. & Kompała A. 2001a: Ekologiczny potencjał nieużytków poprzemysłowych jako podstawa ich biologicznej regeneracji [Ecological potential of post-industrial wastelands as the basis for its biological regeneration]. Materiały Sympozjum Warsztaty 2000 nt. Zagrożeń naturalnych w górnictwie, Wieliczka, 223–233.

Woźniak G. & Kompała A. 2001b: Ecology of spontaneous vegetation on post-industrial waste lands (Upper Silesia – Poland). [In:] Sarsby R.W. & Meggyes T. (eds.): Green 3: The Exploitation of Natural Resources and the Consequences. Thomas Telford, London, 567–573.

Woźniak G. & Rostański A. 2000: Rola traw w spontanicznej sukcesji roślinnej na osadnikach ziemnych wód kopalnianych na Górnym Śląsku. Łąkarstwo w Polsce, 3, 159–167.

Woźniak G. & Sierka E. 2004: Importance of spontaneous succession in reclamation processes. Zeszyty Naukowe Uniwersytetu Zielonogórskiego, Inżynieria Środowiska, 131, 391–398.

Woźniak G. 1992: Osadniki wód kopalnianych jako słabo poznany element środowiska geograficzno-przyrodniczego Górnego Śląska. Kształtowanie środowiska geograficznego i ochrona przyrody na obszarach uprzemysłowionych i zurbanizowanych, 6, UŚ, Katowice-Sosnowiec, 18–26.

Woźniak G. 1998a: Uwarunkowania sukcesji na osadnikach ziemnych wód kopalnianych [The primary succession on coalmine sedimentation pools on the Upper Silesia]. Ph.D. Thesis mscr, praca doktorska wykonana w Katedrze Botaniki i Ochrony Przyrody pod kierunkiem prof. dr. hab. Stanisława Wiki.

Woźniak G. 1998b: Plant succession on sedimentation pools in Upper Silesia. [In:] Sarsby R.W. (ed.): Contaminated and Derelict Land. The Proceeding of Green 2: The Second International Symposium on Geotechnics Related to the Environment. Thomas Telford, London, 60–62.

Woźniak G. 1998c: Primary succession on the sedimentation pools of coal mine. [In:] Faliński J.B., Adamowski W. & Jackowiak B. (eds.): Synanthropization of Plant Cover in New Polish Research. Phytocoenosis, Vol. 10 (N.S.), Supplementum Cartographiae Geobotanica, 9, 189–198.

Woźniak G. 2001a. Flora roślin naczyniowych osadników ziemnych wód kopalnianych – nieużytków poeksploatacyjnych na Górnym Śląsku [Vascular plants flora of the coal mine underground water sedimentation pools – post-industrial wastelands in the Upper Silesia]. Centrum Dziedzictwa Przyrody Górnego Śląska, Materiały i Opracowania, 6, 3–46.

Woźniak G. 2001b: Invasive plants involved in primary succession on post-industrial areas Upper Silesia (Poland). [In:] Brundu G., Brock J., Camarda I., Child L. & Wade M. (eds.): Plant Invasions: Species Ecology and Ecosystem Management. Backhuys Publishers, Leiden, 263–270.

Woźniak G. 2003: Grass-dominated plant communities of the post-industrial waste sites in the Upper Silesia. [In:] Frey L. (ed.): Problems of Grass Biology. W. Szafer Institute of Botany, Polish Academy of Sciences, Kraków, 531–545.

Woźniak G. 2005: Problems of *Calamagrostis epigejos* synecology on post-industrial sites. [In:] Frey L. (ed.): Biology of Grasses. W. Szafer Institute of Botany, Polish Academy of Sciences, Kraków, 353–361.

Woźniak G. 2010: Zróżnicowanie roślinności na zwałach pogórniczych Górnego Śląska [Diversity of vegetation on coal-mine heaps of the Upper Silesia (Poland)]. W. Szafer Institute of Botany, Polish Academy of Sciences, Kraków.

Zgórska A., Trząski L. & Wiesner M. 2016: Environmental risk caused by high salinity mine water discharge from active and closed mines located in the Upper Silesia Coal Basin (Poland). [In:] Drebensted C. & Paul M. (eds.): Proceedings of the International Mine Water Association: Mining Meets Water – Conflicts and Solutions, 85–92.

2.5. Post-industrial Habitats as Centres of Biodiversity – a Case Study on a Sample of Sedimentation Pools

Agnieszka Kompała-Bąba
Institute of Biology, Biotechnology and Environmental Protection,
Faculty of Natural Sciences, University of Silesia in Katowice;
28 Jagiellońska Str., 40-032 Katowice, Poland;
agnieszka.kompala-baba@us.edu.pl; ORCID: 0000-0002-4736-4704

Robert Hanczaruk
Institute of Biology, Biotechnology and Environmental Protection,
Faculty of Natural Sciences, University of Silesia in Katowice;
28 Jagiellońska Str., 40-032 Katowice, Poland;
rhanczaruk@us.edu.pl; ORCID: 0000-0002-8337-8687

Wojciech Bąba
Institute for Ecology of Industrial Areas;
6 Kossutha Str., 40-844 Katowice, Poland;
wojciech.baba12@gmail.com; ORCID: 0000-0001-7428-0429

ABSTRACT: The paper presents the diversity of the spontaneous flora of underground mine water sedimentation pools created as a result of purification of saline mine waters before their introduction into rivers. Despite unfavourable abiotic conditions (fine-grained, often compacted, erosion-prone substrate, salinity, impaired air-water relations), the vascular flora of these objects is quite diverse. A distinct participation of native species, mainly non-forest species was recorded. Ruderal, meadow, thicket and forest species predominate among them. The species represent different ecological groups reflecting the spectrum of habitats occurring on the sedimentation pools as well as environmental gradients connected with light, moisture, soil reaction and soil fertility. Competitors dominate in the vascular flora of sedimentation pools as well as species with a mixed type of strategy (CR, CS, CSR). Most species reproduce via seeds or vegetatively which enables them to survive in a given place and occupy new areas. Five species are present that can grow in habitats with higher than typical contents of salt in the substrate, including *Bulboschoenus maritimus*, *Spergularia salina*, *Juncus ranarius*, *Scirpus lacustris* and *Puccinellia distans*.

KEYWORDS: Vascular flora, salinity, ecological groups, diversity, primary succession

INTRODUCTION

Hard coal mining has been one of the most important branches of the mining industry in Upper Silesia since the XIX Century. Its dynamic development caused serious changes and alteration of the natural environment and its elements such as land relief, landscape, pollution of the atmosphere, soil and surface waters (Czaja 2001; Cabała *et al.* 2004a; 2004b; Machowski *et al.* 2006; Pełka-Gościniak 2006; Dulias 2016). Moreover, vascular flora and plant communities underwent significant transformations or were destroyed completely (Sendek 1981; Jędrzejko 1987; Cabała 1990; Celiński & Wika 1992; Kompała-Bąba 2007). The anthropogenic forms resulting from the extraction and processing of hard coal may include: (i) heaps (above-ground spoil heaps of carboniferous origin), subsidence troughs (land subsidence caused by subsidence of the area lying under exploited coal seams), settling reservoirs (created at the settling point above the exploited reservoir) and sedimentation pools (large reservoirs in which underground mine waters are purified of coal dust suspension). These post-industrial, post-excavation areas are considered as a serious environmental problem and according to Polish Law should be subject to reclamation and subsequent management. Forest reclamation (understood as tree planting) and water reservoir approaches have been used as the most frequent methods of land reclamations (Maciak 2003; Pietrzykowski & Krzaklewski 2018). However, many post-excavation areas were not reclaimed at all and were left to primary natural succession that enabled the colonisation of a variety of species and the development of novel ecosystems *sensu* Hobbs *et al.* (2006), as a result (Woźniak 2010; Prach *et al.* 2011; Kompała-Bąba 2013; Halabowski *et al.* 2018; Błońska *et al.* 2019a).

Most of the scientific research conducted in Poland and other major industrial regions of Europe was devoted to brown or hard coal mining. On these objects the flora and vegetation, mechanisms and rate of their overgrowth, and physical and chemical properties of the soil substrate, as well as relationships between flora and microorganisms were studied (Cabała & Sypień 1987; Patrzałek & Rostański 1992; Rebele & Dettmar 1996; Rostański 1998; 2006; Jochimsen 2001; Woźniak & Błońska 2009; Pecharová *et al.* 2011; Frouz 2014; Woźniak *et al.* 2015; Cudlín *et al.* 2016; Hanczaruk & Gołąb 2016; Błońska *et al.* 2019a; 2019b; Kompała-Bąba *et al.* 2019). However, not many research projects have focused on the diversity of vegetation that spontaneously develops on sedimentation pools (Woźniak 1998; Woźniak & Rostański 2000; Woźniak & Śliwińska-Wyrzychowska 2002; Woźniak 2006). It seems extremely important to acquire knowledge about the diversity of vascular flora of these areas to aid their future management, particularly including natural processes.

THE CHARACTERISTICS OF THE STUDY OBJECTS

Sedimentation pools are objects reflecting very strong impacts of human activity during establishment, combined with periodic destruction of resident vegetation by re-flooding, subsidence or collapse of the area. In the sedimentation pools underground mine waters are cleaned of coal dust (Fig. 1).

Taking into account species responses to increasing disturbance, such habitats can be classified as polyhemerobic (Kowarik 1999; Sukkop 2008). However, the sedimentation pools of mine waters are a specific case of the anthropogenic habitats that do not have a counterpart in natural habitats. Small material (coal dust) is susceptible to drying out during the summer months (then the substrate often breaks into smaller pieces or lumps). In contrast, after intensive rainfall, the surface of the sedimentation pool forms sludge. Without initial soil forming processes, these sites are usually poor in nutrients and have small humus contents. Such sites have poor air-water relations, frequently they may be saline, and the fine-grained soil substrate may be subject to water or wind erosion. However, in comparison to lead and zinc spoil heaps, contents of heavy metals are absent in the soil. There is also a risk of land collapse below the surface of the sedimentation pool. Substrates of secondary origin are a frequent component of these habitats. There is also a lack of seed bank in the soil substrate (Fig. 2).

Two types of sedimentation pools can be distinguished, taking into account the materials used in their construction as well as the type of management: (i) earth sedimentation pools (left to spontaneous succession, after being completely filled with dust) or (ii) concrete sedimentation pools, from which the sludge is removed after filling and collected in sludge tanks (Fig. 2). Such objects are currently of different age, occupy an area ranging from a few to several hectares and are usually located in different places, sometimes also in forests. One mine could have had from 1 to even 11 settlers (on average 3 sedimentation pools) (Woźniak 1998).

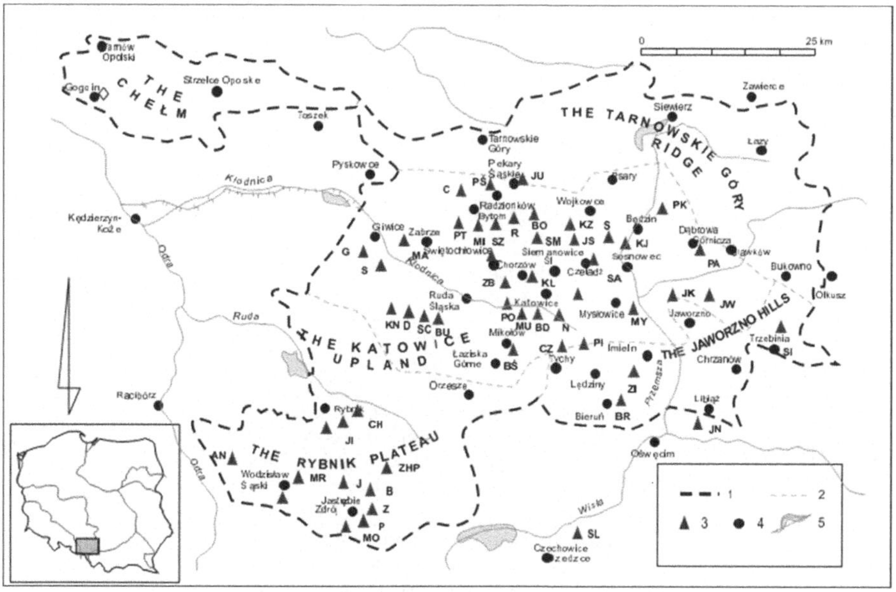

Figure 1. Distribution of coal mine sedimentation pools in the Silesian Upland:
1 – borders of the Silesian Uplands, 2 – borders of the mesoregions, 3 – sedimentation pools,
4 – towns, 5 – lakes and rivers

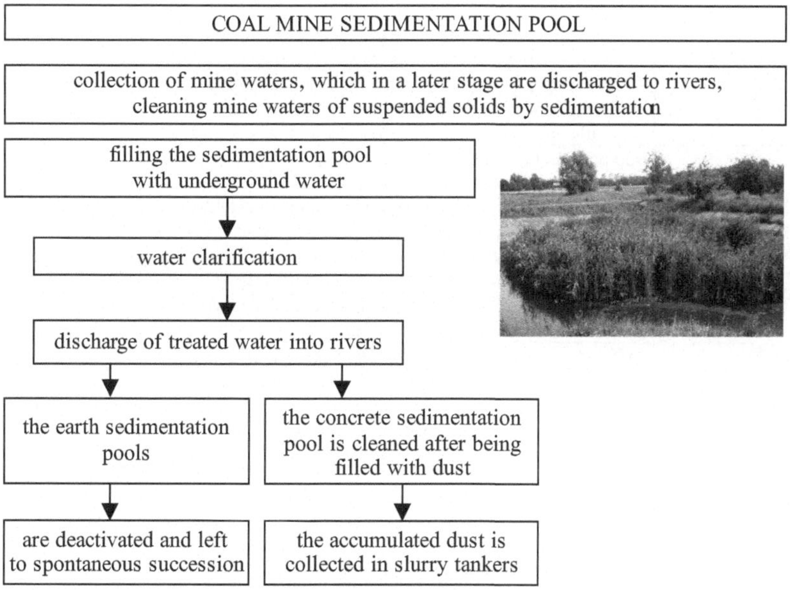

Figure 2. The sedimentation process taking part in sedimentation pools

METHODS USED TO DESCRIBE THE DIVERSITY OF VASCULAR FLORA OCCURRING IN SEDIMENTATION POOLS

Floristic research was conducted on 137 coal mine sedimentation pools located within the area of 50 mines occurring in Upper Silesia, one of the most urbanized and industrialized regions in Poland (Fig. 1). In the field, floristic data were collected in a designated grid of squares or on permanent study plots. Floristic data were collected directly on coal sediment (dust), separately on each of the examined inactive, mainly earth sedimentation pools. For each sedimentation pool species richness and Shannon-Wiener diversity index were calculated. In order to show distribution of sample plots along environmental gradients (Tab. 1) multivariate analyses were applied (ter Braak & Šmilauer 2002). For sample units Ellenberg's indicator (L, T, C, M, SR, N) values were calculated.

Table 1. Traits used in describing the diversity of vascular flora of sedimentation pools

ATTRIBUTES	SHORT CHARACTERISTSICS (EXPLANATIONS)	REFERENCES
Origin of species	apophytes, anthropophytes (archaeophytes, ephemerophytes, kenophytes)	Rutkowski (1998) Mirek *et al.* (2002)
Family		Rutkowski (1998)
Ecological groups of species	forest species (L); grassland species (LK) psammophilous species (MP); xerothermic grassland species (MK); species of clearings and thermophilous fringes (O, ON); ruderal species (RD); salt marsh species (S); peatbog and bog species (TR); rush and aquatic species (WN, W); segetal (SG)	Frank & Klotz (1990) Oberdorfer *et al.* (1990) Klotz *et al.* (2002) Matuszkiewicz (2012)
Life strategies	C – competitive species; CR – competitive ruderals; S – stress-tolerant species; SR – stress tolerant ruderals; R – ruderals; species with mixed strategies: CSR	Grime (1979) Frank & Klotz (1990)
Life forms	hemicryptophytes; therophytes; geophytes; megaphanerophytes; nanophanerophytes; chamaephytes; hydrophytes	Frank & Klotz (1990) Zarzycki *et al.* (2002)
Value of resistance to NaCl content in soil or water	0 – not salt tolerant, glycophytes; 1 – salt tolerant, mostly on low-salt to salt-free soils, but occasionally on slightly salty soils; 2 – often on soils with very low salt content; 3 – mostly on soils with low salt content; 4 – mostly on soils with low to moderate salt content; 5 – mostly on soils with a moderate salt content; 6 – on soils with moderate to high salt content; 7 – on soils with a high salt content; 8 – on soils with a very high salt content; 9 – on soils with a very high, and in dry periods extremely high salt content	Chytrý *et al.* (2018)
Habitat preferences of species according to moisture, organic matter content, light and productivity Ellenberg's indicator values (EIV)	soil without humus and organic matter; 2.0 – soil with very little humus content; 3.0 – mineral-humus soils; 4.0 – peat soils with significant participation of mineral particles; 5.0 – organogenic soils; 2.0 to 3.0 (2.3 – species, which grow on habitats of fertility – values of the H indicator from 2.0 to 3.0); 6 – large range of the H indicator value 1–3, 2–5, 3–5; 7 – lack of data about the humus content preferences of the species	Ellenberg *et al.* (1992) Zarzycki *et al.* (2002)
Share of species indicators of elevated heavy metal content in the substrate	0.0 – species which have not been observed on localities with heavy metals in the soil; 1.0 – species which can tolerate increased contents of heavy metals; 2.0 – species which occur almost exclusively on soils with increased heavy metal contents	Zarzycki *et al.* (2002)
Protected and rare species		Babczyńska-Sendek *et al.* (2012) Sierka (2013)
Frequency of species occurrences on the sedimentation pools	Number of sedimentation pools where a given species occurs	Woźniak (1998; 2005)

DIVERSITY OF VASCULAR FLORA OF SEDIMENTATION POOLS

The vascular flora of a given area become established as a result of both deterministic and stochastic processes. Taxa from a given regional (local) species pool pass through an environmental sieve, which eliminate species that are not able to reach a given area on one hand and/or adapt to the difficult conditions that prevail in mine water sedimentation pools on the other hand (Zobel 2016). Only those species which have the suitable combination of persistence, dispersal and regeneration traits can survive in a given area, effectively use resources and compete with other species. Inter-species interactions (both positive and negative) also play an important role, and they change during the process of succession in the settling ponds.

The species richness of the sedimentation pool is very diversified and depends on many factors, including the area of a given sedimentation pool, its surroundings, and the time that has elapsed since its formation. On individual sedimentation pools from 11 to 129 species of vascular plants were recorded. Sedimentation pools also differed in terms of species diversity (measured by the Shannon-Wiener index) (Fig. 3). The vegetation of some of the objects had relatively low species diversity (Zofiówka, Knurów, Chwałowice, Pstrowski), while the most diverse were Paryż, Powstańców Śląskich, and Niwka Modrzejów.

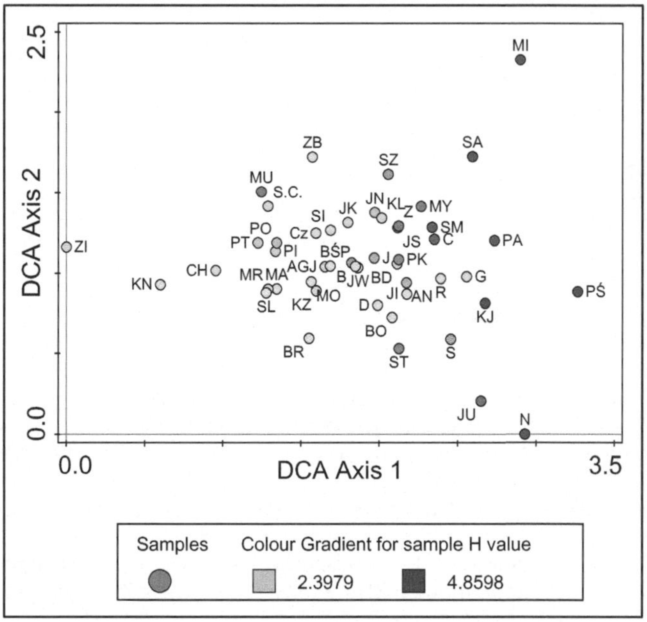

Abbreviations of coal mine spoil heaps (localities):
AN – Anna; AGJ – Andaluzja, Grodziec, Julian; B – Borynia; BD – Boże Dary; BO – Bobrek; BR – Brzeszcze; BŚ – Bolesław Śmiały; C – Centrum; CZ – Czeczott; CH – Chwałowice; D – Dębieńsko; G – Gliwice; J – Jastrzębie; JI – Jankowice; JK – Jan Kanty; JN – Janina; JS – Juliusz; JU – Julian; JW – Jaworzno; KJ – Kazimierz Juliusz; KL – Kleofas; KN – Knurów; KZ – Kazimierz; MA – Makoszowy; MI – Miechowice; MO – Moszczenica; MR – Marcel; MU – Murcki; MY – Mysłowice; N – Niwka Modrzejów; P – Pniówek; PA – Paryż; PI – Piast; PK – Porąbka Klimontów; PO – Pokój; PŚ – Powstańców Śląskich; PT – Pstrowski; R – Rozbark; S – Sosnowiec; SA – Saturn; SC – Szczygłowice; SI – Siersza; SL – Silesia; SM – Siemianowice; ST – Staszic; SZ – Szombierki; W – Wujek; Z – Zofiówka; ZB – Zabrze-Bielszowice; ZI – Ziemowit

Figure 3. DCA analysis of the diversity of vascular flora of the sedimentation pools in Upper Silesia

Ecological groups of species that constitute vascular flora of sedimentation pools

Species that colonised sedimentation pools represent a vast spectrum of ecological groups that reflects the spectrum of habitats and gradients connected with light, moisture, soil reaction and soil fertility (Fig. 4).

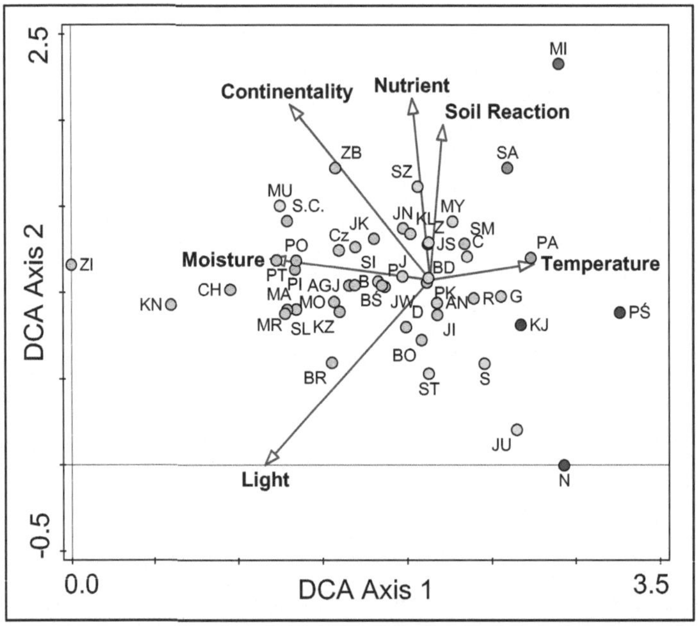

Abbreviations of coal mine spoil heaps (localities):
AN – Anna; AGJ – Andaluzja, Grodziec, Julian; B – Borynia; BD – Boże Dary; BO – Bobrek; BR – Brzeszcze; BŚ – Bolesław Śmiały; C – Centrum; CZ – Czeczot; CH – Chwałowice; D – Dębieńsko; G – Gliwice; J – Jastrzębie; JI – Jankowice; JK – Jan Kanty; JN – Janina; JS – Juliusz; JU – Julian; JW – Jaworzno; KJ – Kazimierz Juliusz; KL – Kleofas; KN – Knurów; KZ – Kazimierz; MA – Makoszowy; MI – Miechowice; MO – Moszczenica; MR – Marcel; MU – Murcki; MY – Mysłowice; N – Niwka Modrzejów; P – Pniówek; PA – Paryż; PI – Piast; PK – Porąbka Klimontów; PO – Pokój; PŚ – Powstańców Śląskich; PT – Pstrowski; R – Rozbark; S – Sosnowiec; SA – Saturn; SC – Szczygłowice; SI – Siersza; SL – Silesia; SM – Siemianowice; ST – Staszic; SZ – Szombierki; W – Wujek; Z – Zofiówka; ZB – Zabrze-Bielszowice; ZI – Ziemowit

Figure 4. DCA analysis of the whole vascular flora of the sedimentation pools

Two main environmental gradients can be detected that are responsible for the differentiation of the vascular flora. One of them is connected with moisture. Species that colonize the bare surface of a sedimentation pool can come from water habitats (*Batrachium circinnatum*, *Batrachium trichophyllum*), rush (*Phragmites australis*, *Typha angustifolia*, *Typha latifolia*), through peat bog (*Parnassia palustris*, *Triglochin palustre*), wet (*Angelica sylvestris*, *Cirsium palustre*), and fresh (*Arrhenatherum elatius*, *Campanula patula*, *Crepis biennis*, *Pastinaca sativa*) to dry habitats (*Jasione montana*, *Corynephorus canescens*, *Hieracium pilosella*). The second gradient is connected with soil pH and fertility as well as light (Fig. 4).

The preferences of species in terms of salinity of the substrate

Because the underground water from coal mines can contain high amounts of salt, it seems that species that can tolerate higher content of salt were found on these habitats. In the vascular flora of sedimentation pools, 5 species were recorded that can grow mostly on soils with a moderate to a very high salt content (sometimes extremely high during dry periods): *Spergularia salina*, *Juncus ranarius*, *Bulboschoenus maritimus*, *Puccinellia distans*, *Scirpus lacustris* (categories 5–7 in Tab. 1). Species that are known to be able to grow on soil substrates with high salt content were not very frequent. *Bulboschoenus maritimus* occurred in rush communities (Julian sedimentation pool), whereas *Spergula salina* was recorded in a sedimentation pool in Knurów, only on one locality in the *Puccinellio-Spergularietum salinae* plant community. More than 50% of species are not salt tolerant and are classified as glycophytes (category 0) (Chytrý *et al.* 2018).

The spectrum of families that constitute the vascular flora of sedimentation pools

Despite harsh habitat conditions that occur on coal mine sedimentation pools the vascular flora of such objects is rich in species and diverse in comparison to other post-industrial sites (coal mine spoil heaps, lead and zinc spoil heaps, sandpits) (Rostański 2006; Woźniak 2010; Hanczaruk & Kompała-Bąba 2019a; 2019b) (Tab. 2).

Table 2. The comparison of vascular flora of sedimentation pools with other post-industrial sites

VARIABLE	SAND-PITS	COAL MINE SEDIMENTATION POOLS	COAL MINE SPOIL HEAPS	SOURCE
Family	*Asteraceae* (13%) *Poaceae* (12%) *Fabaceae* (7%)	*Asteraceae* (16%) *Poaceae* (12.4%) *Fabaceae* (6.6%) *Brassicaceae* (6.1%)	*Asteraceae* (15%) *Poaceae* (12.1%) *Fabaceae* (6.5%) *Brassicaceae* (6.1%)	Bąba & Kompała (2003)
Life forms	Hemicryptophytes (49%) Geophytes (10%) Megaphanerophytes (5%) Nanophanerophytes (5%) Therophytes (14%)	Hemicryptophytes (50%) Geophytes (7%) Megaphanerophytes (5%) Nanophanerophytes (5%) Therophytes (24%) Chamaephytes (4.7%) Hydrophytes (4.2)%	Hemicryptophytes (50%) Geophytes (10%) Megaphanerophytes (5%) Nanophanerophytes (5%) Therophytes (30%) Hydrophytes (2%)	Woźniak (2001) Bąba & Kompała (2003)
Origin	Apophytes (87.5%) Anthropophytes (11%)	Apophytes (75%) Anthropophytes (25%)	Apophytes (70%) Archaeophytes (10%) Kenophytes (12%)	Woźniak (2001) Bąba & Kompała (2003)
Ecological groups of species	Meadow species (22%) Perennial ruderal species (11%) Short-lived ruderal species (8%) Xerothermic grassland species (7%) Rush species (7%) Peat bog and bog species (6%) Psammophilous species (5%)	Meadow species (16.3%) Ruderal species (23.4%) Xerothermic grassland species (5.7%) Rush species (11.1%) Peat bog and bog species (3.5%) Psammophilous species (5.9%)	Meadow species (20%) Ruderal species (22%) Forest species (15%) Waterside habitat species (8%) Segetal species (20%) Xerothermic grassland species (4%) Psammophilous species (5%)	Woźniak (2001) Bąba & Kompała (2003)
ECOLOGICAL INDICATORS				
Moisture	Dry soils (32.4%) Fresh soils (24.4%)	Dry soils (19%) Fresh soils (38%)	Mesophilous species (55%)	Woźniak (2001)

VARIABLE	SAND-PITS	COAL MINE SEDIMENTATION POOLS	COAL MINE SPOIL HEAPS	SOURCE
	Wet soils (21.6%)	Moist soils (20%)	Species preferring moist soils (22%)	Bąba & Kompała (2003)
Soil reaction	Moderately acidic to acidic soils (43.2%) Acidic soils (20.1%)	Moderately acidic to slightly acidic soils (39%) Slightly acidic to slightly basic conditions (41.1%)	Species with wide tolerance range (28%) Acidophilous species (12%) Alkaline species (35%)	Woźniak (2001) Bąba & Kompała (2003)
Productivity	Species moderately poor in nutrients (38%) rich in nutrients soils (43%)	Species of poor habitats (26%) Species with intermediate preferences in terms of productivity (35%) Nitrophilous species (14%)	Species of poor habitats (22%) Species with intermediate preferences in terms of productivity (35%) Nitrophilous species (22%)	Bąba & Kompała (2003)
Humus content	Mineral-humus soils (52.5%) Soils with very little humus content (27.8%)	Mineral-humus soils (52.5%) Soils with a very little humus content (15%)	Mineral-humus soils (50%) Soils very poor in humus (15%)	Bąba & Kompała (2003)
Light	Species preferring places with full light, only periodically shaded (65%)	Species preferring places with full light, only periodically shaded (60%) Full light (19%)	Heliophilous species (50%) Light loving plants or living in full light species (33%)	Woźniak (2001) Bąba & Kompała (2003) Rostański (2006)
Protected, threat, endangered species	*Orchidaceae* *Lycopodiaceae* *Cyperaceae* *Droseraceae*	*Orchidaceae* *Gentianaceae*	*Orchidaceae*	Bąba & Kompała (2003) Sierka (2013)
Soil dispersion preference indicator	Mainly sandy soils with clay admixture	Fine-grained substrate	Species with wide range of granulometric composition (18%) Species preferring sandy soils (45%) Heavy soils (15%) Skeletal soils (10%)	Bąba & Kompała (2003) Sierka (2013)
Obligatory and facultative halophytes	0	18% of species that tolerate higher content of salt in the soils; 1.5% obligatory halophytes	17% tolerate higher content of salt in the soils	Woźniak (2001) Bąba & Kompała (2003) Sierka (2013)
Obligatory and facultative metalophytes	0	1	5% of the vascular flora	Bąba & Kompała (2003) Sierka (2013)

Herbaceous species prevail over shrub and forest species in the floristic composition of sedimentation pools (Fig. 5). Similarly to other post-industrial wastelands the most numerous were species from the *Asteraceae* family, graminoids (species belonging to *Poaceae*, *Cyperaceae*, *Juncaceae* family) and legumes representing the *Fabaceae* family (Rostański 2006). The positive role of grasses in colonization of post-industrial sites was underlined by many studies that also examined other post-industrial sites connected with excavation and processing of mineral resources such as coal mine spoil heaps, quarries, sandpits, lead and zinc spoil heaps, as well as iron spoil heaps

(Kompała & Woźniak 2001; Bąba *et al.* 2003; Kompała-Bąba *et al.* 2005; Kompała-Bąba & Bąba 2011; 2013; Hanczaruk 2017; Błońska *et al.* 2019b). Some grasses are dominants in plant communities (e.g. *Calamagrostis epigejos*, *Phragmites australis*, *Festuca arundinacea*, *F. ovina*, *F. rubra*, *Molinia caerulea*, *Poa compressa*, *Deschampsia caespitosa*, *Agrostis capillaris*), can produce large amounts of biomass, and protect these heaps surface against wind and water erosion. They can also stabilise the fine-grained substrate owing to the well-developed root system. The flora of the sedimentation pools also includes species from the *Fabaceae* family. They are often used in the reclamation process to enrich poor substrates with humus as well as nitrogen. Shrub and tree species have a small share in the flora of sedimentation pools.

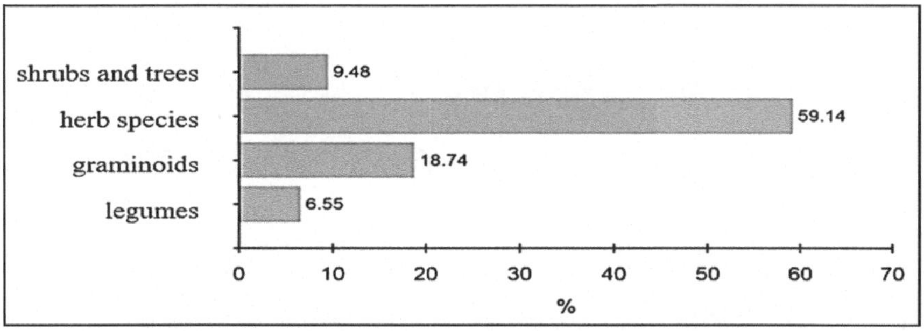

Figure 5. Spectrum of functional groups of species represented in the flora on the coal-mine water sedimentation pools

Origin of species that constitute the vascular flora of sedimentation pools

Similarly to other post-industrial sites (coal mine heaps, sand-pits, lead and zinc spoil heaps), the vast majority of species that constitute vascular species are of native origin (apophytes) (Rostański 1998; 2006; Bąba & Kompała 2003; Woźniak 2010; Sierka 2013; Hanczaruk 2017). Kenophytes predominate among the rare alien species. However, most of them occur only sporadically. The most frequent among them were *Solidago canadensis*, *S. gigantea*, *Matricaria maritima* ssp. *inodora*, *Chamomilla suaveolens*, and *Atriplex prostrata* ssp. *prostrata*. Archaeophytes are represented mainly by segetal weeds (*Papaver rhoeas*, *P. somniferum*, *Vicia angustifolia*, *V. tetrasperma*) as well as short-lived ruderal plant species (Fig. 5, 6).

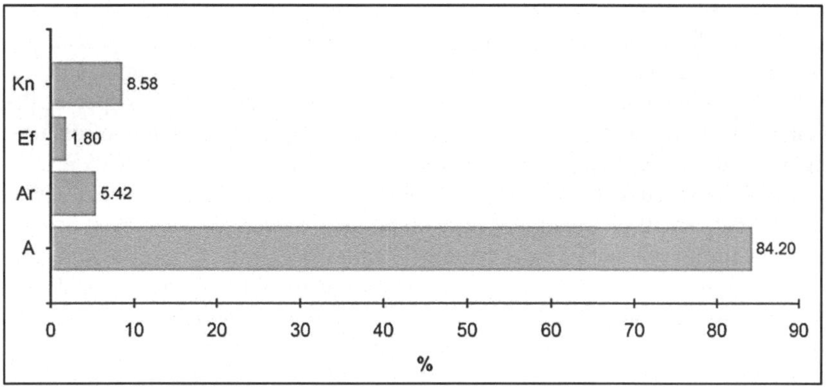

Figure 6. Origin of species that constitute the vascular flora of coal-mine sedimentation pools. Explanations: A – apophytes, Ar – archaeophytes, Ef – ephemerophytes, Kn – kenophytes

Ecological groups of species that occur in the vascular flora of sedimentation pools

The highest participation in the vascular flora was mainly the perennial ruderal species, some meadow species, shrub and forest species, as well as rush species. Because the fine-grained substrate can become very dry during the summer period, xerothermic and psammophilous species are also represented in the vascular flora (Fig. 7).

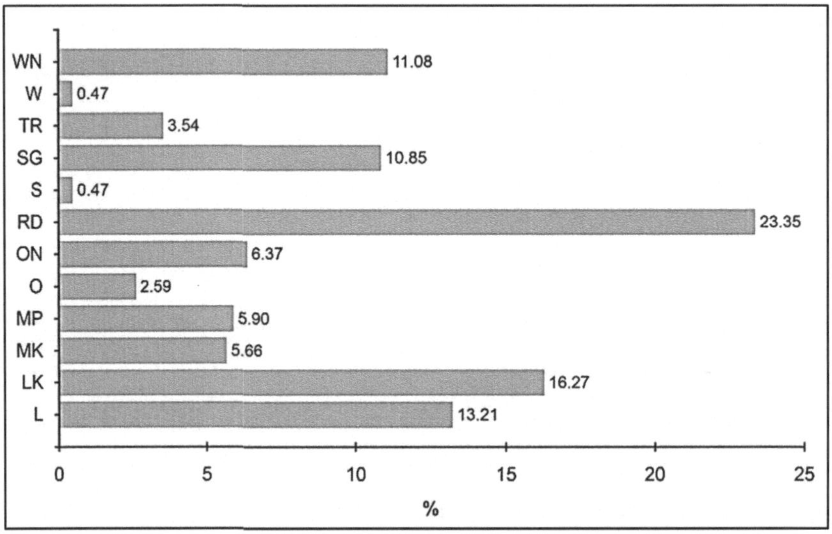

Figure 7. The spectrum of socio-ecological groups in the vascular flora of the coal-mine sedimentation pools. Explanations: L – shrub and forest species, LK – meadow species, MK – xerothermic grassland species, MP – psammophilous species, O – species of clearings and thermophilous fringes, ON – nitrophilous fringe communities, RD – ruderal species, S – species of saline habitats, SG – segetal species, TR – peat bog and bog species, W – aquatic species, WN – rush species

Spectrum of life strategies of vascular flora of sedimentation pools

With respect to the life strategies, the species that represent the competitor (C) strategy dominate the vascular flora of sedimentation pools along with species with mixed types of strategies (CR, CS, CSR) (Fig. 8). Since disturbances with different intensities and frequencies can occur on such sites a group of species with a ruderal type of strategy can be also found. This group of species comprises segetal weeds (*Setaria pumila*, *Vicia angustifolia*, *V. tetrasperma*), as well as species that are components of short-lived ruderal plant communities and are adapted to a variety of disturbances such as trampling, moving, soil transportation, flooding (*Poa annua*, *Polygonum aviculare*, *Senecio vulgaris*). Species in the vascular flora which can tolerate stress and stress-ruderal tolerators (S, SR) composed a relatively small part of the flora. Among them there are species that can grow in sites with higher concentration of salt in the substrate (*Atriplex prostrata*, *Chenopodium glaucum*, *Puccinellia distans*, *Salsola kali*, *Spergula salina*) that appear periodically on the banks of reservoirs that dry up in summer, and species that appear on muddy ground on the bottom and edges of reservoirs, that can be destroyed by flooding, but can reproduce and regrow in the following year (*Ranunculus sceleratus*, *Rumex maritimus*). They are affected by the fact that the coal-mine sedimentation pools have different environmental conditions depending on the time that passed since they were formed. This variability in age of pools enables the presentation of a wider spectrum of the spontaneous flora of specific habitats provided by the coal-mine sedimentation pools.

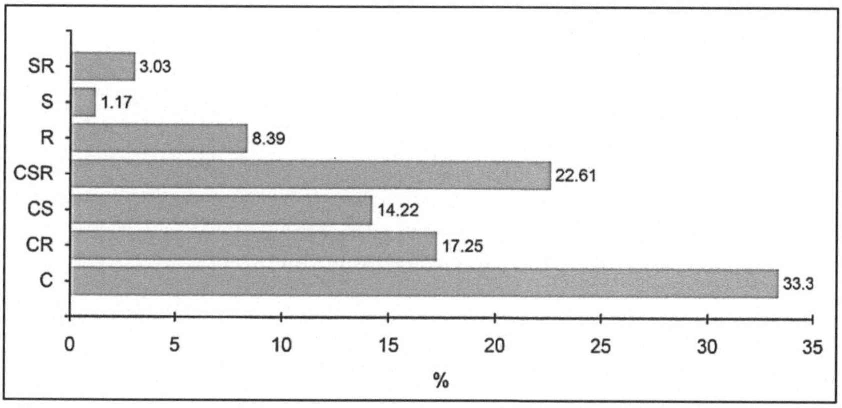

Figure 8. The spectrum of the life strategies of species that constitute the vascular flora of sedimentation pools. Explanations: C – competitors, S – stress-tolerators, R – ruderals, species within mixed type of strategy: CR – competitors-ruderals, CS – competitors-stress tolerators, SR – stress tolerators-ruderals, CSR – competitors-stress tolerators-ruderals

Type of reproduction

Most species that constitute the vascular flora of coal-mine sedimentation pools can reproduce via seeds (Fig. 9). Some of the species are using two types of reproduction strategies, by seeds and vegetatively. Among species representing the first group are mainly ruderal species that constitute both short-lived and perennial ruderal communities. Most of these species are adapted to disturbance and can occupy new areas by producing many, light seeds that can be dispersed by wind, animals or humans. Species that can reproduce both by seeds and vegetatively represent perennial ruderal species and meadow species. Those species are growing on wet substrates as well as psammophilous and xerothermic grassland species.

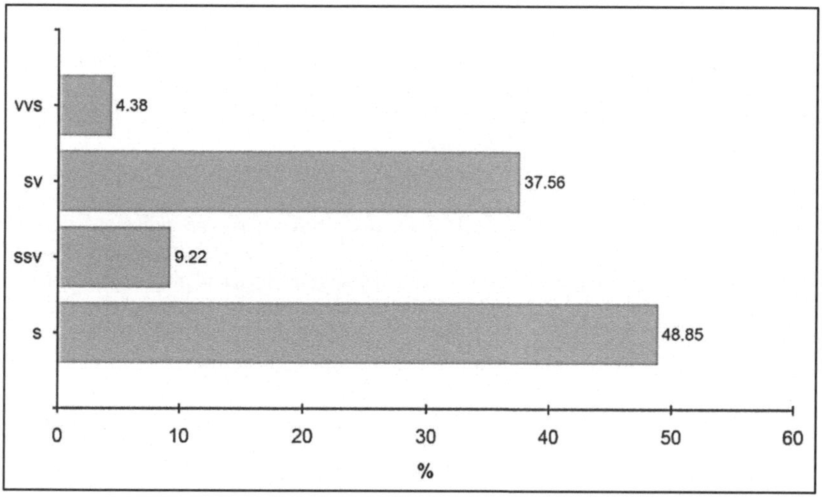

Figure 9. The participation of reproduction types in the vascular flora of coal-mine sedimentation pools. Explanations: s – species that reproduce mostly by seeds, sv – species that reproduce by seeds and vegetatively, ssv – species that reproduce mostly by seeds and vegetatively, vvs – species that reproduce mostly vegetatively and by seeds

CONCLUSIONS

It seems extremely important to carry out detailed studies on vascular flora that develop as a result of colonisation of sedimentation pools, on the mechanisms of formation of plant communities, on the adaptation of plants to extreme conditions which often differ from natural habitat conditions, as well as on the relations between plants and microorganisms, in terms of planning economically justified reclamation activities in the future. In recent years, it has also become more and more important to restore the existing ecosystem services or to create new ones such as provisioning, regulating (e.g. sequestration of carbon, protection against water and wind erosion), and cultural ones.

REFERENCES

Bąba W. & Kompała A. 2003: Piaskownie jako centra bioróżnorodności. Środowisko i Rozwój, 7 (1), 85–102.

Bąba W., Błońska A. & Kompała-Bąba A. 2003: Grasses in plant communities of the sand-pits. [In:] Frey L. (ed.): Problems of Grass Biology. W. Szafer Institute of Botany, Polish Academy of Sciences, Kraków, 547–562.

Babczyńska-Sendek B., Błońska A., Bula R., Cabała S., Michalska-Hejduk D., Hereźniak J., Kołodziejek J., Krajewski Ł., Krause R., Mitka J., Nowak T., Parusel J., Piwowarczyk R., Posz E., Siciński J.T., Spałek K., Szendera W., Urbisz A., Węgrzynek B. & Gałosz J. 2012: The red list of vascular plants of Silesian Voivodship. [In:] Parusel J.B. (ed.): The red lists of chosen groups of mushrooms and plants of Silesian Voivodship. Upper Silesian Nature Heritage Centre, Katowice, 105–177.

Błońska A., Kompała-Bąba A., Sierka E., Besenyei L., Magurno F., Frydecka K., Bierza W. & Woźniak G. 2019a: Impact of selected plant species on enzymatic activity of soil substratum on post-mining heaps. Journal of Ecological Engineering, 20 (1), 138–144.

Błońska A., Kompała-Bąba A., Sierka E., Bierza W., Magurno F., Besenyei L., Ryś K. & Woźniak G. 2019b: Diversity of vegetation dominated by selected grass species on coal-mine spoil heaps in terms of reclamation of post-industrial areas. Journal of Ecological Engineering, 20 (2), 209–217.

Cabała J., Ćmiel S.R. & Idziak A. 2004a: Environmental impact of mining activity in the Upper Silesian Coal Basin. Geologica Belgica, 7 (3–4), 225–230.

Cabała J., Ćmiel S.R. & Idziak A. 2004b: The management of former mining areas in the north-eastern part of the Upper Silesian Coal Basin (Poland). [In:] Hardygóra M., Paszkowska G. & Sikora M. (eds.): Mine Planning and Equipment Selection. Taylor & Francis Group, London, 749–754.

Cabała S. & Sypień B. 1987: Rozwój szaty roślinnej na wybranych zwałowiskach kopalń węgla kamiennego GOP. Archiwum Ochrony Środowiska, 3–4, 169–184.

Cabała S. 1990: Zróżnicowanie i rozmieszczenie zbiorowisk leśnych na Wyżynie Śląskiej. Prace Naukowe Uniwersytetu Śląskiego w Katowicach, Katowice.

Celiński F. & Wika S. 1992: Zagrożenia żywych zasobów przyrody województwa katowickiego. Biblioteczka Fundacji Ekologicznej „Silesia", 5, 43.

Chytrý M., Tichý L., Dřevojan P., Sádlo J. & Zelený D. 2018: Ellenberg-type indicator values for the Czech flora. Preslia, 90, 83–103.

Cudlín O., Řehák Z. & Cudlín P. 2016: Development of soil characteristics and plant communities on reclaimed and unreclaimed spoil heaps after coal mining. IOP Conference Series, Earth and Environmental Science, 44, 052030.

Czaja S. 2001: Mining and hydrological transformations in Upper Silesia from the fifteenth to the nineteenth century. The Geographical Journal, 167 (1), 57–71.

Dulias R. 2016: The Impact of Mining on the Landscape. A study of Upper Silesian Coal Basin in Poland. Springer, Cham.

Ellenberg H., Weber H.E., Dull R., Wirth W., Werner W. & Paulissen D. 1992: Zeigewerte von Pflanzen in Mitteleuropa. 2nd ed. Scripta Geobotanica, 18, 1–258.

Frank D. & Klotz S. 1990: Biologisch-ökologische Daten zur Flora der DDR. Martin Luther Universität, Wissenschaftliche Beiträge, Haale (Saale).

Frouz J. (ed.) 2014: Soil Biota and Ecosystem Development in Post Mining Sites. CRC Press, Taylor & Francis Group, Boca Raton.

Grime J.P. 1979: Plant Strategies and Vegetation Processes. John Wiley & Sons, Chichester-New York-Brisbane-Toronto.
Halabowski D., Sowa A. & Krodkiewska M. 2018: Inland coal mine settling pond as a habitat for the brackish-water plant *Ruppia maritima*. Polish Journal of Ecology, 66 (3), 301–308.
Hanczaruk R. & Gołąb N. 2016: Wybrane problemy rekultywacji zwałowiska odpadów pogórniczych na przykładzie zwałowiska „Ruda" w Zabrzu-Biskupicach. Prace Naukowe Uniwersytetu Ekonomicznego we Wrocławiu, 461, 65–75.
Hanczaruk R. & Kompała-Bąba A. 2019a: Changes in the vascular flora of a postflotation zinc-lead ore spoil heap of the "Orzeł Biały" mining and smelting works in Bytom (Silesian Upland) after 15 years. Acta Agrobotanica, 72 (1), 1762.
Hanczaruk R. & Kompała-Bąba A. 2019b: Effect of thermal activity on the differentiation of the vegetation of the "Ruda" post-mining heap in Zabrze (Poland). Acta Agrobotanica, 72 (3), 1783.
Hanczaruk R. 2017: Spektrum ekologiczno-siedliskowe flory naczyniowej zwałowiska odpadów pogórniczych „Ruda" w Zabrzu-Biskupicach. Prace Naukowe Uniwersytetu Ekonomicznego we Wrocławiu, 494, 50–63.
Hobbs J.R., Arico S., Aronson J., Baron J.S., Bridgewater P., Cramer V.A., Epstein P.R., Ewel J.J., Klink C.A., Lugo A.E., Norton D., Ojima D., Richardson D.M., Sanderson E.W., Valladares F., Vilà M., Zamora R. & Zobel M. 2006: Novel ecosystems: theoretical and management aspects of the new ecological world order. Global Ecology and Biogeography, 15 (1), 1–7.
Jędrzejko K. 1987: Szata roślinna w krajobrazie terenów przemysłowych Wyżyny Śląskiej. [In:] Szczypek T. & Wach J. (eds.): Problemy geograficzne Górnośląsko-Ostrawskiego Regionu Przemysłowego. Oddział Doskonalenia Nauczycieli Instytutu Kształcenia Nauczycieli, Wydział Nauk o Ziemi Uniwersytetu Śląskiego, Katowice-Sosnowiec, 49–56.
Jochimsen M.E. 2001: Vegetation development and species assemblages in a long-term reclamation project on mine spoil. Ecological Engineering, 17 (2–3), 187–198.
Klotz S., Kühn I. & Durka W. (eds.) 2002: BIOLFLOR – Eine datenbank zu biologisch-ökologischen merkmalen der gefäßpfanzen in Deutschland. Schrifenreihe für Vegetationskunde. Bundesamt für Naturschutz, Bonn.
Kompała A. & Woźniak G. 2001: The role of grasses in chosen anthropogenic plant communities in the Upper Silesian industrial district. [In:] Frey L. (ed.): Studies on Grasses in Poland. W. Szafer Institute of Botany, Polish Academy of Sciences, Kraków, 329–353.
Kompała-Bąba A. & Bąba W. 2011: Udział traw w zbiorowiskach roślinnych wykształcających się na obszarze wybranych kamieniołomów Wyżyny Śląskiej. Fragmenta Floristica et Geobotanica Polonica, 18 (2), 359–374.
Kompała-Bąba A. & Bąba W. 2013: Udział traw (*Poaceae*) w zbiorowiskach roślinnych wykształcających się nieużytkach hutnictwa żelaza występujących na obszarze Wyżyny Śląskiej. Fragmenta Floristica et Geobotanica Polonica, 20 (2), 267–284.
Kompała-Bąba A. 2007: The ruderal plant communities of the Dąbrowska Basin (Silesian Uplands). [In:] Wika S. & Woźniak G. (eds.): Threats, Protection and Transformation of Vegetation of the Upper Silesia and Adjacent Areas. University of Silesia in Katowice, Katowice, 45–65.
Kompała-Bąba A. 2013: Abiotic and biotic factors affecting the diversity of ruderal vegetation (Silesian Uplands Poland). Sorus, Poznań.
Kompała-Bąba A., Bierza W., Błońska A., Sierka E., Magurno F., Chmura D., Besenyei L., Radosz Ł. & Woźniak G. 2019: Vegetation diversity on coal mine spoil heaps – how important is the texture of the soil substrate? Biologia, 74, 419–436.
Kompała-Bąba A., Błońska A., Bąba W. & Czyba M. 2005: Grasses in plant communities which develop on the waste sites of zinc-lead industry (Upper Silesia, South of Poland). [In:] Frey L. (ed.): Biology of Grasses. W. Szafer Institute of Botany, Polish Academy of Sciences, Kraków, 269–282.
Kowarik I. 1999: Zum menschlichen Einfluss auf Flora und Vegetation. Schriftenr. Fachber. Landschaftsentwicklung. Technische Universität, Berlin.
Machowski R., Jankowski A. & Gawenda M. 2006: Changes in land use and landscape complexes in strongly industrialized areas (a case study of Ruda Śląska town, Upper Silesian Industrial Region, Southern Poland). [In:] The 14th International Symposium of Landscape Ecology Research: Implementation of Landscape Ecology in new and changing conditions, 4–7 October 2006, Stara Leśna, Slovakia. Institute of Landscape Ecology SAS, Slovak Association for Landscape Ecology, Constantine the Philosopher University in Nitra, 247–254.
Maciak F. 2003: Ochrona i rekultywacja środowiska. Wydawnictwo SGGW, Warszawa.

Matuszkiewicz W. 2012: Przewodnik do oznaczania zbiorowisk roślinnych Polski. Wydawnictwo Naukowe PWN, Warszawa.
Mirek Z., Piękoś-Mirkowa H., Zając A. & Zając M. 2002: Flowering Plants and Pteridophytes of Poland. W. Szafer Institute of Botany, Polish Academy of Sciences, Kraków.
Oberdorfer E., Müller T., Korneck D., Lippert W., Markgraf-Dannenberg I., Patzke E. & Weber H.E. 1990: Pflanzensoziologische Excursionflora. 6 Auflage. Ulmer, Stuttgart.
Patrzałek A. & Rostański A. 1992: Procesy glebotwórcze i zmiany roślinności na skarpie rekultywowanego biologicznie zwałowiska odpadów po kopalnictwie węgla kamiennego. Archiwum Ochrony Środowiska, 3–4, 157–168.
Pecharová E., Broumová-Dušáková K., Novotná K. & Svoboda I. 2011: Function of vegetation in new landscape units after brown coal mining. International Journal of Mining, Reclamation and Environment, 25 (4), 367–376.
Pełka-Gościniak J. 2006: Restoring nature in mining areas of the Silesian Upland (Poland). Earth Surface Processes and Landforms, 31 (13), 1685–1691.
Pietrzykowski M. & Krzaklewski W. 2018: Reclamation of mine lands in Poland. [In:] Prasad M.N.V., de Campos Favas P.J., Maiti S.K. (eds.): Bio-Geotechnologies for Mine Site Rehabilitation. Elsevier, Cambridge, 493–513.
Prach K., Řehounková K., Řehounek J. & Konvalinková P. 2011: Ecological restoration of Central European mining sites: A summary of a multi-site analysis. Landscape Research, 36, 263–268.
Rebele F. & Dettmar J. 1996: Industriebrachen Ökologie und Management E.U. Verlag, Eugen Ulmer GmbH & Co., Stuttgart.
Rostański A. 1998: Anthropophytes and apophytes in colonization process on the post-industrial heaps in Upper Silesia Region. Phytocoenosis, 10, 199–201.
Rostański A. 2006: Spontaniczne kształtowanie się pokrywy roślinnej na zwałowiskach po górnictwie węgla kamiennego na Górnym Śląsku [Spontaneous plant cover on colliery spoil heaps in Upper Silesia (Southern Poland)]. University of Silesia, Katowice.
Rutkowski L. 1998: Klucz do oznaczania roślin naczyniowych Polski niżowej. PWN, Warszawa.
Sendek A. 1981: Analiza antropogenicznych przemian w szacie roślinnej Górnośląskiego Okręgu Przemysłowego. Prace Naukowe Uniwersytetu Śląskiego, Katowice.
Sierka E. 2013: Biodiversity of the Silesian voivodship postindustrial areas – resources, protection, threats. [In:] Parusel J.B. (ed.): Biodiversity of the Silesian voivodship postindustrial areas – resources, protection, threats. Upper Silesian Nature Heritage Centre, Katowice, 1–58.
Sukkop H. 2008: The city as a subject for ecological research. [In:] Sukopp H., Marzluff J.M., Shulenberger E., Endlicher W., Alberti M., Bradley G., Ryan C., Simon U. & ZumBrunnen C. (eds.): Urban Ecology. An International Perspective on the Interaction Between Humans and Nature. Springer, Boston, 281–299.
ter Braak C.J.K. & Šmilauer P. 2002: CANOCO reference manual and CanoDraw for Windows User's Guide. Software for Cannonical Community Ordination (version 4.5). Ithaca, NY: Microcomputer Power.
Woźniak G. & Błońska A. 2009: Bioindicative applicability of vegetation on post coal mine sites. Polish Journal of Environmental Studies, 6, 127–131.
Woźniak G. & Rostański A. 2000: Rola traw w spontanicznej sukcesji roślinnej na osadnikach ziemnych wód kopalnianych na Górnym Śląsku. Łąkarstwo w Polsce, 3, 159–167.
Woźniak G. & Śliwińska-Wyrzychowska A. 2002: Trwałe zadamawianie się gatunków obcych (antropofitów) w spontanicznych zbiorowiskach roślinnych na nieużytkach poprzemysłowych (osadniki ziemne wód kopalnianych) Górnego Śląska. Środowisko i Rozwój, 5–6 (1–2), 145–162.
Woźniak G. 1998: Primary succession on the sedimentation pools of coal mine. Synanthropization of plant cover in new Polish research. Phytocoenosis, 10 (9), 189–198.
Woźniak G. 2001. Flora roślin naczyniowych osadników ziemnych wód kopalnianych – nieużytków poeksploatacyjnych na Górnym Śląsku [Vascular plants flora of the coal mine underground water sedimentation pools – post-industrial wastelands in the Upper Silesia]. Centrum Dziedzictwa Przyrody Górnego Śląska, Materiały i Opracowania, 6, 3–46.
Woźniak G. 2005: Problems of *Calamagrostis epigejos* synecology on post-industrial sites. [In:] Frey L. (ed.): Biology of Grasses. W. Szafer Institute of Botany, Polish Academy of Sciences, Kraków, 353–361.
Woźniak G. 2006: Colonization process on coal mine sedimentation pools (Upper Silesia, Poland). Polish Botanical Studies, 22, 561–568.

Woźniak G. 2010: Zróżnicowanie roślinności na zwałach pogórniczych Górnego Śląska [Diversity of vegetation on coal-mine heaps of the Upper Silesia (Poland)]. W. Szafer Institute of Botany, Polish Academy of Sciences, Kraków.
Woźniak G., Markowicz A., Borymski S., Piotrowska-Seget Z., Chmura D. & Besenyei L. 2015: The relationship between successional vascular plant assemblages and associated microbial communities on coal mine spoil heaps. Community Ecology, 16, 23–32.
Zarzycki K., Trzcińska-Tacik H., Różański W., Szelag Z., Wołek J. & Korzeniak U. 2002: Ecological indicator values of vascular plants of Poland. W. Szafer Institute of Botany, Polish Academy of Sciences, Kraków.
Zobel M. 2016: The species pool concept as a framework for studying patterns of plant diversity. Journal of Vegetation Science, 27, 8–18.

2.6. Mineral Soil Substrate Seed Banks: Understanding Their Role in Primary Succession and Enhancement of Habitat Recovery

Joanna Czarnecka
Department of Botany, Mycology and Ecology,
Institute of Biological Sciences, Maria Curie-Skłodowska University;
Akademicka 19 Str., 20-033 Lublin, Poland;
joanna.czarnecka@mail.umcs.pl; ORCID: 0000-0001-6358-2915

Wiktoria Hryń
Department of Botany, Mycology and Ecology,
Institute of Biological Sciences, Maria Curie-Skłodowska University;
Akademicka 19 Str., 20-033 Lublin, Poland;
wiktoria.hryn@gmail.com; ORCID: 0000-0003-3053-0144

Jaco Vangronsveld
Department of Plant Physiology and Biophysics,
Institute of Biological Sciences, Maria Curie-Skłodowska University;
Akademicka 19 Str., 20-033 Lublin, Poland;
Environmental Biology, Centre for Environmental Sciences, Hasselt University;
Agoralaan Building D, 3590 Diepenbeek, Belgium;
jaco.vangronsveld@mail.umcs.pl; ORCID: 0000-0003-4423-8363

ABSTRACT: One of the key factors determining the pattern of plant colonization is seed availability. Insufficient amounts of diaspores may be an obstacle for plant cover development, especially on sites undergoing primary succession, where all the seed sources are initially situated outside the newly developing ecosystem. Seeds that do not germinate soon after reaching the soil constitute a seed bank, providing an important component of life strategies for many pioneer plant species. This chapter examines the characteristics of soil seed banks found on mine waste heaps, their relationship with heap conditions, as well as their role in the spontaneous colonization of such objects. We also explored the importance of environmental variables influencing germination, in the context of heap-specific conditions.

KEYWORDS: Seed reserves, mine spoil heaps, species richness, germination, seed rain

SEED BANK CONCEPT, METHODOLOGY AND IMPORTANCE

Species distribution depends on dispersal processes at various spatial scales, availability of suitable microsites, and capacity of a population to persist after establishment (Eriksson & Ehrlén 1992; Ehrlén & Eriksson 2000). The ability of species to disperse as well as the seed rain intensity can influence the composition of plant communities (Münzbergová 2012). The importance of seed limitation has been often underestimated; however, experimental studies suggest that the local population size and distribution at the regional scale are sometimes highly limited by seed availability, and sometimes seeds have low efficiency in reaching potential patches for recruitment (Eriksson & Ehrlén 1992; Ehrlén & Eriksson 2000). Local species richness at a given site is usually directly proportional to the regional species pool. Distance and isolation from the pool of diaspores usually excludes many of species, while others reach the microsite but find the habitat unsuitable. Species interactions can also be important in structuring communities, but in the majority of cases (ecological systems) they appear to be less important (Lawton 1999). One of the frequently studied consequences of seed dispersal is formation of seed banks of different sizes (Tab. 1), which influence seedling spatial structure and, consequently, adult plant composition and distribution (Wang & Smith 2002).

Table 1. Examples of soil seed bank density (seeds/m^2) in different vegetation types in connection with the succession type and successional stage

VEGETATION TYPE	SEED BANK DENSITY	SOURCE
Primary succession		
Dune slacks (western coastline of Belgium)		
5-year slacks	339	Bossuyt & Hermy (2004)
55-year slacks	9160	Bossuyt & Hermy (2004)
Reclaimed coal mine: open-pit filled with coal waste and fine-textured sediments and covered with a layer of cattle manure (northern Spain)		
Before hydroseeding	17	González-Aiday *et al.* (2009)
2–3 years after hydroseeding	1813	González-Aiday *et al.* (2009)
Zinc mine spoil heaps (Olkusz, Poland)		
Young heap (< 20 years)	3000–81 000	Grodzińska *et al.* (2000)
Old heap (> 20 years)	6000–20 000	Grodzińska *et al.* (2000)
Secondary succession		
Arable field, cereal crop (Lublin, Poland)	13 056	Czarnecka (2006)
Abandoned fields on poor soils (Białowieża, Poland)		
6 years after farming cessation	30 000	Symonides (1986)
50 years after farming cessation	3000	Kwiatkowska-Falińska *et al.* (2011)
Brushwood overgrowing xerothermic grassland, Central Roztocze Region, Poland	32 459	Czarnecka (2005)
Abandoned *Cirsietum rivularis* meadow (Białowieża, Poland)		
5 years after abandonment	4052	Falińska (1999)
10 years after abandonment	4182	Falińska (1999)
Late successional vegetation		
Primeval temperate deciduous forest (Białowieża, Poland)	2142	Jankowska-Błaszczuk & Grubb (1997)
Well-preserved natural thermophilous oak wood forest (Białowieża, Poland)	2268	Kwiatkowska-Falińska *et al.* (2013)

The soil seed bank is a reserve of viable seeds present on or in the soil (Harper 1977; Simpson *et al.* 1989). The size and species composition of the bank is determined by the rate at which viable seeds are supplied (seed rain) and lost in the soil (Staaf *et al.* 1987) (Fig. 1).

A crucial trait of species is the longevity of their seeds in a seed bank. A few classifications have been proposed, but that suggested by Thompson & Grime (1979) and modified by other authors, e.g. Bakker *et al.* (1996) and Thompson *et al.* (1997), is most commonly used:

- transient – seeds are able to persist in soil less than one year;
- short-term persistent – seeds persist in soil longer than one year but less than five years;
- long-term persistent – seeds persist in soil longer than five years.

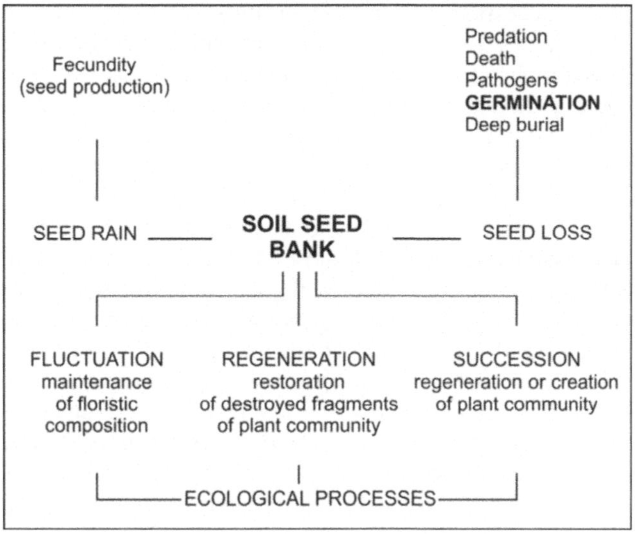

Figure 1. Seed bank role in ecosystem dynamics (after Simpson *et al.* (1989) modified by Czarnecka (2004), ecological processes according to Faliński (2001))

It is broadly accepted that the persistent component of the seed bank is an important reserve of the genetic potential accumulated over a long time (Simpson *et al.* 1989), persistent seeds are often small with a compact shape (close to spherical) and usually photoblastic (Harper 1977; Thompson 1987; Simpson *et al.* 1989; Thompson *et al.* 1998; Jankowska-Blaszczuk & Daws 2007). A consequence of these seed characteristics is the constant presence of seeds of species with a persistent seed bank in deeper soil layers. The LEDA database is a good source of information about the seed bank type (Kleyer *et al.* 2008). It is important to note that there is not a close relationship between seed dormancy (morpho-physiological, physical or physiological) as a characteristic of seeds, with seed longevity in the soil (Thompson *et al.* 2003).

There are two possible pathways for the formation of seed bank assembly during succession. The seed bank can be composed of early successional species during the whole succession process. The explanation is that early successional species produce many small and persistent seeds, which remain in the soil after these species disappear from the vegetation. Their proportion is high during the whole successional process because later successional species usually produce small amounts of seeds with lower persistence. Another possibility is that the seed bank composition can vary during the successional process according to changes in the vegetation: the seed bank structure simply follows the changes in species richness of the vegetation. In this case, early successional species disappear from the seed bank after they are lost from vegetation and their seeds are replaced by seeds of species appearing in the vegetation later in succession when their seeds are gradually incorporated in the soil. The type of pathway depends on the ratio of persistent seeds and the whole seed production by early and late successional species, as well as the type of succession, i.e. primary or secondary (Bossuyt & Hermy 2004). The first pathway was observed during secondary succession on abandoned arable lands when forest communities were formed (e.g. studies of mixed deciduous forests in central Belgium (Bossuyt *et al.* 2002) and fresh coniferous forests overgrowing abandoned fields on poor soils in north-eastern Poland (Kwiatkowska-Falińska *et al.* 2011). The majority of species present in the seed bank in these cases originated from earlier successional phases. The effects of historical land use (grassland and heathland cases) decrease after 50 years due to senescence of persistent seeds of the former vegetation (Bossuyt & Hermy 2001). Forest seed banks on former arable lands recover poorly and the process can last more than 150 years (Plue *et al.* 2010).

In the case of primary succession on nutrient-poor substrates, early successional species must be adapted to high stress conditions to be successful in germination, and the seed bank structure usually follows changes observed in the vegetation (the second pathway). Usually seed density and species richness is very low during the first successional stages and increases over time as a result of higher seed production. It seems to be a rule that the mean seed longevity decreases during primary succession, which is accompanied by an increase in species richness and density, especially in the upper soil layer, and a positive correlation between the seed bank-vegetation similarity index and the study site age is observed; examples include studies of dune slack sites (Bekker *et al.* 1999; Bossuyt & Hermy 2004) and lake islands (Grandin & Rydin 1998). This proves that the occurrence of species in the seed bank usually follows their presence in the vegetation (Bossuyt & Hermy 2004).

Seed bank analysis is usually performed using two widely recommended methods (Gross 1990; Brown 1992; Thompson *et al.* 1997; Gonzalez & Ghermandi 2012): seed extraction under magnification from soil samples concentrated by washing and sieving (the recommended mesh size of 0.25 mm guarantees that even the smallest seeds will remain on the sieve) and the seedling emergence method. Although seedling emergence was the method usually applied in studies of the soil seed bank (e.g. Staaf *et al.* 1987; Jankowska-Błaszczuk 1998; Falińska 1999; Bossuyt *et al.* 2002; Plue *et al.* 2010; Kwiatkowska-Falińska *et al.* 2011), the seed extraction method, although time consuming, facilitates detection of all seeds that have reached the soil in the studied ecosystems. The seedling emergence method fails to determine presence of dormant seeds and seeds with specific germination requirements (Gross 1990; Brown 1992).

SEED BANKS OF SPOIL HEAPS

Seed rain and seed banks are crucial for plant community development at the initial stages of succession (Zhao *et al.* 2012). They constitute a reserve of propagules from which plant individuals may recruit into the plant cover (Grandin & Rydin 1998). As objects that are initially devoid of vegetation, spoil heaps create an opportunity to examine the beginning of successional processes (Markowicz *et al.* 2015), including formation of the seed bank and its relationship to vegetation. The revegetation of such highly transformed areas is a great challenge, and the presence of the initial plant cover is a very important step in the process of stabilization of the substrate. Although it is broadly known that the soil seed bank is an important component of ecosystem resilience (Wang *et al.* 2013) and is recognized as a source of genetic novelty for many plant species (Levin 1990), and a base of plant regenerative strategies in habitats subjected to seasonal or unpredictable disturbances (Grime 1989), there are few studies of the seed rain and seed bank at early stages of restoration of coal wastes (Tab. 1).

In the case of seed banks of spoil heaps, a distinction should first be made between recultivated and non-recultivated spoil heaps. Technical reclamation of spoil heaps often includes levelling of the terrain and covering it with a layer of topsoil, usually containing some amount of seeds (Zhang *et al.* 2001). The reason is that a huge problem can be posed by the limitation of revegetation caused by the absence of a seed bank in coal mining wastes and the reduced opportunities for dispersal from nearby colonization sources (González-Alday *et al.* 2009). Sometimes additional seeds of desired plant species are sown directly and therefore supply topsoil reserves (Gonzáles-Alday *et al.* 2008). The observation of seed bank formation in coal wastes filling open pits created during coal mining in northern Spain showed that, at the initial stage, only a few seeds were detected in the substrate (17 seeds m^{-2} representing 6 plant species). Hydro-seeding was applied (seed mixture of commercial and crop sources with grasses and legumes, mixed with soluble chemical fertilizer) to improve revegetation in that area. After two growing seasons post hydro-seeding, the soil seed bank structure was studied again. Although the seed density increased significantly (1813 seeds m^{-2}), it was composed mainly of native species coming from plants germinating from the initial seed bank and seeds colonizing the substrate from the surrounding vegetation. Seed density of native species increased when the distance to the nearest natural community acting as a seed source decreased. The seed bank varied significantly between different areas of the same mine, and the topography of the surface appeared to be another important factor affecting its structure. The development of the seed bank turned out to be an important part of the site restoration and it was concluded that it should be taken into account during management and planning. If a commercial seed mixture is planned to be applied (hydro-seeding), species must be chosen with care, as they can become a permanent part of the seed bank, which may cause long-term problems in the future (González-Alday *et al.* 2009). Application of a commercial seed mixture without earlier studies of the potential of local colonization processes appeared to be ineffective (Martínez-Ruiz *et al.* 2007 and literature cited therein). The expected results of these treatments were observed two years after the application on only some of the treated microhabitats (north-facing slopes with better moisture conditions). It was found that the commercial mixture of seeds performed worse than the native species (Martínez-Ruiz *et al.* 2007), which also confirmed results obtained earlier (Pinaya *et al.* 2000; Tinsley *et al.* 2006). Additionally, the species from the surrounding plant communities had enormous capacity to displace the initial vegetation and maintain high diversity of the developing vegetation. This emphasizes the potential of natural colonization in maintaining and preserving biodiversity. It was also hypothesised that on some degraded habitats the role of humanity in the process of site restoration can be limited to aiding and enhancing of natural colonization. The resistance of communities assembled from a large local species pool also seems to be higher than those coming from an introduced exotic seed pool (Martínez-Ruiz *et al.* 2007).

While some long-living seeds may survive burdening with gangue material, a majority of seeds composing the seed bank of non-recultivated heaps come from the surrounding vegetation (Horáčková *et al.* 2016). The composition of the seed bank therefore depends on the species composition of plant communities located approximately within 5 km of the heap (Woch *et al.* 2013). Crucial factors for successful diaspore transportation include the ability to disperse over a long distance (*via* anemochory or zoochory) and the abundance of diaspores in the surrounding habitats (Grime *et al.* 1986). As reported by Salisbury (1953), plants that successfully disperse and colonize new habitats need to produce large numbers of seeds; this is possible only at sufficiently high density and fertility of the original population. Most seeds that travel to the heap successfully represent species that are capable of long-distance dispersal and production of large numbers of seeds and reach high abundances in adjacent ecosystems. Such features are typical for annual and perennial herbs representing the ruderal life strategy (Grime *et al.* 1986). They are typically fast-growing, light-seeded, photoblastic, anemochorous species adapted to colonizing open spaces and dominating the seed banks during the first stages of succession (Grime 1977; Šerá & Šerý 2004). Many of such species require high nutrient content (e.g. nitrogen) in the soil to germinate successfully, which may be problematic on the heap (Lambers *et al.* 2008). Many of the incoming seeds may therefore not germinate, and accumulate in the soil forming the seed bank. Whilst a majority of ruderal species have long-lived seeds (Klimkowska *et al.* 2010), their mortality in the heap substrate may be increased due to unfavourable edaphic conditions such as low pH, high salinity, oxidative stress, high accumulation of heavy metals, and so on (Kompała-Bąba *et al.* 2019). Stress-tolerant species, which are better adapted to harsh environmental conditions, have a limited participation in the initial seed pool of spoil heaps, due to their limited fertility, and usually low abundances in the adjacent areas that are usually eutrophicated as a result of various human activities (Grime *et al.* 1986). They may appear at the later stages of succession: for example, there are plants growing on zinc mine ore and smelter heaps with seed tolerance to heavy metals *via* accumulation thereof outside of the embryo tissues (Grodzińska *et al.* 2000).

At the initial stages of succession, the seed banks of non-recultivated spoil heaps are small and consist mainly of anemochorous species. Another important vector of dispersal in such habitats is rain and water runoff (Horáčková *et al.* 2016). Some colonizing plants have no visible adaptations for dispersal but are transported by humans (purposefully as a recultivation means – e.g. sowing *Fabaceae* species for enrichment of the substrate with nitrogen, or involuntarily – on shoes, clothes and cars) (Grime *et al.* 1986; Koszelnik-Leszek *et al.* 2013). There is evidence that even mining activities themselves might be an accidental source of seed introduction (Prach *et al.* 2013). Seeds that travel to the heap *via* anthropochory accumulate in the soil and may establish in the vegetation with time. Once the plant cover is established, the local dispersal supplies the seed pool further and the seed densities increase although some species, e.g. *Tussilago farfara* and *Calamagrostis epigejos*, seem to rely more on vegetative expansion, rather than on a seed bank, as a regeneration strategy (Ogden 1974; Grime 1989; Baasch *et al.* 2012).

One of the important aspects affecting seed banks forming *de novo* on heaps is spatial habitat heterogeneity (Frouz *et al.* 2018). The terrain of non-recultivated spoil heaps is usually rough due to uneven deposition of gangue material (Martín-Moreno *et al.* 2016; Frouz *et al.* 2018). There are both slopes with various shapes and degrees of steepness and plateaus located in close proximity. Steep slopes are prone to erosion and both seeds and substrate material accumulate at the foot of the slope (Martín-Moreno *et al.* 2016). Areas with flat topography have better chances of accumulating large densities of seeds; however, when they are surrounded by artificial site elevations (due to gangue deposition), the neighbourhood might act like a physical barrier preventing seed migration. Another source of spatial heterogeneity of seed bank densities is extreme substrate texture (Tesnerová *et al.* 2017). Immediately after deposition, the substrate texture tends to be coarse, with gravel as a dominant soil fraction. This coarse substrate enables water to permeate through its top layer, which may transport seeds to a depth from which they are unable to germinate. In such places, new material is also sometimes deposited again after some time, which buries the seed reserves deep below the ground level. The loose, bare substrate is a subject to fast weathering and, within ten or more years after the deposition, it is usually heavily dominated by the silt fraction. Such transformed substrate does not allow many seeds to travel below the topsoil level, but may also increase their mortality by possible hypoxia and anoxia due to water stagnation during wet seasons. To summarize, post coal-mine heap seed banks are characterized by both vertical and horizontal structural heterogeneity, with some sites (such as the bottoms of the slopes) having higher seed abundances than adjacent areas. This should be kept in mind when conducting surveys on the seed bank structure of post coal-mine heaps.

At the late stages of succession, post coal-mine heap vegetation often develops into patches strongly dominated by one species of competitive, perennial grass with quick vegetative spread. Such grasses as *Calamagrostis epigejos* expand aggressively and may even halt succession by prevention of woody vegetation establishment (Prach & Pyšek 2001). Dense above-ground biomass and quick necromass accumulation of such species might have a detrimental effect upon seed bank formation, preventing access of seeds from seed rain to the soil (Roubíčková *et al.* 2012; Młynkowiak *et al.* 2014). On the other hand, at lower densities, such grasses may also act as "nurse plants" (Groenveld & Rochefort 2002; Kostrakiewicz-Gierałt 2014) stabilizing the soil and preventing nutrient, water, and seed runoff, thus enhancing the rate of germination.

There are not many quantitative studies targeted specifically on seed banks of soil substrates of heaps. Grodzińska *et al.* (2000) studied both seed rain and seed banks on young (up to 20 years old) and older zinc mine heaps in the Olkusz region (southern Poland). The seed rain peaked in the autumn months and was one to two orders of magnitude higher on the older spoils than on the younger ones. In both cases, it was strongly dominated by a few species, which was more apparent on the younger heaps. 70–80% of species from the vegetation had a significant participation in the seed rain on the old heaps, while only 30% were recorded in the younger sites. *Festuca ovina* or *Thymus pulegioides* were abundant in the plant cover of both the old and young heaps, but they produced significant amounts of seeds only on the former ones. The most dominant species in the seed rain on the older sites were *Euphrasia stricta*, *Festuca ovina*, and *Thymus pulegioides*, whilst the seed rain in the younger sites was dominated mainly by *Cardaminopsis arenosa* and *Silene vulgaris*. On both the older and younger heaps, four species identified from the seed rain were not present in the plant cover, but could be found in the adjacent vegetation. The seed bank structure was screened using the seedling emergence method. The older heaps had seed bank abundances in the range of 6000–20 000 seeds m^{-2}; the variance was higher on the younger heaps where the range of density was 3000–81 000 seeds m^{-2}. Perhaps such a wide range of the results obtained is due to the aforementioned spatial heterogeneity of such habitats. A range of 5–11 species were identified in the seed banks of the younger spoils and 11–14 species were found on the older heaps. The dominant species in the seed bank on the older heaps were *Cardaminopsis arenosa*, *Euphrasia stricta*, *Agrostis stolonifera*, *Campanula rotundifolia*, and *Leontodon hispidus*. The dominance was not very strong and a large proportion of the seed banks consisted of species represented by single seedlings. On the younger heaps, there were three strongly dominant species in the bank: *Cardaminopsis arenosa*, *Agrostis stolonifera*, and *Herniaria glabra*; together they constituted over 90% of seeds. 30% of all species found in the vegetation or the seed bank were found in both the vegetation and the soil on both the old and young heaps, as long as the ecosystems were developing from local material. However, if the topsoil of a different place of origin was applied on the surface of the heap, the similarity between the vegetation and the seed bank dropped to 3%, as many seeds from the seed bank failed to establish in the plant cover. This again highlights the need to choose local reserves of diaspores for introduction.

"SOŚNICA" SPOIL HEAP CASE STUDY

The aim of the study was to determine the seed bank potential and its relationship with the plant cover of different microhabitats on one of the active colliery spoil heaps in the Upper Silesia district region. The "Sośnica" spoil heap is located within the Sośnica estate of the Gliwice city (exact geolocation: 50°16'N, 18°44'E), southern Poland (Fig. 2). It is formed mainly from carboniferous rocks extracted from the depth of about 0.5 to 1 km beneath the surface and deposited at the spoil site (Kompała-Bąba *et al.* 2019). New material is still transported to the top of the heap by a railway connecting the mine and the heap. This causes a vertical gradient of age in different points on the heap (sites of recent deposition are located closer to the top of the heap and old sites are situated at its foot). The vegetation is characterized by spatial heterogeneity, with an apparent mosaic of plant communities with varying degrees of complexity and most probably representing different successional series. They usually have a simplified structure, with one species strongly dominating the others. Five types of microhabitats were chosen for the study. Empty spots were sites with succession halted for an unknown reason, although adjacent places had developed vegetation. Communities with dominance of *Tussilago farfara* and with dominance of *Daucus carota* grew on more initial sites with coarse texture of the substrate – the first one occupied mainly the slopes, while *D. carota* communities seemed not to have affiliation to a certain terrain form and occurred on both the slopes and plateaus. Communities with dominance of perennial grasses – *Calamagrostis epigejos* and *Poa compressa* typically had a more complex and dense structure than the previous two types of vegetation and were located mainly on the lower parts of the heap. During our study, we wanted to address the following questions:

- What is the structure (qualitative and quantitative) of the seed banks of the different types of microhabitats?
- What is the level of similarity between the seed bank and vegetation? Does the seed bank structure follow the structure of vegetation?

The seed bank analysis was performed using the seedling emergence method (samples taken in March 2018) (Price *et al.* 2010). To compare the seed bank structure to the vegetation, we conducted a vegetation study in July 2018 (species diversity and species abundance were noted).

Figure 2. "Sośnica" spoil heap – top of the heap with sites of recent deposition (Photo: Joanna Czarnecka)

The most abundant seed bank was recorded in microhabitats where the vegetation structure was already relatively complex and the percent of area covered by plants was high (Tab. 2). The lowest seed bank densities were registered in the "empty" microhabitat. On average, the number of species was slightly higher in the plant cover than in the seed bank.

More thorough insight into the species structure of the seed banks vs. vegetation revealed different regenerative strategies of the dominant species of the microhabitats. The only microhabitat with the same species dominating strongly both the seed pool and the vegetation was the community with *D. carota*. The other dominants seemed to produce low amounts or, as in the case of *T. farfara*, no seeds at all. The most expansive plant among the dominants examined, i.e. *C. epigejos*, has a limited share in the seed pool in plots where it is abundant. Instead, *Erigeron annuus* contributes significantly to the density of seeds. *C. epigejos* seeds were more frequently found in the soils from the *T. farfara* microhabitat, in spite of their limited success in its plant cover.

Analysis of the share of anemochorous species showed that they contributed greatly to the vegetation and seed bank of the examined phytocoenoses (Tab. 3). Their greatest number was detected in the "empty" and *T. farfara* microhabitats, where the vegetation had the most initial character. The percentage of anemochorous species and their share in the seed pool was smaller than their share in the vegetation.

Table 2. Structure of the plant cover and the seed bank of examined microhabitats; the values are means from all examined plots; N = 8 except in the case of the plant cover of *D. carota* plots, where N = 6 (two plots were lost due to overburdening with gangue material)

MICROHABITAT	PLANT COVER STRUCTURE		SEED BANK STRUCTURE	
	NUMBER OF SPECIES	TOTAL AREA COVERED BY PLANTS (%)	NUMBER OF SPECIES	SEED DENSITY (SEEDS m^{-2})
"Empty"	5.4	4.7	3.7	337
T. farfara	9.5	30.0	8.1	1075
D. carota	12.0	33.3	6.4	925
P. compressa	15.1	47.5	11.5	1537
C. epigejos	14.6	63.7	13.4	3944

Table 3. Mean share of anemochorous species in the plant cover and seed bank of the examined microhabitats: N = 8 except in the case of plant cover of *D. carota* plots where N = 6 (two plots were lost due to overburdening with gangue material)

MICROHABITAT	PLANT COVER		SEED BANK	
	% SHARE OF ALL SPECIES	% SHARE IN SUM OF PLANT COVER	% SHARE OF ALL SPECIES	% SHARE IN SEED POOL
"Empty"	92	90	77	80
T. farfara	72	90	70	76
D. carota	52	43	39	25
P. compressa	57	73	47	47
C. epigejos	62	85	48	59

SEEDLING GERMINATION CONDITIONS ON SPOIL HEAPS AND THEIR CONSEQUENCES TO PLANT DISTRIBUTION

Germination is the most vulnerable stage of the plant life cycle (Ji *et al.* 2018). The process determines the establishment of an individual in the plant cover. Spatial distribution of individual plants in the population depends on the availability of suitable germination microsites constituting a regeneration niche (Falińska 1996). Spatial heterogeneity of spoil heap conditions (for example, water regimes) therefore affects plant distribution and community structure.

The most important factors that affect germination are water regimes, light and temperature, soil chemical properties, and biotic interactions (Lambers *et al.* 2008; Baskin & Baskin 2014). Some of these factors will be presented below with consideration of spoil heap-specific conditions.

Water regimes

Sufficient moisture is necessary to stimulate germination of seeds of all species. The imbibition phase in which the seed rapidly absorbs water from the surroundings is required for the occurrence of enzymatic processes during the concomitant germination activation phase (Fenner & Thompson 2005). Alternating water regimes in the close vicinity of seeds may also modulate (induce or break) their dormancy (Baskin & Baskin 2014), and drought is believed to accelerate seed senescence (Adams & Rinne 1980). Two aspects of water availability should be considered crucial on the spoil heap substrate: substrate texture and soil salinity. Substrate texture affects its permeability to water

(Kompała-Bąba *et al.* 2019). On spoil heaps, the texture usually ranges from very coarse to very fine, often exhibiting extreme values, which in turn affects water regimes (Tesnerová *et al.* 2017). In sites where the time from the last deposition is short, the gravelly fraction dominates the substrate and enables water runoff. The availability of water for seeds is therefore limited. In older sites, the most dominant fraction is silt, which prevents water from permeating to the deeper layers of the substrate. This leads to temporary water stagnation, which increases water availability for seeds, but may also lead to increased seed mortality due to their vulnerability to microbial attacks (Schafer & Kotanen 2003) and oxygen stress (Sarlistyaningsih *et al.* 1996).

A different factor influencing the ability of seeds to absorb water from their environment is the salinity of the soil solution. High salt content decreases the osmotic potential in the substrate solution, thus making it less accessible for plants to absorb (Howell *et al.* 1984). High salinity of the substrate solution, often occurring in spoil heaps (Kompała-Bąba *et al.* 2019), may therefore inhibit the imbibition of seeds, impeding their germination even in conditions of high moisture.

Light

Plants that require specific light regimes to germinate are called photoblastic (Lambers *et al.* 2008). In the temperate zone, light is usually utilized as a germination cue by r-strategists, with a small amount of storage material in seeds. Survival of these plants at the seedling stage depends on the fast establishment of photosynthesis and this in turn requires availability of light at the time of germination. Light regimes are therefore used to initiate germination only when there is a gap in vegetation and the competition for light is low. Light cues are also helpful in inhibiting germination from deeper layers of soil, which is again important for small-seeded species unable to emerge from a depth more than just a few centimetres. Plants sense not only differences in light intensity, but also its spectral composition – known as the red:far-red ratio. The plant canopy absorbs more red than far-red light; therefore, the more a place is shaded by plants, the lower the red:far-red ratio is. Similarly, far-red penetrates soil more deeply than red light: higher values of the red:far-red ratios therefore indicate better light conditions for germination (Šerá & Šerý 2004; Fenner & Thompson 2005; Jankowska-Blaszczuk & Daws 2007; Jankowska-Błaszczuk 2008; Lambers *et al.* 2008; Tiansawat & Dalling 2013).

At the initial stages of spoil heap vegetation, the competition for light is low, and the light regimes are favourable for small-seeded species of early heap colonists. It is therefore safe to assume that light is not a germination-limiting factor in developing ecosystems of spoil heaps.

Temperature

The material forming the heap is often dark, which enhances heat conduction and accelerates the rate of warming and cooling. The lack of dense vegetation allows most sunbeams to reach the ground, which further contributes to this phenomenon. The temperature of the spoil heap surface may even reach 50°C (Piekarska-Stachowiak *et al.* 2014). Sometimes endogenous thermal activity is recorded on such objects, and then the heap surface temperatures may exceed 60°C and are much higher (up to over 300°C) below-ground (Hanczaruk & Kompała-Bąba 2019). It can therefore be suspected that temperature and its fluctuations will have significant influences on germination patterns in spoil heap ecosystems, as temperature is also an important environmental cue for dormancy and germination.

Temperature fluctuations along with light are a germination cue for the depth of soil burial. This is especially important in the case of larger seeds that are able to germinate from a depth where light does not permeate easily (Jankowska-Blaszczuk & Daws 2007). Sometimes temperature fluctuations are required to break seed dormancy. A common example is the stratification process, where seeds of the temperate climate zone must be subjected to cold temperatures before they can germinate (this inhibits their germination before winter). There are also data that suggest that high and highly fluctuating daily temperatures (such as those observed on the spoil heaps) may play a role in breaking the physical dormancy of some species (Baskin & Baskin 2014). Temperatures which are too high can be dangerous to seeds and contribute to their mortality (Egley 1990); they can also decrease their germinability by inactivating their core microbiome (Holland 2019).

Nutrient content

Low trophy of the heap substrate (Piekarska-Stachowiak *et al.* 2014) can pose problems not only with plant survival in the plant cover, but also with germination, especially for pioneer species. Nitrate concentrations in soils are a factor stimulating germination of some species, mainly those with the ruderal life strategy (Fenner & Thompson 2005; Hryń *et al.* 2020). It is not associated with the role of nitrates in protein synthesis, as no nitrate reductase activity is detected in seeds. It is hypothesized that it serves a role as an environmental cue helpful in detecting a gap in vegetation (as the plant cover is destroyed, the nitrogen uptake by plants decreases and its concentration in soil locally increases) and signalling a nutrient-rich habitat (Fenner & Thompson 2005). Little is known about whether other inorganic nutrients, such as phosphates, are able to trigger germination.

A role for seed endophytes in colonization of hostile environments by plants

Seeds that can persist for years in a dormant state in the seed bank represent a remarkable phase in the life cycle of spermatophytes. Only when it meets suitable conditions, a seed can germinate and develop into a new plant (Nelson 2004). When germination starts, a seed imbibes water and begins to secrete exudates that attract bacteria from the surroundings, which colonize the spermosphere, rhizosphere, and the internal tissues of the seedling itself (Truyens *et al.* 2015). These bacteria can directly and indirectly promote plant growth and health (Nelson 2004; Weyens *et al.* 2009a; 2009b). Direct plant growth promoting mechanisms involve mobilization of nutrients such as phosphorus and iron by production of organic acids and siderophores, fixation of atmospheric nitrogen, production of plant hormones such as auxins and cytokinins, and lowering the levels of stress ethylene by producing the enzyme 1-aminocyclopropane-1-carboxylate (ACC) deaminase (Weyens *et al.* 2009b). Indirect growth promoting effects prevent the development or activity of phytopathogens through competition for space and nutrients, antibiosis, production of hydrolytic enzymes, inhibition of toxins, and stimulation of plant defense mechanisms (Weyens *et al.* 2009b).

However, besides bacteria recruited from the seed surroundings, those that are already present inside the seed are at least as important for the development of the microbial community of the seedling (Johnston-Monje & Raizada 2011). Indeed, many seeds contain a diversity of both bacterial and fungal endophytes (Truyens *et al.* 2015). Such seed endophytes are of particular interest for plants, as they can be transmitted from generation to generation (Sánchez-López *et al.* 2018). Many seed endophytic bacteria provide benefits to their host comparable to those that were described above and can increase the tolerance of their host to metals, salts, and other environmental stressors (Mastretta *et al.* 2009; Weyens *et al.* 2009a; Sánchez-López *et al.* 2018). Therefore, it was hypothesized that the establishment of plants in hostile environments may be connected to the presence and transmission of specific seed endophytic bacteria that have important roles during germination and seedling development (Johnston-Monje & Raizada 2011; Hardoim *et al.* 2012; Truyens *et al.* 2013; 2014; 2016). Moreover, seed endophytes, particularly those transmitted across different seed generations, can play an important role in the adaptation of plant populations to stressful environments (Truyens *et al.* 2013; 2016; Sánchez-López *et al.* 2018).

Comparison of seed endophytes of *Arabidopsis thaliana* plants that were exposed for several generations to cadmium (Cd) with seed endophytes of plants that were never exposed to Cd, revealed obvious differences between the two types of seeds in both the bacterial species present and the phenotypic characteristics of the bacterial strains (Truyens *et al.* 2016). These results support the hypothesis that certain endophytes are selected for transfer to the next generation and that their presence might be important for subsequent germination of seeds and early seedling development. For instance, increased metal tolerance and 1-aminocyclopropane-1-carboxylate deaminase activity were predominantly found in strains from seeds of Cd-exposed plants, while the production of siderophores, indole-3-acetic acid, and organic acids were more frequently reported in endophytes of seeds of plants that had never been exposed to Cd.

By being seed-borne, such "adapted" endophytes assure their occurrence in the new plant generation. Such vertical transmission should select against pathogenicity and favour mutualism, as these endosymbionts totally depend on their host for survival and reproduction (Ewald 1989; Rudgers *et al.* 2009). Concurrently, the host plant benefits by providing its progeny with beneficial endosymbionts (Ewald 1989).

An interesting illustration of the vertical transmission of endophytes in plants growing in hostile conditions through seeds was provided by Sánchez-López *et al.* (2018), who investigated the transgenerational transfer of bacterial seed endophytes across three consecutive seed generations of *Crotalaria pumila* (low rattlebox), i.e. a pioneer species colonizing metal-polluted mining residues in the semi-arid mining region of Zimapan (central Mexico). This annual plant species completes its full life cycle on this highly polluted and very dry substrate and produces viable seeds, hence increasing the numbers of plants colonizing the bare mine tailings year after year (Sánchez-López *et al.* 2015). The bacterial endophytic communities across three consecutive generations of seeds were shown to be very similar. The results indicated the existence of a core microbiome that was present across successive seed generations. The most representative endophytes of this core microbiome belonged to the genus *Methylobacterium* and were shown to possess several plant growth promotion traits and increased metal tolerance. The results presented by Truyens *et al.* (2016) and Sánchez-López *et al.* (2018) demonstrate that bacterial genera present in seeds and preserved across generations are functional traits that can be beneficial for their host plant when growing in hostile environments. Such results highlight the relevance and drive future avenues in exploiting seed-endophytes occurring in natural systems for increasing the efficiency of establishing plant cover on hostile substrates.

REFERENCES

Adams C.A. & Rinne R.W. 1980: Moisture component as a controlling factor in seed development and germination. International Review of Cytology, 68, 1–8.

Baasch A., Kirmer A. & Tischew S. 2012: Nine years of vegetation development in a post mining site: effects of spontaneous and assisted site recovery. Journal of Applied Ecology, 49, 261–260.

Bakker J.P., Bakker E.S., Rosén E., Verweij G.L. & Bekker R.M. 1996: Soil seed bank composition along a gradient from dry alvar grassland to *Juniperus* shrubland. Journal of Vegetation Science, 7, 165–176.

Baskin C.C. & Baskin J.M. 2014: Seeds. Ecology, Biogeography, and Evolution of Dormancy and Germination. 2nd ed. Academic Press, San Diego.

Bekker R.M., Lammerts E.J., Schutters A. & Grootjans A.P. 1999: Vegetation development in dune slacks: the role of persistent seed banks. Journal of Vegetation Science, 10, 745–754.

Bossuyt B. & Hermy M. 2001: Influence of land use history on seed banks in European temperate forest ecosystems: a review. Ecography, 24 (2), 225–238.

Bossuyt B. & Hermy M. 2004: Seed bank assembly follows vegetation succession in dune slack. Journal of Vegetation Sciences, 15 (4), 449–456.

Bossuyt B., Heyn M. & Hermy M. 2002: Seed bank and vegetation composition of forest stands of varying age in central Belgium: consequences for regeneration of ancient forest vegetation. Plant Ecology, 162, 33–48.

Brown D. 1992: Estimating the composition of a forest seed bank: a comparison of the seed extraction and seedling emergence methods. Canadian Journal of Botany, 70 (8), 1603–1612.

Czarnecka J. 2004: Glebowy bank nasion – powstawanie i znaczenie w ekologii roślin. Edukacja Biologiczna i Środowiskowa, Kwartalnik dla Nauczycieli, 4 (12), 5–11.
Czarnecka J. 2005: Seed dispersal effectiveness in three adjacent plant communities: xerothermic grassland, brushwood and woodland. Annales Botanici Fennici, 42, 161–171.
Czarnecka J. 2006: Roadsides as plant diversity refuges in agricultural landscape. Ecological Questions, 7, 37–46.
Egley G.H. 1990: High-temperature effects on germination and survival of weed seeds in soil. Weed Science, 38 (4/5), 429–435.
Ehrlén J. & Eriksson O. 2000: Dispersal limitation and patch occupancy in forest herbs. Ecology, 81 (6), 1667–1674.
Eriksson O. & Ehrlén J. 1992: Seed and microsite limitation of recruitment in plant populations. Oecologia, 91 (3), 360–364.
Ewald P.W. 1989: Transmission modes and evolution of the parasitism-mutualism continuum. Annals of the New York Academy of Sciences, 503, 295–306.
Falińska K. 1996: Ekologia roślin. Wydawnictwo Naukowe PWN, Warszawa.
Falińska K. 1999: Seed bank dynamics in abandoned meadows during a 20-year period in the Białowieża National Park. Journal of Ecology, 87 (3), 461–475.
Faliński J. 2001: Przewodnik do długoterminowych badań ekologicznych. Wydawnictwo Naukowe PWN, Warszawa.
Fenner M. & Thompson K. 2005: The Ecology of Seeds. Cambridge University Press, Cambridge.
Frouz J., Mudrák O., Reitschmiedová E., Walmsley A., Vachová P., Šimáčková H., Albrechtowa J., Moradi J. & Kučera J. 2018: Rough wave-like shaped heaped overburden promotes establishment of woody vegetation while levelling promotes grasses during unassisted post-mining site development. Journal of Environmental Management, 205, 50–58.
Gonzáles-Alday J., Marris R.H. & Martínez-Ruiz C. 2008: The influence of aspect on the early growth dynamics of hydroseeded species in coal reclamation areas. Applied Vegetation Science, 11 (3), 405–412.
Gonzalez S.L. & Ghermandi L. 2012: Comparison of methods to estimate soil seed banks: the role of seed size and mass. Community Ecology, 13 (2), 238–242.
González-Alday J., Marrs R.H. & Martínez-Ruiz C. 2009: Soil seed bank formation during early revegetation after hydroseeding in reclaimed coal wastes. Ecological Engineering, 35 (7), 1062–1069.
Grandin U. & Rydin H. 1998: Attributes of the seed bank after a century of primary succession on islands in Lake Hjalmaren, Sweden. Journal of Ecology, 86 (2), 293–303.
Grime J.P. 1977: Evidence for the existence of three primary strategies in plants and its relevance to ecological and evolutionary theory. The American Naturalist, 111 (982), 1169–1194.
Grime J.P. 1989: Seed Banks in Ecological Perspective. [In:] Leck M.A., Parker T.V. & Simpson R.L. (eds.). Ecology of Soil Seed Banks. Academic Press, Inc. San Diego, California, XV-XXII (Introduction).
Grime J.P., Crawley M.J., Southwood R., Harper J.L., Mack R.N., Gray A.J. & Gibbs A. 1986: The circumstances and characteristics of spoil colonization within a local flora. Philosophical Transactions of the Royal Society of London, Series B, Biological Sciences, 314 (1167), 637–654.
Grodzińska K., Korzeniak U., Szarek-Łukaszewska G. & Godzik B. 2000: Colonization of zinc mine spoils in southern Poland – preliminary studies on vegetation, seed rain and seed bank. Fragmenta Floristica et Geobotanica, 45 (1–2), 123–145.
Groenveld E.V.G. & Rochefort L. 2002: Nursing plants in peatland restoration: on their potential use to alleviate frost heaving problems. Suoseura – Finnish Peatland Society, 53 (3–4), 73–85.
Gross K.L. 1990: A comparison of methods for estimating seed numbers in the soil. Journal of Ecology, 78 (4), 1079–1093.
Hanczaruk R. & Kompała-Bąba A. 2019: Effect of thermal activity on the differentiation of the vegetation of the "Ruda" postmining heap in Zabrze (Poland). Acta Agrobotanica, 72 (3), 1783.
Hardoim P.R., Hardoim C.C.P., van Overbeek L.S. & van Elsas J.D. 2012: Dynamics of seed-borne rice endophytes on early plant growth stages. PLoS ONE, 7 (2), e30438.
Harper J.L. 1977: Population Biology of Plants. Academic Press, New York.
Holland M.A. 2019: Thinking about PPFM bacteria as a model of seed endophytes: who are they? Where did they come from? What are they doing for the plant? What can they do for us? [In:] Verma S.K. & White J.F. Jr (eds.): Seed Endophytes. Biology and Biotechnology. Springer, Cham.

Horáčková M., Řehounková K. & Prach K. 2016: Are seed and dispersal characteristics of plants capable of predicting colonization of post-mining sites? Environmental Science and Pollution Research, 23 (14), 13617–13625.

Howell T.A., Hatfield J.L., Rhoades J.D. & Meron M. 1984: Response of Cotton Water Stress Indicators to Soil Salinity. Irrigation Science, 5, 25–36.

Hryń W., Czarnecka J., Kitowski I., Franczak M. & Bartmiński P. 2020: How does the engineering species affect forest ecosystems? Changes in germination and development rate of non-forest species under the nests of *Ardea cinerea* L. Ekológia (Bratislava), 39 (3), 250–259.

Jankowska-Blaszczuk M. & Daws M.I. 2007: Impact of red: far red ratios on germination of temperate forest herbs in relations to shade tolerance, seed mass and persistence in the soil. Functional Ecology, 21, 1055–1062.

Jankowska-Błaszczuk M. & Grubb P.J. 1997: Soil seed banks in primary and secondary deciduous forest in Białowieża, Poland. Seed Science Research, 7 (3), 281–292.

Jankowska-Błaszczuk M. 1998: Variability of the soil seed banks in the natural deciduous forest in the Białowieża National Park. Acta Societatis Botanicorum Poloniae, 67 (3–4), 313–324.

Jankowska-Błaszczuk M. 2008: Bank nasion w lasach liściastych strefy umiarkowanej – ewolucyjne i ekologiczne aspekty badań. Wiadomości Botaniczne, 52 (3/4), 25–41.

Ji M., Qin H., Chen H. & Wen B. 2018: Seed germination requirements of *Ficus virens* (Moraceae) as adaptation to its hemi-epiphyte life form. Polish Journal of Ecology, 66 (1), 14–22.

Johnston-Monje D. & Raizada M.N. 2011: Conservation and diversity of seed associated endophytes in *Zea* across boundaries of evolution, ethnography and ecology. PLoS ONE, 6 (6), e20396.

Kleyer M., Bekker R.M., Knevel I.C., Bakker J.P., Thompson K., Sonnenschein M., Poschlod P., Van Groenendael J.M., Klimes L., Klimesová J., Klotz S., Rusch G.M., Hermy M., Adriaens D., Boedeltje G., Bossuyt B., Dannemann A., Endels P., Götzenberger L., Hodgson J.G., Jackel A-K., Kühn I., Kunzmann D., Ozinga W.A., Römermann C., Stadler M., Schlegemilch J., Steendham H.J., Tackenberg O., Wilmann B., Cornelissen J.H.C., Eriksson O., Garnier E. & Peco B. 2008: The LEDA Traitbase: a database of life-history traits of Northwest European flora. Journal of Ecology, 96 (6), 1266–1274.

Klimkowska A., Bekker R.M., Van Diggelen R. & Kotowski W. 2010: Species trait shifts in vegetation and soil seed bank during fen degradation. Plant Ecology, 206, 59–82.

Kompała-Bąba A., Bierza W., Błońska A., Sierka E., Magurno F., Chmura D., Besenyei L., Radosz Ł. & Woźniak G. 2019: Vegetation diversity on coal mine spoil heaps – how important is the texture of soil substrate? Biologia, 74 (4), 419–436.

Kostrakiewicz-Gierałt K. 2014: Are *Deschampsia caespitosa* (L.) Beauv. tussocks safe sites for seedling recruitment in the succession of wet meadows? Polish Journal of Ecology, 62, 707–721.

Koszelnik-Leszek A., Podlaska M. & Tomaszewska K. 2013: Diversity of vascular flora of waste dumps and dumping grounds in Lower Silesia. Archives of Environmental Protection, 39 (2), 81–85.

Kwiatkowska-Falińska A.J., Jankowska-Błaszczuk M. & Wódkiewicz M. 2011: The pattern of seed banks during secondary succession on poor soil. Acta Societatis Botanicorum Poloniae, 80 (4), 269–274.

Kwiatkowska-Falińska A.J., Panufnik-Mędrzycka D., Wódkiewicz M., Sondej I. & Jaroszewicz B. 2013: Ancient forest species and the diversity of vegetation and seed bank indicate the aptitude of transformed thermophilous oak wood patches for restoration. Polish Journal of Ecology, 61 (1), 65–80.

Lambers H., Chapin F.S.III. & Pons T.L. 2008: Plant Physiological Ecology. 2nd ed. Springer Science + Business Media, LLC, New York.

Lawton J.H. 1999: Are there general laws in ecology? Oikos, 84, 177–192.

Levin D.A. 1990: The Seed Bank as a Source of Genetic Novelty in Plants. American Naturalist, 135 (4), 563–572.

Markowicz A., Woźniak G., Borymski S., Piotrowska-Seget Z. & Chmura D. 2015: Links in the functional diversity between soil microorganisms and plant communities during natural succession on coal mine spoil heaps. Ecological Research, 30 (6), 1005–1014.

Martínez-Ruiz C., Fernández-Santos B., Putwain P.D. & Fernández-Gómez M.J. 2007: Natural and man--induces revegetation on mining wastes: Changes in the floristic composition during early succession. Ecological Engineering, 30 (3), 286–294.

Martín-Moreno C., Duque J.F.M., Ibarra J.M.N., Rodríguez N.H., Santos M.A.S. & Castillo L.S. 2016: Effects of topography and surface soil cover on erosion for mining reclamation: the experimental heap at El Machorro Mine (Central Spain). Land Degradation & Development 27, 145–159.

Mastretta C., Taghavi S., van der Lelie D., Mengoni A., Galardi F., Gonnelli C., Barac T., Boulet J., Weyens N. & Vangronsveld J. 2009: Endophytic bacteria from seeds of *Nicotiana tabacum* can reduce cadmium phytotoxicity. International Journal of Phytoremediation, 11 (3), 251–267.
Młynkowiak E., Malinowska K. & Kutyna I. 2014: Występowanie *Calamagrostis epigejos* L. (Roth) w zbiorowiskach roślinnych różnych biotopów. Folia Pomeranae Universitatis Technologiae Stetinensis. Agricultura, Alimentaria, Piscaria et Zootechnica, 309 (29), 85–100.
Münzbergová Z. 2012: Seed density significantly affects species richness and composition in experimental plant communities. PLoS ONE, 7 (10), e46704.
Nelson E.B. 2004: Microbial dynamics and interactions in the spermosphere. Annual Review of Phytopathology, 42, 271–309.
Ogden J. 1974: The reproductive strategy of higher plants. II. The reproductive strategy of *Tussilago farfara* L. Journal of Ecology, 62 (1), 291–324.
Piekarska-Stachowiak A., Szary M., Ziemer B., Besenyei L. & Woźniak G. 2014: An application of the plant functional group concept to restoration practice on coal mine spoil heaps. Ecological Restoration, 29, 843–853.
Pinaya I., Soto B., Arias M. & Dias-Fierros F. 2000: Revegetation of burnt areas: Relative effectiveness of native and commercial seed mixtures. Land Degradation & Development, 11 (1), 93–98.
Plue J., Verheyen K., Van Calster H., Marage D., Thompson K., Kalamees R., Jankowska-Blaszczuk M., Bossuyt B. & Hermy M. 2010: Seed banks of temperate deciduous forests during secondary succession. Journal of Vegetation Science, 21 (5), 965–978.
Prach K. & Pyšek P. 2001: Using spontaneous succession for restoration of human-disturbed habitats: Experience from Central Europe. Ecological Engineering, 17 (1), 55–62.
Prach K., Lencová K., Řehounková K., Dvořáková H., Jírová A., Konvalinková P., Mudrák O., Novák J. & a comparison across seres. Environmental Science and Pollution Research, 20 (11), 7680–7685.
Price J.N., Wright B.R., Gross C.L. & Whalley W.R.D.B. 2010: Comparison of seedling emergence and seed extraction techniques for estimating the composition of soil seed banks. Methods in Ecology and Evolution, 1 (2), 151–157.
Roubíčková A., Mudrák O. & Frouz J. 2012: The effect of belowground herbivory by wireworms (Coleoptera: Elateridae) on performance of *Calamagrostis epigejos* (L) Roth in post-mining sites. European Journal of Soil Biology, 50, 51–55.
Rudgers J.A., Afkhami M.E., Rúa M.A., Davitt A.J., Hammer S. & Huguet V.M. 2009: A fungus among us: broad patterns of endophyte distribution in the grasses. Ecology, 90 (6), 1531–1539.
Salisbury E.J. 1953: A changing flora as shown in the study of weeds of arable lands and waste places. [In:] Lousley J.E. (ed.): The Changing Flora of Britain. Arbroath, Buncle, 130–139.
Sánchez-López A.S., González-Chávez M.D.C.A., Carrillo-González R., Vangronsveld J. & Díaz-Garduño M. 2015: Wild flora of mine tailings: perspectives for use in phytoremediation of potentially toxic elements in a semi-arid region in Mexico. International Journal of Phytoremediation, 17 (5), 476–484.
Sánchez-López A.S., Thijs S., Beckers B., González-Chávez M.C., Weyens N., Carrillo-González R. & Vangronsveld J. 2018: Community structure and diversity of endophytic bacteria in seeds of three consecutive generations of *Crotalaria pumila* growing on metal mine residues. Plant and Soil, 422 (1–2), 51–66.
Sarlistyaningsih L., Sivasithamparam K. & Setter T.L. 1996: The role of limited oxygen supply and soil microorganisms or germination and survival of lupin seeds (*Lupinus angustifolius* L. cv. Gungurru) in waterlogged soil. Australian Journal of Experimental Agriculture, 36 (3), 323–329.
Schafer M. & Kotanen P.M. 2003: The influence on soil moisture on losses of buried seeds to fungi. Acta Oecologica, 24 (5–6), 255–263.
Šerá B. & Šerý M. 2004: Number and weight of seeds and reproductive strategies of herbaceous plants. Folia Geobotanica, 39, 27–40.
Simpson R.L., Leck M.A. & Parker T. 1989: Seed banks: general concepts and methodological issues. [In:] Leck M.A., Parker T.V., Simpson R.L. (eds.): Ecology of Soil Seed Banks. Academic Press, Inc. San Diego, California, 3–21.
Staaf H., Jonsson M. & Olsén L. 1987: Buried germinative seeds in mature beech forests with different vegetation and soil types. Holarctic Ecology, 10, 268–277.
Symonides E. 1986: Seed bank in old-field successional ecosystems. Ekologia Polska, 34 (1), 3–29.
Tesnerová C., Zadinová R., Pikl M., Zemek F., Kadochová Š., Matějíček L., Mihajlevič M. & Frouz J. 2017: Predicting the toxicity of post-mining substrates, a case study based on laboratory tests, substrate chemistry, geographic information systems and remote sensing. Ecological Engineering, 100, 56–62.

Thompson K. & Grime J.P. 1979: Seasonal variation in the seed banks of herbaceous species in ten contrasting habitats. Journal of Ecology, 67 (3), 893–921.
Thompson K. 1987: Seeds and seed banks. New Phytologist, 106 (Suppl. 1), 23–34.
Thompson K., Bakker J.P. & Bekker R.M. 1997: The Soil Seed Banks of North West Europe: Methodology, Density and Longevity. University Press, Cambridge.
Thompson K., Bakker J.P., Bekker R.M. & Hodgson J.G. 1998: Ecological correlates of seed persistence in soil in the north-west European flora. Journal of Ecology, 86 (1), 163–169.
Thompson K., Ceriani R., Bakker J.P. & Bekker R.M. 2003: Are seed dormancy and persistence in soil related? Seed Science Research, 13 (2), 97–100.
Tiansawat P. & Dalling J.W. 2013: Differential seed germination responses to the ratio of red to far-red light in temperate and tropical forests. Plant Ecology, 214 (5), 751–764.
Tinsley M.J., Simmons M.T. & Windhager S. 2006: The establishment success of native versus non-native herbaceous seed mixes on revegetated roadside in Central Texas. Ecological Engineering, 26 (3), 231–240.
Truyens S., Beckers B., Thijs S., Weyens N., Cuypers A. & Vangronsveld J. 2016: Cadmium-induced and trans-generational changes in the cultivable and total seed endophytic community of *Arabidopsis thaliana*. Plant Biology, 18 (3), 376–381.
Truyens S., Jambon I., Croes S., Janssen J., Weyens N., Mench M., Carleer R., Cuypers A. & Vangronsveld J. 2014: The effect of long-term Cd and Ni exposure on seed endophytes of *Agrostis capillaris* and their potential application in phytoremediation of metal-contaminated soils. International Journal of Phytoremediation, 16 (7–8), 643–659.
Truyens S., Weyens N., Cuypers A. & Vangronsveld J. 2013: Changes in the population of seed bacteria of transgenerationally Cd-exposed *Arabidopsis thaliana*. Plant Biology, 15 (6), 971–981.
Truyens S., Weyens N., Cuypers A. & Vangronsveld J. 2015: Bacterial seed endophytes: genera, vertical transmission and interaction with plants. Environmental Microbiology Reports, 7 (1), 40–50.
Wang B.C. & Smith T.B. 2002: Closing the seed dispersal loop. Trends in Ecology & Evolution, 17 (8), 379–385.
Wang Y., Jiang D., Toshio O. & Zhou Q. 2013: Recent advances in soil seed bank research. Contemporary Problems of Ecology, 6 (5), 520–524.
Weyens N., van der Lelie D., Taghavi S. & Vangronsveld J. 2009a: Phytoremediation: plant–endophyte partnerships take the challenge. Current Opinion in Biotechnology, 20 (2), 248–254.
Weyens N., van der Lelie D., Taghavi S., Newman L. & Vangronsveld J. 2009b: Exploiting plant–microbe partnerships to improve biomass production and remediation. Trends in Biotechnology, 27 (10), 591–598.
Woch M.W., Radwańska M. & Stefanowicz A.M. 2013: Flora of spoil heaps after hard coal mining in Trzebinia (southern Poland): effect of substratum properties. Acta Botanica Croatica, 72 (2), 237–256.
Zhang Z.Q., Shu W.S., Lan C.Y. & Wong M.H. 2001: Soil seed bank as an input seed source in revegetation of lead/zinc mine tailings. Restoration Ecology, 9 (4), 378–385.
Zhao Z., Bai Z., Zhang Z., Guo D., Li J., Xu L. & Pan Z. 2012: Population structure and spatial distributions of 17 years old plantation in a reclaimed spoil of Pingshuo opencast mine, China. Ecological Engineering, 44, 147–151.

Green Scenarios..., Dyczko, Jagodziński & Woźniak (eds.) © 2022 Taylor & Francis Group, London, UK

2.7. Bryoflora of Post-industrial Areas in Poland

Anna Salachna
Institute of Environmental Protection and Engineering,
Faculty of Materials, Civil and Environmental Engineering, University of Bielsko-Biala;
2 Willowa Str., 43-309 Bielsko-Biała, Poland;
asalachna@ath.bielsko.pl; ORCID: 0000-0002-3789-2765

ABSTRACT: A list of bryophyte species occurring in different types of post-industrial areas in Poland is presented. The bryoflora of these habitats consists of 166 species, including 147 species of mosses and 19 species of liverworts. Most of them are common, hemerophilous species, but more valuable species also occur, such as species protected in Poland (28 species), threatened in Poland (eight species) and threatened in Europe (one species). Moreover, the most important biological features of bryophytes that determine their role in natural and transformed ecosystems are described.

KEYWORDS: Bryophyte biology, ecological role of bryophytes, mosses, liverworts, post-industrial habitats

INTRODUCTION

The development of industries is inextricably linked to land and soil degradation. Post-industrial areas currently represent a significant landform in the landscape of many regions in Europe. In Poland the main reason for land transformation is mining and quarrying of raw materials, i.e. brown coal, hard coal and rock raw materials (Gonda-Soroczyńska & Kubicka 2016). These areas cover almost 62,000 ha, which is about 0.2% of the total area of our country (Ochrona środowiska 2018). Their reclamation aims to restore utility or natural values by properly shaping the terrain, improving physical and chemical properties, regulating water relations, and restoring soils and vegetation. The selection of a suitable reclamation direction should be preceded by geobotanical research to determine the individual potential of the habitat and the possibility of using natural processes such as ecological succession. The specific environmental conditions of transformed areas causes them to be colonized by various well-adapted groups of organisms. Spontaneous development of vegetation in these types of habitats is a well-known and documented phenomenon (Borgegård 1990; Rostański 2006; Řehounková 2007; Woźniak 2010). However, most of these studies concern the diversity and dynamics of vascular plants communities.

This work discusses the biological features of bryophytes associated with their ecological function in natural and transformed ecosystems. Based on published data, as well as information shared by other authors, a list of bryophyte species occurring in selected types of post-industrial areas in Poland was compiled. The characteristics of the distinguished habitat groups of bryophytes (epigeic, epiphytic, epixylic, epilithic, paludicolous, aquatic) was also carried out.

BRYOPHYTE BIOLOGY

Bryophytes are a group of spore plants that appeared on Earth about 500 million years ago in the Ordovician Period. Current phylogenetic and paleobotanical research confirms that they were the first plants which colonized the land (Kenrick & Crane 1997; Wellman *et al.* 2003; Dombrovska & Qiu 2004). For this reason, bryophytes occupy a very important position in the world of living organisms, as they hold the key to understanding the evolution of vascular plants.

The intensification of molecular and phylogenetic studies at the beginning of the 21st Century led to revolutionary changes in bryophyte systematics and taxonomy. Nowadays, the term bryophytes does mean a systematic unit, but it is only used to indicate a group of plants characterized by several common features, namely: dominance of haploid gametophytes in the life cycle (Fig. 1), small body size and (compared to vascular plants) relatively simple morphological and anatomical structures. In many modern classification systems, the traditionally distinguished group of bryophytes now belongs to the subkingdom *Bryobiotina*, which is divided into three separate phyla: liverworts (*Marchantiophyta*), mosses (*Bryophyta*) and hornworts (*Anthocerotophyta*) (Frey & Stech 2009; Goffinet *et al.* 2009). The representatives of *Bryobiotina* have some common features with vascular plants, i.e. the formation of an embryo inside the multicellular reproductive organ and the presence of sporopolenine and flavonoids (Glime 2017). It is worth noting that there are still new reports regarding phylogenesis of bryophytes, suggesting, inter alia, that mosses and liverworts belong to the same clade "Setaphyta" and together with hornworts form a monophyletic group (Puttick *et al.* 2018), which gives a new look to bryophyte systematics.

In terms of species richness, bryophytes occupy the second place after vascular plants. In the world, 18 000–23 000 bryophyte species have been catalogued and described, of which 11 000–13 000 are mosses, 7000–9000 liverworts and 200–250 species are hornworts (Magill 2010; Villareal *et al.* 2010; Söderström *et al.* 2016). In Poland, the list of bryophyte species contains about 700 species of mosses (Ochyra *et al.* 2003), 250 species of liverworts and four species of hornworts (Szweykowski 2006).

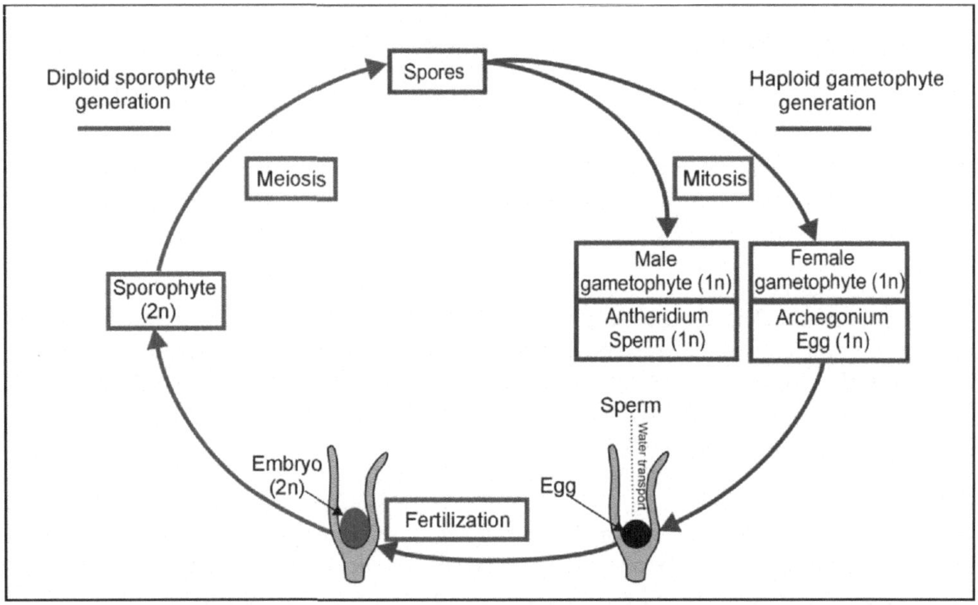

Figure 1. Life cycle of bryophytes (in accordance with Hodgetts *et al.* (2019), changed)

The life cycle of bryophytes is characterized by dominance of haploid gametophyte which distinguishes bryophytes from vascular plants (Fig. 1). The gametophyte-dependent sporophyte dies after maturation of the spores. The gametophyte may be in the form of a leafy stem (mosses and some liverworts) or thallus (hornworts and some liverworts) (Tab. 1). Male and female sexual reproductive structures (antheridia and archegonia) are located at the top of the shoot or on the upper surface of the thallus. For fertilization, water is needed that allows sperm to enter the egg. This fact explains the abundant occurrence of bryophytes in wet places, i.e. peat bogs, stream banks, and spring areas. The sporophyte developing from the zygote forms a seta terminated with a sporangium in which haploid spores are formed. They are small and light, so they can be spread by the wind even over considerable distances, which allows bryophytes to colonize new, often distant places. Vegetative reproduction plays a very important role in the bryophyte life cycle. It is carried out with the help of special propagules (gemmae) or with fragments of stems, leaves and thalli. The resulting new individuals are genetically identical with their parents. This method of reproduction determines the occurrence of bryophytes in the colonies, whose shape is ultimately modified by environmental factors, mainly water and light. Among other various life forms observed in bryophytes which are the most adapted to adverse environmental conditions, are cushions and balls. These forms preserve individuals against water and heat loss and minimize wind turbulence. The space between the densely arranged stems acts as capillaries that conduct water from the environment and retain it inside. In addition, this process supports concave leaves, lack of cuticle and presence of paraphyllia (reduced leaf) and tomentum (abundant rhizoids on the stem). The ability to retain water inside cushions is very significant for bryophytes because most of them are ectohydric, which means that they absorb water with the entire surface.

Table 1. Chosen morphological and anatomical differences among bryophytes

CHARACTER	MOSSES	LIVERWORTS	HORNWORTS
GAMETOPHYTE			
Gametophyte form	Leafy shoot	Leafy shoot or thallus; thallus simple or with air chamber	Simple thallus
Leaf arrangement	Leaves in spirals	Leaves in three rows	Not applicable
Leaf form	Leaves undivided, midvein present	Leaves divided into 2 + lobes, no midvein	Not applicable
Special organelles	None	Oil bodies	Single plastids with pyrenoids
SPOROPHYTE			
Capsule	Complex with operculum, theca and neck; dehiscence at operculum and peristome teeth	Undifferentiated, spherical or elongate; dehiscence into 4 valves	Undifferentiated, horn shaped; dehiscence into 2 valves
Seta	Photosynthetic	Hyaline	Absent

Only a few endohydric species develop an internal conductive system, which includes hydroids (water-conducting cells) and leptoids (sugar-conducting cells). In addition to morpho-anatomical adaptations bryophytes evolved physiological mechanism allowing them to survive in conditions of prolonged lack of water. During desiccation bryophytes limit metabolic activity which they resume relatively quickly after rehydration. Return to normal functioning is possible by the action of numerous rehydration proteins and rapid membrane repair (Glime 2017). Longevities ranging from a few days in some species to even several years in other species have been recorded (Stark *et al.* 2017). Desiccation tolerance of bryophytes distinguishes them from vascular plants and enables survival in extreme climatic conditions, i.e. polar regions and deserts.

ECOLOGICAL ROLE OF BRYOPHYTES

Specific morphological and anatomical features of bryophytes and their physiology mean that these plants perform very important roles in ecosystems. One of them is the previously mentioned water retention. This is most clearly seen in peat bogs, where *Sphagnum* mosses dominate. Hummocks of *Sphagnum* quickly absorb water like a sponge, holding it inside and very slowly releasing it, which regulates water flows in nature and minimizes the risk of flooding. In terrestrial ecosystems bryophytes form soil-covering mats, which help maintain the humidity regime and prevent soil erosion (Hodgetts *et al.* 2019). In arid and semi-arid landscapes they are a component of cryptogamic crusts formed apart from them by lichens, cyanobacteria, green algae and fungi. These associations of organisms perform biological protection of the soil by binding fine particles into more stable aggregates (Vanderpoorten & Goffinet 2009). Some species of mosses, e.g. of the genus *Atrichum* and *Pohlia* are recognized as erosion inhibitors due to their resistance to trampling and regenerative abilities (Bahuguna *et al.* 2013).

Another major function of bryophytes is to influence the carbon and nutrient cycles of ecosystems. Annual bryophyte production (g/m^2) in some types of terrestrial ecosystems (e.g. Atlantic deciduous woodland, arctic wet meadow and peat-bog) is a significant part of total ecosystem photosynthesis. This means that a large portion of the CO_2 consumed is stored in bryophyte communities. It was found that the aforementioned mosses *Sphagnum* have more ability for long-term carbon storage than other vascular and non-vascular plants. This is due to presence of lignin-like polymers in their cell walls and physico-chemical parameters occurring in peat bogs (low temperature, low pH, and anaerobic soil) which significantly inhibit degradation. As a consequence, we observe a much lower decomposition rate on peatlands (about 15%) than in, for example, shrubs and grasses (25–70%) (Vanderpoorten & Goffinet 2009).

Bryophytes absorb nutrients with water over their entire surface and used them in metabolic processes to increase own biomass. After decomposition these elements are available to other organisms. Moreover, some species of bryophytes e.g. the liverwort *Blasia pusilla* and the moss *Pleurozium schreberi* form symbioses with colonies of the cyanobacteria *Nostoc*, which are capable of fixing atmospheric nitrogen. This is a valuable source of this nutrient in some ecosystems, for example in boreal forests (Glime 2017; Hodgetts *et al.* 2019). Another way of bryophyte nutrient uptake is their mycorrhizal associations with fungi. This relationship is very common especially in liverworts. It was found that almost half of known liverwort species establish this symbiotic association and they are mostly obligatory. *Glomeromycota* fungal partners create characteristic arbuscules in cells of rhizoids or ventral epidermis of bryophytes and support nutrient and water absorption. Bryophytes can be also colonized by representatives of *Ascomycota* and, rarely, *Basidiomycota* (Vanderpoorten & Goffinet 2009).

Bryophytes are considered pioneering plants because they promote soil formation on areas transformed by natural processes (natural erosion, volcano eruption) or human activity (post-industrial areas). In such habitats bryophytes appear as one of the first organisms, fertilize the soil by their own decomposition, secretion of physiological products and trapping wind-blow organic and inorganic materials. In this way they create an organic layer on the barren ground which initiates the succession of vascular plants and reconstruction of the ecosystem. Some species, e.g. *Pohlia andalusica*, *Mielichhoferia mielichhoferiana*, and *Cephaloziella massalongi*, referred to as "copper mosses", show affinity to increased content of copper, zinc and iron in the substratum (Širka *et al.* 2018).

THE DIVERSITY OF BRYOPHYTES IN POST-INDUSTRIAL HABITATS

Bryophyte flora of the areas transformed due to industrial activities has recently been the subject of research in many countries (e.g. Duda 1989; Holyoak & Lockhart 2011; Širka *et al.* 2018). In Poland these comprehensive studies were carried out for a few industrial facilities such as coal and lignite mines, limestone quarry and zinc ore mines (Stebel 2012; Ochyra & Godzik 2015; Stebel *et al.* 2015; Jagodziński *et al.* 2018; Wierzcholska unpublished; Woźniak unpublished). Information on the occurrence of bryophytes in these types of habitats are also included in bryological studies of the large industrial regions (Jędrzejko 1985; 1990; Stebel 1997) and in the works on the ecology of some species (Gugała 1996; Stebel 2006; Krywult *et al.* 2013). Knowledge on this subject is still insufficient and requires further comprehensive research.

Analysis of available data showed that bryophyte flora of post-industrial areas in Poland consist of 166 species, including 147 species of mosses and 19 species of liverworts (Tab. 2). They are mostly hemerophilous species adapted to exist in very difficult habitat conditions.

Table 2. Bryophyte species occurring in selected post-industrial habitats in Poland

SPECIES NAME	TYPE OF POST--INDUSTRIAL SITE					HABITAT					
	I	II	III	IV	V	I	II	III	IV	V	VI
MOSSES											
!*Abietinella abietina* (Hedw.) M. Fleisch	x	x	x	x			x				
Amblystegium juratzkanum Schimp.		x	x	x		x		x			
!**Amlystegium radicale* (P. Beauv.) Schimp.				x		x					
Amblystegium serpens (Hedw.) Schimp.		x		x		x	x	x	x		
!*Anomodon viticulosus* (Hedw.) Hook &Taylor		x					x				
Atrichum undulatum (Hedw.) P. Beauv.	x	x	x	x	x	x					
!*Aulacomnium palustre* (Hedw.) Schwägr.				x		x					
Barbula convoluta Hedw.	x	x	x	x		x			x		
Barbula unguiculata Hedw.		x	x	x		x					
Brachytheciastrum velutinum (Hedw.) Ignatov & Huttunen	x	x	x	x		x	x				
Brachythecium albicans (Hedw.) Schimp.	x	x	x	x		x	x				
Brachythecium glareosum (Bruch *ex* Spruce) Schimp.		x	x	x		x		x	x		
Brachythecium mildeanum (Schimp.) Schimp.		x		x		x					
Brachythecium rivulare Schimp.		x				x					
Brachythecium rutabulum (Hedw.) Schimp.	x	x	x	x		x	x	x	x		
Brachythecium salebrosum (Hoffm. *ex* F. Weber & D. Mohr)	x	x	x	x		x	x	x			
Bryoerythrophyllum recurvirostrum (Hedw.) P.C. Chen		x		x		x					
Bryum argenteum Hedw.	x	x		x		x		x	x		
Bryum bicolor Dicks.		x							x		
Bryum caespiticium Hedw.	x	x	x	x		x			x		
Bryum dichotomum Hedw.				x		x					
Bryum intermedium (Brid.) Blandow		x				x					
Bryum pallescens Schwägr.			x	x					x		
Bryum pseudotriquetrum (Hedw.) P. Gaertn.		x		x		x				x	
Buxbaumia aphylla Hedw.		x				x					
Calliergon giganteum (Schimp.) Kindb.				x		x				x	
!*Calliergonella cuspidata* (Hedw.) Loeske		x	x	x		x		x	x	x	
Campyliadelphus chrysophyllus (Brid.) R.S. Chopra		x	x						x		
Campylophyllopsis calcarea (Crundwell & Nyholm) Ochyra		x	x	x		x			x		
Campylium polygamum (Schimp.) Lange & C.E.O. Jensen			x			x				x	

SPECIES NAME	TYPE OF POST--INDUSTRIAL SITE					HABITAT					
	I	II	III	IV	V	I	II	III	IV	V	VI
Campylium protensum (Brid.) Kindb.			x			x				x	
Campylium stellatum (Hedw.) Lange & C.E.O. Jensen				x		x				x	
Campylopus introflexus (Hedw.) Brid.		x				x					
Ceratodon purpureus (Hedw.) Brid.	x	x	x	x	x	x	x	x	x		
Cirriphyllum piliferum (Hedw.) Grout		x				x		x			
!*Climacium dendroides* (Hedw.) F. Weber & D. Mohr		x	x	x		x		x		x	
Cratoneuron filicinum (Hedw.) Spruce			x	x							x
Dicranella heteromalla (Hedw.) Schimp.	x	x		x		x					
Dicranella rufescens (Dicks.) Schimp.				x		x					
Dicranella staphylina H. Whitehouse				x		x					
Dicranella varia (Hedw.) Schimp.		x	x	x		x					
!*Dicranum polysetum* Sw. *ex* anon.		x		x		x					
!*Dicranum scoparium* Hedw.		x		x		x	x	x			
Dicranoweisia cirrata (Hedw.) Lindb.				x			x				
Didymodon fallax (Hedw.) R.H. Zander		x	x	x		x					
Didymodon rigidulus Hedw.		x		x		x			x		
Didymodon tophaceus (Brid.) Lisa				x		x					
Ditrichum pusillum (Hedw.) Hampe				x		x					
Drepanocladus aduncus (Hedw.) Warnst.		x	x	x	x						x
Drepanocladus polycarpos (Blandow *ex* Voit) Warnst.		x	x	x		x			x	x	
! • **Drepanocladus sendtneri* (Schimp. *ex* H. Müll.) Warnst.				x		x				x	
Dryptodon pulvinatus (Hedw.) Brid.		x	x	x		x			x		
Encalypta streptocarpa Hedw.		x	x	x		x	x		x		
Eurhynchiastrum pulchellum (Hedw.) Ignatow & Huttunen				x		x			x		
!*Eurhynchium angustirete* (Broth.) T.J. Kop.		x	x	x		x	x	x	x		
Eurhynchium striatum (Schreb. *ex* Hedw.) Schimp.		x	x			x					
Fissidens taxifolius Hedw.		x	x	x		x					
Funaria hygrometrica Hedw.	x	x		x		x					
Hedwigia ciliata (Hedw.) P. Beauv.			x						x		
Herzogiella seligeri (Brid.) Z. Iwats.				x				x			
Homalothecium lutescens (Hedw.) H. Rob.			x						x		
Homalothecium sericeum (Hedw.) Schimp.		x							x		
!*Hylocomiadelphus triquetrus* (Hedw.) Ochyra & Stebel		x				x					
!*Hylocomium splendens* (Hedw.) Schimp.	x	x	x	x		x		x			
Hypnum cupressiforme Hedw.		x	x	x	x	x	x	x	x		
Hypnum pallescens (Hedw.) P. Beauv.		x		x			x				
Isothecium alopecuroides (Lam. *ex* Dubois) Isov.		x					x				
Kindbergia praelonga (Hedw.) Ochyra		x		x		x			x		
Leptobryum pyriforme (Hedw.) Wilson				x		x					
Leptodictyum riparium (Hedw.) Loeske			x	x							x
Leskea polycarpa Hedw.		x					x	x			
Leskeella nervosa (Brid.) Loeske		x					x	x			
Mnium hornum Hedw.		x			x	x					

SPECIES NAME	TYPE OF POST- -INDUSTRIAL SITE					HABITAT					
	I	II	III	IV	V	I	II	III	IV	V	VI
Niphotrichum canescens (Hedw.) Bednarek-Ochyra & Ochyra		x		x		x			x		
Niphotrichum elongatum (Ehrh. *ex* Frisvoll) Bednarek-Ochyra & Ochyra		x				x			x		
Orthodicranum montanum (Hedw.) Loeske		x		x		x	x	x			
Orthotrichum affine Schrad. *ex* Brid.		x	x	x			x	x			
Orthotrichum anomalum Hedw.		x	x	x			x		x		
Orthotrichum diaphanum Schrad. *ex* Brid.		x	x	x			x	x	x		
!*Orthotrichum lyellii* Hook. & Taylor		x					x				
Orthotrichum obtusifolium Brid.		x					x				
Orthotrichum pallens Brid.			x				x		x		
**Orthotrichum patens* Brid.			x				x				
Orthotrichum pumilum Sw. *ex* anon.		x	x				x	x			
Orthotrichum speciosum Nees		x	x	x			x	x			
**Orthotrichum stramineum* Hornsch. *ex* Brid.				x			x				
Oxyrrhynchium hians (Hedw.) Loeske	x	x	x	x		x					
Paraleucobryum longifolium (Ehrh. *ex* Hedw.) Loeske					x	x					
!*Philonotis calcarea* (Bruch & Schimp.) Schimp.			x			x				x	
!**Philonotis marchica* (Hedw.) Brid.			x			x					
Physcomitrella patens (Hedw.) Bruch & Schimp.		x				x					
Plagiomnium affine (Blandow *ex* Funck) T.J. Kop.		x	x	x	x	x			x		
Plagiomnium cuspidatum (Hedw.) T.J. Kop.	x	x	x	x		x	x	x	x		
Plagiomnium elatum (Bruch & Schimp.) T.J. Kop.		x		x		x					
Plagiomnium rostratum (Schrad.) T.J. Kop.				x		x			x		
Plagiomnium undulatum (Hedw.) T.J. Kop.		x		x		x		x			
Plagiothecium curvifolium Schlieph. *ex* Limpr.				x		x	x	x			
Plagiothecium denticulatum (Hedw.) Schimp.				x	x	x	x				
Plagiothecium laetum Schimp.		x		x		x	x	x			
Plagiothecium nemorale (Mitt.) A. Jaeger		x				x	x				
Plagiothecium ruthei Limpr.				x		x					
Platygyrium repens (Brid.) Schimp.		x					x	x			
!*Pleurozium schreberi* (Willd. *ex* Brid.) Mitt.		x		x		x	x	x			
Pogonatum urnigerum (Hedw.) P. Beauv.					x	x					
Pohlia melanodon (Brid.) A.J. Shaw			x			x					
Pohlia nutans (Hedw.) Lindb.	x	x		x		x	x	x	x		
Pohlia wahlenbergii (F. Weber & D. Mohr) A.L. Andrews			x	x							
Polytrichastrum formosum (Hedw.) G.L. Sm.		x		x	x	x					
!*Polytrichum commune* Hedw.				x		x					
Polytrichum juniperinum Hedw.	x	x		x	x	x	x				
Polytrichum piliferum Hedw.	x	x		x	x	x			x		
Pseudocrossidium hornschuchianum (Schulz) R.H. Zander				x		x					
!*Pseudoscleropodium purum* (Hedw.) M. Fleisch.		x	x	x		x		x			
!*Ptilium crista-castrensis* (Hedw.) De Not.		x		x		x					
Pylaisia polyantha (Hedw.) Schimp.		x	x	x			x				

SPECIES NAME	TYPE OF POST--INDUSTRIAL SITE					HABITAT					
	I	II	III	IV	V	I	II	III	IV	V	VI
Rhizomnium punctatum (Hedw.) T.J. Kop.		x			x	x	x	x			
Rhodobryum roseum (Hedw.) Limpr.		x				x					
Rhynchostegium megapolitanum (Blandow ex F. Weber & D. Mohr) Schimp.		x				x					
Rhynchostegium murale (Hedw.) Schimp.		x	x	x		x					
!*Rhytidiadelphus squarrosus* (Hedw.) Warnst.				x		x					
Rhytidiadelphus subpinnatus (Lindb.) T.J. Kop.				x		x					
Rosulabryum capillare (Hedw.) J.R. Spence		x	x	x		x					
Rosulabryum moravicum (Podp.) Ochyra & Stebel		x		x			x	x			
Sanionia uncinata (Hedw.) Loeske		x		x		x		x			
Schistidium apocarpum (Hedw.) Bruch & Schimp.		x		x					x		
Schistidium crassipilum H.H. Blom		x	x	x					x		
Schistidium dupretii (Thér.)W.A. Weber				x					x		
Sciuro-hypnum oedipodium (Mitt.) Ignatov & Huttunen		x		x		x		x			
Sciuro-hypnum plumosum (Hedw.) Ignatov & Huttunen				x							x
Sciuro-hypnum populeum (Hedw.) Ignatov & Huttunen		x		x		x			x		
Sciuro-hypnum reflexum (Starke) Ignatov & Huttunen		x		x		x					
!*Syntrichia papillosa* (Wilson) Jur.		x					x				
Syntrichia ruralis (Hedw.) F. Weber & D. Mohr		x		x		x	x		x		
Tetraphis pellucida Hedw.				x				x			
!*Thuidium assimile* (Mitt.) A. Jaeger			x	x		x					
!*Thuidium delicatulum* (Hedw.) Schimp.		x		x		x					
!*Thuidium tamariscinum* (Hedw.) Schimp.		x				x					
Torrentaria riparioides (Hedw.) Ochyra				x							x
Tortella inclinata (R. Hedw.) Limpr.		x	x						x		
Tortella tortuosa (Hedw.) Limpr.		x	x	x					x		
Tortula muralis Hedw.		x	x	x					x		
Tortula subulata Hedw.			x						x		
Tortula truncata Hedw.		x				x					
!*Ulota bruchii* Hornsch *ex* Brid.		x					x				
!**Ulota crispa* (Hedw.) Brid.		x		x			x				
Warnstorfia fluitans (Hedw.) Loeske		x				x				x	
Weissia controversa Hedw.				x		x					
LIVERWORTS											
Aneura pinguis (L.) Dumort.		x	x	x		x				x	
Barbilophozia barbata (Schmidel *ex* Schreb.) Loeske				x		x					
Cephaloziella divaricata (Sm.) Schiffn.		x		x		x					
Cephaloziella rubella (Nees) Warnst.				x		x					
Chiloscyphus pallescens (Ehrh. *ex* Hoffm.) Dumort.				x		x				x	
!*Frullania dilatata* (L.) Dumort.		x					x				
**Leiocolea badensis* (Gottsche) Jörg.			x			x					
Lophocolea heterophylla (Schrad.) Dumort.	x	x	x	x		x	x	x			
Lophocolea bidentata (L.) Dumort.		x		x		x		x			
Lophocolea minor Nees				x		x					

SPECIES NAME	TYPE OF POST-INDUSTRIAL SITE					HABITAT					
	I	II	III	IV	V	I	II	III	IV	V	VI
Lophozia bicrenata (Schmidel *ex* Hoffm.) Dumort		x				x					
Lophozia ventricosa (Dicks.) Dumort.				x		x					
Marchantia aquatica (Nees) Burgeff			x								x
Marchantia polymorpha L. *emend.* Burgeff				x	x	x			x		
Pellia endiviifolia (Dicks.) Dumort.		x		x		x				x	
!!*Porella platyphylla* (L.) Pfeiff.		x	x				x				
Ptilidium pulcherrimum (Weber) Vain.		x		x			x	x			
Radula complanata (L.) Dumort.		x					x	x			
Riccardia palmata (Hedw.) Carruth.				x				x			

KEY OF THE TABLE 2
TYPE OF POST-INDUSTRIAL SITE:
I – spoil heap of coal mine; II – reclaimed spoil heap of lignite mine; III – limestone quarry;
IV – zinc ore mine; V – spoil heap of uranium
HABITATS:
I – epigeic; II – epiphytic; III – epixylic; IV – epilithic; V – paludicolous; VI – aquatic; !! – strictly protected species;
! – partially protected species; • – species threatened in Europe; * – species threatened in Poland

Noteworthy are the occurrences of less common bryoflora elements, which are mainly related to areas subjected to forest reclamation. There are 28 species protected in Poland (Anonymous 2014), eight species threatened in Poland (Żarnowiec *et al.* 2004; Klama & Górski 2018) and one species threatened in Europe (Hodgetts *et al.* 2019).

The most species-rich bryophyte group of the studied sites is the epigean flora consisting of 120 species. On bare soil (spoil heaps, wasteland) the most common species are *Ceratodon purpureus* (Fig. 2), *Bryum caespiticium*, *Brachythecium rutabulum* (Fig. 2). However, common forest species such as *Atrichum undulatum*, *Hylocomium splendens* (Fig. 2) and *Dicranella heteromalla* also occur. Other habitat groups are comparable in terms of species richness. The epiphytic flora is represented by 49 species of bryophytes. Among them the most frequent are multi-substrate species such as *Amblystegium serpens*, *Brachythecium rutabulum*, *B. salebrosum*, *Eurhynchium angustirete*, *Hypnum cupressiforme* and *Lophocolea heterophylla*. To the typical epiphytes belong *Orthotrichum lyellii* (Fig. 3), *O. pallens*, *O. patens*, *O. stramineum*, *Syntrichia papillosa*, *Ulota bruchii*, *U. crispa* and *Frullania dilatata* (Fig. 3).

Epilithic habitats such as rocks, boulders and stones are colonized by 44 species including the exclusive epilithic species *Hedwigia ciliata*, *Tortula muralis*, *T. subulata*, *Homalothecium lutescens* and *H. sericeum*. Bryoflora associated with decaying wood consist of 40 species and most often they are common, multi-substrate bryophytes like *Orthodicranum montanum*, *Tetraphis pellusida* and *Plagiomnium cuspidatum*.

The least numerous but interesting habitat groups of bryophytes in post-industrial areas are the paludicolous species which prefer wet soil in depressions and ditches (15 species) and aquatic species growing in water ponds and canals (6 species). Particularly noteworthy is the mountain species *Sciuro-hypnum plumosum*, found in water of the Baba Canal near the flotation tailings heap of the Olkusz mine (Stebel *et al.* 2015).

Figure 2. Common species of bryophytes on post-industrial areas: a. moss *Brachythecium rutabulum*; b. moss *Ceratodon purpureus* with sporophytes; c. moss *Hylocomium splendens*; d. thaloid liverwort *Aneura pinguis* (Photo: Robert Zubel)

Figure 3. Epiphytic species of bryoflora on post-industrial areas: a. moss *Orthotrichum lyellii*; b. moss *Ulota crispa* with sporophyte; c. liverwort *Porella platyphylla*; d. liverwort *Frullania dilatata* (Photo: Robert Zubel)

SUMMARY

Bryophytes are small spore plants considered in the light of recent research to be one of the first terrestrial plant groups. Through evolution, they have developed a number of adaptations that allow them to exist in extreme habitat conditions. Growth in colonies, uptake of water and nutrients over the entire body surface and desiccation tolerance are the most important traits of bryophyte biology. Specific properties of these plants determine their ecological functions including water retention, prevention of soil erosion, effects on carbon and nutrient cycles and soil formation. For these reasons they are used in soil and landscape restoration (Greenwood *et al.* 2019). Areas transformed due to industrial development represent very unfavourable conditions for all living organisms. The soil substratum of these places is characterized by low humidity and fertility, coarse grain size and very often increased heavy metal content. Bryophyte species inhabiting post-industrial areas are mostly hemerophilous species well adapted to difficult environmental conditions. However, rare and more valuable elements of the bryoflora also occur, which are related mainly with afforested areas. Appropriate forest restoration increases species diversity of bryophytes, in particular woodland species including specialized epiphytes (Stebel 2012; Jagodziński *et al.* 2018).

Acknowledgements

The author would like to thank Prof. Gabriela Woźniak (University of Silesia in Katowice) for providing data concerning bryophytes on the post-mining heaps of Upper Silesia and Dr. Sylwia Wierzcholska (Wrocław University of Environmental and Life Sciences) for sharing information on epiphytes in reclaimed external spoil heap of the Bełchatów Lignite Mine.

REFERENCES

Anonymous 2014: Rozporządzenie Ministra Środowiska z dnia 9 października 2014 r. w sprawie ochrony gatunkowej roślin. Dz. U. z 2014 r., poz. 1409.

Bahuguna Y.M., Gairola S., Semwal D.P., Uniual P.L. & Bhatt A.B. 2013: Bryophytes and Ecosystem. [In:] Gupta R.K. & Kumar M. (eds.): Diversity of Lower Plants. IK International Publishing House Pvt. Ltd., New Delhi, 279–296.

Borgegård S.-O. 1990: Vegetation development in abandoned gravel pits: effects of surrounding vegetation, substrate and regionality. Journal of Vegetation Science, 1 (5), 675–682.

Dombrovska O. & Qiu Y.-L. 2004: Distribution of introns in the mitochondrial gene *nad1* in land plants: phylogenetic and molecular evolutionary implications. Molecular Phylogenetics and Evolution, 32 (1), 246–263.

Duda J. 1989: Mechorosty Ostravsko-Karvinského Revíru. Časopis Slezského Muzea Opava, Série A, 38, 83–95.

Frey W. & Stech M. 2009: Marchantiophyta, Bryophyta, Anthocerotophyta. [In:] Frey W. (ed.): Syllabus of Plant Families. A Engler's Syllabus der Pflanzenfamilien. Part 3, Bryophytes and Seedless Vascular Plants. Gebrüder Borntraeger Verlagsbuchhandlung, Stuttgart, 13–263.

Glime J.M. 2017: Bryophyte Ecology. Volume 1. Physiological Ecology. E-book sponsored by Michigan Technological University and the International Association of Bryologists. http://www.bryoecol.mtu.edu

Goffinet B., Buck W.R. & Shaw A.J. 2009: Morphology, anatomy and classification of the Bryophyta. [In:] Goffinet B. & Shaw A.J. (eds.): Bryophyte Biology. 2nd ed. Cambridge University Press, Cambridge, 55–138.

Gonda-Soroczyńska E. & Kubicka H. 2016: Znaczenie rekultywacji i zagospodarowania gruntów w Polsce w kontekście ochrony środowiska. Infrastruktura i Ekologia Terenów Wiejskich, 1, 163–175.

Greenwood J.L., Stark L.R. & Chiquoine L.P. 2019: Effects of rate of drying, life history phase, and ecotype on the ability of the moss *Bryum argenteum* to survive desiccation events and the influence on conservation and selection of material for restoration. Frontiers in Ecology and Evolution, 7, 388.

Gugała A. 1996: Ekologia mchów występujących na hałdach kopalnianych uranu w Wojcieszycach i Kopańcu koło Jeleniej Góry. Acta Unversitatis Wratislaviensis 1835, Prace Botaniczne, 68, 105–115.

Hodgetts N., Cálix M., Englefield E. *et al.* 2019: A miniature world in decline. European Red List of Mosses. Liverworts and Hornworts. IUCN, Brussels. https://doi.org/10.2305/IUCN.CH.2019.ERL.2.en
Holyoak D.T. & Lockhart N. 2011: A survey of bryophytes and metallophyte vegetation of metalliferous mine spoil in Ireland. Journal of the Mining Heritage Trust of Ireland, 11, 3–16.
Jagodziński A.M., Wierzcholska S., Dyderski M.K., Horodecki P., Rusińska A., Gdula A.K. & Kasprowicz M. 2018: Tree species effects on bryophyte guilds on a reclaimed post-mining site. Ecological Engineering, 110, 117–127.
Jędrzejko K. 1985: The Liverworts (Hepaticopsida) in the Upper Silesian Industrial District and the Forest Protective Belt in the area of the Silesian Upland and its relation to anthropopressure. Śląska Akademia Medyczna, Katowice.
Jędrzejko K. 1990: Mosses (Bryopsida) in the Upper Silesian Industrial District and the Forest Protective Belt and their relation to anthropopressure. Prace i Studia, 39, 1–264.
Kenrick P. & Crane P.R. 1997: The Origin and Early Diversification of Land Plants. A Cladistics Study. Smithsonian Institution Press, Washington–London.
Klama H. & Górski P. 2018: Red List of Liverworts and Hornworts of Poland (4th edition, 2018). Cryptogamie, Bryologie, 39 (4), 415–441.
Krywult M., Salachna A., Chmura D. & Żarnowiec J. 2013: Nitrate reductase activity in chosen mosses: a case of study in Skalny colliery waste tip. Chemistry–Didactics–Ecology–Metrology, 18 (1–2), 61–68.
Magill R.E. 2010: Moss diversity: New look at old numbers. Phytotaxa, 9, 167–174.
Ochrona środowiska 2018: Główny Urząd Statystyczny, Warszawa.
Ochyra R. & Godzik B. 2015: Bryophytes of selected habitat types in the Olkusz Ore-bearing Region. [In:] Godzik B. (ed.): Natural and historical values of the Olkusz Ore-bearing Region. W. Szafer Institute of Botany, Polish Academy of Science, Kraków, 173–199.
Ochyra R., Żarnowiec J. & Bednarek-Ochyra H. 2003: Census Catalogue of Polish Mosses. Biodiversity of Poland. Volume 3. W. Szafer Institute of Botany, Polish Academy of Sciences, Kraków.
Puttick M.N., Morris J.L., Williams T.A., Cox C.J., Edwards D., Kenrick P., Pressel S., Wellman C.H., Schneider H., Pisani D. & Donoghue P.C.J. 2018: The interrelationships of land plants and the nature of the ancestral embryophyte. Current Biology, 28 (5), 733–745.
Řehounková K. 2007: Variability of spontaneous vegetation succession in disused gravel-sand pits: importance of environmental factors and surrounding vegetation. PhD Thesis, University of South Bohemia, České Budějovice, Czech Republic, 1–100.
Rostański A. 2006: Spontaniczne kształtowanie się pokrywy roślinnej na zwałowiskach po górnictwie węgla kamiennego na Górnym Śląsku [Spontaneous plant cover on colliery spoil heaps in Upper Silesia (Southern Poland)]. Katowice, University of Silesia.
Širka P., Kubešowá S. & Mišiková K. 2018: Bryophytes of spoil heaps rich in toxic metals in Central Slovakia. Thaiszia – Journal of Botany, Košice, 28 (1), 59–77.
Söderström L., Hagborg A., von Konrat M. *et al.* 2016: World checklist of hornworts and liverworts. PhytoKeys, 59, 1–828.
Stark L.R., Greenwood J.L. & Brinda J.C. 2017: Desiccated *Syntrichia ruralis* shoots regenerate after 20 years in the herbarium. Journal of Bryology, 39 (1), 85–93.
Stebel A. 1997: Mszaki Rybnickiego Okręgu Węglowego. Fragmenta Floristica et Geobotanica, Series Polonica, 4, 121–233.
Stebel A. 2006: New bryophyte data for mineral workings in Upper Silesia (Poland). [In:] Nowak A. & Hebda G. (eds.): Biodiversity of Quarries and Pits. Opole Scientific Society, 3rd Department of Natural Science, Opole–Górażdże, 71–81.
Stebel A. 2012: Bryophytes as indicators of natural values on mineral workings areas – a case study from limestone quarry "Górażdże" (Opole Province, Poland). Quarry Life Award – Final Project Report. https://www.quarrylifeaward.pl/node/261
Stebel A., Ochyra R., Godzik B. & Bednarek-Ochyra H. 2015: Bryophytes of the Olkusz Ore-bearing Region (southern Poland). W. Szafer Institute of Botany, Polish Academy of Sciences, Kraków.
Szweykowski J. 2006: An annotated checklist of Polish liverworts and hornworts [Krytyczna lista wątrobowców i glewików Polski]. Biodiversity of Poland. Volume 4. W. Szafer Institute of Botany, Polish Academy of Sciences, Kraków.
Vanderpoorten A. & Goffinet B. (eds.) 2009: Introduction to Bryophytes. Cambridge University Press, Cambridge.

Villareal J.C., Cargill D.C., Hagborg A., Söderström S. & Renzaglia K.S. 2010: A synthesis of hornwort diversity: Patterns, causes and future work. Phytotaxa, 9, 150–166.
Wellman C., Osterloff P. & Mohiuddin U. 2003: Fragments of the earliest land plants. Nature, 425, 282–285.
Woźniak G. 2010: Zróżnicowanie roślinności na zwałach pogórniczych Górnego Śląska [Diversity of vegetation on coal-mine heaps of the Upper Silesia (Poland)]. W. Szafer Institute of Botany, Polish Academy of Sciences, Kraków.
Żarnowiec J., Stebel A. & Ochyra R. 2004: Threatened moss species in the Polish Carpathians in the light of a new red-list of mosses in Poland. [In:] Stebel A. & Ochyra R. (eds.): Bryological Studies in the Western Carpathians. Sorus, Poznań, 9–28.

Part Three

Modern Methods Approach to Novel Ecosystem Studies

3.1. Immunochemistry of Cell Wall – a Tool for Evaluation of the Response of Plants to Changed Habitat

Katarzyna Sala
Institute of Biology, Biotechnology and Environmental Protection,
Faculty of Natural Sciences, University of Silesia in Katowice;
28 Jagiellońska Str., 40-032 Katowice, Poland;
katarzyna.sala@us.edu.pl; ORCID: 0000-0002-5978-3623

Kamila Godel-Jędrychowska
Institute of Biology, Biotechnology and Environmental Protection,
Faculty of Natural Sciences, University of Silesia in Katowice;
28 Jagiellońska Str., 40-032 Katowice, Poland;
kamila.godel-jedrychowska@us.edu.pl; ORCID: 0000-0002-0136-2746

Ewa Kurczyńska
Institute of Biology, Biotechnology and Environmental Protection,
Faculty of Natural Sciences, University of Silesia in Katowice;
28 Jagiellońska Str., 40-032 Katowice, Poland;
ewa.kurczynska@us.edu.pl; ORCID: 0000-0003-0772-8961

ABSTRACT: The chapter presents the definition, molecular basis of the immunohistochemistry methods and the use of these methods in examining plant cell walls. Recommended protocols for immunostaining of plant tissues and visualization together with examination on light and electron microscopy levels are presented. The structural and chemical composition of the cell wall is thoroughly described. Moreover, this chapter shows the current state of knowledge about the relationship between the chemical structure of the wall and environmental conditions, and how immunohistochemical methods can be used to monitor changes in the environment.

KEYWORDS: Cell wall modification, immunohistochemistry, immunocytochemistry, tissue preparation, abiotic stress

CELL WALL DEFINITION

Robert Hooke discovered the cell wall in 1665. At present, it is known that cell wall is a part of the cells of plants, fungi, algae, most bacteria, and archaea, and is an outer layer deposited outside of the cell membrane.

Cell walls are an integral, metabolically active part of the cell, reacting dynamically to stimuli from the inside of the cell or the environment. The chemical composition of the walls is used as a source of information defining the future direction of cell differentiation. Cell walls are also an origin of signals that regulate the developmental processes of plant cells.

Primary walls are involved in determination and maintenance of the cell shape and structure. As such, the wall must be sufficiently strong, flexible and plastic to prevent the cell from changes (Cosgrove 2003). Numerous literature data indicate that cell walls play an important role in plant cell differentiation and development (Showalter 1993; Cassab 1998; for review see Lamport *et al.* 2018). Increasing knowledge of the structure and function of the wall components, their synthesis and modifications during cell growth, including reaction to biotic and abiotic factors, is required for understanding the mechanisms which control plant development.

The walls of angiosperm and gymnosperm cells are composed of cellulose, hemicelluloses, and pectic polysaccharides, amounts and composition of which change during plant growth and development. This indicates that analysis of the spatio-temporal changes in the presence and distribution of wall chemical components can be used as an indicator of the influence of environment on plants.

There are two general types of wall based on the relative amounts of pectic polysaccharides and the structure and amounts of hemicellulosic polysaccharides (Carpita & Gibeaut 1993). Type I walls typically contain xyloglucan and/or glucomannan, structural proteins and 20–35% pectin. These walls are found in all dicotyledons, the non-graminaceous monocotyledons, and gymnosperms. Type II walls, characteristic for commelinolid monocotyledons (e.g. grasses), are rich in glucuronoarabinoxylans (instead of xyloglucan), and they also contain small amounts of structural proteins and <10% pectin (Vogel 2008).

The structure of wall polymers is diversified. Some wall polysaccharides are linear polymers that are composed of a single type of glycosyl residue (e.g. cellulose is composed of 1,4-linked β-glucosyl residues whereas homogalacturonan is composed of 1,4-linked α-galacturonic acid residues). Other wall polysaccharides (e.g. xyloglucan and rhamnogalacturonan II) have a regular branching pattern. In contrast, the polysaccharide backbone of rhamnogalacturonan I and the protein backbone of arabinogalactan proteins are substituted with a diverse range of arabinosyl and galactosyl-containing oligosaccharide side chains. Despite the progress that was made in understanding the structures of these polysaccharides and proteoglycans, determining their primary structures remains a major challenge (O'Neill & York 2003). This is due in large part to the fact that the biosynthesis of polysaccharides is, in contrast to proteins and nucleic acids, not template-driven. Thus, the primary sequence of a polysaccharide cannot be inferred from genomic data.

STRUCTURE, COMPOSITION AND FUNCTION

Walls as a specialized structures, bind together the plant cells and form tissues. Cells from various tissues have different functions, thus, the structure and wall composition can vary among particular cell types. In general, the wall can be considered as a two-phase system, in which the cellulose microfibryls (the fibrillar part) are surrounded and connected by amorphous matrix, composed mainly of hemicelluloses, pectins, proteins, phenolic compounds and water with minerals (Fig. 1). It is postulated that molecules of microfibrils and matrix material are bound together either covalently or by hydrogen bonds. The main component of the cell wall, cellulose, can constitute up to 20–90% of cell wall components, depending on the wall type (primary, secondary) and tissue (parenchyma cells, collenchyma, fibres). Wall proteins can be divided as enzymatic or structural.

Among structural proteins, the most abundant include hydroxyproline-rich glycoproteins (HRGP), also called extensins (EXT), arabinogalactan proteins (AGP), glycine-rich proteins (GRP), and proline-rich proteins (PRP). With the exception of glycine-rich proteins, all the previously mentioned proteins are glycosylated and contain hydroxyproline (Hyp). Each class of glycoprotein is defined by a characteristic, highly repetitive protein sequence. Chimeric proteins contain two or more different domains, each with a sequence from a different class of glycoprotein. Most cell wall proteins are cross-linked to the cell wall components and may have structural functions.

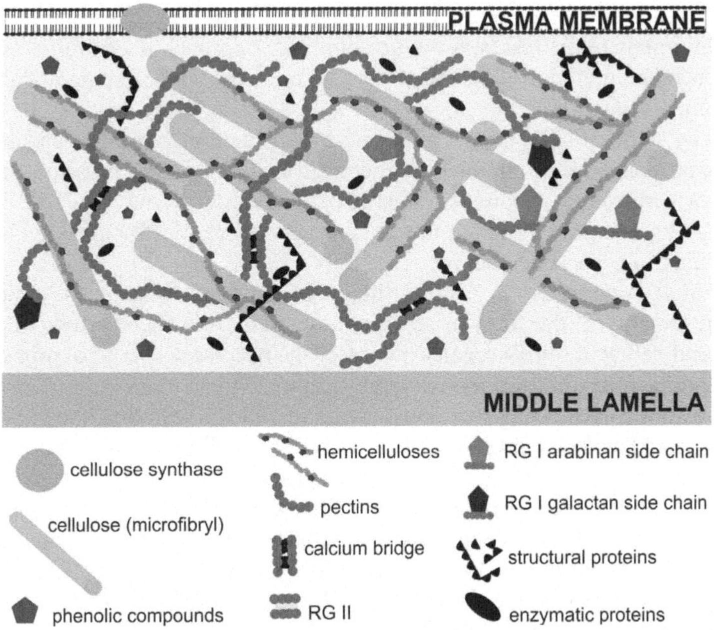

Figure 1. Schematic representation of the main components and their arrangement within plant cell walls

The functions of cell walls are diverse. Wall determines the shape, growth and development of plant cells as well as protects the cytoplasm from invasion by viral, bacterial and fungal pathogens (Karthikeyan *et al.* 2006; Filippi *et al.* 2014; da Silva *et al.* 2018). In this reaction, an increase of wall peroxidases and deposition of phenolic compounds takes place, which enhances the mechanical strength of the host wall. Cell wall is also involved in the transport of materials and metabolites into and out of cells (Byrt *et al.* 2018; Canarini *et al.* 2019). Moreover, wall withstands the turgor pressure and due to this, regulates cell growth (Hernández-Hernández *et al.* 2020).

HABITATS AND CELL WALL MODIFICATION

Plants stay in a dynamic balance with the environment, constantly responding at the organ, tissue and cellular level to any environmental changes (Kuhn 1988; Le Gall *et al.* 2015). Extreme changes that hinder plant growth or development are called stress (Kuhn 1988). Plants developed strategies to adapt to various biotic (caused by pathogens, insects or herbivores) or abiotic (such as drought, flooding, heat, cold, salinity, heavy metals, light, air pollutants) stresses (Le Gall *et al.* 2015). Responses of plant cells, their adaptive strategy to stress or to any changes in environment, may be reflected by the cell wall composition. Also, specific wall modifications can be an expression of a particular plant habitat.

The responses of the plant cell wall to the abiotic stresses mentioned above are intensively investigated at the morphological, physiological, transcriptomic, proteomic and biochemical levels. Based on many studies it appears that plant responses are highly complex and dependent on the plant species, genotype, age of the plant, duration of the stress application, and intensity of this stress. Thus, it is difficult to present a common pattern of stress response in cell wall architecture that is involved in adaptation and/or resistance to abiotic stress (Le Gall *et al.* 2015). The common reaction of cell walls involves an elevation in xyloglucan endotransglucosylase/hydrolase (XTH) and expansion action and an increased cell wall thickening by incorporation of hemicellulose and lignin into the wall (Le Gall *et al.* 2015). Such results indicate that to answer the questions regarding how plants adapt to the environmental stresses, a multidisciplinary approach needs to be used, including the immunohistochemical methods which allow to determine the spatio-temporal changes in chemical composition of the cell wall.

Despite many studies on the impact of the environment on plant growth and development, there is very little data concerning immunohistochemical analyses of the presence and distribution of pectic, AGPs or hemicellulose epitopes. Therefore, knowledge about the correlation between wall composition and its changes in relation to the environment are still fragmentary. Some examples of the current findings are presented below.

In the leaf of *Myrothamnus flabellifolius* high concentration of arabinose which was associated with pectin presumably in the form of arabinan polymers, was found. Authors proposed that the arabinan-rich leaf cell wall of this species possesses the necessary structural properties to be able to undergo repeated periods of desiccation and rehydration (Moore *et al.* 2006). The reaction of plants to desiccation also involves changes in composition of cell wall. Studies showed that levels of arabinan-rich pectin and arabinogalactan proteins increase in cell walls and may act as "pectic plasticizers" (Moore *et al.* 2006). Up-regulation of wall proteins and osmoprotectants has also been described. Moreover, the hemicellulose-rich walls of the grass-like *Xerophyta* sp. and the resurrection grass *E. nindensis* were found to contain highly arabinosylated xylans and arabinogalactan proteins (Moore *et al.* 2006). According to Moore *et al.* (2013) such data support a general mechanism of "plasticising" the cell walls during desiccation and implicate arabinose-rich polymers as the major contributors in maintaining the wall flexibility.

Immunohistochemical analysis of *Tilia* × *euchlora* leaves showed that pectins, rhamnogalacturonan I (RG-I) and homogalacturonan (HG), and some AGP proteins, were differentially distributed in leaves from healthy trees, in contrast to leaves from injured trees which were subjected to salt stress (Milewska-Hendel *et al.* 2017). The differences were especially visible for RG-I arabinans, which are required for the maintenance of cell wall flexibility, and RG-I galactans, which can be associated with the stiffening of the wall and firmness of the cell wall. Studies on the roots of crop plants showed the effects of salt stress on cell wall modifying enzymes, cellulose microfibril orientation and non-cellulosic polysaccharide deposition in root elongation zones (Byrt *et al.* 2018). It was determined that structural and chemical cell wall changes were linked to tolerance to salt stressed and water limited roots.

In *Brachypodium distachyon* leaves, it was shown that hydroxyproline-rich glycoproteins may be markers of temperature stress (Pinski *et al.* 2019), and the reaction also involved altering the cell wall architecture, which is facilitated by the AGP and EXT.

An important aspect of pollution concerns the presence of trace metals in the soil (Krzesłowska 2011). Such metals include Cu, Fe, Zn, Pb, Al, Hg and Cd and all of these influence plant developmental processes. As a cell wall is the main barrier for their entrance into the plant, analysis of the chemical composition of the wall is essential for understanding the interaction of the trace metals and cell wall modification. In this case the most important are pectins which can bind divalent and trivalent metal cations (Krzesłowska 2011). For example, immunohistochemical analysis of barley roots showed that methyl-esterified HG can be a marker for newly arising cell walls under Al exposure. Moreover, the response of root cells to Al can be expressed by the specific distribution of pectins in the cell wall and may be an explanation of the mechanism by which Al inhibits root elongation (Jaskowiak *et al.* 2019). Another example of the habitat influence was found in three grass species that were growing in post-industrial or control habitats (Milewska-Hendel *et al.* 2021). Among others, presence of arabinan and galactan (pectins) in leaves from the same species increased in plants collected from polluted areas.

Changes in wall structure were also described for root tip cells of *Biscutella laevigata* growing in a zinc-lead waste heap and pointed out that detoxification of metals in this species involves the accumulation of metals in vacuoles, cell walls and intercellular spaces (Wierzbicka *et al.* 2017), indicating that the role of the cell wall and its chemical composition is important.

The importance of the analysis of wall modifications under biotic and abiotic factors is confirmed by the analysis of gene expression. As was pointed out by Houston *et al.* (2016), cell wall and polysaccharide-related activities may fulfil a common function in promoting remodelling of the cell wall (for example modification of particular wall components), as a direct response to abiotic stress, that leads to stress tolerance.

All of these studies show that spatio-temporal analysis of the wall chemical components with the use of immunohistochemical methods is significant for understanding the influence of environment and the protection or adaptation mechanisms that plants activate in their responses. Moreover, knowledge of the chemical composition of cell walls can be used for selection of plants for phytoremediation of coal mine dumps (Rana & Maiti 2018) or other changed habitats (Kushwaha *et al.* 2016).

IMMUNOCYTOCHEMISTRY/IMMUNOHISTOCHEMISTRY

Immunochemistry (ICH) includes immunocytochemistry (ICC) which examines cells and immunohistochemistry (IHC) which examines the tissues. It is a method that allows for identification or localisation of specific antigens within cells or tissues. The method utilises an antigen-specific antibody which binds an antigen in a cell/tissue section, fluorochromes (at the light microscope level) and colloidal-gold particles (at the electron microscopy level) to visualize where the antibody was bound. ICH, nowadays, is a common laboratory technique that is used to visualize the presence of a specific antigen in cells/tissues by use of a specific primary antibody that binds to it. In a two-step reaction, the primary antibody allows for visualization of the antigen under a fluorescence microscope only when it is bound by a secondary antibody that has a conjugated fluorophore (Fig. 2). ICH helps to evaluate whether or not cells/tissues in a particular sample express the antigen in question. In cases where an immunopositive signal is found, we are able to determine which sub-cellular compartments are expressing the antigen.

It can be concluded that immunohistochemistry is a method of detecting various antigenic substances in microscopic sections, involving the use of a specific antibody directed against the searched specimen, and then a detection system that creates a coloured insoluble substance visible in a bright field, fluorescence or electron microscope (fluorochromes, colloidal metals, 3'3 diaminobenzidine enzymes).

ICH GLOSSARY

To better understand the principle of immunohistochemistry, below are the definitions of the basic terms that are used to describe the procedure (see also Fig. 2):

- *Antigens* – molecules capable of stimulating an immune response. Each antigen has distinct surface features, or epitopes, resulting in specific responses.
- *Epitope* – (antigenic determinant) is the part of an antigen that is recognized by antibodies.
- *Antibodies* (immunoglobins) – Y-shaped proteins produced by B cells of the immune system in response to exposure to antigens. Each antibody contains a paratope which recognizes a specific epitope on an antigen, acting like a lock and key binding mechanism. This binding helps to eliminate antigens from the body, either by direct neutralization or by "tagging" for other arms of the immune system.

- *Primary antibody* – is the one that binds directly to the antigen. The variable region of the primary antibody recognizes an epitope on the antigen. It is produced by a host organism that is of a different species than the specimen.
- *Secondary antibody* – is cross-adsorbed against a serum antibody protein from another species or is adsorbed against a mixture of serum antibody protein from several species, available commercially as conjugate with e.g. fluorochrome.
- *Fluorochrome* – any of various fluorescent substances used in biological staining to produce fluorescence in a specimen. For antibody staining, fluorochromes like AMCA, Cy® or Alexa Fluor® dyes, FITC, PE, PerCP, Rhodamine and Texas Red® are most commonly used.
- *Fluorescence microscopy* – a special form of light microscopy which uses the ability of fluorochromes to emit light after being excited with light of a certain wavelength. Proteins of interest can be marked with fluorochromes *via* antibody staining or tagging with fluorescent proteins. It allows for the determination of the distribution of a single molecule species, its amount and its localization inside a cell.

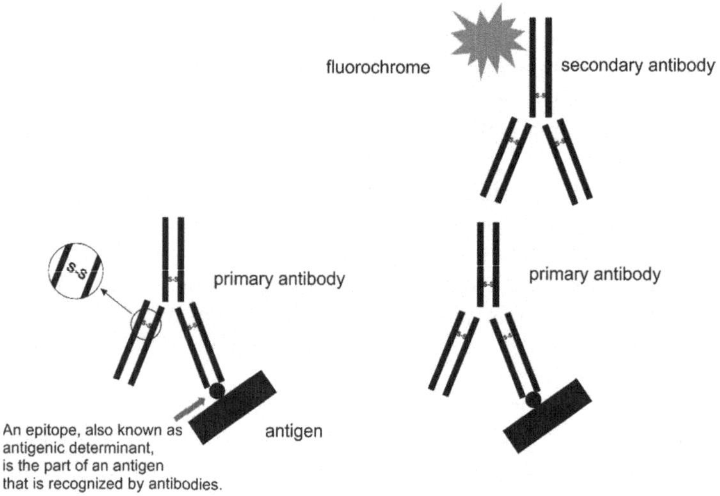

Figure 2. Schematic representation of the epitope, antibody, antigen and fluorochrome (S-S – disulfide bond)

Principles of ICH for cell wall composition

As plant cell walls are composed of polysaccharides and (glyco)proteins, that form complex networks and cross-links *in muro* (O'Neill & York 2003), preparation of antibodies directed to such wall components is not direct as in the case of animal cells. In the case of plant cell wall chemical components, the generation of monoclonal antibodies is not straightforward. For example, limitations in producing the antibodies for glycans include the low immunogenicity of this compound which forces the preparation of neoglycoprotein immunogens. This involves the coupling of a defined oligosaccharide to an immunogenic protein such as BSA (Paul Knox Cell Wall Lab, University of Leeds). The generated probes are generally rat hybridoma or phage display monoclonal antibodies which allow for generation of a series of monoclonal antibodies that recognize diverse carbohydrate structures within plant cell walls. Nowadays, characterization of the epitope structures recognized by antibodies that each bind to unique glycosyl structures within the plant cell wall is available. During preparation of such, different techniques are used, including an array of plant cell wall-derived carbohydrates which are presented to the animals' immune system. The details of procedures used for obtaining the antibodies directed to diverse cell wall components can be found, e.g. in Willats & Knox (2003).

Monoclonal antibodies as tools for cell wall characterization

Monoclonal antibodies are specific and sensitive reagents that have been used to monitor changes in primary wall composition and organization at the cellular and sub-cellular level (Knox 1997; Bush & McCann 1999; Freshour *et al.* 2003). For example, monoclonal antibodies have been used to show that wall composition differs amongst cell types and even in the walls surrounding a single cell. Antibodies have provided evidence that within a single wall there are sub-domains containing different glycoconjugates. In addition, studies utilising antibodies suggest that occurrence of carbohydrate epitopes of glycoproteins and polysaccharides is changed during plant development. However, the lack of well-characterized antibodies is a major hindrance to advances in these types of studies. To date, ~40 monoclonal antibodies have been generated that bind to plant cell wall carbohydrate structures. Only some of these antibodies have had their binding sites (epitopes) characterized in detail (Willats & Knox 2003).

The most frequently applied antibodies and epitopes they recognise are listed in Table 1.

Table 1. List of commonly used primary rat monoclonal antibodies

ANTIBODY	EPITOPE	REFERENCES
	PECTIC POLYSACCHARIDES	
	Homogalacturonan (HG) – methyl-esterification degree	
LM19 JIM5	non-methyl-esterified, partially methyl-esterified HG	Verhertbruggen *et al.* (2009) Willats *et al.* (2000)
LM20 JIM7	methyl-esterified HG	Verhertbruggen *et al.* (2009) Willats *et al.* (2000)
	Rhamnogalacturonan-I – side chains	
LM5	linear tetrasaccharide in (1-4)-β-D-galactans	Jones *et al.* (1997)
LM6	linear pentasaccharide in (1-5)-α-L-arabinans	Willats *et al.* (1998)
	NON-CELLULOSIC, NON-PECTIC POLYSACCHARIDES	
	Hemicelluloses	
LM25	XLLG, XXLG and XXXG oligosaccharides of xyloglucan	Pedersen *et al.* (2012)
LM21	heteromannan	Marcus *et al.* (2010)
LM10	(1-4)-β-D-xylan	McCartney *et al.* (2005)
	PLANT CELL WALL PROTEOGLYCANS/GLYCOPROTEINS	
	Arabinogalactan protein (AGP) glycan	
JIM8	Arabinogalactan	Pennel *et al.* (1991)
JIM13	Arabinogalactan/Arabinogalactan protein	Knox *et al.* (1991)
JIM16	Arabinogalactan/Arabinogalactan protein	Knox *et al.* (1991)
MAC207	Arabinogalactan protein	Pennell *et al.* (1989)
	Extensin	
LM1	Extensin/HRGP	Smallwood *et al.* (1995)
JIM11	Extensin/HRGP glycoprotein	Smallwood *et al.* (1994)
JIM12	Extensin/HRGP glycoprotein	Smallwood *et al.* (1994)
JIM20	Extensin/HRGP glycoprotein	Smallwood *et al.* (1994)

PROTOCOLS FOR IMMUNOCYTOCHEMISTRY OF PLANT MATERIAL

In this part, the recommended methods of fixation and preparation of the plant material for immunostaining, based on our expertise in such methods, is described (Sala *et al.* 2013; 2017; 2019). The first and most important step in this procedure is tissue fixation. This process is affected by: time of fixation, size of tissue, tissue aeration, fixatives and solvents, and temperature. The purpose of fixation is to stop life processes and preserve the tissue in an appropriate chemical environment, while maintaining the composition and structure of plant cells to the maximum possible extent. Under the influence of fixation, structures giving erroneous histological images should not be formed. The principle of fixation is to create cross-links (mainly, but not only) between the amino groups of proteins present in the cell. Such reaction should permanently stabilize the cell structure, ultrastructure and physiological state. The fixative solutions are characterized by relatively slow diffusion, but small objects are usually impregnated with the fixating agent faster. In contrast, thick and large pieces require more time to be fixed. A good fixative solution should quickly kill cells, but gradually coagulate proteins.

Fixation and tissue preparation

The fixation process takes place at room temperature; however, a lower temperature (0–4°C) or fixing on ice, is used in some cases, for example, to study enzymes. The first step is a selection of fresh, well-cleaned tissue which we cut it into small pieces (5 mm × 5 mm × 5 mm maximum size for light microscopy, and 1 mm × 1 mm × 1 mm maximum size for electron microscopy) with a sharp blade and put the material into the fixative solution. The fixative should be used in excess of the fixed material (over 20 times). Also, volume of fixative depends on the amount of material and type of utensil in which the material will be fixed (Eppendorf, falcon or beaker). Because it is necessary to set the fixative final pH at 7.2 (with HCl when too high, with NaOH when too low), the minimum volume should be at least 25 ml so that the electrode can be submerged. The tissue can be left in the fixation solution for a few to several hours (for extended time of fixation it is recommended to store the material at 0–4°C).

Fixatives used for immunohistochemical research include glutaraldehyde (GA) and paraformaldehyde (PFA). GA stabilizes protein structures by forming cross-links between them and penetrates slowly into the cells. Subsequent action of GA causes even more permanent cross-linking of cellular proteins, and as a consequence, the tissue is better fixed and gives positive results regarding cell ultrastructure. PFA quickly fixes and penetrates into cells and is responsible for structure pre-stabilization. An extremely practical advantage of formaldehyde is its high ability to penetrate biological material, which allows it to effectively fix relatively larger fragments. A mixture of PFA + GA at particular concentrations is used in examination of cell ultrastructure, as well as in fluorescence microscopy analyses (see Scheme 1).

Fixative solution

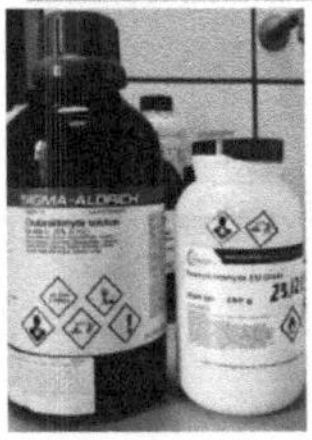

4% PFA + 2% GA
pH recommended: 6.9-7.2

final volume of fixation solution = proper buffer + PFA (dissolve on heating block) + GA (add when solution is cooled down)

Scheme 1. Composition of the basic fixative

As there is no universal fixative for all plant tissues, the given values should be treated as initial, which could be modified depending on the type of plant material. For example, for fixing "difficult" plant tissues it is recommended to add other components to fixative such as sucrose, $CaCl_2$ and detergent (Chen & Baldwin 2007).

The best medium for fixation solution is an appropriate buffer, which provides the proper environment for fixation, namely, maintains a constant pH during fixation, ensures adequate osmolality (to preserve the natural shape of cells and organelles) and ionic composition (to prevent extraction or precipitation of fixed cell components). For fluorescence immunostaining it is preferred to use a phosphate or PBS buffer at pH = 7.2 (the composition of PBS buffer is presented in Table 2), for electron microscopy analysis the cacodylate buffer (pH 7.2) is recommended.

Table 2. Composition of Phosphate Buffered Saline (PBS buffer), pH 7.2 (all components are from Merck)

SALT	CONCENTRATION (mMol/l)	CONCENTRATION (g/l)
NaCl	137	8.0
KCl	2.7	0.2
Na_2HPO_4	10	1.44
KH_2PO_4	1.76	0.24

Cacodylate buffer pH = 7.2, 0.15 M formula:

1. Dissolve 21.4 g of sodium cacodylate in 900 ml of demineralized water.
2. Adjust the pH to 7.2 and add water to 1000 ml.

Fixative preparation

1. Measure out the appropriate volume of buffer.
2. Add PFA and slowly dissolve at 50°C (you can add a few drops of NaOH).
3. Cool the solution and filter through a filter paper.
4. Add appropriate volume of GA.
5. Add buffer, leaving 2–3 ml to final volume.
6. Establish the pH to appropriate value.
7. Add buffer to final volume of the fixative.

Tissue preparation

1. Put tissue fragments into an Eppendorf tube containing 2 ml of fixation mixture. The time between excising tissue and placing it in the fixative solution must be as short as possible!
2. Vent the samples under a vacuum in a desiccator until the samples sink completely to the bottom of the Eppendorf tube (in some cases it is better to repeat the action several times).
3. Incubate material in fixative solution at room temperature or transfer the samples into the fridge (4°C) for overnight incubation.
4. Carefully remove the fixative with a pipette and make 3 washes in a buffer (PBS or cacodylate buffer), each one for 5–20 minutes, depending on the size of the samples.
5. Go to the next step, which is dehydration of the sample by replacing the buffer with an increasing ethanol series.
   Remember to work under the fume hood during preparation of fixative and fixation process.

Dehydration

Dehydration is a process of removing water from the fixed tissue. It is necessary step before infiltration and embedding in a Steedman's wax or in LR-White Resin. Dehydration consists of treating material in an ethanol series from low concentration (10%, 30%, 50%) to high concentration (70%, 90%, 100%). There should be two changes at each concentration for 15–30 min (depending on the size and type of the material). For electron microscopy softer dehydration steps are used prior to a higher ethanol series (5%, 10%, 20%, 30%, 40%, 50%, 60%, 70%, 80%, 90%, 100%). In this case one change in each concentration for 30 min is sufficient. Material can be stored in 70% ethanol at 0–4°C.

Embedding

Before embedding the biological material, it should be incubated with solutions of gradually increasing concentration of the substance in which it will be embedded (Scheme 2). As a result of certain chemical processes, the probe will change the state of aggregation from liquid to solid. Such a substance should be characterized by good solubility in solvents, relatively low density in the liquid state, uniform polymerization, the smallest change in volume during polymerization and optimal hardness which will enable cutting after polymerization. Embedding media used for immunostaining are Steedman's wax and LR-White acrylic resins. For immunostaining at the electron microscopy level only LR-White acrylic resin is used. LR-White acrylic resin is available commercially and it polymerizes at 50°C (if the probes are soft and locally "jelly", keep the probes in an incubator for one more day).

The Steedman's wax is prepared from individual components as follows: 90 g polyethylene glycol distearate and 10 g 1-hexadecanol. Weigh out the appropriate amount into a glass beaker and put it on the heating plate (do not exceed 55–60°C during melting). The solutions of Steedman's wax and pure Steedman's wax should be kept at 39°C, as well as the incubated material. The polymerization of Steedman's wax takes place at room temperature (rt).

Sectioning the material

Prepare a wax block with material embedded in Steedman's wax (Fig. 3A), then select the area where the sample is located and trim the block (Fig. 3B). The block with plant material has to be attached to a wooden stub. To do this, warm the remaining wax block (without material) on the blade over the burner (Fig. 3C) and place it warmed up on a wooden stub (Fig. 3D). Then place the block with material on the wooden stub (Fig. 3E) and avoid touching the top of the wax block. To fasten the polymerization, the wooden stub can be placed in the freezer for 5–10 min. The last step is clamping the block into the specimen holder of a rotary microtome and correct orientation of the material to the cutting surface (Fig. 3F). As soon as the wax block starts being cut and ribbons begin to form, pick them up and place the ribbons on slides coated with poly-L-lysine (PolySciences) or Mayer's adhesive. Add drops of water on the slide to unfold the sections. The thickness of sections recommended for immunostaining and examination at the fluorescence microscope level is 5–10 μm. It is necessary to de-wax the slides with ribbons of Steedman's wax before immunostaining (Scheme 3).

Material embedded in LR-White must be cut with the use of an ultramicrotome. The thickness of ribbons recommended for immunostaining and examination at the fluorescence microscope level is 1,5 μm. Ribbons are placed on slides coated with poly-L-lysine (PolySciences) and no additional action is needed prior to immunostaining. The thickness of a section for immunolabelling and detection at the electron microscopy level ranges between 70–90 nm. The sections are cut with the use of an ultra-diamond knife (Diatome, Switzerland). Sections are collected on formvar-coated nickel grids (300 meshes; PolySciences). Next, grids undergo the immunolabelling procedures.

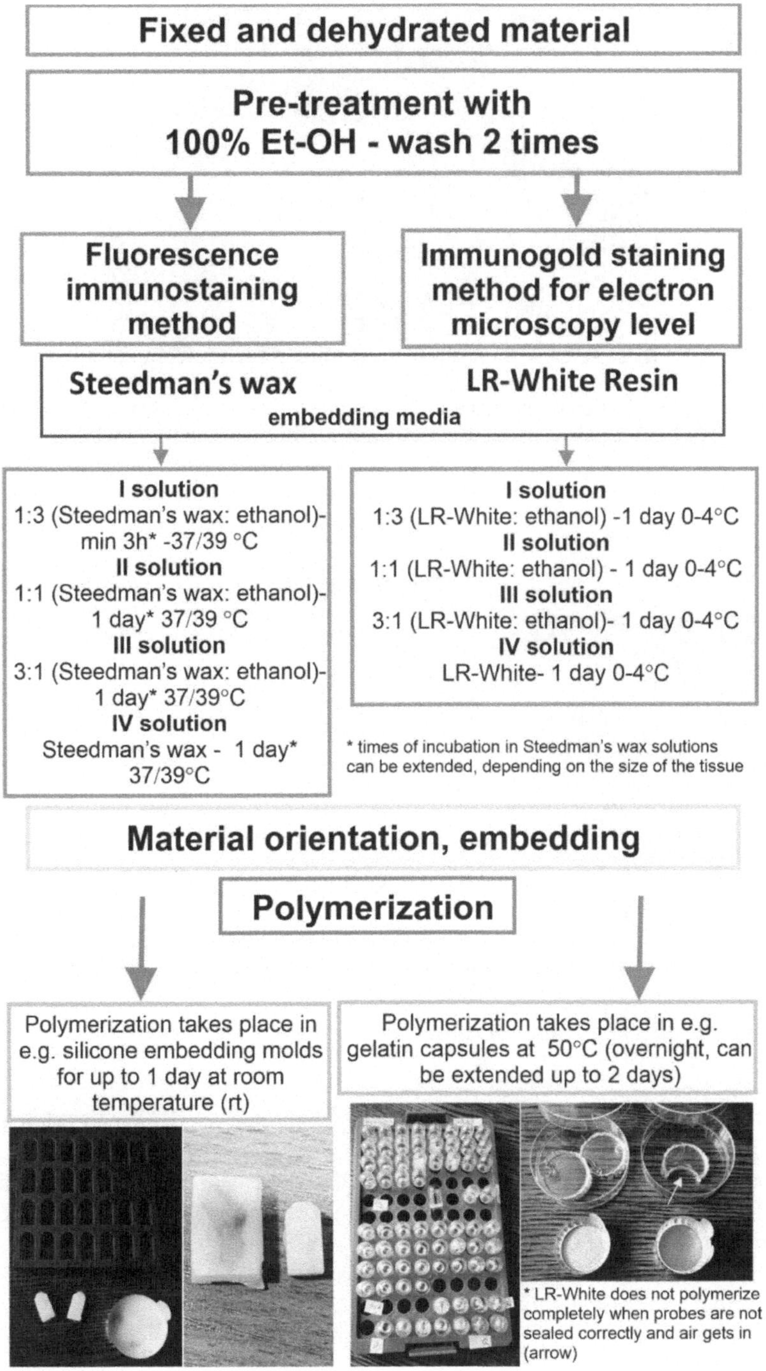

Scheme 2. Procedure of embedding material in the proper medium
for fluorescence immunostaining (left) and for immunogold staining (right)

Figure 3. Step by step preparation of the material for cutting on a rotary microtome. A – Material embedded in a wax block. B – Selected areas of the wax block and the wooden stub. C – Melting the wax (without material). D – Slide off the warm wax on the wood stub. E – Stick the wax block with the material onto the wood stub. F – Clamping the block into the specimen holder of a rotary microtome

DE-WAXING

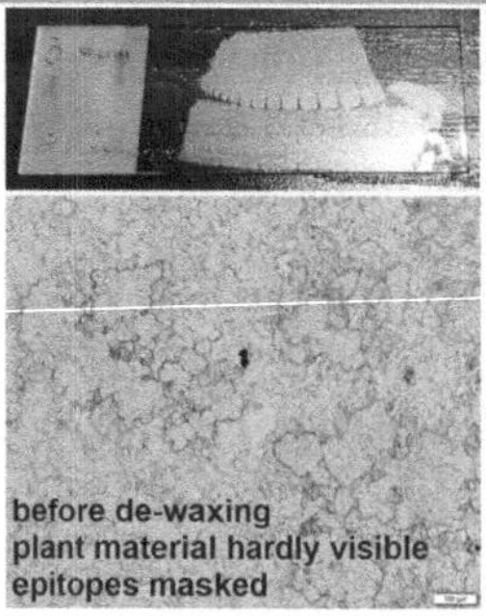

1 step: 100% ethanol (3 changes, 10 min each), rt*

2 step: 90% ethanol (10 min), rt**
3 step: 50% ethanol (10 min), rt**
4 step: PBS (10 min), rt

↓

IMMUNOSTAINING

*** when preparations are more than 4 weeks after sectioning it is better to heat the ethanol up to 39°C**
**** (v/v) ethanol solution in PBS**

Scheme 3. De-waxing procedure for Steedman's wax preparation

Immunostaining (Immunolabelling procedure)

DAY 1

1. Outline the sections on slides with a hydrophobic marker (PAP pen: Merck; Fig. 4A-D) and put the slides in a moist chamber.
2. Apply the blocking buffer on slides in the outlined areas or on formvar-coated nickel grids and incubate 30 min at room temperature (blocking buffer composition: 2% BSA (Bovine Serum Albumin), 2% FCS (Fetal Calf Serum) in PBS buffer, pH = 7.2).
3. Incubate slides/grids with the primary antibody overnight at 4°C (dilute the antibody 1:20 in blocking buffer – 1 µl antibody per 19 µl blocking buffer).

NOTE: the whole procedure can be performed within one day, in which case the incubation time with primary antibody is 1–1.5 h.

DAY 2

4. Wash slides/grids in blocking buffer (3 times, 10 min each).
5. Incubate slides/grids with secondary antibody (for example: goat anti-rat IgG H+L, fluorochrome AlexaFluor®488 conjugated, or for electron microscopy goat anti-rat secondary antibody conjugated with 12 nm gold – Jackson ImmunoResearch Laboratories), 1.5–2 h at

room temperature in the dark; dilution 1:100 in blocking buffer (1 µl antibody per 99 µl blocking buffer).
6. Wash slides/grids in blocking buffer (3 times, 10 min each).
7. Wash slides/grids in PBS (3 times, 5 min each).
8. Optionally for immunofluorescence detection to visualize cell walls: stain sections with a Calcofluor (Fluorescence Brightener 28; Merck; 0.01% solution in PBS, filtered through a 0.22 µm syringe filter) for 5–10 min. Alternatively, to perform bright field observations or to quench autofluorescence, a toluidine blue O solution can be used (TBO, Merck; 0.05% solution in PBS). Wash slides in PBS (3 times, 10 min each).
9. Wash slides/grids in sterile ultrapure water (3 times, 5 min each).
10. Close the slides in Fluoromount (anti-fading medium; Merck; Fig. 4E, 4F). After about 1 hour or the next day, outline the edges of the coverslip with clear nail polish. Store immunostained slides at 0–4°C.

NOTE: Remember to perform the negative controls (for slides/grids) for each antibody used (omit the primary antibody step, to check secondary antibody binding specificity). The lack of fluorescence is the desired result.

NOTE: After immunolabelling the sections on the formvar-coated nickel grids, stain sections with 2% aqueous uranyl acetate for 45 minutes and Reynolds lead citrate for 7 minutes. Store at rt in TEM grid storage boxes.

Visualization

For immunostaining visualization with examination at the light microscopy level: observe in a fluorescent microscope or confocal microscope at the indicated wavelength of light for the fluorochrome used, conjugated with the secondary antibody (example: for AlexaFluor®488 observe in blue light; maximum absorption 493 nm, maximum emission 518 nm).

Detection at the light and electron microscopy level

In probes that were stained with antibodies conjugated with fluorochrome, the presence and distribution of different wall epitopes can be analyzed with the use of fluorescence microscopy, including a confocal microscope (Fig. 5). At the level of electron microscope, immunogold labelling is used, where cell wall components are detected using secondary antibodies tagged with electron-dense colloidal gold particles (Fig. 6).

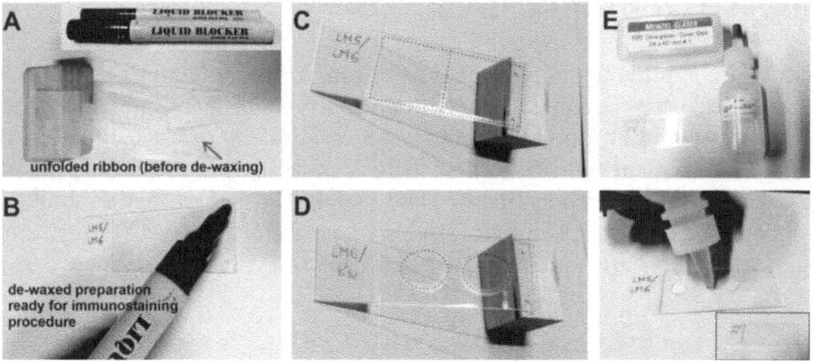

Figure 4. Actions associated with immunolabelling. A – Slides with ribbons (before de-waxing) and hydrophobic markers – PAP pens used for immunostaining. B – De-waxed sections ready for immunolabelling. C, D – two options for outlining the sections on a slide. E – Fluromount and cover slips prepared for closing the slides. F – Application of the Fluoromount on slides and closing with coverslips

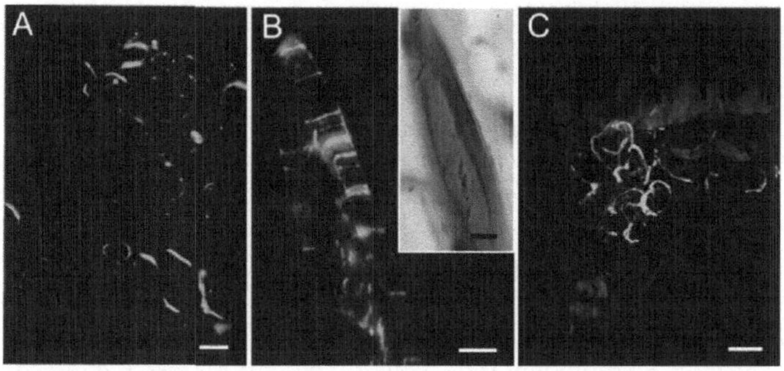

Figure 5. Examples of immunostaining viewed with the use of epifluorescence (A and B) and confocal microscope (C). A – leaf blade of *Phragmites australis* showing the distribution of the AGP epitope recognised by JIM13 antibody, cross section (green colour indicates the distribution of epitope). B – fibre from *Solanum lycopersicum* L. stem stained with LM10 primary antibody, longitudinal section, inset – the same cell stained with toluidine blue O and observed in bright field microscopy. C – graft union from Arabidopsis hypocotyl immunolabelled with the JIM20 antibody, whole mount (green colour indicates the distribution of epitope, blue colour indicates the presence of cellulose); scale bars: A, B, B inset = 10 μm, C = 20 μm

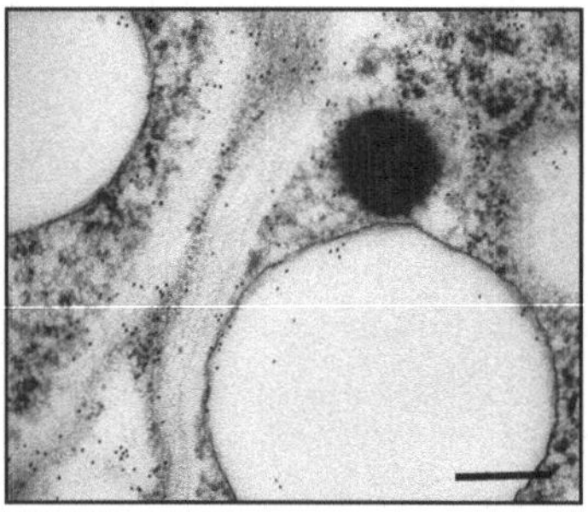

Figure 6. Example of immunogold stained explant cells of *Arabidopsis thaliana* (antibodies were directed against the lipid transfer proteins; black dots are gold particles). Scale bar = 100 nm (Photo: courtesy of Dr. Izabela Potocka)

CONCLUDING REMARKS AND OUTLOOK

Many techniques are applied to analyse the plants from different habitats or exposed to abiotic stresses caused by industrialization of the environment. Among them, immunohistochemical technique can be very helpful as it gives insight into the spatio-temporal and structural organization of walls from individual plant cells. Such knowledge can be used to understand the plant defence mechanisms and may result in production of plant varieties more resistant to abiotic factors, and which in turn, could be used for planting in places where factors such as salinity, drought, heavy metals or high temperatures are a problem for e.g. crops or plants, which can be used for phytoremediation.

REFERENCES

Bush M.S. & McCann M.C. 1999: Pectic epitopes are differentially distributed in the cell walls of potato (*Solanum tuberosum*) tubers. Physiologia Plantarum, 107 (2), 201–213.

Byrt C.S., Munns R., Burton R.A., Gilliham M. & Wege S. 2018: Root cell wall solutions for crop plants in saline soils. Plant Science, 269, 47–55.
Canarini A., Kaiser C., Merchant A., Richter A. & Wanek W. 2019: Root exudation of primary metabolites: mechanisms and their roles in plant responses to environmental stimuli. Frontiers in Plant Science, 10, 157.
Carpita N.C. & Gibeaut D.M. 1993: Structural models of primary cell walls in flowering plants: consistency of molecular structure with the physical properties of the walls during growth. The Plant Journal, 3 (1), 1–30.
Cassab G.I. 1998: Plant cell wall proteins. Annual Review of Plant Physiology and Plant Molecular Biology, 49, 281–309.
Chen W. & Baldwin T.C. 2007: An improved method for the fixation, embedding and immunofluorescence labeling of resin-embedded plant tissue. Plant Molecular Biology Reporter, 25, 27–35.
Cosgrove D. 2003: Expansion of the plant cell. [In:] Rose J.K.C. (ed.): Annual Plant Reviews. Volume 8. The Plant Cell Wall. Blackwell Publishing/CRC Press, Oxford–Boca Raton, 237–263.
da Silva J.M., Medeiros M.B.C.L., Oliveira J.T.C., de Medeiros E.V., de Souza-Motta C.M. & Moreira K.A. 2018: Resistance inducers and biochemical mechanisms in the control of anthracnose in cowpea. Ciencia e Investigación Agraria, 45 (3), 290–300.
Filippi M.C., da Silva G.B., da Silva-Lobo V.L., Viana H.F., Côrtes M.V.C.B. & Prabhu A.S. 2014: Induction of resistance to rice leaf blast by a virulent isolates of *Magnaporthe oryzae*. Revista de Ciências Agrárias, 57 (4), 388–395.
Freshour G., Bonin C.P., Reiter W.-D., Albersheim P., Darvill A.G. & Hahn M.G. 2003: Distribution of fucose-containing xyloglucans in the cell walls of the *mur1* mutant of Arabidopsis. Plant Physiology, 131, 1602–1612.
Hernández-Hernández V., Benítez M. & Boudaoud A. 2020: Interplay between turgor pressure and plasmodesmata during plant development. Journal of Experimental Botany, 71 (3), 768–777.
Houston K., Tucker M.R., Chowdhury J., Shirley N. & Little A. 2016: The plant cell wall: a complex and dynamic structure as revealed by the responses of genes under stress conditions. Frontiers in Plant Science, 7, 984.
Jaskowiak J., Kwasniewska J., Milewska-Hendel A., Kurczynska E.U., Szurman-Zubrzycka M. & Szarejko I. 2019: Aluminium alters the histology and pectin cell wall composition of barley roots. International Journal of Molecular Sciences, 20, 3039.
Jones L., Seymour G.B. & Knox J.P. 1997: Localization of pectic galactan in tomato cell walls using a monoclonal antibody specific to (1→4)-β-D-galactan. Plant Physiology, 113 (4), 1405–1412.
Karthikeyan M., Radhika K., Mathiyazhagan S., Bhaskaran R., Samiyappan R. & Velazhahan R. 2006: Induction of phenolics and defense-related enzymes in coconut (*Cocos nucifera* L.) roots treated with biocontrol agents. Brazilian Journal of Plant Physiology, 18 (3), 367–377.
Knox J.P. 1997: The use of antibodies to study the architecture and developmental regulation of plant cell walls. International Review of Cytology, 171, 79–120.
Knox J.P., Linstead P.J., Peart J., Cooper C. & Roberts K. 1991: Developmentally-regulated epitopes of cell surface arabinogalactan proteins and their relation to root tissue pattern formation. The Plant Journal, 1 (3), 317–326.
Krzesłowska M. 2011: The cell wall in plant cell response to trace metals: polysaccharide remodeling and its role in defense strategy. Acta Physiologiae Plantarum, 33, 35–51.
Kuhn D.N. 1988: Plant stress responses: discussion of models for race-specific resistance. [In:] Conn E.E. (ed.): Opportunities for Phytochemistry in Plant Biotechnology. Plenum Press, New York, 127–161.
Kushwaha A., Rani R., Kumar S. & Gautam A. 2016: Heavy metal detoxification and tolerance mechanisms in plants: Implications for phytoremediation. Environmental Reviews, 24 (1), 39–51.
Lamport D.T.A., Tan L., Held M., Marcia J. & Kieliszewski M.J. 2018: The role of the primary cell wall in plant morphogenesis. International Journal of Molecular Science, 19 (9), 2674.
Le Gall H., Philippe F., Domon J.-M., Gillet F., Pelloux J. & Rayon C. 2015: Cell wall metabolism in response to abiotic stress. Plants (Basel), 4 (1), 112–166.
Marcus S.E., Blake A.W., Benians T.A.S., Lee K.J.D., Poyser C., Donaldson L., Leroux O., Rogowski A., Petersen H.L., Boraston A., Gilbert H.J., Willats W.G.T. & Knox J.P. 2010: Restricted access of proteins to mannan polysaccharides in intact plant cell walls. The Plant Journal, 64 (2), 191–203.
McCartney L., Marcus S.E. & Knox J.P. 2005: Monoclonal antibodies to plant cell wall xylans and arabinoxylans. Journal of Histochemistry & Cytochemistry, 53 (4), 543–546.
Milewska-Hendel A. Baczewska A.H., Sala K., Dmuchowski W., Brągoszewska P., Gozdowski D., Jozwiak A., Chojnacki T., Swiezewska E. & Kurczynska E. 2017: Quantitative and qualitative characteristics of cell

wall components and prenyl lipids in the leaves of *Tilia* × *euchlora* trees growing under salt stress. PLoS ONE, 12 (2): e017268.
Milewska-Hendel A., Chmura D., Wyrwał K., Kurczyńska E.U., Kompała-Bąba A., Jagodziński A.M. & Woźniak G. 2021: Cell wall epitopes in grasses of different novel ecosystem habitats on post-industrial sites. Land Degradation & Development, 32 (4), 1680–1694.
Moore J.P., Nguema-Ona E., Chevalier L., Lindsey G.G., Brandt W.F., Lerouge P., Farrant J.M. & Driouich A. 2006: Response of the leaf cell wall to desiccation in the resurrection plant *Myrothamnus flabellifolius*. Plant Physiology, 141 (2), 651–662.
Moore J.P., Nguema-Ona E.E., Vicré-Gibouin M., Sørensen I., Willats W.G.T., Driouich A. & Farrant J.M. 2013: Arabinose-rich polymers as an evolutionary strategy to plasticize resurrection plant cell walls against desiccation. Planta, 237 (3), 739–754.
O'Neill M. & York W. 2003: The composition and structure of primary cell walls. [In:] Rose J.K.C. (ed.): Annual Plant Reviews. Volume 8. The Plant Cell Wall. Blackwell Publishing/CRC Press, Oxford–Boca Raton, 1–54.
Pedersen H.L., Fangel J.U., McCleary B., Ruzanski C., Rydahl M.G., Ralet M.-C., Farkas V., von Schantz L., Marcus S.E., Andersen M.C.F., Field R., Ohlin M., Knox J.P., Clausen M.H. & Willats W.G.T. 2012: Versatile high resolution oligosaccharide microarrays for plant glycobiology and cell wall research. Journal of Biological Chemistry, 287 (47), 39429–39438.
Pennell R.I., Janniche L., Kjellbom P., Scofield G.N., Peart J.M. & Roberts K. 1991: Developmental regulation of a plasma membrane arabinogalactan protein epitope in oilseed rape flowers. The Plant Cell, 3 (12), 1317–1326.
Pennell R.I., Knox J.P., Scofield G.N., Selvendran R.R. & Roberts K. 1989: A family of abundant plasma membrane-associated glycoproteins related to the arabinogalactan proteins is unique to flowering plants. Journal of Cell Biology, 108 (5), 1967–1977.
Pinski A., Betekhtin A., Sala K., Godel-Jedrychowska K., Kurczynska E. & Hasterok R. 2019: Hydroxyproline-rich glycoproteins as markers of temperature stress in the leaves of *Brachypodium distachyon*. International Journal of Molecular Science, 20 (10), 2571.
Rana V. & Maiti S.K. 2018: Differential distribution of metals in tree tissues growing on reclaimed coal mine overburden dumps, Jharia coal field (India). Environmental Science and Pollution Research, 25, 9745–9758.
Sala K., Karcz J., Rypień A. & Kurczyńska E.U. 2019: Unmethyl-esterified homogalacturonan and extensins seal *Arabidopsis* graft union. BMC Plant Biology, 19, 151.
Sala K., Malarz K., Barlow P.W. & Kurczyńska E.U. 2017: Distribution of some pectic and arabinogalactan protein epitopes during *Solanum lycopersicum* (L.) adventitious root development. BMC Plant Biology, 17, 25.
Sala K., Potocka I. & Kurczynska E. 2013: Spatio-temporal distribution and methyl-esterification of pectic epitopes provide evidence of developmental regulation of pectins during somatic embryogenesis in *Arabidopsis thaliana*. Biologia Plantarum, 57 (3), 410–416.
Showalter A.M. 1993: Structure and function of plant cell wall proteins. The Plant Cell, 5 (1), 9–23.
Smallwood M., Beven A., Donovan N., Neill S.J., Peart J., Roberts K. & Knox J.P. 1994: Localization of cell wall proteins in relation to the developmental anatomy of the carrot root apex. The Plant Journal, 5 (2), 237–246.
Smallwood M., Martin H. & Knox J.P. 1995: An epitope of rice threonine- and hydroxyproline-rich glycoprotein is common to cell wall and hydrophobic plasma-membrane glycoproteins. Planta, 196 (3), 510–522.
Verhertbruggen Y., Marcus S.E., Haeger A., Ordaz-Ortiz J.J. & Knox J.P. 2009: An extended set of monoclonal antibodies to pectic homogalacturonan. Carbohydrate Research, 344 (14), 1858–1862.
Vogel J. 2008: Unique aspects of the grass cell wall. Current Opinion in Plant Biology, 11 (3), 301–307.
Wierzbicka M., Pielichowska M., Bemowska-Kałabun O. & Wąsowicz P. 2017: Microevolution on anthropogenically changed areas on the example of *Biscutella laevigata* plants from calamine waste heap in Poland. Journal of Environmental & Analytical Toxicology, 7 (4), 471.
Willats W.G.T. & Knox J.P. 2003: Molecules in context: probes for cell wall analysis. [In:] Rose J.K.C. (ed.): Annual Plant Reviews. Volume 8. The Plant Cell Wall. Blackwell Publishing/CRC Press, Oxford–Boca Raton, 92–110.
Willats W.G.T., Limberg G., Buchholt H.C., van Alebeek G.-J., Benen J., Christensen T.M.I.E., Visser J., Voragen A., Mikkelsen J.D. & Knox J.P. 2000: Analysis of pectic epitopes recognised by hybridoma and

phage display monoclonal antibodies using defined oligosaccharides, polysaccharides and enzymatic degradation. Carbohydrate Research, 327 (3), 309–320.
Willats W.G.T., Marcus S.E. & Knox J.P. 1998: Generation of a monoclonal antibody specific to (1→5)-α-L-
-arabinan. Carbohydrate Research, 308 (1–2), 149–152.

Green Scenarios..., Dyczko, Jagodziński & Woźniak (eds.) © 2022 Taylor & Francis Group, London, UK

3.2. Soil Enzymes – a Tool to Monitor Soil-forming Processes in Coal Mine Spoil Heaps

Wojciech Bierza
Institute of Biology, Biotechnology and Environmental Protection,
Faculty of Natural Sciences, University of Silesia in Katowice;
28 Jagiellońska Str., 40-032 Katowice, Poland;
wojciech.bierza@us.edu.pl; ORCID: 0000-0002-1789-5512

ABSTRACT: Coal mine spoils, which are deposited above-ground in the form of huge heaps, change aesthetic values and ecological functions of the landscape. This material consists mainly of carboniferous minerals of gravel consistency, and minor quantities of low-grade coal. Deposit of spoil material causes massive disturbance of ecosystems, which are often completely destroyed. The spoil material excavated from great depths typically has very low biological activity. Soil formation and re-establishment of soil biological functions are basic preconditions of recovery of ecologically sensitive ecosystems on spoil heaps, which can be monitored using soil enzymes as indicators of the condition of the soil environment. While the processes of soil formation in reclaimed post-mining sites have been extensively studied, data concerning the development of soils expressed by increases in soil enzymatic activity under natural spontaneous succession, are comparatively rare.

KEYWORDS: Soil enzymatic activity, soil forming processes, plant-soil interactions, soil health indicators

SOIL ENZYMES. THE STATE OF THE ART

Enzymes are organic catalysts produced by an organism's cells. They are both low molecular and high molecular weight proteins produced in living organisms, which catalyse the processes of metabolic reactions. Enzymes are key to understanding soil biochemical processes and the role of soil in the global carbon cycle. Enzymes are active both in the cells of living soil organisms and as independent extracellular proteins, actively secreted into the soil by plant roots, bacteria and fungi.

Soil enzymes play a key role in the organic matter cycle in soil ecosystems. The activity of these enzymes is inseparably related to the soil organic matter content, soil physicochemical properties, and the activity and biomass of soil microorganisms. Soil enzymes participate in catalysing many important reactions necessary in the life processes of soil microorganisms, which help to stabilize the soil structure, take part in the decomposition of organic residues, the formation of organic matter and the circulation of nutrients. The analysis of the activity of soil enzymes, due to their connection with biological processes in soil, provides information on the specific metabolic activities and functions of the soil microorganism community. Soil enzyme activity responds rapidly to stress factors, e.g. heavy metals, drought or salinity, and the magnitude of changes in enzymatic activity is related to the intensity of these factors. Therefore, the assessment of soil enzyme activity is considered to be a good bioindicator, reflecting natural and anthropogenic soil disturbances (Hinojosa *et al.* 2004; Khan *et al.* 2007). The activity of soil enzymes reflects the physiologically active biomass of soil microorganisms and is a sensitive indicator of changes in microbial viability under the influence of biotic and abiotic factors (Kieliszewska-Rokicka 2001; Koper *et al.* 2004; Lasota 2006).

The main types of soil enzymes that are used as soil condition indicators are presented in Table 1. The activity of these soil enzymes can be used to evaluate the speed of important soil processes, soil productivity, the influence of vegetation on soil microbiological activity, and inhibiting impacts of stress factors such as drought, salinity, and heavy metal contamination.

Table 1. The most important soil enzymes and their functions (Rao *et al.* (2017))

ENZYME	SOURCE	REACTION CATALYZED	END PRODUCT	SOIL FUNCTION INDICATED	FACTORS INFLUENCING ENZYME ACTIVITY
α-amylase	Plants, animals and microorganisms	Starch hydrolysis	Glucose and/or oligosaccharides	C-cycling	Type of vegetation, environment and soil types
β-amylase	Mainly plants	Starch hydrolysis	Maltose		
Dehydrogenase	Microorganisms	Oxidation of organic compounds	Transfer of H to NAD or NADP (electron transport system)	C-cycling, microbial oxidative activity	Soil water content, temperature, pesticides, trace elements, pollution, etc.
Endo-1, 4-β--glucanase	Microorganism, protozoa and termites	Cellulose endohydrolysis	Oligosaccharides	C-cycling	Temperature, pH, water, O_2 contents, quality and location of organic matter, mineral elements and fungicides
Exo-1, 4-β--glucanase		Cellulose cleavage at ends	Glucose and cellobios		
β-glucosidase		Cellobiose hydrolysis	Glucose (sugar)		
Phenol oxidase	Plants and microorganisms	Lignin hydrolysis	C compounds (humic substances)	C-cycling	Soil pH, mean annual precipitation and temperature, SOM content, N enrichment etc.
Urease	Microorganisms, plants, and some invertebrates	Urea hydrolysis	Ammonia (NH_3) and CO_2	N-cycling	Organic matter content, soil depth, heavy metals, temperature, pH, etc.
Alkaline phosphatase	Mainly bacteria	Hydrolysis of esters and anhydrides of phosphoric acid	Phosphate (PO_4)	P-cycling	Organic matter content, pH, pollution
Acid phosphatase	Plants, fungi and bacteria				
Arylsulfatase	Microorganisms, plants, and animals	Hydrolysis of sulfate esters	Sulfate (SO_4^{2-})	S-cycling	Heavy metal pollution, pH, organic matter content and composition and availability of organic sulfate esters
Protease	Microorganisms and plants	N mineralization	Plant available N	N-cycling	Humic acid concentration, availability of C and N, etc.
Chitinase	Plants and microorganisms	Degradation and hydrolysis of chitin	Carbohydrates and inorganic nitrogen	C- and N-cycling	Availability of N, soil depth, atmospheric CO_2 levels, etc.

NAD: nicotinamide adenine dinucleotide; NADP: nicotinamide adenine dinucleotide phosphate

The location of enzymes in the soil and their considerable range of individual properties make the microbial community able to significantly improve the ability to decompose very diverse exogenous substrates (Wallenstein & Burns 2011). Soil enzymes can be divided into four categories according to their location:

1. Enzymes associated with living, metabolically active cells of microorganisms present in the soil. They occur in cell cytoplasm, as cell wall-bound or as extracellular enzymes, recently released into the environment.
2. Enzymes associated with non-proliferating cells (e.g. spores).
3. Enzymes associated with dead cells or their debris.
4. Enzymes that are "permanently" immobilised on soil clay or on humic colloids. This group of enzymes can remain active for a long time.

The detailed classification of soil enzymes according to their location in soil is presented in Figure 1.

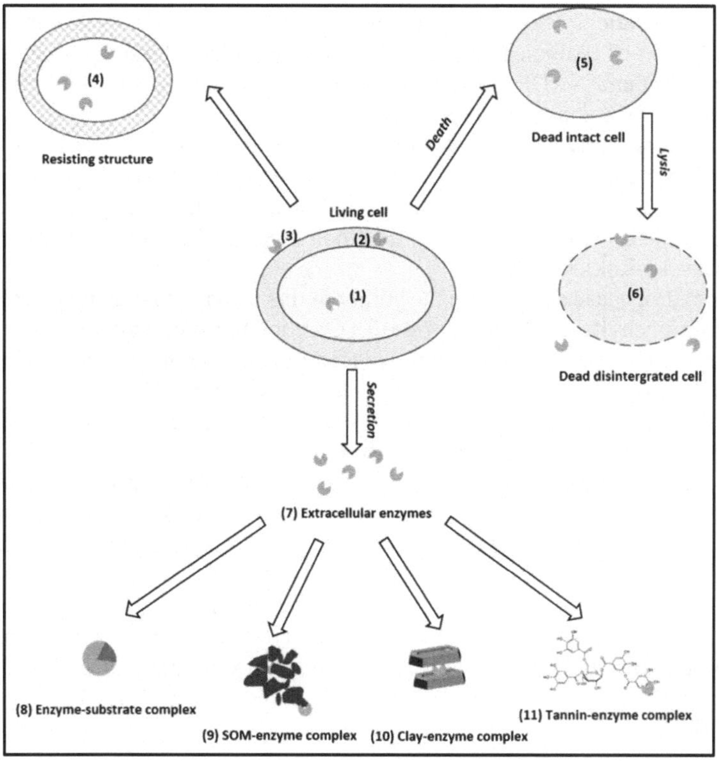

Figure 1. Locations of enzymes in soil (Burns *et al.* 2013): (1) functioning within the cytoplasm of microbial cells; (2) restricted to the periplasmatic space of Gram-negative bacteria; (3) located on the outer surface of cells with their active sites extending into the soil environment, contained within polysomes or retained by biofilms; (4) situated within resting cells including fungal spores, protozoal cysts and bacterial endospores; (5) attached to entire dead cells and cell debris; (6) leaking from intact cells or released from lysed cells; (7) free in soil water; (8) associated with enzyme substrate complexes; (9) complexed with soil organic matter by absorption, entrapment or co-polymerization; (10) sorbed to the external and internal surfaces of clay minerals; (11) bound to condensed tannins

Most commonly studied soil enzymes

The most commonly investigated soil enzymes include dehydrogenase, urease and phosphatase. This is because these enzymes determine the general condition of soil microorganisms, and thus the rate of carbon transformation (dehydrogenase), as well as the rate of mineralization of the most important nutrients like N (urease) and P (phosphatases).

Dehydrogenase [EC 1.1]

Dehydrogenase belongs to the group of oxidoreductases, enzymes located in cytoplasm or specific structures formed from cytoplasmic membranes and catalyses the oxidation of organic compounds by breaking off two hydrogen atoms (two protons and two electrons) from the organic molecule and transferring them to the final acceptor (Moeskops *et al.* 2010; Zhao *et al.* 2010; Yuan & Yue 2012). Under aerobic conditions, this transport takes place through a series of respiratory chain conveyors on O_2. Under anaerobic conditions, the function of the final acceptor is performed by available, oxidized inorganic forms NO^{3-}, Mn (IV), Fe (III), SO_4^{2-}, CO_2 (anaerobic breathing) or organic compounds (fermentation) (Alef & Nannipieri 1995; Brzezińska & Włodarczyk 2005). Regardless of the state of soil oxygenation, dehydrogenases are an element of respiratory metabolism, closely related to the production of biological energy in the form of ATP. Through coenzymes of dehydrogenases (mainly NAD and NADP), protons separated from oxidized substrates also participate in anabolic processes (Subhani *et al.* 2001).

Total dehydrogenase activity is an indicator of the redox system and a measure of the respiratory activity of soil microorganisms (Gu *et al.* 2009; Salazar *et al.* 2011). The substrates in the reactions catalysed by dehydrogenases are usually simple hydrocarbons. Dehydrogenases, unlike other soil enzymes, are active only within living cells, and their death is followed by their degradation very quickly (Kieliszewska-Rokicka 2001).

Studies of dehydrogenase activity use the ability of this enzyme to transfer electrons to synthetic acceptors such as triphenyltetrazolium chloride (TTC) or iodonitrotetrazolium chloride (INT), which in oxidized form are almost colourless and in reduced form are a coloured compound – tetrazolium formate (TTF) (Wolińska & Stępniewska 2012). The activity of dehydrogenases is commonly used to evaluate factors adversely affecting soil microorganisms (Gu *et al.* 2009; Salazar *et al.* 2011). Close correlation between dehydrogenase activity and organic matter content, soil fertility, number and biomass of soil microorganisms, C_{org}, N_{org}, proteolytic activity, nitrification, denitrification, respiration and activity of other enzymes present in soil, including acid and alkaline phosphatase, sulfatase, β-glucosidase, catalase, amylase, saccharase was observed (Brzezińska & Włodarczyk 2005; Błońska *et al.* 2016). The usefulness of the measurement of dehydrogenase activity in soil as an ecotoxicological test was confirmed by numerous studies on the changes in biological activity of soil contaminated with heavy metals, pesticides, oil derivatives and polycyclic aromatic hydrocarbons (Kieliszewska-Rokicka 2001; Brzezińska & Włodarczyk 2005).

Phosphatases [EC 3.1.3]

The term "phosphatase" is used to refer to a wide range of enzymes that catalyses the hydrolysis of esters of orthophosphoric acid (Condron *et al.* 2005). Phosphatases are enzymes that play an important role in soil as they stimulate the transformation of organic phosphorus compounds into inorganic phosphates (HPO_4^{2-} and H_2PO^{4-}), which are directly available to plants and soil organisms (Bielińska 2005). Lack of phosphorus in soil stimulates plant roots and microorganisms to increase the secretion of phosphatase to intensify the solubilization and remobilization of phosphate, thereby influencing the ability of the plant to cope with phosphorus-stressed conditions (Kai *et al.* 2002). The division of phosphatases is based on the number of ester bonds of the respective substrates and they can be divided into: hydralases of phosphate monoesters (phosphomonoesterases), hydralases of phosphate diesters (phosphodiesterases) and hydralases of phosphate triesters (phosphotriesterases) (Margesin & Schinner 1994; Makoi & Ndakidemi 2008). In the soil environment, the following most frequently occur: phosphomonoesterases (e.g. phytases, glycerophosphatases, nucleases), phosphodiesterases, phosphotriesterases, polyphosphatases (e.g. ATP-ases, pyrophosphatases) and P-N hydrolyses (e.g. phosphatidases; Bielińska 2005).

Phosphomonoesterases are the most studied among the phosphatases present in the soil (Adetunji *et al.* 2017). Phosphomonoesterases were divided by Hoffmann (1968) into three types depending on the optimum pH. According to this division, we distinguish between acidic, neutral and alkaline phosphatases. An acidic reaction (pH 4–6) is optimal for acidic phosphonoesterase, commonly known as acid phosphatase, and an alkaline reaction (pH 9–11) for alkaline phosphonoesterase, commonly known as alkaline phosphatase (Dodor & Tabatabai 2003). Neutral phosphatases with optimum activity at pH 6.5–7 can also be distinguished in the soil (Bielińska 2005).

The main sources of phosphatases in the soil environment are mainly soil microorganisms, as well as plant roots and soil fauna (Adetunji *et al.* 2017). The activity of phosphatases in the soil environment reflects the activity of enzymes associated with soil colloids and humus substances, free phosphatases in soil solution and phosphatases associated with living and dead cells of plants and microorganisms (Bielińska 2005). Phosphatases can be a good indicator of the potential of organic phosphorus mineralisation and soil biological activity (Piotrowska-Długosz & Charzyński 2015; Adetunji *et al.* 2017).

The activity of phosphatases starts to play an important role after the initial decomposition of soil organic matter, which is catalysed simultaneously by multicomponent enzyme systems produced by different groups of microorganisms (Bielińska 2005).

Urease [EC 3.5.1.5]

Urease is an enzyme from the group of deaminases, catalysing hydrolysis of urea to ammonia and carbon dioxide (Das & Varma 2010). Urease also catalyses the hydrolysis of hydroxyurea, dihydroxyurea and semicarbazide with nickel as a co-factor (Alef & Nannipieri 1995). Ureases are widely distributed in nature and originate from bacteria, yeasts, fungi, algae, animal wastes and plants (Follmer 2008). Studies on soil urease activity showed that this enzyme can be used as a good soil quality indicator, because of the role of urease in the regulation of N supply to plants after fertilization with urea (Piotrowska-Długosz & Charzyński 2015). Urease activity in soils can be influenced by many factors. These include organic matter content of the soil, soil pH, soil amendments, heavy metals and other environmental factors such as temperature (Nannipieri 1995; Yang *et al.* 2006; Alef & Machuca *et al.* 2015). The highest urease activity was recorded under soil pH 5.8 while the lowest activity was at pH 4.2 (Błońska & Lasota 2014). The reduction of soil moisture by 10% and 21% led to reductions of 10–67% and 42–62% in urease activity, respectively, which explains the link between drought and slower nutrient turnover (Sardans & Peñuelas 2005).

VEGETATION IMPACTS ON ACTIVITIES OF SOIL ENZYMES

Vegetation can directly or indirectly change soil properties and maintain soil fertility due to the close relationship between plants and microorganisms (Liu *et al.* 2020). Studies of Kompała-Bąba *et al.* (2021) on hard coal spoil heaps have partially confirmed this statement since they detected a significantly higher enzyme activity in plots with vegetation in comparison to plots that had no vegetation cover.

Many research results indicate that the composition of soil microorganisms can fundamentally modify microbiological processes, regardless of environmental factors such as water availability or temperature. Therefore, scientists are increasingly interested in better understanding the factors determining the composition and activities of soil microorganisms (Zogg *et al.* 1997; Balser & Firestone 2005). Previous studies indicated that the composition of soil microorganisms depends on such factors that can be regulated by vegetation such as soil pH (Fierer & Jackson 2006; Högberg *et al.* 2007), phenolic compounds (Hättenschwiler & Vitousek 2000) or nutrients and carbon availability (Fierer *et al.* 2007; Högberg *et al.* 2007). Moreover, many authors indicate that the composition of microorganisms in soil may vary depending on the species of plants growing on it, the diversity of plant cover and the length of vegetation period (Grayston & Campbell 1996; Westover *et al.* 1997; Priha *et al.* 1999; Waldrop *et al.* 2000; Grayston *et al.* 2001; Porazinska *et al.* 2003; Bartelt-Ryser *et al.* 2005; Liang *et al.* 2007; Kao-Kniffin & Balser 2007).

In view of the above, differences in microorganism communities may also result from differences in the number of fine roots and their metabolic activity specific for a given plant species (Coleman *et al.* 2000), as well as in the number and chemical composition of the litter produced. Under the influence of the dominance of different plant species, the functional structure of soil microorganism communities also changes, which is reflected in the activity of enzymes involved in the mineralisation of nutrients (Waldrop *et al.* 2000).

Litter influence on soil enzymatic activity

Biomass and type of litter determine the composition of mineral compounds in soil. Although the amount and composition of litter produced by a particular plant species may vary, there are general differences in the structure and ability to decompose litter from one species to another. Reich *et al.* (2005) found that different plant species produce different amounts of litter with different nitrogen, calcium and lignin contents. These authors postulated that the quantity and quality of litter are reflected in significant differences in soil pH, fertility and the activity of soil enzymes produced by microorganisms. High lignin content in the litter reduces the rate of its decomposition by microorganisms and the ratio of lignin to nitrogen is used as a measure of litter quality (Finzi & Canham 1998). In addition to the lignin content, the chemical structure of litter determines its ability to decompose easily by microorganisms (Berg 1986). For example, besides lignin and waxes, a large number of other polyphenolic compounds make conifer needles more difficult to decompose, while leaves of angiosperms often contain more easily degradable and water-soluble compounds such as sugars, amino-acids and aliphatic acids (Johansson 1995; Harris & Safford 1996). The difficulty for microorganisms to decompose conifer needles is due to the fact that phenolic compounds inhibit the activity of enzymes that take part in litter decomposition and/or precipitate nutrient proteins (Hättenschwiler & Vitousek 2000).

Influence of root exudates on soil enzymatic activity

Root exudates are one of the key factors increasing the number and activity of microorganisms in the rhizosphere compared to bare soil (Shi *et al.* 2011). Low molecular carbon compounds present in root exudates such as sugars, organic acids and amino acids are easily assimilable by microorganisms and play a significant role in regulating the dynamics of soil microbial communities (Bais *et al.* 2006; Weisskopf *et al.* 2008). Root exudates also affect the level of soil contamination by stimulating bacterial growth, which results in a more effective distribution of environmental pollutants (Kuiper *et al.* 2004). Root activity depends on the plant species. Smith (1976) showed that the composition of the root exudates of yellow birch (*Betula alleghaniensis*), American beech (*Fagus grandifolia*) and sugar maple (*Acer saccharum*) differs significantly. American beech secreted the largest amounts of amino-acids and organic acids, while yellow birch released large amounts of sugars into the soil. Sandnes *et al.* (2005) showed that the roots of the silver birch (*Betula pendula*) tree secreted more diverse and more organic acids than the Norway spruce (*Picea abies*).

CASE STUDY: THE INFLUENCE OF DOMINANT PLANT SPECIES ON THE ENZYMATIC ACTIVITY OF SPOIL HEAP SOIL SUBSTRATUM BASED ON BŁOŃSKA *ET AL.* (2019)

The aim of this study was to compare the activity of chosen soil enzymes in the vegetation patches dominated by two grass species (*Calamagrostis epigejos* and *Poa compressa*) and two herbaceous plants (*Daucus carota* and *Tussilago farfara*). For this purpose three heaps "Sośnica" Heap (Gliwice) (50°16'22"N, 18°44'43"E), "Wesoła" Heap (Mysłowice) 50°10'28"N, 19°5'44"E) and "Kostuchna" Heap (Katowice) (50°11'4"N, 19°0'33"E) were chosen (Fig. 2). Selected heaps were of the same age classes (over 30 years), on which herbaceous vegetation appeared spontaneously (Woźniak 2010). In total, 75 research plots were established, with dominance of the four above-mentioned plant species and plots without vegetation (control). Soil samples were taken from each plot at depths between 0 and 10 cm. Samples were sieved through a 2 mm mesh sieve and stored at 4°C until being analysed. Four soil enzymes, namely: dehydrogenase, acid and alkaline phosphatases and urease were investigated.

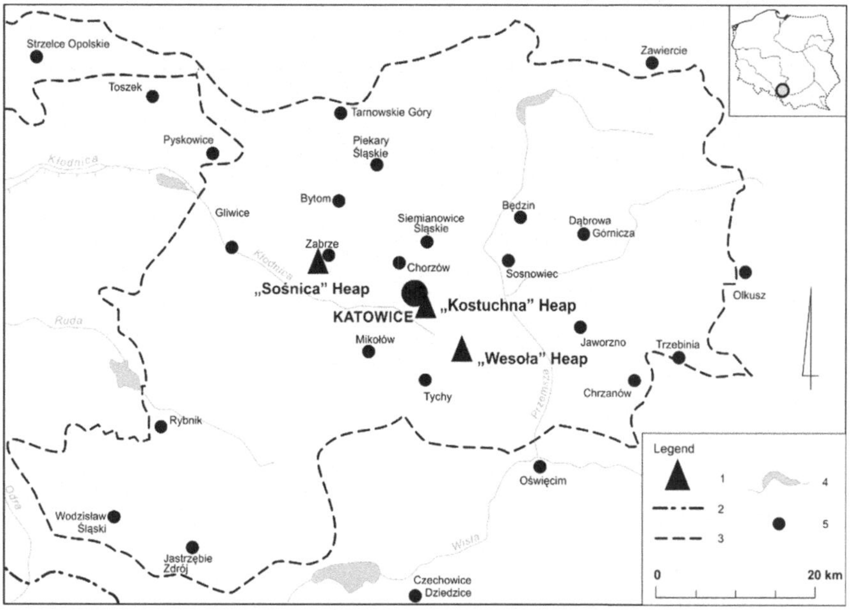

Figure 2. Location of investigated coal mine spoil heaps (marked in colour). Explanations: 1 – investigated heaps, 2 – state border, 3 – area of Silesian Upland macroregion, 4 – water reservoirs, 5 – cities

Dehydrogenase [EC 1.1] activity was determined by reduction of 2, 3, 5-triphenyltetrazolium chloride (TTC) to triphenylformazan (TPF) by the method developed by Schinner *et al.* (1996). Briefly, 2.5 mL of 0.1 M TRIS-HCl buffer and 2.5 mL of 1% aqueous solution of TTC were added to 2.5 g of fresh soil samples. All the samples were incubated in the dark at 37°C for 24 h. After that, 12.5 mL of acetone was added to all the samples, shaken for 2 h in darkness and subsequently filtered. The assay of acid phosphatase [EC 3.1.3.2] and alkaline phosphatase [EC 3.1.3.1] activity was determined by measuring the p-nitrophenol (PNP) released by phosphatase activity when soil is incubated with buffered (pH 6.0 for acid phosphatase and pH 11 for alkaline phosphatase) 115 mM sodium p-nitrophenyl phosphate solution at 37°C for 1 h (Schinner *et al.* 1996). The urease [EC 3.5.1.5] activity in the soil was determined by a modification of the method proposed by Hoffmann and Teicher (1961). Briefly, 5 mL of a 10% urea solution and 10 mL of citrate buffer (pH 6.7) were added to 5 g of fresh soil, and then incubated for 3 h at 37°C. The ammonium content was determined by a modified indophenol-blue reaction. The activity of dehydrogenases was expressed in μg TPF g^{-1} d.m. h^{-1}, urease in μg $N\text{-}NH_4$ g^{-1} d.m. h^{-1} and acid phosphatase and alkaline phosphatase in [μg PNP g^{-1} d.m. h^{-1}]. All enzymes were determined with a DR 5000 HACH-Lange spectrophotometer. The activity of dehydrogenases was measured at the wavelength of $\lambda = 546$ nm, urease at $\lambda = 580$ nm and that of acid and alkaline phosphatase at $\lambda = 410$ nm. The differences between investigated soil enzyme activity were tested using ANOVA with Welsh correction, and wherever significant differences were identified, a *post hoc* Tukey test was performed.

The soils from the control plots and those covered by the studied plant species differed significantly in terms of dehydrogenase ($F = 7.358$; $p = 0.000$) (Fig. 3), as well as the acid ($F = 21.770$; $p = 0.000$) (Fig. 4) and alkaline ($F = 35.831$; $p = 0.000$) (Fig. 5) phosphatase activity. Only in the case of urease ($F = 3.293$; $p = 0.150$) (Fig. 6), were no statistically significant differences in the activity found. Higher activity for alkaline and acid phosphatases was recorded for the patches dominated by grasses, compared to patches dominated by herbaceous species. A similar situation was observed for the dehydrogenase activity.

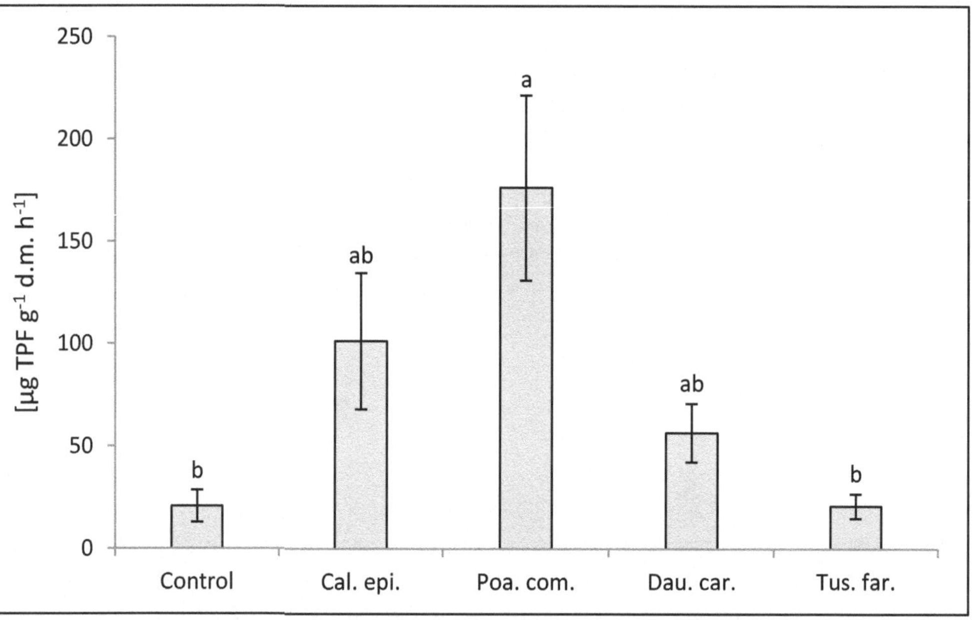

Figure 3. Comparison of the dehydrogenase activity in substrate samples of plots dominated by *Calamagrostis epigejos* (Cal. epi.), *Poa compressa* (Poa. com.), *Daucus carota* (Dau. car.) and *Tussilago farfara* (Tus. far.) with substrate control samples. Results marked with the same letter do not differ statistically significantly ($p<0.05$); the activity of dehydrogenases was expressed in μg triphenylformazan (TPF) g^{-1} d.m. h^{-1}; d.m. – dry mass of soil

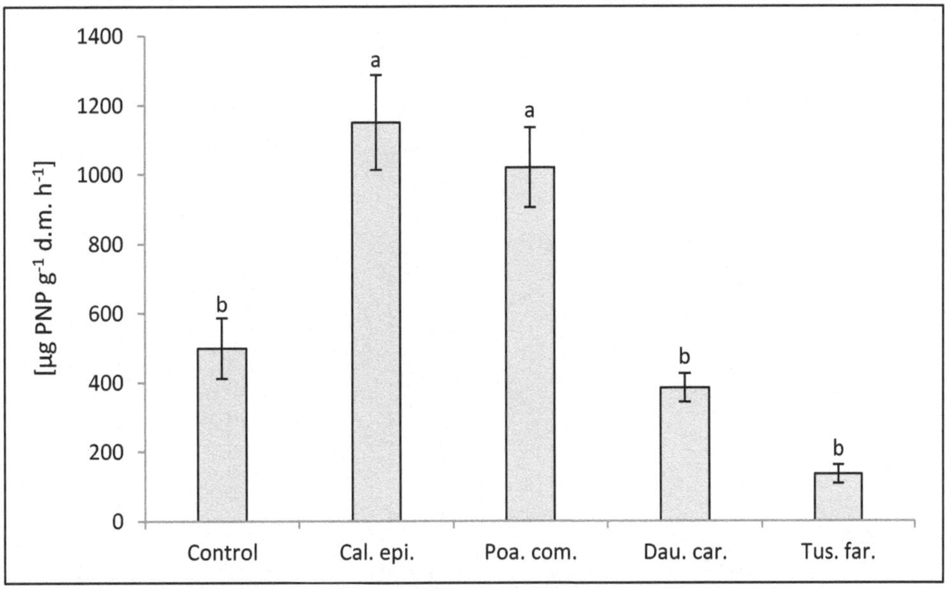

Figure 4. Comparison of the acid phosphatase activity in substrate samples of plots dominated by *Calamagrostis epigejos* (Cal. epi.), *Poa compressa* (Poa. com.), *Daucus carota* (Dau. car.) and *Tussilago farfara* (Tus. far.) with substrate control samples. Results marked with the same letter do not differ statistically significantly ($p<0.05$); the activity of acid phosphatase was expressed in µg p-nitrophenol (PNP) g^{-1} d.m. h^{-1}; d.m. – dry mass of soil

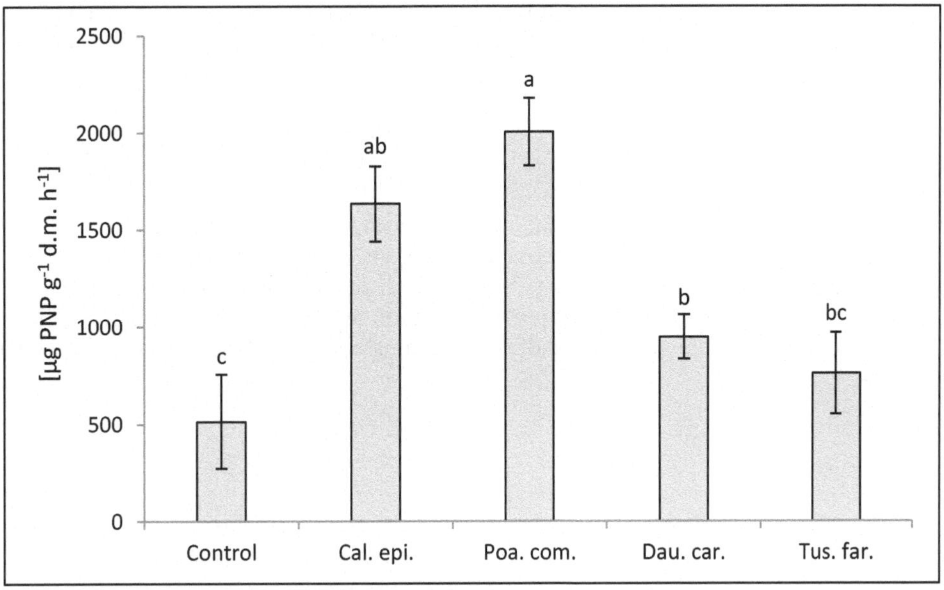

Figure 5. Comparison of the alkaline phosphatase activity in substrate samples of plots dominated by *Calamagrostis epigejos* (Cal. epi.), *Poa compressa* (Poa. com.), *Daucus carota* (Dau. car.) and *Tussilago farfara* (Tus. far.) with substrate control samples. Results marked with the same letter do not differ statistically significantly ($p<0.05$); the activity of alkaline phosphatase was expressed in µg p-nitrophenol (PNP) g^{-1} d.m. h^{-1}; d.m. – dry mass of soil

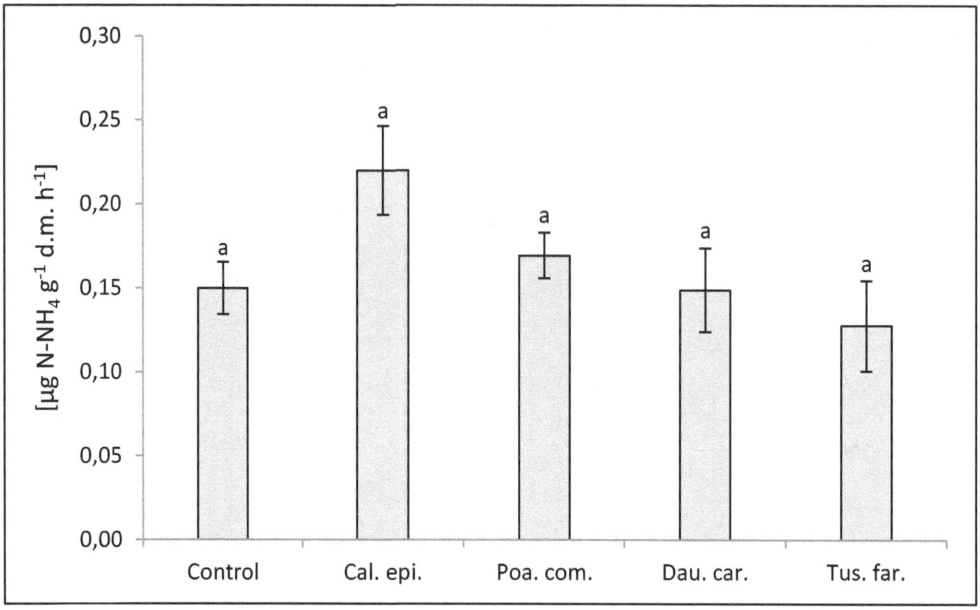

Figure 6. Comparison of the urease activity in substrate samples of plots dominated by *Calamagrostis epigejos* (Cal. epi.), *Poa compressa* (Poa. com.), *Daucus carota* (Dau. car.) and *Tussilago farfara* (Tus. far.) with substrate control samples. Results marked with the same letter do not differ statistically significantly ($p<0.05$); the activity of urease was expressed in µg $N\text{-}NH_4$ g^{-1} d.m. h^{-1}; d.m. – dry mass of soil

Higher soil enzyme activity can be connected to the stimulating effect of the extensive root system of grasses, and thus a large amount of root exudates stimulating the development and activity of microorganisms. Lower enzyme activity on the plots of *Daucus carota* and *Tussilago farfara* may also be associated with lower abundance (biomass) of these species, in comparison to grasses.

CONCLUSIONS

Soil enzymes can be used as a powerful and relatively cheap indicator of changes in soil substrate microbiological activity under the influence of changing vegetation during spontaneous succession on coal mine spoil heaps. Understanding the way in which spontaneous vegetation and activities of soil enzymes are developed together will provide a tool to assess the environmental conditions of coal mine spoil heaps and the possibilities of strengthening reclamation processes based on natural mechanisms.

REFERENCES

Adetunji A.T., Lewu F.B., Mulidzi R. & Ncube B. 2017: The biological activities of β-glucosidase, phosphatase and urease as soil quality indicators: a review. Journal of Soil Science and Plant Nutrition, 17 (3), 794–807.

Alef K. & Nannipieri P. 1995: Methods in applied soil microbiology and biochemistry. Academic Press, London.

Bais H.P., Weir T.L., Perry L.G., Gilroy S. & Vivanco J.M. 2006: The role of root exudates in rhizosphere interactions with plants and other organisms. Annual Review of Plant Biology, 57, 233–266.

Balser T.C. & Firestone M.K. 2005: Linking microbial community composition and soil processes in a California annual grassland and mixed-conifer forest. Biogeochemistry, 73, 395–415.

Bartelt-Ryser J., Joshi J., Schmid B., Brandl H. & Balser T. 2005: Soil feedbacks of plant diversity on soil microbial communities and subsequent plant growth. Perspectives in Plant Ecology, Evolution and Systematics, 7 (1), 27–49.

Berg B. 1986: Nutrient release from litter and humus in coniferous forest soils – a mini review. Scandinavian Journal of Forest Research, 1 (1–4), 359–369.

Bielińska E.J. 2005: Methods of determination of photosphatase activity. Acta Agrophysica, 3, 63–74.

Błońska A., Kompała-Bąba A., Sierka E., Besenyei L., Magurno F., Frydecka K., Bierza W. & Woźniak G. 2019: Impact of selected plant species on enzymatic activity of soil substratum on post-mining heaps. Journal of Ecological Engineering, 20 (1), 138–144.

Błońska E. & Lasota J. 2014: Biological and biochemical properties in evaluation of forest soil quality. Folia Forestalia Polonica, 56 (1), 23–29.

Błońska E., Lasota J. & Gruba P. 2016: Effect of temperate forest tree species on soil dehydrogenase and urease activities in relation to other properties of soil derived from loess and glaciofluvial sand. Ecological Research, 31 (5), 655–664.

Brzezińska M. & Włodarczyk T. 2005: Enzymy wewnątrzkomórkowych przemian redoks (oksydoreduktazy). Acta Agrophysica, 3, 11–26.

Burns R.G., DeForest J.L., Marxsen J., Sinsabaugh R.L., Stromberger M.E., Wallenstein M.D., Weintraub M.N. & Zoppini A. 2013: Soil enzymes in a changing environment: Current knowledge and future directions. Soil Biology and Biochemistry, 58, 216–234.

Coleman M.D., Dickson R.E. & Isebrands J.G. 2000: Contrasting fine-root production, survival and soil CO_2 efflux in pine and poplar plantations. Plant and Soil, 225 (1/2), 129–139.

Condron L.M., Turner B.L., Cade-Menun B.J., Sims J. & Sharpley A. 2005: Chemistry and dynamics of soil organic phosphorus. [In:] Sims J.T., Sharpley A.N. (eds.): Phosphorus: Agriculture and the Environment. Agronomy Monographs, 46, 87–121.

Das S.K. & Varma A. 2010: Role of enzymes in maintaining soil health. [In:] Shukla G. & Varma A. (eds.): Soil Enzymology. Soil Biology 22. Springer-Verlag, Berlin–Heidelberg, 25–42.

Dodor D.E. & Tabatabai M.A. 2003: Effect of cropping systems on phosphatases in soils. Journal of Plant Nutrition and Soil Science, 166 (1), 7–13.

Fierer N. & Jackson R.B. 2006: The diversity and biogeography of soil bacterial communities. Proceedings of the National Academy of Sciences of the United States of America, 103 (3), 626–631.

Fierer N., Bradford M.A. & Jackson R.B. 2007: Toward an ecological classification of soil bacteria. Ecology, 88 (6), 1354–1364.

Finzi A.C. & Canham C.D. 1998: Non-additive effects of litter mixtures on net N mineralization in a southern New England forest. Forest Ecology and Management, 105 (1–3), 129–136.

Follmer C. 2008: Insights into the role and structure of plant ureases. Phytochemistry, 69 (1), 18–28.

Grayston S.J. & Campbell C.D. 1996: Functional biodiversity of microbial communities in the rhizospheres of hybrid larch (*Larix eurolepis*) and Sitka spruce (*Picea sitchensis*). Tree Physiology, 16 (11–12), 1031–1038.

Grayston S.J., Griffith G.S., Mawdsley J.L., Campbell C.D. & Bardgett R.D. 2001: Accounting for variability in soil microbial communities of temperate upland grassland ecosystems. Soil Biology and Biochemistry, 33 (4–5), 533–551.

Gu Y., Wag P. & Kong C.H. 2009: Urease, invertase, dehydrogenase and polyphenoloxidase activities in paddy soils influenced by allelopathic rice variety. European Journal of Soil Biology, 45 (5–6), 436–441.

Harris M.M. & Safford L.O. 1996: Effects of season and four tree species on soluble carbon content in fresh and decomposing litter of temperate forests. Soil Science, 161 (2), 130–135.

Hättenschwiler S. & Vitousek P.M. 2000: The role of polyphenols in terrestrial ecosystem nutrient cycling. Trends in Ecology & Evolution, 15 (6), 238–243.

Hinojosa M.B., Carreira J.A. & García-Ruíz R. & Dick R.P. 2004: Soil moisture pre-treatment effects on enzyme activities as indicators of heavy metal-contaminated soils. Soil Biology and Biochemistry, 36 (10), 1559–1568.

Hoffmann G. 1968: Phosphatases in the enzyme system of cultivated soils (in Germany) and possibilities of determining their activity. Zeitschrift für Pflanzenernährung, Düngung, Bodenkunde, 118, 153–160.

Hoffmann G.G. & Teicher K. 1961: Ein kolorimetrisches Verfahren zur Bestimmung der Ureaseaktivität in Böden. Zeitschrift für Pflanzenernährung, Düngung, Bodenkunde, 95 (1), 55–63.

Högberg M.N., Högberg P. & Myrold D.D. 2007: Is microbial community composition in boreal forest soils determined by pH, C-to-N ratio, the trees or all three? Oecologia, 150, 590–601.

Johansson M.-B. 1995: The chemical composition of needle and leaf litter from Scots pine, Norway spruce and white birch in Scandinavian forests. Forestry, 68 (1), 49–62.
Kai M., Takazumi K., Adachi H., Wasaki J., Shinano T. & Osaki M. 2002: Cloning and characterization of four phosphate transporter cDNAs in tobacco. Plant Science, 163 (4), 837–846.
Kao-Kniffin J. & Balser T.C. 2007: Elevated CO_2 differentially alters belowground plant and soil microbial community structure in reed canary grass-invaded experimental wetlands. Soil Biology and Biochemistry, 39 (2), 517–525.
Khan S., Cao Q., Hesham A.E.-L., Xia Y. & He J. 2007: Soil enzymatic activities and microbial community structure with different application rates of Cd and Pb. Journal of Environmental Sciences, 19 (7), 834–840.
Kieliszewska-Rokicka B. 2001: Enzymy glebowe i ich znaczenie w badaniach aktywności mikrobiologicznej gleby. [In:] Dahm H. & Pokojska-Burdziej A. (eds.): Drobnoustroje środowiska glebowego: aspekty fizjologiczne, biochemiczne, genetyczne. Wydawnictwo Adam Marszałek, Toruń, 37–47.
Kompała-Bąba A., Bierza W., Sierka E., Błońska A., Besenyei L. & Woźniak G. 2021: The role of plants and soil properties in the enzyme activities of substrates on hard coal mine spoil heaps. Scientific Reports, 11, 5155.
Koper J., Lemanowicz J. & Igras J. 2004: The effect of fertilization on the phosphatase activity and the content of some phosphorus forms. Annales Universitatis Mariae Curie-Sklodowska, Sectio E, Agricultura (Poland), 59 (2), 679–686.
Kuiper I., Lagendijk E.L., Bloemberg G.V. & Lugtenberg B.J.J. 2004: Rhizoremediation: a beneficial plant–microbe interaction. Molecular Plant-Microbe Interactions, 17 (1), 6–15.
Lasota J. 2006: Enzymatic activity of soils in forest sites of the Western Beskid Mountains. Polish Journal of Soil Science, 39 (1), 43–53.
Liang C., Fujinuma R., Wie L. & Balser T.C. 2007: Tree species-specific effects on soil microbial residues in an upper Michigan old-growth forest system. Forestry, 80 (1), 65–72.
Liu Y., Zhu G., Hai X., Li J., Shangguan Z., Peng C. & Deng L. 2020: Long-term forest succession improves plant diversity and soil quality but not significantly increase soil microbial diversity: Evidence from the Loess Plateau. Ecological Engineering, 142, 105631.
Machuca A., Cuba-Díaz M. & Córdova C. 2015: Enzymes in the rhizosphere of plants growing in the vicinity of the Polish Arctowski Antarctic Station. Journal of Soil Science and Plant Nutrition, 15 (4), 833–838.
Makoi J.H.J.R. & Ndakidemi P.A. 2008: Selected soil enzymes: Examples of their potential roles in the ecosystem. African Journal of Biotechnology, 7 (3), 181–191.
Margesin R. & Schinner F. 1994: Phosphomonoesterase, phosphodiesterase, phosphotriesterase, and inorganic pyrophosphatase activities in forest soils in an alpine area: effect of pH on enzyme activity and extractability. Biology and Fertility of Soils, 18 (4), 320–326.
Moeskops B., Sukristiyonubowo, Buchan D., Sleutel S., Herawaty L., Husen E., Saraswati R., Setyorini D. & De Neve S. 2010: Soil microbial communities and activities under intensive organic and conventional vegetable farming in West Java, Indonesia. Applied Soil Ecology, 45 (2), 112–120.
Piotrowska-Długosz A. & Charzyński P. 2015: The impact of the soil sealing degree on microbial biomass, enzymatic activity, and physicochemical properties in the Ekranic Technosols of Toruń (Poland). Journal of Soils and Sediments, 15 (1), 47–59.
Porazinska D.L., Bardgett R.D., Blaauw M.B., Hunt W.H., Parsons A.N., Seastedt T.R. & Wall D.H. 2003: Relationships at the aboveground–belowground interface: plants, soil biota, and soil processes. Ecological Monographs, 73 (3), 377–395.
Priha O., Grayston S.J., Pennanen T. & Smolander A. 1999: Microbial activities related to C and N cycling and microbial community structure in the rhizospheres of *Pinus sylvestris*, *Picea abies* and *Betula pendula* seedlings in an organic and mineral soil. FEMS Microbiology Ecology, 30 (2), 187–199.
Rao S.C., Grover M., Kundu S. & Desai S. 2017: Soil enzymes. [In:] Lal R. (ed.): Encyclopedia of Soil Science. CRC Press, London, 2100–2107.
Reich P.B., Oleksyn J., Modrzynski J., Mrozinski P., Hobbie S.E., Eissenstat D.M., Chorover J., Chadwick O.A., Hale C.M. & Tjoelker M.G. 2005: Linking litter calcium, earthworms and soil properties: a common garden test with 14 tree species. Ecology Letters, 8 (8), 811–818.
Salazar S., Sánchez L.E., Alvarez J., Valverde A., Galindo P., Igual J.M., Peix A. & Santa-Regina I. 2011: Correlation among soil enzyme activities under different forest system management practices. Ecological Engineering, 37 (8), 1123–1131.

Sandnes A., Eldhuset T.D. & Wollebæk G. 2005: Organic acids in root exudates and soil solution of Norway spruce and silver birch. Soil Biology & Biochemistry, 37 (2), 259–269.
Sardans J. & Peñuelas J. 2005: Drought decreases soil enzyme activity in a Mediterranean *Quercus ilex* L. forest. Soil Biology & Biochemistry, 37 (3), 455–461.
Schinner F., Öhlinger R., Kandeler E. & Margesin R. 1996: Methods in soil biology. Springer-Verlag, Berlin.
Shi S., Richardson A.E., O'Callaghan M., DeAngelis K.M., Jones E.E., Stewart A., Firestone M.K. & Condron L.M. 2011: Effects of selected root exudate components on soil bacterial communities. FEMS Microbiology Ecology, 77 (3), 600–610.
Smith W.H. 1976: Character and significance of forest tree root exudates. Ecology, 57 (2), 324–331.
Subhani A., Changyong H., Zhengmiao X., Min L. & El-ghamry A.M. 2001: Impact of soil environment and agronomic practices on microbial/dehydrogenase enzyme activity in soil. A review. Pakistan Journal of Biological Sciences, 4 (3), 333–338.
Waldrop M.P., Balser T.C. & Firestone M.K. 2000: Linking microbial community composition to function in a tropical soil. Soil Biology and Biochemistry, 32 (13), 1837–1846.
Wallenstein M.D. & Burns R.G. 2011: Ecology of extracellular enzyme activities and organic matter degradation in soil: a complex community-driven process. Methods of Soil Enzymology, 9, 35–55.
Weisskopf L., Le Bayon R.-C., Kohler F., Page V., Jossi M., Gobat J.-M., Martinoia E. & Aragno M. 2008: Spatio-temporal dynamics of bacterial communities associated with two plant species differing in organic acid secretion: A one-year microcosm study on lupin and wheat. Soil Biology & Biochemistry, 40 (7), 1772–1780.
Westover K.M., Kennedy A.C. & Kelley S.E. 1997: Patterns of rhizosphere microbial community structure associated with co-occurring plant species. Journal of Ecology, 85, 863–873.
Wolińska A. & Stępniewska Z. 2012: Dehydrogenase activity in the soil environment. [In:] Canuto R.A. (ed.): Dehydrogenases. IntechOpen, Rijeka, 183–210.
Woźniak G. 2010: Zróżnicowanie roślinności na zwałach pogórniczych Górnego Śląska [Diversity of vegetation on coal-mine heaps of the Upper Silesia (Poland)]. W. Szafer Institute of Botany, Polish Academy of Sciences, Kraków.
Yang Z.-X., Liu S.-Q., Zheng D.-W. & Feng S.-D. 2006: Effects of cadmium, zinc and lead on soil enzyme activities. Journal of Environmental Sciences, 18 (6), 1135–1141.
Yuan B.-C. & Yue D.-X. 2012: Soil microbial and enzymatic activities across a chronosequence of Chinese pine plantation development on the Loess Plateau of China. Pedosphere, 22 (1), 1–12.
Zhao B., Chen J., Zhang J. & Qin S. 2010: Soil microbial biomass and activity response to repeated drying--rewetting cycles along a soil fertility gradient modified by long-term fertilization management practices. Geoderma, 160 (2), 218–224.
Zogg G.P., Zak D.R., Ringelberg D.B., White D.C., MacDonald N.W. & Pregitzer K.S. 1997: Compositional and functional shifts in microbial communities due to soil warming. Soil Science Society of America Journal, 61 (2), 475–481.

3.3. Functional Diversity of Bacteria as an Important Element of Post-mining Ecosystem Functioning – the Use of the BIOLOG® Method in Environmental Research

Monika Malicka
Institute of Biology, Biotechnology and Environmental Protection, Faculty of Natural Sciences, University of Silesia in Katowice; 28 Jagiellońska Str., 40-032 Katowice, Poland; monika.malicka@us.edu.pl; ORCID: 0000-0002-1034-0995

Monika Szalbot
Institute of Plant Protection – National Research Institute, Sośnicowice Branch; 29 Gliwicka Str., 44-153 Sośnicowice, Poland; M.Szalbot@iorpib.poznan.pl; ORCID: 0000-0001-9180-0828

Wojciech Bierza
Institute of Biology, Biotechnology and Environmental Protection, Faculty of Natural Sciences, University of Silesia in Katowice; 28 Jagiellońska Str., 40-032 Katowice, Poland; wojciech.bierza@us.edu.pl; ORCID: 0000-0002-1789-5512

ABSTRACT: Many studies on the development and succession of plant communities along habitat gradients focus only on abiotic factors; however, the structural and functional interactions between plants and soil microorganisms are equally important. The species composition of plants has a strong effect on the metabolic activity of soil microbial communities and the production of soil organic matter. Spontaneous vegetation occurs on coal mine spoil heaps that provide an opportunity to study *de novo* ecosystem development and the establishment of relationships between plants and rhizosphere microorganisms and soil formation. The functional diversity of microorganisms might easily be analysed using the Biolog™ method based on the assessment of physiological activity of microorganisms. The physiological profiles of microbial communities are determined by their ability and intensity of degradation of various carbon substrates. Soil physicochemical parameters and plant associations shape these profiles. Their analysis enables the prediction of changes in plant and microbial growth and biochemical processes occurring in the soil environment. The Biolog™ method is a cheap, quick, and sensitive method of comparative analysis of microbial communities from different habitats, which enables conclusions about qualitative and quantitative changes in the soil more efficiently than using plant-based systems.

KEYWORDS: Community level physiological profiles, soil microbial communities, soil organic matter, rhizosphere, vegetation

Many studies on the ecology of plant communities are focused on explaining the occurrence of spontaneous development of plant species and patches of vegetation along habitat gradients. Most of them consider only the analysis of abiotic factors. Over time, however, more attention has been paid to structural and functional interactions between plants and soil microorganisms. Changes in the structure, biomass, and metabolic activity of microbial communities are reliable indicators of changes in the whole ecosystem. That reliability lies in the fact that microorganisms react quickly to changes in their environment because of their high surface-to-volume ratio and low homeostasis (Boivin *et al.* 2002).

The relationship between plants and microorganisms has strong structural and functional relationships in natural (Faliński 2003; Urbanová *et al.* 2011), semi-natural, and reclaimed areas (Liu *et al.* 2008; Zhang *et al.* 2011). It has been postulated that the links between vegetation and soil microbial communities are determined by the development of vegetation during succession and by the activity and abundance of soil microorganisms (Ehrenfeld *et al.* 2005; Schaaf *et al.* 2011).

The species composition of plants has a strong effect on soil microbial communities (Grayston *et al.* 2001). Some studies have revealed that particular plant species support soil-forming processes in post-mining sites (Frouz *et al.* 2009; Chodak & Niklińska 2010). It has also been found that different plant species are involved in the production of different types of organic matter, depending on the source of carbon (litter or roots) or the amount of exudates and plant matter that individual tree species add to soil, thus affecting the pH or contents of macronutrients (Rahmonov 2009; Urbanová *et al.* 2011).

Coal mine spoil heaps are sites where the spoil material extracted together with the coal is stored. The spoil material does not contain organic carbon, and the amounts of nitrogen and phosphorus are low (Ledin & Pedersen 1996). Spontaneous colonization by living organisms on post-industrial sites represents a model situation of *de novo* ecosystem development. It provides an opportunity to study the structure of spontaneously established vegetation and interactions between plants and rhizosphere microorganisms in successional processes over a long time. Increases in plant diversity during vegetation development have been recorded in many studies (Whittaker 1965; Matthews 1979; Pineda *et al.* 1981), but the results of these studies have not consistently explained how plant diversity changes during spontaneous succession.

Post-mining areas are a good model for studying the mechanisms of soil organic matter (SOM) accumulation, which is an essential component of soil formation. The accumulation of SOM affects a broad spectrum of soil characteristics, such as structure, water retention, and sorption of nutrients (Lavelle & Spain 2003). SOM is also of great practical importance, because in post-mining areas where the extracted material has not been covered by a soil layer, the accumulation of SOM has a significant impact on the success of remedial actions (Frouz *et al.* 2007).

In addition to plants, which are the main source of SOM, the soil microbiota also plays an important role in the accumulation of SOM. Microorganisms are mainly responsible for the decomposition and chemical transformation of SOM. Soil fauna has a negligible impact on the direct mineralization of SOM; however, it significantly affects the distribution of SOM in the soil by fragmenting the litter and mixing it with the mineral soil fraction, which facilitates the degradative activity of soil microorganisms (Frouz *et al.* 2007).

In recent years attention has been paid to study the relationship between structural and functional diversity of rhizosphere microorganisms and plant communities. The analyses of phospholipid fatty acids (PLFA) (Dickens & Anderson 1999; Enami *et al.* 2001) and denaturing/temperature gradient gel electrophoresis (DGGE/TGGE) (Smalla *et al.* 1998; Pennanen *et al.* 2001; Yrjälä *et al.* 2004) are examples of biochemical and molecular methods used to study the structural diversity of microbial communities. The functional diversity of microorganisms can be easily studied using the Biolog™ method based on the assessment of the physiological activity of microorganisms, which enables sensitive detection of changes in qualitative and quantitative soil microbial parameters (Perucci 1992; Islam *et al.* 2011). The Biolog EcoPlate™ test is used in ecological studies to approximate the metabolic potential of microbial communities and create community level physiological profiles (CLPP) of rhizosphere bacteria. The CLPP profile is determined by the ability and intensity of degradation of various carbon substrates by microbial communities (Insam 1997). Based on the

CLPP profile, changes in microbial growth and biochemical processes occurring in the soil environment can be understood as the effects of particular physicochemical parameters that are shaped by environmental factors and plant associations (Chaerun *et al.* 2011; Islam *et al.* 2011). Garland & Mills (1991), who used the Biolog™ method for the first time, have demonstrated that this method is appropriate to characterize the differences between habitats and between samples in the same habitat. The authors showed that the separation of samples according to the main components was directly related to differences in the use of particular carbon sources.

Biolog EcoPlates™ are 96-well plates containing 31 different carbon substrates representing compounds commonly found in the soil environment (Fig. 1; Stefanowicz 2006). These substrates are classified into five biochemical groups. The largest group is composed of 10 different carbohydrates, including mannitol, xylose and cellobiose. The other groups include carboxylic acids, ketones, amino acids and polymers, including glycogen, α-cyclodextrin, Tween 40 and Tween 80. The least numerous group of substrates are amines represented by phenylethylamine and putrescine (Furtak 2017). To get the correct result the plate system allows the analyst to perform a triple check for each carbon source. Negative controls are three wells containing water (Tab. 1).

Each well with the medium contains tetrazolium violet as an indicator dye. Dye reduction to a coloured product in a given well indicates substrate degradation by microorganisms present in a sample. Depending on microbial species composition, the substrates on a plate are metabolized at variable rates, which is indicated by the intensity of colour development in a well (Rutgers *et al.* 2016). Bacteria are the only microorganisms which contribute to the colour reaction because the dye used in the plate wells is not metabolized by fungi (Preston-Mafham *et al.* 2002).

Analysis of the parameters of functional microbial diversity using the Biolog™-CLPP method (Martínez-Iñigo *et al.* 2009) allows one to perform quick and cheap comparative analyses of microbial communities. Obtained CLPP profiles are the basis for calculation of biodiversity indices and statistical multi-factorial analyses of the studied microbial communities (Haack *et al.* 1995). The method has been widely used to characterize microbial communities of different habitats, from natural, native soil environments to organic soil, composts, and sewage sludge (Doran *et al.* 1996; Sharma *et al.* 1998).

Monitoring and understanding the causes of changes in soil microbial communities and their activity may be the key to explanation of basic issues of ecology, especially functional ecology. Rapid changes in the quantity and quality of microbial indicators in soils, enables much easier studies of interactions between communities of organisms than can be done in the plant system (Garland 1997). Soil biological parameters, including functional biodiversity of microorganisms are highly sensitive indicators of ecological stress that can be studied using convenient and accurate methods, like Biolog™ (Dick & Tabatabai 1993; Gil-Sotres *et al.* 2005).

Table 1. Carbon sources in Biolog EcoPlate™ wells. The codification matrix and groups of carbon sources are specified (the real configuration is in triplicate of this table)

BIOLOG ECOPLATE	1	2	3	4
A	Water (blank)	β-Methyl-D-Glucoside	D-Galactonic Acid γ-Lactone	L-Arginine
B	Pyruvic Acid Methyl Ester	D-Xylose	D-Galacturonic Acid	L-Asparagine
C	Tween 40	i-Erythritol	2-Hydroxy-Benzoic Acid	L-Phenylalanine
D	Tween 80	D-Mannitol	2-Hydroxy-Benzoic Acid	L-Serine
E	α-Cyclodextrin	N-Acetyl-D-Glucosamine	γ-Hydroxy-Butyric Acid	L-Threonine
F	Glycogen	D-Glucosaminic Acid	Itaconic Acid	Glycyl-L-Glutamic Acid
G	D-Cellobiose	Glucose-1-Phosphate	α-Keto-Butyric Acid	Phenylethylamine
H	α-D-Lactose	D,L-α-Glycerol-Phosphate	D-Malic Acid	Putrescine

Amines
Carbohydrates
Polymers
Carboxylic acids
Amino acids
Phosphate-carbon

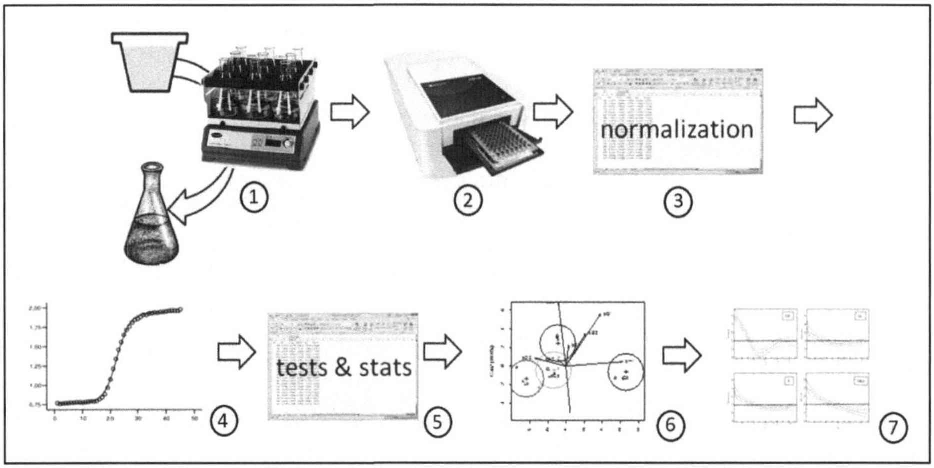

Figure 1. Diagram showing experimental design and data processing: (1) Soil samples collected from an environment are suspended in saline solution by orbital shaker; (2) A suspension containing microorganisms is inoculated into a Biolog EcoPlate™; (3) The results of colour development after 72 h incubation are calculated and normalized; (4) Results are fitted to a sigmoidal model; (5) Parameters are calculated (slope, time to reach 50% of the maximum colour intensity); (6) Multivariate analysis; (7) Calculation of diversity indices and estimation of changes in the diversity of substrate utilisation

REFERENCES

Boivin M.-E.Y., Breure A.M., Posthuma L. & Rutgers M. 2002: Determination of field effects of contaminants – significance of pollution-induced community tolerance. Human and Ecological Risk Assessment, 8 (5), 10–35.

Chaerun S.K., Pangesti N.P.D., Toyota K. & Whitman W.B. 2011: Changes in microbial functional diversity and activity in paddy soils irrigated with industrial wastewaters in Bandung, West Java Province, Indonesia. Water, Air, & Soil Pollution, 217, 491–502.

Chodak M. & Niklińska M. 2010: Effect of texture and tree species on microbial properties of mine soils. Applied Soil Ecology, 46 (2), 268–275.

Dick W.A. & Tabatabai W.A. 1993: Significance and potential use of soil enzymes. [In:] Metting F.B. Jr. (ed.): Soil Microbial Ecology: Application in Agricultural and Environmental Management. Marcel Dekker Inc., New York, 95–130.

Dickens H.E. & Anderson J.M. 1999: Manipulation of soil microbial community structure in bog and forest soils using chloroform fumigation. Soil Biology and Biochemistry, 31 (14), 2049–2058.

Doran J.W., Sarrantonio M. & Liebig M.A. 1996: Soil health and sustainability. Advances in Agronomy, 56, 1–54.

Ehrenfeld J.G., Ravit B. & Elgersma K. 2005: Feedback in the plant-soil system. Annual Review of Environment and Resources, 30, 75–115.

Enami Y., Okano S., Yada H. & Nakamura Y. 2001: Influence of earthworm activity and rice straw application on the soil microbial community structure analyzed by PLFA pattern. European Journal of Soil Biology, 37 (4), 269–272.

Faliński J.B. 2003: Long-term studies on vegetation dynamics: some notes on concepts, fundamentals and conditions. Community Ecology, 4 (1), 107–113.

Frouz J., Pižl V. & Tajovský K. 2007: The effect of earthworms and other saprophagous macrofauna on soil microstructure in reclaimed and un-reclaimed post-mining sites in Central Europe. European Journal of Soil Biology, 43 (Supplement 1), S184–S189.

Frouz J., Pižl V., Cienciala E. & Kalčík J. 2009: Carbon storage in post-mining forest soil, the role of tree biomass and soil bioturbation. Biogeochemistry, 94, 111–121.

Furtak K. 2017: Analiza profilu metabolicznego populacji mikroorganizmów z zastosowaniem techniki ECO-plate Biolog®. [In:] Panfil M. (ed.): Badania i rozwój młodych naukowców w Polsce. Nauki Przyrodnicze, Część I. Wydawnictwo Młodzi Naukowcy, Poznań, 31–37.

Garland J.L. & Mills A.L. 1991: Classification and characterization of heterotrophic microbial communities on the basis or patterns of community-level sole-carbon-source utilization. Applied and Environmental Microbiology, 57 (8), 2351–2359.

Garland J.L. 1997: Analysis and interpretation of community-level physiological profiles in microbial ecology. FEMS Microbiology Ecology, 24 (4), 289–300.

Gil-Sotres F., Trasar-Cepeda C., Leirós M.C. & Seoane S. 2005: Different approaches to evaluating soil quality using biochemical properties. Soil Biology and Biochemistry, 37 (5), 877–887.

Grayston S.J., Griffith G.S., Mawdsley J.L., Campbell C.D. & Bardgett R.D. 2001: Accounting for variability in soil microbial communities of temperate upland grassland ecosystems. Soil Biology and Biochemistry, 33 (4–5), 533–551.

Haack S.K., Garchow H., Klug M.J. & Forney L.J. 1995: Analysis of factors affecting the accuracy, reproducibility and interpretation of microbial community carbon source utilization patterns. Applied and Environmental Microbiology, 61 (4), 1458–1468.

Insam H. 1997: A new set of substrates proposed for community characterization in environmental samples. [In:] Insam H. & Rangger A. (eds.): Microbial Communities. Functional Versus Structural Approaches. Springer, Berlin– Heidelberg, 259–260.

Islam M.R., Chauhan P.S., Kim Y., Kim M. & Sa T. 2011: Community level functional diversity and enzyme activities in paddy soils under different long-term fertilizer management practices. Biology and Fertility of Soils, 47, 599–604.

Lavelle P. & Spain A.V. 2003: Soil Ecology. Kluwer Academic Publishers, New York–Boston–Dordrecht–London–Moscow.

Ledin M. & Pedersen K. 1996: The environmental impact of mine wastes – Roles of microorganisms and their significance in treatment of mine wastes. Earth-Science Reviews, 41 (1–2), 67–108.

Liu Z., Liu G., Fu B. & Zheng X. 2008: Relationship between plant species diversity and soil microbial functional diversity along a longitudinal gradient in temperate grasslands of Hulunbeir, Inner Mongolia, China. Ecological Research, 23, 511–518.
Martínez-Iñigo M.J., Pérez-Sanz A., Ortiz I., Alonso J., Alarcón R., García P. & Lobo M.C. 2009: Bulk soil and rhizosphere bacterial community PCR–DGGE profiles and β-galactosidase activity as indicators of biological quality in soils contaminated by heavy metals and cultivated with *Silene vulgaris* (Moench) Garcke. Chemosphere, 75 (10), 1376–1381.
Matthews J.A. 1979: A study of the variability of some successional and climax plant assemblage-types using multiple discriminant analysis. Journal of Ecology, 67 (1), 225–271.
Pennanen T., Paavolainen L. & Hantula J. 2001: Rapid PCR-based method for the direct analysis of fungal communities in complex environmental samples. Soil Biology and Biochemistry, 33 (4–5), 697–699.
Perucci P. 1992: Enzyme activity and microbial biomass in a field soil amended with municipal refuse. Biology and Fertility of Soils, 14, 54–60.
Pineda F.D., Nicolas J.P., Pou A. & Galiano E.F. 1981: Ecological succession in oligotrophic pastures of central Spain. Vegetatio, 44 (3), 165–176.
Preston-Mafham J., Boddy L. & Randerson P.F. 2002: Analysis of microbial community functional diversity using sole-carbon-source utilisation profiles – a critique. FEMS Microbiology Ecology, 42 (1), 1–14.
Rahmonov O. 2009: The chemical composition of plant litter of black locust (*Robinia pseudoacacia* L.) and its ecological role in sandy ecosystems. Acta Ecologica Sinica, 29 (4), 237–243.
Rutgers M., Wouterse M., Drost S.M., Breure A.M., Mulder C., Stone D., Creamer R.E., Winding A. & Bloem J. 2016: Monitoring soil bacteria with community-level physiological profiles using Biolog™ ECO-plates in the Netherlands and Europe. Applied Soil Ecology, 97, 23–35.
Schaaf W., Bens O., Fischer A., Gerke H.H., Gerwin W., Grünewald U., Holländer H.M., Kögel-Knabner I., Mutz M., Schloter M., Schulin R., Veste M., Winter S. & Hüttl R.F. 2011: Patterns and processes of initial terrestrial-ecosystem development. Journal of Plant Nutrition and Soil Science, 174 (2), 229–239.
Sharma S., Rangger A. & Insam H. 1998: Effects of decomposing maize litter on community level physiological profiles of soil bacteria. Microbial Ecology, 35 (3), 301–310.
Smalla K., Wachtendorf U., Heuer H., Liu W.-T. & Forney L. 1998: Analysis of BIOLOG GN substrate utilization patterns by microbial communities. Applied and Environmental Microbiology, 64 (4), 1220–1225.
Stefanowicz A. 2006: The Biolog plates technique as a tool in ecological studies of microbial communities. Polish Journal of Environmental Studies, 15 (5), 669–676.
Urbanová M., Kopecký J., Valášková V., Ságová-Marečková M., Elhottová D., Kyselková M., Moënne-Loccoz Y. & Baldrian P. 2011: Development of bacterial community during spontaneous succession on spoil heaps after brown coal mining. FEMS Microbiology Ecology, 78 (1), 59–69.
Whittaker R.H. 1965: Dominance and diversity in land plant communities: Numerical relations of species express the importance of competition in community function and evolution. Science, 147 (3655), 250–260.
Yrjälä K., Katainen R., Jurgens G., Saarela U., Saano A., Romantschuk M. & Fritze H. 2004: Wood ash fertilization alters the forest humus *Archaea* community. Soil Biology and Biochemistry, 36 (1), 199–201.
Zhang C., Liu G., Xue S. & Song Z. 2011: Rhizosphere soil microbial activity under different vegetation types on the Loess Plateau, China. Geoderma, 161 (3–4), 115–125.

3.4. Novel Ecosystems Establishment in Environmental Management with Geoinformatics Tools: Remote Sensing and Machine Learning Application

Agata Lisiecka
WMB.EXPERT Sp. z o.o.;
137/31 Pełczyńskiego Str., 01-437 Warszawa, Poland;
agata.lisiecka@wmb.expert; ORCID: 0000-0002-1279-4691

Remigiusz Gałka
WMB.EXPERT Sp. z o.o.;
137/31 Pełczyńskiego Str., 01-437 Warszawa, Poland;
remigiusz.galka@wmb.expert; ORCID: 0000-0001-7688-1113

ABSTRACT: This section defines and establishes geoinformatic tools used for investigating the potential of Novel Ecosystems, which directly affect the efficiency of urban services and the quality of life in urbanized and industrialized areas. Human activities influence the Earth's surface and result in the formation of areas with unique biological properties and environmental conditions. In recent years, geoinformatics techniques, such as remote sensing and machine learning, have undergone significant development.

KEYWORDS: Ecology, vegetation indices, ecosystem services, teledetection, LiDAR, GIS

INTRODUCTION

Human beings influence the Earth's ecosystems in remarkable ways. The high concentration of population in urban areas, the increase in the urban population, and thus the development of the economy's industrial sector, contribute to fundamental ecological changes. As a result of human activities, the land is transformed on a large scale and with resources consumed very fast. The relationships between natural and anthropogenic phenomena and processes as well as their impact on biodiversity are very complex. Changes in ecosystems at the level of bio-physical processes are manifested in the reduction of biodiversity, change of species habitats, disruption of energy flow, change in soil characteristics and many others (McKinney 2002; Alberti *et al.* 2003; Milgrom 2008).

Despite serious modifications in the ecosystem, under certain conditions in some environments, and resulting from natural processes, changes in species composition and structure of biocoenoses occur. Spontaneous succession on post-industrial sites enables the development of organisms better adapted to the presence of new environmental conditions and enhances the formation of stable novel ecosystems. The species composition of flora and fauna of such habitats differ significantly from the known original natural, semi-natural and synanthropic vegetation ecosystems (Kowarik 2011).

So called Novel Ecosystems (NE) comprise new combinations of species that appeared as a result of human activities, environmental changes, habitat fragmentation, biodiversity loss and the deliberate or inadvertent introduction of species from other parts of the world (Hobbs *et al.* 2006).

The growing demand for ecosystem services in communities related to highly industrialized regions makes research on novel ecosystems more and more significant, as their abilities and importance for human well-being in crowded urban areas are underestimated.

NOVEL ECOSYSTEMS

For a better understanding of the kind of challenges humans face nowadays, we need to understand exactly what Novel Ecosystems mean for the future ecosystem mosaic and broader scale environmental functioning. The definition evolves, but Morse *et al.* (2014) give the most up-to-date version. Morse's proposed definition exposes novel ecosystems as a unique combination of biological and environmental conditions. These conditions arose directly following intentional or unintentional human activities and they differ from the previously known original conditions. Moreover, attempts to restore systems to their original state are highly unlikely. Characteristics of novel ecosystems are defined as self-sustaining in terms of species composition, structure, functioning, biochemistry and biogeochemical cycles, and therefore also their ecosystem services. So novel ecosystems can be easily characterized by two indicators: novelty as in species groups not seen before and resulting from human intervention, but independent of it to be properly maintained (Morse *et al.* 2014).

CLASSIFICATION

Ecosystems that appear in areas that have been transformed as a result of human activity usually have vegetation with specific species composition, making it very difficult to classify, as they have specific plant communities that do not fit into the conceptual framework developed from the accepted phytosociological methods (Box 1999). Phytosociological classification takes into account the division of vegetation into natural, semi-natural and anthropogenic vegetation types, according to the characteristic combination of species in a given habitat type (Matuszkiewicz 2001). Due to the disadvantages and inconsistencies associated with the traditional phytosociological approach, and the difficulty in classifying habitats arising in post-industrial areas, the latest research aims to develop a new functional approach that would allow for classification of vegetation based on the study of the species traits and composition (Botta-Dukát 2005; Morse *et al.* 2014). Identification of vegetation types based on species composition and/or their functional features is also possible by using geoinformatics techniques.

GEOINFORMATICS

Given the pace of changes taking place in the development of novel ecosystems, as well as the speed of creation of these new ecosystems, the issues of speed and low costs of obtaining information concerning their environments is becoming extremely important. Geoinformatics techniques have proven to be a very useful tool. This science combines other various branches of science, including physical, biological and social sciences. Methods such as satellite Remote Sensing, Global Positioning Systems, Geographical Information Systems, and information systems, together with ground-based research, form the basis of the field called geoinformatics (Murthy *et al.* 2003). The main purpose of geoinformatics is to support the process of collecting spatially oriented measurement data, their advanced analysis and final presentation of the results, including spatial visualizations.

So, what challenges are the scientists facing? First of all, species composition needs to be determined, to indicate chemical plant structures to predict the directions of natural succession or to model recommended changes for obtaining the desired ecosystem services. But the biggest challenge for Novel Ecosystems is that standard phytosociological methods cannot be used (Braun-Blanquet 1965).

Remote Sensing (RS), also called teledetection, is used to observe the surface of the Earth. It is used in broadly understood landscape research. Remote sensing allows to detect and monitor the physical features of an area. It enables the identification and measurement of numerous properties such as soil and water temperatures, biomass, crop and timber-stand vigour. It is also used for determining soil characteristics and soil moisture conditions (DeFries 2013; Ahlmer *et al.* 2018). Several remote sensing systems can be distinguished: aerial photographs, manned satellite images, Landsat, thermal infrared images, and radar images (Sabins 1987).

Obtaining vegetation composition and structure data by remote sensing is based on the properties of plants to absorb and reflect light waves of different lengths, which is caused by the presence of pigments, cell structures and water content (Mazur & Chojnacki 2017). Measuring the amount of reflected light in a given range allows to determine the spectral curve which illustrates the state of plant health (Mazur & Chojnacki 2017). Remote sensing enables us to calculate Vegetation Indices (VIs) which are algorithms that allow for quick quantitative and qualitative evaluation of many biotic environmental properties (Xue & Su 2017). Vegetation Indices play important roles in such processes as characterization of vegetation, canopy biomass, and absorbed radiation, and studies on drought, agricultural activities, climate and hydrology (Candiago *et al.* 2015). Figure 1 shows the changes in the spectral curve in individual wavelength ranges for different states of the plant and soil.

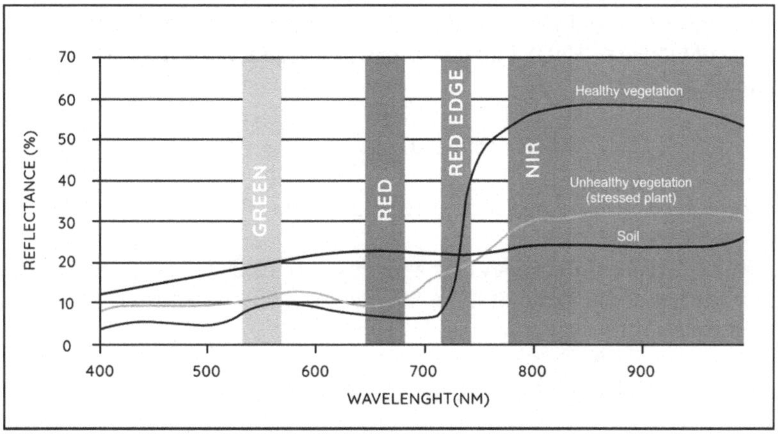

Figure 1. Green Vegetation Reflectance

Dr. Gregory P. Asner of the Carnegie Institution of Washington (Department of Global Ecology) has selected the most important vegetation categories, described and calculated the most characteristic indices from 150 Vegetation Indices that have been published in the literature (Asner 1998).

Vegetation Indices can be divided into a few categories: Basic Vegetation Indices, Vegetation Indices considering Atmospheric Effects, Adjusted-Soil Vegetation Index, and Tasseled Cap Transformation of Greenness Vegetation Index. Basic Vegetation Indices are Ratio Vegetation Index (RVI), Difference Vegetation Index (DVI), Perpendicular Vegetation Index (PVI), Normalized Difference Vegetation Index (NDVI). NDVI, GNDVI and SAVI are the most frequently used Vegetation Indices (Mazur & Chojnacki 2017; Xue & Su 2017).

Ratio Vegetation Index (RVI), also called the simple ratio vegetation index (SR), corresponds to the relationship between the amount of reflected near-infrared and red radiation. RVI ranges from

0 to 20, and this value will increase with increasing biomass (Wójtowicz *et al.* 2005). RVI is calculated using the following formula:

$$\mathrm{RVI} = (\mathrm{NIR}/\mathit{Red}),$$

where: NIR corresponds to the light reflectance values in the near infrared; *Red* means the light reflectance values within the red band (Mazur & Chojnacki 2017; Xue & Su 2017).

Normalized Difference Vegetation Index (NDVI) based on the multispectral data and characterize the photosynthetic activity of the plants. NDVI is calculated from reflectance measurements in the red and near infrared (NIR) part of the spectrum, the formula is:

$$\mathrm{NDVI} = (\mathrm{NIR} - \mathit{Red})/(\mathrm{NIR} + \mathit{Red}),$$

(Mazur & Chojnacki 2017; Xue & Su 2017).

NDVI index values varies from –1 (maximum reflection of RED, which means absence of vegetation) to 1 (highest level of vegetation). Reliable NDVI index determination is based on satellite research and is influenced by many non-vegetative factors such as atmospheric conditions, geometry and calibration of the satellite, soil backgrounds, biomass and crop canopy. Calculated NDVI index enables the determination of canopy height and vegetation vigour (Cabrera-Bosquet *et al.* 2011).

Green Normalized Difference Vegetation Index (GNDVI) is calculated similarly to NDVI except that the green band is used instead of the red one. GNDVI is used for determining the photosynthetic activity characterizing the water or nitrogen uptake into the canopy. The GNDVI index is sensitive to the concentration of chlorophyll. That is why it is linearly correlated with the leaf area index (LAI) and volume of biomass. GNDVI ranges from 0 to 1 (Candiago *et al.* 2015; Mazur & Chojnacki 2017). The formula is:

$$\mathrm{GNDVI} = (\mathrm{NIR} - \mathit{Green})/(\mathrm{NIR} + \mathit{Green}),$$

where: *Green* corresponds to the light reflectance values within the green band (Gitelson *et al.* 1996).

Soil-Adjusted Vegetation Index (SAVI) is an index corrected for soil brightness in areas where vegetative cover reaches a maximum of 40%, and it allows for correct assessment of the condition and performance the plants in sparse vegetation. The formula is:

$$\mathrm{SAVI} = (1 + \mathrm{L})(\mathrm{NIR} - \mathit{Red})/(\mathrm{NIR} + \mathit{Red} + \mathrm{L}),$$

where: L is the soil cover factor, which ranges from 0 to 1 (0 – total vegetation cover, then SAVI = = NDVI; 1 – no vegetation cover (Mazur & Chojnacki 2017).

Difference Vegetation Index (DVI) belongs to the group of the distance-based vegetation indices. Calculations of these indices assume that the coefficient "a" is the slope of the soil line. Thus, the DVI index takes into account the slope of the ground in the near infrared range (0 – bare soil, negative values – water, positive values – vegetation) (Mróz & Sobieraj 2004; Silleos *et al.* 2006). The formula is:

$$\mathrm{DVI} = a\mathrm{NIR} - \mathit{Red},$$

where: *a* is the slope of the soil line (Silleos *et al.* 2006).

Transformed Vegetation Index (TVI) allows to analyze NDVI histograms as a normal distribution, by excluding negative values. It cannot be calculated when NDVI is less than 0.5. The formula is:

$$\mathrm{TVI} = \sqrt{\mathrm{NDVI} + 0.5},$$

(Mróz & Sobieraj 2004).

The Corrected Transformed Vegetation Index (CTVI) includes a correction to the TVI to ultimately prevent negative values:

$$\mathrm{CTVI} = \frac{(\mathrm{NDVI}+0.5)}{|(\mathrm{NDVI}+0.5)|} * \sqrt{|(\mathrm{NDVI} + 0.5)|},$$

(Silleos *et al.* 2006).

Thiam's Transformed Vegetation Index (TTVI). In 1997, Thiam proposed another amendment to improve the quality of the CTVI image. He suggested removing the first term from the equation. The formula is:

$$\mathrm{TVI} = \sqrt{|(\mathrm{NDVI} + 0.5)|},$$

(Silleos *et al.* 2006).

Until recently, satellite imaging and aircraft-based methods of obtaining data about vegetation composition and structure were the only available methods. With the development of drone technology (UAV), it turned out to be an excellent and inexpensive way of obtaining images. Remote sensing performed by using satellites or aircraft consists of creating an image in a stationary or mobile position at a distance remote from the surface. Remote sensing allows measurement of some properties from a distance. It is one of the most important methods of obtaining vegetation data. This method enables the monitoring of Earth's surface in a feasible way. Drone availability solves the problem of obtaining specific data "on demand". The key criterion for the selection of the imaging method is the extent area to be examined (Ruwaimana *et al.* 2018). For small areas up to 5 ha, UAV drones seem to be ideal due to the very low costs, high resolution and the possibility of performing a flight mission at any time. The application of satellites facilitates obtaining free-of-charge images, because such data is available for the satellites like Landsat 7–8, Sentinel-1 and 2, CBERS-4, and MODIS. The main disadvantages are the low resolution of the images, the lack of choice of the time period when the images were obtained and dependence on the amount of cloud cover. The methods that employ airplane flights involve considerable costs and this option should be considered for larger areas.

The time when the image is taken also has a very significant impact on further data processing. The spectral response changes in particular periods of the phenological cycle. For example, for species separation/identification, the best months are April and July (Cole *et al.* 2014).

Another remote sensing method, in addition to multispectral passive scanning, is the use of active laser scans. LiDAR is an acronym for Airborne Laser Scanning (ALS). The main function of data processing with ALS for vegetated areas is to separate the impulses reflected from the tree layer from those reflected from the ground surface (Hyyppä *et al.* 2005; Wężyk *et al.* 2015). LiDAR remote sensing finds its application in measurement of the vertical forest structure. This method enables the determination of such features as canopy heights, and above-ground stand volume and biomass. High spatial resolution of the data from LiDAR are used for land cover classification, mapping of habitats and identification of many ecosystem services (Dubayah & Drake 2000; Lim *et al.* 2003).

Another IT concept is the Internet of Things (IOTs). By using small sensors, it can even be placed on the leaves, or measure plant growth on a micrometer scale. To respond to the need of monitoring microclimatic conditions in crops at the ground level, Khan & Hussain (2019) proposed a monitoring method based on using an ultra-lightweight flexible and transparent sensory platform. Such a platform would be placed on the leaf of a plant. Scientists have also developed a solution for the deployment system of multisensory platforms for remote, large-scale macroclimate investigations (Khan & Hussain 2019). By combining those methods (with multispectral LiDAR in particular) huge support can be given to the scientific basis of plant phenotyping and classification (Lin 2015; Budei *et al.* 2018) and further on, to the modelling of ecosystem services derived from the identification of ecosystem functioning processes such as the biomass amount, quality and dynamics.

The data obtained by using methods described above can be processed by means of computer algorithms based on machine learning. Machine learning (ML) is a field of artificial intelligence that uses examples, data, and experience which allows the computer to learn (Maheshwari *et al.* 2020). By using machine learning in geoinformatics, the calculation of vegetation indices can be extended, combined with other data like LiDAR, weather data, and calculate derived indicators to find correlations, as well as the use of time series for data prediction. Machine learning models also give powerful tools for preparing various simulations of possible ecosystem changes, which makes it possible to give advice to decision makers on how to fulfil and improve the needs for ecosystem services.

The development of the idea of Convolution Neural Networks (CNN) attempts to perform remote plant identification (Dangtongdee & Kurfess 2018; Wäldchen *et al.* 2018). CNN is a deep learning algorithm. The capabilities of self-learning in terms of the Convolution Neural Network enables effective analysis of the contextual signal of multiple neighbouring pixels. This allows to determine the essential features of the leaf and canopy traits required for the identification of vegetation communities or species (Kattenborn *et al.* 2019). Maggiori *et al.* (2017) used a Convolution Neural Network as a classification tool to create a framework for categorization of satellite imagery. Kattenborn *et al.* (2019) proposed a method of segmentation and mapping of vegetation and plant communities based on a Convolution Neural Network with 84% accuracy. They showed how remote sensing (high-resolution satellite imagers or unmanned aerial vehicles (UAV)) simplifies accessibility to high quality optical Earth observation data and enables revealing of the vegetation patterns (Kattenborn *et al.* 2019).

GEOGRAPHIC INFORMATION SYSTEM (GIS)

Geographic Information System (GIS) is a framework for gathering, organizing, analysing and visualisation of geo-referenced data. It is applicable to all sciences related to the Earth's surface. According to the National Aeronautics and Space Administration (NASA), a Geographic Information System comprises hardware, software and qualified people, and combines geographically referenced resource data from topography, demography and imaging. GIS finds application in mapping, mo-delling and computing statistics essential for studies of distribution of vegetation types. According to Dr. Marguerite Madden from the University of Georgia, remote sensing data (such as aerial photographs, multispectral and hyperspectral airborne scanner images, and satellite images) combined with a Geographic Information System and Global Positioning System, provides a tremendously powerful tool for generating vegetation maps and databases (Madden 2004).

ECOSYSTEM SERVICES

Ecosystem services are most often defined as the overall benefits that humans gain from the natural environment and healthy ecosystems. The definition formulated by scientists Burkhard and Maes in their publication from 2017 says that ecosystem services are the inputs of ecosystem structures and functions to human well-being (Costanza et al. 2014; Burkhard & Maes 2017). Ecosystem services should be used by scientists, policy makers, and managers to manage the novel ecosystems and create specific goals. Until now, researchers and managers have been arguing about the classification and meaning of novel ecosystems for society. In their publications Evers *et al.* (2018) emphasized the importance of developing an appropriate novel ecosystems management model that will regulate the issues of biodiversity and ecosystem services. The formation of novel ecosystems depends on decisions and actions taken by organizations, social values, and personal preferences (Simberloff & Vitule 2014; Evers *et al.* 2018). Novel ecosystems are estimated to cover about 40% of the non-ice-covered land and a similar percentage of the marine environment (Bridgewater *et al.* 2011). It is already predictable that novel ecosystems will eventually cover most of the Earth's surface. According to Bridgewater *et al.* (2011), adequate intervention in rapidly changing novel ecosystems is limited due to the fact that such ecosystems are not treated with due attention. They are not adequately taken into account in the global discussion of environmental management and sustainable development (Bridgewater *et al.* 2011). The management of these specific ecosystems, which in their present state differ from their original form, must take into consideration the new values and functions, and not limit its activities to traditional conservation and restoration measures (Hobbs *et al.* 2014).

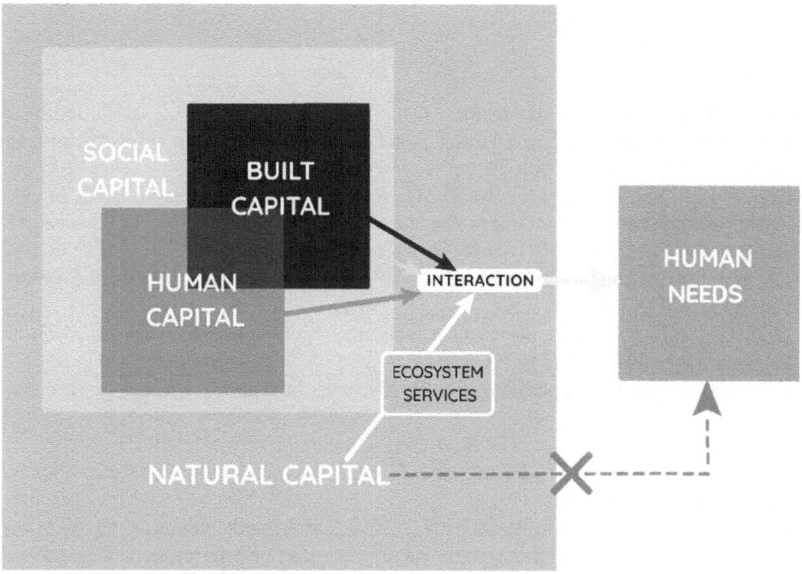

Figure 2. Interactions between built, social, human and natural capital required to provide ecosystem services necessary for human well-being (Costanza *et al.* (2014))

According to The Economics of Ecosystems and Biodiversity (TEEB), four main types of ecosystem services can be distinguished. Those are Provisioning, Regulating, Habitat and Cultural Services. And yet, novel ecosystems are of key importance when it comes to genetic, species and ecosystem biodiversity. Thus, they provide vital ecosystem services such as carbon sequestration, the circulation of nutrients, elements and energy flow in nature, and nourishment. Sites such as regulated rivers, managed wetlands and even secondary forests, are nothing more than a manifestation of the novel ecosystems' cultural ecosystem services. These are recreational places where people can jointly appreciate the beauty of nature and gain knowledge (Bridgewater *et al.* 2011).

SUMMARY

The impact of novel ecosystems on water retention, CO_2 sequestration, biodiversity or creation of new habitats, has become a significant problem. The most important challenge of the current time is to model the dynamics of changes in novel ecosystems, both spontaneous and affected by humans. These kinds of information can be provided to owners of these areas and relevant state institutions, to understand the mechanisms of how the novel ecosystems are functioning, and in this way to enhance the management of ecosystem services, which will lead to improvements in the quality of environmental health and life of the population.

Despite the growing importance of novel ecosystems in the global ecosystems mosaic, and the increasing area they occupy, they are still relatively poorly explored (Hobbs *et al.* 2006). But the most important issue is to change the perception and recognition of novel ecosystems from sites that are devastated, useless and even destructive, to something that can give new opportunities for providing new services to mankind.

Detailed knowledge about the operation of novel ecosystems can allow individuals to direct the current process of natural succession on a path based on natural processes at a low cost. It can be done by the reinforcement of current natural succession in the direction of the desirable ecosystem services.

REFERENCES

Ahlmer A.-K., Cavalli M., Hansson K., Koutsouris A.J., Crema S. & Kalantari Z. 2018: Soil moisture remote-sensing applications for identification of flood-prone areas along transport infrastructure. Environmental Earth Sciences, 77, 533.

Alberti M., Marzluff J.M., Shulenberger E., Bradley G., Ryan C. & Zumbrunnen C. 2003: Integrating humans into ecology: Opportunities and challenges for studying urban ecosystems. BioScience, 53 (12), 1169–1179.

Asner G.P. 1998: Biophysical and biochemical sources of variability in canopy reflectance. Remote Sensing of Environment, 64 (3), 234–253.

Botta-Dukát Z. 2005: Rao's quadratic entropy as a measure of functional diversity based on multiple traits. Journal of Vegetation Science, 16 (5), 533–540.

Box J. 1999: Nature conservation and post-industrial landscapes. Industrial Archaeology Review, 21 (2), 137–146.

Braun-Blanquet J. 1965: Plant Sociology: The Study of Plant Communities. Hafner, London.

Bridgewater P., Higgs E.S., Hobbs R.J. & Jackson S.T. 2011: Engaging with novel ecosystems. Frontiers in Ecology and the Environment, 9 (8), 423.

Budei B.C., St-Onge B., Hopkinson C. & Audet F.-A. 2018: Identifying the genus or species of individual trees using a three-wavelength airborne lidar system. Remote Sensing of Environment, 204, 632–647.

Burkhard B. & Maes J. (eds.) 2017: Mapping Ecosystem Services. Pensoft Publishers, Sofia.

Cabrera-Bosquet L., Molero G., Stellacci A.M., Bort J., Nogués S. & Araus J.L. 2011: NDVI as a potential tool for predicting biomass, plant nitrogen content and growth in wheat genotypes subjected to different water and nitrogen conditions. Cereal Research Communications, 39 (1), 147–159.

Candiago S., Remondino F., De Giglio M., Dubbini M. & Gattelli M. 2015: Evaluating multispectral images and vegetation indices for precision farming applications from UAV images. Remote Sensing, 7 (4), 4026–4047.

Cole B., McMorrow J. & Evans M. 2014: Spectral monitoring of moorland plant phenology to identify a temporal window for hyperspectral remote sensing of peatland. ISPRS Journal of Photogrammetry and Remote Sensing, 90, 49–58.

Costanza R., de Groot R., Sutton P., van der Ploeg S., Anderson S.J., Kubiszewski I., Farber S. & Turner R.K. 2014: Changes in the global value of ecosystem services. Global Environmental Change, 26 (1), 152–158.

Dangtongdee K.D. & Kurfess F. 2018: Plant identification using Tensorflow. California Polytechnic State University, Luis San Obispo, CA, USA.

DeFries R. 2013: Remote sensing and image processing. [In:] Levin S.A. (ed.): Encyclopedia of Biodiversity. 2nd ed. Academic Press, 389–399.
Dubayah R.O. & Drake J.B. 2000: Lidar remote sensing for forestry. Journal of Forestry, 98 (6), 44–46.
Evers C.R., Wardropper C.B., Branoff B., Granek E.F., Hirsch S.L., Link T.E., Olivero-Lora S. & Wilson C. 2018: The ecosystem services and biodiversity of novel ecosystems: A literature review. Global Ecology and Conservation, 13, e00362.
Gitelson A.A., Kaufman Y.J. & Merzlyak M.N. 1996: Use of a green channel in remote sensing of global vegetation from EOS-MODIS. Remote Sensing of Environment, 58 (3), 289–298.
Hobbs R.J., Arico S., Aronson J., Baron J.S., Bridgewater P., Cramer V.A., Epstein P.R., Ewel J.J., Klink C.A., Lugo A.E., Norton D., Ojima D., Richardson D.M., Sanderson E.W., Valladares F., Vilà M., Zamora R. & Zobel M. 2006: Novel ecosystems: theoretical and management aspects of the new ecological world order. Global Ecology and Biogeography, 15 (1), 1–7.
Hobbs R.J., Higgs E., Hall C.M., Bridgewater P., Chapin III F.S., Ellis E.C., Ewel J.J., Hallett L.M., Harris J., Hulvey K.B., Jackson S.T., Kennedy P.L., Kueffer C., Lach L., Lantz T.C., Lugo A.E., Mascaro J., Murphy S.D., Nelson C.R., Perring M.P., Richardson D.M., Seastedt T.R., Standish R.J., Starzomski B.M., Suding K.N., Tognetti P.M., Yakob L. & Yung L. 2014: Managing the whole landscape: historical, hybrid, and novel ecosystems. Frontiers in Ecology and the Environment, 12 (10), 557–564.
Hyyppä H., Yu X., Hyyppä J., Kaartinen H., Kaasalainen S., Honkavaara E. & Rönnholm P. 2005: Factors affecting quality of DTM generation in forested areas. The International Archives of the Photogrammetry, Remote Sensing and Spatial Information Sciences, 36 (3/W), 85–90.
Kattenborn T., Eichel J. & Fassnacht F.E. 2019: Convolutional neural networks enable efficient, accurate and fine-grained segmentation of plant species and communities from high-resolution UAV imagery. Scientific Reports, 9, 17656.
Khan S. & Hussain M.M. 2019: IoT enabled plant sensing systems for small and large scale automated horticultural monitoring. 2019 IEEE, 5th World Forum on Internet of Things (WF-IoT), 303–308.
Kowarik I. 2011: Novel urban ecosystems, biodiversity and conservation. Environmental Pollution, 159 (8–9), 1974–1983.
Lim K., Treitz P., Wulder M., St-Onge B. & Flood M. 2003: LiDAR remote sensing of forest structure. Progress in Physical Geography, 27 (1), 88–106.
Lin Y. 2015: LiDAR: An important tool for next-generation phenotyping technology of high potential for plant phenomics? Computers and Electronics in Agriculture, 119, 61–73.
Madden M. 2004: Remote sensing and geographic information system operations for vegetation mapping of invasive exotics. Weed Technology, 18, 1457–1463.
Maggiori E., Tarabalka Y., Charpiat G. & Alliez P. 2017: Convolutional neural networks for large-scale remote sensing image classification. IEEE Transactions on Geoscience and Remote Sensing, 55 (2), 645––657.
Maheshwari K., Ruetzler K. & Saugel B. 2020: Perioperative intelligence: applications of artificial intelligence in perioperative medicine. Journal of Clinical Monitoring and Computing, 34 (4), 625–628.
Matuszkiewicz W. 2001: Przewodnik do oznaczania zbiorowisk roślinnych Polski. Vademecum Geobotanicum 3. PWN, Warszawa.
Mazur P. & Chojnacki J. 2017: Wykorzystanie dronów do teledetekcji multispektralnej w rolnictwie precyzyjnym. Technika Rolnicza, Ogrodnicza, Leśna, 1, 25–28.
McKinney M.L. 2002: Urbanization, Biodiversity, and Conservation. BioScience, 52 (10), 883–890.
Milgrom T. 2008: Environmental aspects of rehabilitating abandoned quarries: Israel as a case study. Landscape and Urban Planning, 87 (3), 172–179.
Morse N.B., Pellissier P.A., Cianciola E.N., Brereton R.L., Sullivan M.M., Shonka N.K., Wheeler T.B. & McDowell W.H. 2014: Novel ecosystems in the Anthropocene: a revision of the novel ecosystem concept for pragmatic applications. Ecology and Society, 19 (2), 12.
Mróz M. & Sobieraj A. 2004: Comparison of several vegetation indices calculated on the basis of a seasonal SPOT XS time series, and their suitability for land cover and agricultural crop identification. Technical Sciences, 7, 39–66.
Murthy M.S.R., Giriraj A. & Dutt C.B.S. 2003: Geoinformatics for biodiversity assessment. Biological Letters, 40 (2), 75–100.
Ruwaimana M., Satyanarayana B., Otero V., Muslim A.M., Syafiq A.M., Ibrahim S., Raymaekers D., Koedam N. & Dahdouh-Guebas F. 2018: The advantages of using drones over space-borne imagery in the mapping of mangrove forests. PLoS ONE, 13 (7), e0200288.

Sabins F.F. Jr. 1987: Remote Sensing. Principles and Interpretation. 2nd ed. W. H. Freeman & Co., New York–Oxford.
Silleos N.G., Alexandridis T.K., Gitas I.Z. & Perakis K. 2006: Vegetation indices: advances made in biomass estimation and vegetation monitoring in the last 30 years. Geocarto International, 21 (4), 21–28.
Simberloff D. & Vitule J.R.S. 2014: A call for an end to calls for the end of invasion biology. Oikos, 123 (4), 408–413.
Wäldchen J., Rzanny M., Seeland M. & Mäder P. 2018: Automated plant species identification – Trends and future directions. PLOS Computational Biology, 14 (4), e1005993.
Wężyk P., Szostak M., Krzaklewski W., Pająk M., Pierzchalski M., Szwed P., Hawryło P. & Ratajczak M. 2015: Landscape monitoring of post-industrial areas using LiDAR and GIS technology. Geodesy and Cartography, 64 (1), 125–137.
Wójtowicz A., Wójtowicz M. & Piekarczyk J. 2005: Zastosowanie teledetekcji do monitorowania i oceny produktywności plantacji rzepaku. Rośliny Oleiste, 26 (1), 269–276.
Xue J. & Su B. 2017: Significant remote sensing vegetation indices: A review of developments and applications. Journal of Sensors, 2017, 1353691.

Part Four

Natural Capital as the Basis for Ecosystem Services

4.1. "Natural Capital" Concept – a New Approach to Environmental Management and Post-industrial Landscapes

Gabriela Woźniak
Institute of Biology, Biotechnology and Environmental Protection, Faculty of Natural Sciences, University of Silesia in Katowice; 28 Jagiellońska Str., 40-032 Katowice, Poland; gabriela.wozniak@us.edu.pl; ORCID: 0000-0003-1936-2880

Agnieszka Hutniczak
Institute of Biology, Biotechnology and Environmental Protection, Faculty of Natural Sciences, University of Silesia in Katowice; 28 Jagiellońska Str., 40-032 Katowice, Poland; agnieszka.hutniczak@us.edu.pl; ORCID: 0000-0002-6235-6139

Jörg Dettmar
Department of Design and Landscape Architecture, Faculty of Architecture, Technical University of Darmstadt; 1 El-Lissitzky Str., 64287 Darmstadt, Germany; dettmar@freiraum.tu-darmstadt.de

ABSTRACT: We are currently observing significant transformations that affect the quality of human life. Careless use of the environment significantly limits its resources and leads to its destruction. Therefore, responsible management of the natural environment and natural capital is crucial for us to benefit from it in the future. The natural environment, natural processes related to all environmental components, largely determine wild welfare. Ecological economics gives tremendous importance to biological processes (capital) that create everything on which people are dependent. Natural capital is closely related to the state of ecosystems.

Natural capital is a term used for the entity of the living and non-living components of ecosystems. It "includes all forms of ecosystems and natural resources that create human welfare but are not its work". The relationship between ecosystems and processes generates goods and services for people. Natural capital consists mainly of ecosystems, primary producers, species composition, the related above and below-ground organisms, the processes, and resulting ecosystem functioning, which are the fundamentals for providing ecological services. Ecosystem services maintain people's lives and the balance of ecosystems. Undoubtedly, natural capital stocks and services of biological systems are of crucial importance for the functioning of life on the Earth. Moreover, they contribute to the welfare of people, forming "part of the total economic value of the planet".

The issue dealt with in the paper is, among others, the concept of natural capital in post-industrial landscapes. Currently, this issue assumes significant importance due to the environmental crisis and depletion of natural resources. Furthermore, changes in the economy and technology mean that natural capital is currently undergoing further significant transformations. In these circumstances, a new approach to environmental management, including the concept of natural capital, is becoming an urgent need.

KEYWORDS: Ecosystem services, natural processes, spontaneous processes, adaptation, evolution

INTRODUCTION

The rapid development of civilization, including urbanization, the constant demand for natural resources, especially energy resources, the development of chemical industry, fertilizers and herbicides in agriculture, etc., lead to transformation of the environment.

The uncontrolled exploitation of natural capital is causing loss of biodiversity, water retention, and climate change (Jeżowski 2012). According to the Millennium Ecosystem Assessment (MEA 2005), significant threats to many ecosystems and the natural environment are visible in Europe and on a global scale.

Ecological economics gives tremendous importance to elements of natural processes (capital) that create everything on which people are dependent. Natural capital consists mainly of ecosystems, primary producers, species composition, the related above- and below-ground organisms, the processes, and resulting ecosystem functioning, which are the fundamentals for providing ecological services. The ecosystem services maintain people's lives and the balance of ecosystems on the Earth.

Surprisingly, research shows that some areas associated with exploitation of mineral resources are unique and valuable habitats. Urban post-industrial landscapes create a unique, specific structure, becoming an excellent site for colonization of plants and animals, including rare species. Such places require special care, and their protection seems obvious (Box 1999). Apart from their natural value, they are a testimony to the post-industrial activity of man. Therefore, they are worth protecting and preserving for future generations (Tokarska-Guzik 2000). Undoubtedly, these areas, like the whole landscape, create an exceptional natural capital.

Concerning natural capital, domestic and foreign literature raises some questions related to it, among others: ecosystem services (e.g. de Groot *et al.* 2002; 2010; Turner & Daily 2008; Fisher *et al.* 2009; Costanza *et al.* 2014), critical natural capital resources (e.g. Ekins 2003; Ekins *et al.* 2003; de Groot *et al.* 2003; Brand 2009; Pelenc & Ballet 2015), a capability approach (e.g. Ballet *et al.* 2013; Pelenc & Ballet 2015), as well as sustainable development (among others: Ekins *et al.* 2003; Pelenc & Ballet 2015).

THE DEFINITIONS AND CONCEPT OF "NATURAL CAPITAL"

The very word "capital" "comes from (neo-classical) economics and is defined as the stock of real goods, with power of producing further goods (or utilities) in the future" (de Groot *et al.* 2003). According to de Groot *et al.* (2003), the following factors refer to the capital: land, labour, and artificial infrastructure. Costanza and Daly (1992) used the terms natural capital, human and manufactured capital. For the first time, the term "natural capital" was used by E.F. Schumacher in 1973 (Solon *et al.* 2017). This concept gained popularity and slowly became adequately understood and appreciated in the 1990s (Telega 2014). The concept of natural capital is widely discussed by Barbier (2019).

The term "natural capital" is an interdisciplinary term relatively new in natural sciences. It is also related to the term and concept of "ecosystem services" (Telega 2014; 2015; Barbier 2019). Modern economies use a combination of many forms of capital. One of them is natural capital (Binner *et al.* 2017). Natural processes (capital) create everything on which the people are dependent.

The Earth, ecosystems, ecosystem processes, and functioning geology with mineral diversity, water, air resources, etc., are natural capital. On the other hand, knowledge, work and technologies create human capital. Artificial capital is the following: buildings, infrastructure, etc. (Costanza & Daly 1992; de Groot *et al.* 2003; Jeżowski 2012). None of this could emerge without the geological, hydrological, atmospheric, pedospheric and biospheric resources and interlinked relationships between those elements (Fig. 1).

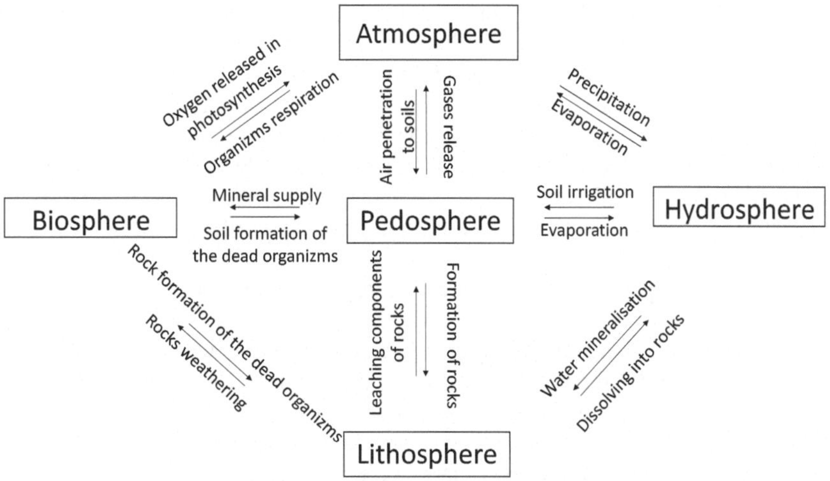

Figure 1. Functional relationships are linking individual elements of the natural environment, including oxygen released by plants in photosynthesis processes; breathing the oxygen; emission of gases; the penetration of gases (air) into the soil or substrate; rainfall; evaporation; providing mineral salts; soil irrigation; the formation of soil from dead fragments of living organisms; soil hydration; evaporation; weathering and rock formation; penetration of elements and minerals; water mineralization; rock dissolution; the shape of soil from rocks; soil formation process; leaching of components from the soil to the rocks; water mineralization; rock solubility

The essential component of natural capital, which completes the generated capital, human and social capital, is the Earth (Goździewicz-Biechońska 2017). Natural capital has become transformed into such systems as agriculture or aquaculture (de Groot *et al.* 2003). As Telega (2015) writes, preserving natural capital intact is unrealistic and unjustified, given the constant pursuit of prosperity.

As defined by Solon *et al.* (2017) "Capital of nature (natural) means real and potential resources, forces, processes and structural elements of nature as well as composition and mutual relationship between various components of the environment, due to which life processes on Earth are maintained. Natural capital is a component of the capital of nature. (…). Natural capital includes all forms of ecosystems and natural resources that participate in the creation of human welfare but are not a product of its work".

Binner *et al.* (2017) defined natural capital in the following way: "Natural capital is the stock of physical assets that generate flows of environmental goods and services that benefit people".

According to Ekins *et al.* (2003): "Natural capital is a metaphor to indicate the importance of elements of nature (e.g., minerals, ecosystems and ecosystem processes) to human society". Many environmental factors define natural ecosystems by characteristics that "determine the ecosystems' capacity to provide goods and services" (Ekins *et al.* 2003).

According to Hodgson and Winram (2017): "Natural capital refers to the stock of natural assets upon which our economy and society is built". Natural processes provide the natural capital that provides value and goods for people, such as timber or minerals. Even more significant are biodiversity increase, water retention, carbon sequestration, climate regulation, and air purification. Sometimes production is simply the result of natural capital combining with natural processes, sometimes people need to come up with some innovations to realize the benefits (Fig. 2). In the publication "Natural Capital Account 2016–2017", Hodgson and Winram (2017) from Forest Enterprise England (FEE) highlight the Natural Capital Account (NCA), which they use to make forest-related decisions binding. From the publication "Forest Enterprise England – Natural Capital Accounts 2015/16", we learn that Forest Enterprise England (FEE), together with environmental economics consultants, have "developed the first organisation-wide Natural Capital Account (NCA)". The natural capital account allows you to quantify the total value of services provided by natural assets under the care of FEE in a relatively orderly and transparent way.

The paper from 1997 emphasizes that natural capital stocks and services of ecological systems are of crucial importance for Earth's functioning. They contribute to the welfare of people, forming "part of the total economic value of the planet" (Costanza *et al.* 1997). Based on published research and original calculations, the authors estimated the economic value of 17 ecosystem services for 16 biomes. According to their accounting the entire biosphere is worth US $ 16–54 trillion per year (Costanza *et al.* 1997).

Dominati *et al.* (2010) pay special attention to ecosystem services and soil natural capital, distinguishing five related elements:

1. "Soil natural capital, characterised by standard soil properties well known to soil scientists.
2. The processes behind soil natural capital formation, maintenance and degradation.
3. Drivers (anthropogenic and natural) of soil processes.
4. Provisioning, regulating, and cultural ecosystem services.
5. Human needs fulfilled by soil ecosystem services".

In turn, Olewiler (2006) defines natural capital as follows: natural capital is becoming the conceptual framework for identification, measurement and assessment of "the role that the natural environment plays in sustaining communities". For example, different organizations in Canada (the National Round Table on the Economy and Environment, or the Nature Conservancy) use the term "natural capital" to analyse and measure environmental sustainability (Olewiler 2006).

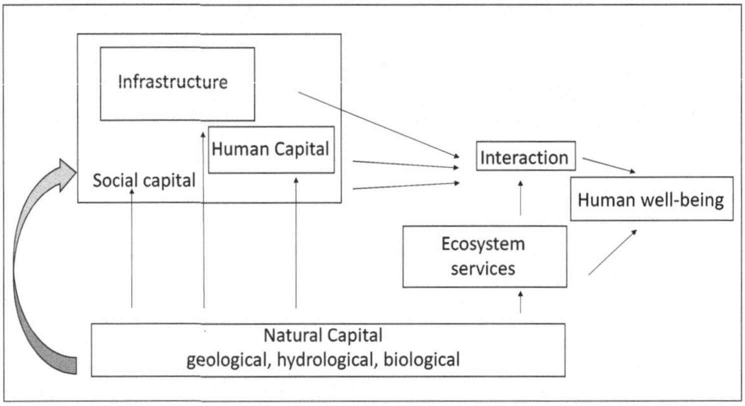

Figure 2. The flow of benefits coming from natural capital

According to Ekins *et al.* (2003), natural capital depends on four environmental factors: air, water, land, and habitats. The biological, ecological, and environmental functions of natural capital are the following: regulatory, production, habitat, and information. Regulatory functions are the crucial ones that regulate ecological processes. For example, biomass production functions such as crops, e.g., food, are partially possible due to natural ecosystem enhancement (pollinators). The habitat functions such as providing natural places of refuge for wild plants and animals by ecosystems or information functions mean providing opportunities for recreation or education.

Natural capital is a term used for the entity of the living and non-living components of eco systems. The relationship between organisms, populations, ecosystems, and related processes provides goods and services for people. Riparian vegetation buffers and holds the soil in place during flooding, preventing erosion and improving water quality. The varied types of vegetation: woodlands, forests, and particular water bodies, lakes, oceans, wetlands and peatlands store the carbon in organic matter as biomass and, even more critical, as soil organic matter (SOM), thereby significantly influencing the regulation of climate. Lakes and mountains provide recreational opportunities. These benefits come from the interactions between biotic and abiotic ecosystem elements (Fisher *et al.* 2009; Guerry *et al.* 2015). Plants as primary producers, animals, and microbes – the decomposers, are the main parts of each ecosystem providing essential services. Some ecosystems are human-engineered and a part or component of the social and ecological mosaic system, which includes, e.g., agroecosystems, urbanized, and industrialized areas.

For a very long time, natural capital was not included in economic calculations. Substantially wrong assumptions caused any natural element which does not bring any financial benefits, to be considered to have no value, which led to an increasing asymmetry in economic systems. This asymmetry necessarily needs to be balanced. Moreover, the asymmetry can be removed due to a better understanding of the essence of natural capital, which is undoubtedly crucial to maintaining all people's needs and society's existence. Humans are ruthlessly losing what they need (natural capital) for something they sometimes do not need. The significance of natural capital and ecosystem services has been strongly emphasized in the Millennium Ecosystem Assessment, stating that human activities are the reasons for the decline in ecosystem services (MEA 2005; Kinzig *et al.* 2011; Guerry *et al.* 2015).

Currently, understanding ecosystems and natural capital resilience on different scales is vitally important. First, it is related to the concerns connected with the responses of ecosystems to climate change (Folke *et al.* 2002; Guerry *et al.* 2015). Second, natural capital resilience gives the ability to adapt and provide ecosystem services for the safety of the social and ecological systems. Therefore, it is essential to understand complex social and environmental systems (Fig. 3; Regan *et al.* 2005; Biggs *et al.* 2012; Reyers *et al.* 2012; Guerry *et al.* 2015; Liu *et al.* 2015; Scheffer *et al.* 2015).

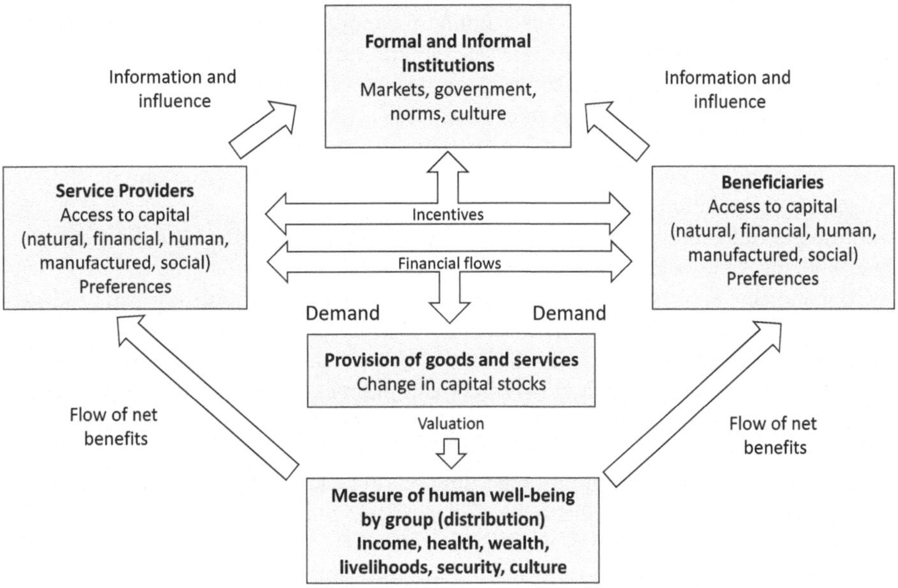

Figure 3. The scheme for natural capital in the context of formal and informal decision-making institutions along with other forms of capital: financial, human, manufactured, and social. Formal and informal institutions influence decisions by both service providers and beneficiaries. Access to various forms of capital ("capabilities") affect the decisions of service suppliers and beneficiaries. The joint actions of service providers and beneficiaries determine the flow of goods and services (including ecosystem services). Natural capital is the only type of capital which is self-recreating. These makes natural capital the basis and source for all other known types of capital

CRITICAL NATURAL CAPITAL

In literature, you can also find definitions referring to the so-called "critical" natural capital (de Groot *et al.* 2003), e.g.:

- "critical natural capital may then be defined as natural capital which is responsible for important environmental functions and which cannot be substituted in the provision of these functions by manufactured capital" (Ekins *et al.* 2003);
- "critical natural capital (life-supporting capital) – capital necessary for life reproduction (the ozone layer, biodiversity, river ecosystems, wetlands swamps, forests as species habitats)", (Telega 2014).

De Groot *et al.* (2003) distinguish two types of criticalities. The first type is based on the ecocentric perspective – focused mainly on maintaining regulatory and habitat functions. The second type is based on an anthropocentric view and shows the ecosystem services that are most important for our survival. This type of criticality focuses mainly on the production and information functions of natural ecosystems. According to the authors, the criticality depends on ecology[1] and economics, political and social background.

As reported by de Groot *et al.* (2003) after MacDonald *et al.* (1999), comprehensive definitions of critical natural capital are rare or focus on the functional aspect of fundamental values of ecosystems for overall functioning of the biosphere.

NATURAL CAPITAL AND ECOSYSTEM SERVICES – – TERMINOLOGY AND RESPONSIBILITY

Sometimes the term "natural capital" has a similar meaning to "ecosystem services", although there exist some significant differences. For example, natural capital is a resource that can survive. At the same time, environmental goods and services are variable (liquid) and appear due to biogeochemical and ecological processes over time (Binner *et al.* 2017).

The concept of "ecosystem services" is inseparably connected with natural capital, from which we benefit. Ecosystem services are "the supply of matter, energy, and information needed for human life provided by biosphere" (Telega 2015).

According to Solon *et al.* (2017), different authors distinguish different definitions of ecosystem services. And so: "(1) Costanza *et al.* (1997); Solon (2008) – a set of products (e.g., wood, forest fruit, wild game) and ecosystem services (e.g., water and air purification, oxygen production, recreation areas) used by society; (2) Boyd & Banzhaf (2007) – nature components consumed directly or experienced to build human needs; (3) TEEB (2010) – elements of ecosystems used (actively or passively) to produce human welfare; (4) direct and indirect contributions of ecosystems to human needs. The concept of goods and benefits is synonymous with ecosystem services" (Solon *et al.* 2017).

There is no unambiguous definition regarding natural capital and ecosystem services, and there is terminological chaos regarding this issue (Telega 2015; Solon *et al.* 2017). Moreover, sometimes expression of ecosystem services is used interchangeably with environmental services. Unfortunately, the word "ecosystem" and the word "environment" have different connotations. People understand the word "ecosystem" as an integrated system of related biotic and abiotic elements related frequently in feedback loops in a dynamic balance. Unfortunately for many people, "environment" means environmental services connected with waste and recycling services provided by local institutions or government (Nahlik *et al.* 2012; Guerry *et al.* 2015). The lack of discipline in using

[1]Ecology is understood as relationships between living organisms and their non-living abiotic habitat. The application of the term ecology in other meanings is wrong and misleading and has no scientific justification.

precisely defined terms in environmental sciences and practical applications worsens the situation on a global scale.

PROGRESS IN DEVELOPING THE CONCEPT OF NATURAL CAPITAL

In recent decades, there has been progress in defining, developing and separating the concept of natural capital and ecosystem services (Goldman-Benner *et al.* 2012). Climate change, biodiversity loss, increasing water deficiency, the underestimated importance of people's dependence on nature, and links between the functioning of ecosystems and society's survival, have attracted human awareness. The close human relationship with nature is recognized. Attention in this subject is still growing, however, not quickly enough (UK National Ecosystem Assessment 2011; Duraiappah *et al.* 2014; Risky Business Project 2014). Andrejczuk (2013) presents concepts of cooperation between people and nature in the landscape. He mentions five visions:

1. "The conception of deification of nature and respect for it.
2. The consumer conception.
3. The conception of nature conservation.
4. The conception of the greening of the economy and life.
5. The sustainable development conception".

He also drew attention to human perception of nature since ancient times.

Our civilization is closely related to the development of cities. Life in cities is more straightforward. Nowadays, people focus on education, life comfort, and rendered services (the city is perceived as a driving force for development). Citizens often move away from nature. Their relationship with nature is less recognizable (Guerry *et al.* 2015).

Issues related to the environment are sometimes considered and discussed at various meetings, including the World Economic Forum, during which environmental problems are considered among the critical issues in business (World Economic Forum 2015). Therefore, introducing new environmentally friendly solutions and the right decisions regarding arable land concerning natural capital are very important. The Ministry of the Environment, Industry, Agriculture, and Finance should adopt and practice the concept of ecosystem services and natural capital. Ecosystem services and ideas for natural capital should cover a wide range of decision-making and alteration in parts of the world where transformations are going on. These solutions should be used and go beyond classic protection. It is essential to help and support issues related to the environment that are sometimes considered and discussed at various meetings, including the World Economic Forum (Goldstein *et al.* 2012; Arkema *et al.* 2015; Guerry *et al.* 2015; Li *et al.* 2015). Environmental problems are regarded among the critical issues in business (World Economic Forum 2015).

Appling available in advanced science, a wide range of biological and biotechnological methods for analysis, management, and decision can support identifying the critical natural capital elements. Developing and creating accessible tools might be possible to prepare maps of natural capital and ecosystem services distribution. It is possible to explore a quantification system and attach values to a particular element of the environment and highlight spatial and temporal aspects. For example, the differences in social and natural capital perceptions of ecosystem services causes and effects, the specific needs in delivering services to various groups of people, and the associated trade-offs between the education intensity and required social appreciation (Guerry *et al.* 2015). The scientific challenges and progress can be described by five groups: 1) understanding the resilience and provision of ecosystem services in industrial and urban contexts; 2) obtaining and developing knowledge about the emergence, persistence, and value of natural capital in highly populated areas; 3) to study the ecosystem services resources in terms of ecosystem and environment functioning; 4) working out the rules of natural capital, ecosystem services management and monitoring the effects of the discussions and actions; 5) it is necessary to understand the mechanisms of impact of policy and management on ecosystem services provisioning systems and natural capital persistence and regeneration (Guerry *et al.* 2015).

For all of the above to be feasible, constant knowledge development using the newest advanced biology, biochemistry, hydrobiology, biophysics, and even biotechnology is necessary. New metrics, data, and tools will make it easier to monitor, assess and account for nature's benefits to human society's survival. In addition, developing knowledge about the synergic feedbacks within the natural capital of novel ecosystems can provide tangible ways to identify and weigh trade-offs derived from different management and policy decisions (Guerry *et al.* 2015).

The study of mapping and looking for relationships among multiple ecosystem service quantifications and biodiversity have resulted in progress (Raudsepp-Hearne *et al.* 2010; Cardinale *et al.* 2012). Those results can support the information needed to predict changes of the critical drivers of the ecosystem and climate, and differences in land use (Lawler *et al.* 2014). In this area, geoinformatics, spectral data, and numerical maps have allowed the development of spatial modelling. Furthermore, the use of the geoinformatics methods can be beneficial in assessing how changes in ecosystems and their essential ecosystem functional processes are likely to cause changes in the provision and persistence of ecosystem services (Guerry *et al.* 2015).

NATURAL CAPITAL AND NOVEL ECOSYSTEMS

In novel ecosystems, the organism's complex adaptive system dynamics are fundamental and still very little understood. The natural power of adaptability and evolutionary processes is an irreplaceable biological power affecting and enhancing the natural capital providing ecosystem services. The organism's adaptations to novel ecosystems include feedback and synergisms with a high potential to compensate for significant disruptions and challenges such as climate change and other functional constraints (Mooney *et al.* 2009; Seppelt *et al.* 2011). A similar mechanism develops in complex natural systems shaping ecosystem resilience and the related ecosystem services (Biggs *et al.* 2012; Reyers *et al.* 2013). Recent progress in understanding the strength of spontaneous and natural ecosystems, including ecosystem services, incorporates social science and natural science has helped to learn how the environmental and social systems respond to maintain or undermine a dynamic balance and sustainable development. As a result, wild, natural, semi-natural, and anthropogenic novel ecosystems, as well as the closely related ecosystem services, are more secure in an uncertain future.

This new environmental management should emerge by combining approaches including ecosystem services, geoinformatics modelling assisted by constant evaluation, and design of new, better solutions to obtain more secure natural capital and ecosystem services in the future (Reyers *et al.* 2015). For decision-makers, managers, and politicians, the process and criteria of ecosystem services validation are not always clear. Monetary valuation, market and nonmarket valuation methods known from economics are sometimes used and help to estimate ecosystem service values (Bateman *et al.* 2015). When considering human health or livelihoods, the monetary valuation is insufficient for robust extrapolation and is highly contested or lacks robustness for some ecosystem services (Plummer 2009; Seppelt *et al.* 2011). For many environmental solutions, the monetary value metrics are not relevant to make the decision. To report the outcomes in biogeochemical terms or directly in terms of impacts on human health or livelihood, we need to consider the newest scientific achievements in biogeochemical, environmental, scientific and applied knowledge (Myers *et al.* 2013; Guerry *et al.* 2015; Ruckelshaus *et al.* 2015).

Sen (1985) and Guerry *et al.* (2015) underline that the costs, rules, and limits in access to various forms of capital ("capabilities") affect decisions of service suppliers and beneficiaries of the provided goods, and the life conditions of different groups in society. Components of the system change on relatively long timescales, including essential regulating services and matching them with the conditions of life and human health metrics. It is a significant research gap to be filled. Natural capital accounting helps avoid the wrong, insufficient approach, focusing only on trends in providing services. Natural capital needs time to be rebuilt and maintained. Natural capital regeneration is costless. Natural capital accounts will point to areas with a "natural capital deficit", and intervention is required. The report called "inclusive wealth" attempts to value social, human, manufactured, and natural assets (Arrow *et al.* 2004). The performance of wealth can provide a helpful gauge of sustainability. The maintenance of natural capital involves a significant time scale and has to include a future time scale. The appropriateness of limiting natural capital is controversial and highly questionable ethically. Regarding climate change, society should divide the natural capital over time (Stern 2008; Guerry *et al.* 2015). Cooperation between scientists (biologists, chemists, hydrologists, geographers, environmentalists) and the economy, markets, and different institutions should provide possibilities to work out a balance between absorbing, conservation, and valuing natural capital asserts (Kinzig *et al.* 2011).

Social and ecological systems are complex and differ from ecosystem services and biogeochemical scales. The governance of social and biogeochemical systems should reflect and account for the complex temporal and spatial intersection (Ostrom 1998; 2010; Shogren & Taylor 2008; Camerer *et al.* 2011; Polasky *et al.* 2014; Bateman *et al.* 2015; Guerry *et al.* 2015). Knowledge of cultural norms of the surrounding nature is a prerequisite for implementing changes in human-nature interactions. Based on the literature, we can learn that the study on evaluating the impacts of protected areas (for example: Andam *et al.* 2010; Ferraro *et al.* 2011; Sala *et al.* 2013) and PES (provision of ecosystem services) programs on social and biogeochemical outcomes can be an example (Arriagada *et al.* 2012; Li *et al.* 2015). The development of a system for education can explain the impacts of natural capital and ecosystem services for health impacts and conservation action enhancement (Ferraro & Pattanayak 2006). The best education and health care systems will not be helpful if there will not be available fresh air, water, and food free from GMOs, herbicides, pesticides, etc. In this respect, the evaluation of ecosystem function reestablishment, habitat restoration, recreation, and conservation actions remains a primary challenge not for the future but for now. The evaluation of impacts has to start with the relevant ecosystem functioning, socio-economic and biogeochemical monitoring measures.

Many of the current monitoring systems and collected data are inadequate because they are not comprehensive. Tracking data collection has to be more relevant to ecosystem function, particularly the functioning of novel ecosystems. Proper monitoring systems and management impact assessment can be the basis for an efficiently working post-implementation program (Ferraro & Pattanayak 2006; Ferraro *et al.* 2015; Guerry *et al.* 2015). Understanding complex multidimensional relationships and links between the above and below-ground biotic and abiotic components are incomplete and dynamically changing with the emerging novel climate and ecosystem conditions, not to mention the social and ecological dynamic processes. Some management scenarios need a long time to observe the results and introduce the impact assessment procedure. When the aim is to restore the transformed site, it can eventually result in biodiversity loss (or increase), but it can take decades or even more (Tilman *et al.* 1994; Schultz *et al.* 2015). In many actions (like vehicle use reduction), the result can be assessed after a much shorter time (Ferraro *et al.* 2015; Guerry *et al.* 2015; Li *et al.* 2015).

NATURAL CAPITAL AND ECOSYSTEM SERVICES – EXAMPLES OF MANAGEMENT AND STRATEGY IN THE WORLD

According to Guerry *et al.* (2015), some governments and international organizations are trying to include natural capital and ecosystem services in management plans, but this is still not a standard practice. Nevertheless, these activities can be observed, among others, in China, South Africa, Costa Rica, Belize or Great Britain.

China plans to become the "ecological civilization of the 21st Century" (Guerry *et al.* 2015). Interest in this topic led to the compilation of a comprehensive book (Pan 2015), which shows the difficulties associated with economic growth and ecological sustainability in China.

According to Arriagada *et al.* (2012), Costa Rican authorities pay attention to payments for environmental services (PES). As a result, the share of forests on farms increased over eight years from 11% to 17%.

In South Africa, you may come across the Grassland Program Initiative. It aims to protect both biodiversity and ecosystem services (Egoh *et al.* 2011).

MANAGEMENT OF POST-INDUSTRIAL AREAS ON THE EXAMPLE OF GREAT BRITAIN – ADMINISTRATIVE AND LEGAL ASPECTS

The development of post-mining areas in Great Britain is related to protecting and preserving these areas' natural values and historical heritage (Besenyei – personal communications). Places of high biodiversity in brownfield areas can be included, according to strictly defined, detailed rules. In the UK, the Sites of Special Scientific Interest (SSSI) are a common form of protection. The government and territorial administration units are obliged to protect these areas, enhance and follow the Wildlife and Countryside Act (1981). This form of protection may cover areas of particular interest due to flora, fauna, and biological processes in the place, taking into account geological, physiographic, or scientific features. Clayhanger in Walsall is an example of a Site of Special Scientific Interest (SSSI), where gangue protection arises from the exploitation of hard coal, water-filled settling troughs, and the remains of unfertilized pastures. Based on strict and transparent criteria, valuable non-status areas (SSSI) are protected as National Nature Reserves (NNR), (O'Halloran 1988). A Local Nature Reserve (LNR) establishment is another form of protection in post-industrial areas, including mining areas. They are designated by local authorities, taking into account their specific regional, natural, educational, and social values (Natural England 2010). Country Parks covering post-industrial areas with exciting elements of animate or inanimate nature are another form of protection. In turn, other valuable areas in West Midlands in the mid-west part of Great Britain, where Black Country is located, are places identified as Sites of Interest for Nature Conservation (SINC) and Sites of Local Interest for Nature Conservation (SLINC). In Staffordshire, Cheshire, and Greater Manchester, one can see Sites of Biological Interest (SBI) and places important due to local biological values, Biological Action Sites (BAS). Post-mining areas resulting from the closure of mines allow the spontaneous spread of organisms that adapt to the conditions which prevail in these areas, thus creating refuges for wildlife. Since the beginning of the 1960s, the Department of the Environment in the United Kingdom has emphasized the importance of protecting the industrial heritage. The natural and historical aspects are considered, implementing a program whose goal was to designate valuable natural areas and cover them with appropriate forms of protection (Besenyei – personal communications).

CONCLUSIONS

Natural capital, including ecosystem services, is an essential element of the economy. It is worth noting that preserving natural capital does not mean giving up the consumption of biological, geological, chemical, and environmental goods, but requires respect for the environment and several changes that will significantly help preserve its value (Telega 2015). Some factors can influence natural capital negatively and be a real threat (Ekins *et al.* 2003). The overuse of natural capital may destabilize a natural balance – upsetting its stability (Islam *et al.* 2019).

Natural capital is significant for the flow of ecosystems and sustainable development (Islam *et al.* 2019). Therefore, its maintenance is one of the most critical problems facing the 21st Century (Clark 2007; Brand 2009; Islam *et al.* 2019). However, despite the consensus on the main idea of sustainable development, which type of capital should be preserved for the present and future generations is controversial. For a long time, the natural capital has been neglected. Finally, cultivated natural capital, artificial, social, and human capital, have been acknowledged as capital that should be preserved (Brand 2009). To achieve the success of sustainable development (Guerry *et al.* 2015) suggest taking the following actions on this issue:

- preparation and development of evidence combined with decisions that would affect natural capital and ecosystem services as well as human well-being;
- establishing cooperation with representatives from various sectors of the economy and society;
- policy changes.

Cooperation with various representatives may turn into action. Joint commitment, knowledge, and mutual understanding will allow co-creation of tools to solve broadly set challenges (Guerry *et al.* 2015).

REFERENCES

Andam K.S., Ferraro P.J., Sims K.R.E., Healy A. & Holland M.B. 2010: Protected areas reduced poverty in Costa Rica and Thailand. Proceedings of the National Academy of Sciences of the United States of America, 107 (22), 9996–10001.

Andrejczuk W. 2013: Koncepcje współdziałania człowieka i natury w krajobrazie. Krajobraz a człowiek w czasie i przestrzeni. Prace Komisji Krajobrazu Kulturowego, 20, 9–16.

Arkema K.K., Verutes G.M., Wood S.A., Clarke-Samuels C., Rosado S., Canto M., Rosenthal A., Ruckelshaus M., Guannel G., Toft J., Faries J., Silver J.M., Griffin R. & Guerry A.D. 2015: Embedding ecosystem services in coastal planning leads to better outcomes for people and nature. Proceedings of the National Academy of Sciences of the United States of America, 112 (24), 7390–7395.

Arriagada R.A., Ferraro P.J., Sills E.O., Pattanayak S.K. & Cordero-Sancho S. 2012: Do payments for environmental services affect forest cover? A farm-level evaluation from Costa Rica. Land Economics, 88 (2), 382–399.

Arrow K., Dasgupta P., Goulder L., Daily G., Ehrlich P., Heal G., Levin S., Mäler K.-G., Schneider S., Starrett D. & Walker B. 2004: Are we consuming too much? Journal of Economic Perspectives, 18 (3), 147–172.

Ballet J., Koffi J.-M. & Pelenc J. 2013: Environment, justice and the capability approach. Ecological Economics, 85, 28–34.

Barbier E.B. 2019: The concept of natural capital. Oxford Review of Economic Policy, 35 (1), 14–36.

Bateman I.J., Coombes E., Fitzherbert E., Binner A., Bad'ura T., Carbone C., Fisher B., Naidoo R. & Watkinson A.R. 2015: Conserving tropical biodiversity via market forces and spatial targeting. Proceedings of the National Academy of Sciences of the United States of America, 112 (24), 7408–7413.

Biggs R., Schlüter M., Biggs D., Bohensky E.L., BurnSilver S., Cundill G., Dakos V., Daw T.M., Evans L.S., Kotschy K., Leitch A.M., Meek C., Quinlan A., Raudsepp-Hearne C., Robards M.D., Schoon M.L., Schultz L. & West P.C. 2012: Toward principles for enhancing the resilience of ecosystem services. Annual Review of Environment and Resources, 37, 421–448.

Binner A., Smith G., Bateman I., Day B., Agarwala M. & Harwood A. 2017: Valuing the social and environmental contribution of woodlands and trees in England, Scotland and Wales. Forestry Commission Research Report, Forestry Commission, Edinburgh.

Box J. 1999: Nature conservation and post-industrial landscapes. Industrial Archaeology Review, 21 (2), 137–146.

Boyd J. & Banzhaf S. 2007: What are ecosystem services? The need for standardized environmental accounting units. Ecological Economics, 63 (2–3), 616–626.

Brand F. 2009: Critical natural capital revisited: Ecological resilience and sustainable development. Ecological Economics, 68 (3), 605–612.

Camerer C.F., Loewenstein G. & Rabin M. (eds.) 2011: Advances in Behavioural Economics. Princeton University Press, Princeton
Cardinale B.J., Duffy J.E., Gonzalez A., Hooper D.U., Perrings C., Venail P., Narwani A., Mace G.M., Tilman D., Wardle D.A., Kinzig A.P., Daily G.C., Loreau M., Grace J.B., Larigauderie A., Srivastava D.S. & Naeem S. 2012: Biodiversity loss and its impact on humanity. Nature, 486 (7401), 59–67.
Clark W.C. 2007: Sustainability science: A room of its own. Proceedings of the National Academy of Sciences of the United States of America, 104 (6), 1737–1738.
Costanza R. & Daly H.E. 1992: Natural capital and sustainable development. Conservation Biology, 6 (1), 37–46.
Costanza R., d'Arge R., de Groot R., Farber S., Grasso M., Hannon B., Limburg K., Naeem S., O'Neill R.V, Paruelo J., Raskin R.G., Sutton P. & van den Belt M. 1997: The value of the world's ecosystem services and natural capital. Nature, 387 (15), 253–260.
Costanza R., de Groot R., Sutton P., van der Ploeg S., Anderson S.J., Kubiszewski I., Farber S. & Turner R.K. 2014: Changes in the global value of ecosystem services. Global Environmental Change, 26 (1), 152–158.
de Groot R., Van der Perk J., Chiesura A. & van Vliet A. 2003: Importance and threat as determining factors for criticality of natural capital. Ecological Economics, 44 (2–3), 187–204.
de Groot R.S., Alkemade R., Braat L., Hein L. & Willemen L. 2010: Challenges in integrating the concept of ecosystem services and values in landscape planning, management and decision making. Ecological Complexity, 7 (3), 260–272.
de Groot R.S., Wilson M.A. & Boumans R.M.J. 2002: A typology for the classification, description and valuation of ecosystem functions, goods and services. Ecological Economics, 41 (3), 393–408.
Dominati E., Patterson M. & Mackay A. 2010: A framework for classifying and quantifying the natural capital and ecosystem services of soils. Ecological Economics, 69 (9), 1858–1868.
Duraiappah A.K., Asah S.T., Brondizio E.S., Kosoy N., O'Farrell P.J., Prieur-Richard A.-H., Subramanian S.M. & Takeuchi K. 2014: Managing the mismatches to provide ecosystem services for human well-being: a conceptual framework for understanding the New Commons. Current Opinion in Environmental Sustainability, 7, 94–100.
Egoh B.N., Reyers B., Rouget M. & Richardson D.M. 2011: Identifying priority areas for ecosystem service management in South African grasslands. Journal of Environmental Management, 92 (6), 1642–1650.
Ekins P. 2003: Identifying critical natural capital: Conclusions about critical natural capital. Ecological Economics, 44 (2–3), 277–292.
Ekins P., Simon S., Deutsch L., Folke C. & de Groot R. 2003: A framework for the practical application of the concepts of critical natural capital and strong sustainability. Ecological Economics, 44 (2–3), 165–185.
Ferraro P.J. & Pattanayak S.K. 2006: Money for nothing? A call for empirical evaluation of biodiversity conservation investments. PLoS Biology, 4 (4), e105.
Ferraro P.J., Hanauer M.M. & Sims K.R.E. 2011: Conditions associated with protected area success in conservation and poverty reduction. Proceedings of the National Academy of Sciences of the United States of America, 108 (34), 13913–13918.
Ferraro P.J., Hanauer M.M., Miteva D.A., Nelson J.L., Pattanayak S.K., Nolte C. & Sims K.R.E. 2015: Estimating the impacts of conservation on ecosystem services and poverty by integrating modeling and evaluation. Proceedings of the National Academy of Sciences of the United States of America, 112 (24), 7420–7425.
Fisher B., Turner R.K. & Morling P. 2009: Defining and classifying ecosystem services for decision making. Ecological Economics, 68 (3), 643–653.
Folke C., Carpenter S., Elmqvist T., Gunderson L., Holling C.S. & Walker B. 2002: Resilience and sustainable development: Building adaptive capacity in a world of transformations. AMBIO: A Journal of the Human Environment, 31 (5), 437–440.
Forest Enterprise England – Natural Capital Accounts 2015/16. 2016 Forestry Commission, England. Report.
Goldman-Benner R.L., Benitez S., Boucher T., Calvache A., Daily G., Kareiva P., Kroeger T. & Ramos A. 2012: Water funds and payments for ecosystem services: Practice learns from theory and theory can learn from practice. Oryx, 46 (1), 55–63.
Goldstein J.H., Caldarone G., Duarte T.K., Ennaanay D., Hannahs N., Mendoza G., Polasky S., Wolny S. & Daily G.C. 2012: Integrating ecosystem-service tradeoffs into land-use decisions. Proceedings of the National Academy of Sciences of the United States of America, 109 (19), 7565–7570.
Goździewicz-Biechońska J. 2017: Nowe paradygmaty ochrony ziemi jako zasobu środowiska w prawie rolnym. Przegląd Prawa Rolnego, 2 (21), 77–92.

Guerry A.D., Polasky S., Lubchenco J., Chaplin-Kramer R., Daily G.C., Griffin R., Ruckelshaus M., Bateman I.J., Duraiappah A., Elmqvist T., Feldman M.W., Folke C., Hoekstra J., Kareiva P.M., Keeler B.L., Li S., McKenzie E., Ouyang Z., Reyers B., Ricketts T.H., Rockström J., Tallis H. & Vira B. 2015: Natural capital and ecosystem services informing decisions: From promise to practice. Proceedings of the National Academy of Sciences of the United States of America, 112 (24), 7348–7355.
Hodgson S. & Winram M. 2017: Natural Capital Account 2016–2017. Forestry Commission, England.
Islam M., Yamaguchi R., Sugiawan Y. & Managi S. 2019: Valuing natural capital and ecosystem services: a literature review. Sustainability Science, 14, 159–174.
Jeżowski P. 2012: Rozwój zrównoważony i jego nowe wyzwania. Kwartalnik Kolegium Ekonomiczno-Społecznego, Studia i Prace, 2, 99–124.
Kinzig A.P., Perrings C., Chapin III F.S., Polasky S., Smith V.K., Tilman D. & Turner II B.L. 2011: Paying for ecosystem services – Promise and peril. Science, 334 (6056), 603–604.
Lawler J.J., Lewis D.J., Nelson E., Plantinga A.J., Polasky S., Withey J.C., Helmers D.P., Martinuzzi S., Pennington D. & Radeloff V.C. 2014: Projected land-use change impacts on ecosystem services in the United States. Proceedings of the National Academy of Sciences of the United States of America, 111 (20), 7492–7497.
Li C., Zheng H., Li S., Chen X., Li J., Zeng W., Liang Y., Polasky S., Feldman M.W., Ruckelshaus M., Ouyang Z. & Daily G.C. 2015: Impacts of conservation and human development policy across stakeholders and scales. Proceedings of the National Academy of Sciences of the United States of America, 112 (24), 7396–7401.
Liu J., Mooney H., Hull V., Davis S.J., Gaskell J., Hertel T., Lubchenco J., Seto K.C., Gleick P., Kremen C. & Li S. 2015: Systems integration for global sustainability. Science, 347 (6225), 1258832.
MacDonald D.V., Hanley N. & Moffatt I. 1999: Applying the concept of natural capital criticality to regional resource management. Ecological Economics, 29 (1), 73–87.
MEA – Millennium Ecosystem Assessment. 2005: Ecosystems and Human Well-being: Biodiversity Synthesis. World Resources Institute, Washington DC.
Mooney H., Larigauderie A., Cesario M., Elmquist T., Hoegh-Guldberg O., Lavorel S., Mace G.M., Palmer M., Scholes R. & Yahara T. 2009: Biodiversity, climate change and ecosystem services. Current Opinion in Environmental Sustainability, 1 (1), 46–54.
Myers S.S., Gaffikin L., Golden C.D., Ostfeld R.S., Redford K.H., Ricketts T.H., Turner W.R. & Osofsky S.A. 2013: Human health impacts of ecosystem alteration. Proceedings of the National Academy of Sciences of the United States of America, 110 (47), 18753–18760.
Nahlik A.M., Kentula M.E., Fennessy M.S. & Landers D.H. 2012: Where is the consensus? A proposed foundation for moving ecosystem service concepts into practice. Ecological Economics, 77, 27–35.
Natural England 2010: Local Nature Reserves in England: A guide to their selection and declaration. www.naturalengland.org.uk/publications, access: 27.11.2021.
O'Halloran D. 1988: Earth Science Conservation. Nature Conservancy Council, 24, 1–59.
Olewiler N. 2006: Environmental sustainability for urban areas: The role of natural capital indicators. Cities, 23 (3), 184–195.
Ostrom E. 1998: A behavioural approach to the rational choice theory of collective action: Presidential Address, American Political Science Association, 1997. The American Political Science Review, 92 (1), 1–22.
Ostrom E. 2010: Beyond markets and states: Polycentric governance of complex economic systems. American Economic Review, 100 (3), 641–672.
Pan J. 2015: China's Environmental Governing and Ecological Civilization. Springer, Berlin–Heidelberg.
Pelenc J. & Ballet J. 2015: Strong sustainability, critical natural capital and the capability approach. Ecological Economics, 112, 36–44.
Plummer M.L. 2009: Assessing benefit transfer for the valuation of ecosystem services. Frontiers in Ecology and the Environment, 7 (1), 38–45.
Polasky S., Lewis D.J., Plantinga A.J. & Nelson E. 2014: Implementing the optimal provision of ecosystem services. Proceedings of the National Academy of Sciences of the United States of America, 111 (17), 6248–6253.
Raudsepp-Hearne C., Peterson G.D. & Bennett E.M. 2010: Ecosystem service bundles for analysing trade-offs in diverse landscapes. Proceedings of the National Academy of Sciences of the United States of America, 107 (11), 5242–5247.

Regan H.M., Ben-Haim Y., Langford B., Wilson W.G., Lundberg P., Andelman S.J. & Burgman M.A. 2005: Robust decision-making under severe uncertainty for conservation management. Ecological Applications, 15 (4), 1471–1477.

Reyers B., Biggs R., Cumming G.S., Elmqvist T., Hejnowicz A.P. & Polasky S. 2013: Getting the measure of ecosystem services: a social-ecological approach. Frontiers in Ecology and the Environment, 11 (5), 268–273.

Reyers B., Nel J.L., O'Farrell P.J., Sitas N. & Nel D.C. 2015: Navigating complexity through knowledge coproduction: Mainstreaming ecosystem services into disaster risk reduction. Proceedings of the National Academy of Sciences of the United States of America, 112 (24), 7362–7368.

Reyers B., Polasky S., Tallis H., Mooney H.A. & Larigauderie A. 2012: Finding common ground for biodiversity and ecosystem services. BioScience, 62 (5), 503–507.

Risky Business Project 2014: Risky Business: The Economic Risks of Climate Change in the United States. https://riskybusiness.org/reports, access: 20.02.2020.

Ruckelshaus M., McKenzie E., Tallis H., Guerry A., Daily G., Kareiva P., Polasky S., Ricketts T., Bhagabati N., Wood S.A. & Bernhardt J. 2015: Notes from the field: Lessons learned from using ecosystem service approaches to inform real-world decisions. Ecological Economics, 115, 11–21.

Sala E., Costello C., Dougherty D., Heal G., Kelleher K., Murray J.H., Rosenberg A.A. & Sumaila R. 2013: A general business model for marine reserves. PLoS ONE, 8 (4), e58799.

Scheffer M., Barrett S., Carpenter S.R., Folke C., Green A.J., Holmgren M., Hughes T.P., Kosten S., van de Leemput I.A., Nepstad D.C., van Nes E.H., Peeters E.T.H.M. & Walker B. 2015: Creating a safe operating space for iconic ecosystems. Science, 347 (6228), 1317–1319.

Schultz L., Folke C., Österblom H. & Olsson P. 2015: Adaptive governance, ecosystem management, and natural capital. Proceedings of the National Academy of Sciences of the United States of America, 112 (24), 7369–7374.

Sen A. 1985: Commodities and Capabilities. Elsevier, Amsterdam, North-Holland.

Seppelt R., Dormann C.F., Eppink F.V., Lautenbach S. & Schmidt S. 2011: A quantitative review of ecosystem service studies: approaches, shortcomings and the road ahead. Journal of Applied Ecology, 48 (3), 630–636.

Shogren J.F. & Taylor L.O. 2008: On behavioural-environmental economics. Review of Environmental Economics and Policy, 2 (1), 26–44.

Solon J. 2008: Koncepcja "Ecosystem Services" i jej zastosowania w badaniach ekologiczno-krajobrazowych. [In:] Chmielewski T.J. (ed.): Struktura i funkcjonowanie systemów krajobrazowych: meta-analizy, modele, teorie i ich zastosowania. Problemy Ekologii Krajobrazu, 21, 25–44.

Solon J., Roo-Zielińska E., Affek A., Kowalska A., Kruczkowska B., Wolski J., Degórski M., Grabińska B., Kołaczkowska E., Regulska E. & Zawiska I. 2017: Świadczenia ekosystemowe w krajobrazie młodoglacjalnym. Ocena potencjału i wykorzystania. Wydawnictwo Akademickie SEDNO, Polska Akademia Nauk, Instytut Geografii i Przestrzennego Zagospodarowania im. Stanisława Leszczyckiego, Warszawa.

Stern N. 2008: The economics of climate change. American Economic Review: Papers & Proceedings, 98 (2), 1–37.

TEEB 2010: The Economics of Ecosystems and Biodiversity: Mainstreaming the Economics of Nature: A synthesis of the approach, conclusions and recommendations of TEEB. Report.

Telega I. 2014: Kapitał naturalny w koncepcji rozwoju trwałego. Rozprawa doktorska. Uniwersytet Ekonomiczny w Krakowie, Wydział Finansów, Katedra Polityki Przemysłowej i Ekologicznej.

Telega I. 2015: Kapitał naturalny a proces ekologizacji gospodarki. [In:] Kożuch M. (ed.): Ekologizacja gospodarki. Wydawnictwo Fundacja Uniwersytetu Ekonomicznego w Krakowie, Kraków, 31–47.

Tilman D., May R.M., Lehman C.L. & Nowak M.A. 1994: Habitat destruction and the extinction debt. Nature, 371 (6492), 65–66.

Tokarska-Guzik B. 2000: Przyrodnicze zagospodarowanie nieużytków miejsko-przemysłowych na przykładzie centrów górniczych Europy. Inżynieria Ekologiczna, 1, 72–80.

Turner R.K. & Daily G.C. 2008: The ecosystem services framework and natural capital conservation. Environmental and Resource Economics, 39, 25–35.

UK National Ecosystem Assessment – Synthesis of the Key Findings. 2011: UNEP-WCMC, Cambridge.

Wildlife and Countryside Act 1981: http://www.legislation.gov.uk/ukpga/1981/69, access: 22.02.2020.

World Economic Forum 2015: Global Risks 2015. 10th ed. https://www3.weforum.org/docs/WEF_Global_Risks_2015_Report15.pdf, access: 20.02.2020.

4.2. Changing the Management of Post-mining Synanthropic Sites

Philip James
School of Science, Engineering and Environment;
University of Salford, G46 Peel Building, Salford, M5 4WT, UK;
P.James@salford.ac.uk; ORCID: 0000-0001-9079-3953

Mark Howard Champion
Wigan Projects Manager, Lancashire Wildlife Trust;
Worsley Mesnes Drive, Wigan WN3 5UF, UK;
mchampion@lancswt.org.uk; ORCID: 0000-0002-5196-1667

ABSTRACT: The world's first industrial city, Manchester, relied on coal extracted from the Lancashire coalfield. The legacy of the coal mining industry is writ large across the post-industrial landscape of the towns within that coalfield. Today there are complex social-ecological processes interacting in this landscape. A case study of Wigan, a township within Greater Manchester in the north-west of England, in general and of the site of former collieries, identifies a range of twenty-first-century values attributable to the sites. Wigan has extensive derelict sites as a result of the closure of the mining industry and associated engineering and textile industries. At the site of the Bickershaw colliery, innovative ecological conservation interventions are bringing forward new uses for this once-scarred landscape. Anti-social behaviour and neglect have been addressed by creating landscapes that are not conducive to the activities that once took place there, and at the same time provide flood prevention for local residential areas and offer biodiversity gains. At the site of the Garswood Hall collieries, now called Three Sisters Local Nature Reserve, there is value associated with the botanical communities that have become established on the unusual, for this area, surface geology which is the result of the deposition of mining spoil. Towards the centre of Wigan, at what is now called Wigan Flashes Nature Reserve, once disparate communities have been joined, and the reserve is used for a range of informal and formal recreational activities, in addition to providing biodiversity gains. The sites discussed in this case study demonstrate how post-industrial landscapes can be re-imagined to address the complex social-ecological problems that face those living in the twenty-first century and provide guidance that may be considered for adoption at other sites.

KEYWORDS: Ecosystem services, urban ecosystems, green infrastructure

INTRODUCTION

In the eighteenth century, industrialization, which began in Great Britain, spread throughout the world. Today humans inhabit a rapidly urbanizing and industrial planet (United Nations 2014; McPhearson *et al.* 2016). The global urban population is expected to rise from around four billion, the level in the second decade of the twenty-first century, to over six billion by 2045 (United Nations 2014). To the majority of people who live in urban areas, that is to say more than 51% of the global population, nearly 80% of European and US citizens, almost 50% of Asians and more than 90% of Latin Americans (United Nations 2012; Haase 2014), urban landscapes are their everyday environment.

With this increase in urbanization comes a demand for more built infrastructure and the associated movement of natural resources from where they are extracted from the earth, or otherwise provisioned, to sites at which they are processed and pass-through manufacturing processes to the towns and cities where they are used. Population growth, climate change, growing slums and informal settlements, changing demographic characteristics, social inequity, economic fluctuations, pollution, the replacement of ageing infrastructure and other stressors are all factors that involve this flow from the extraction of material to end use and disposal. Such complex social-ecological problems belong to the class of problems known as wicked problems, that is to say problems that cannot be solved by the traditional approach – definition, analysis, and solution – arrived at by sequential steps, rather, they are tackled by authoritative, competitive or collaborative processes (Tab. 1).

Table 1. Approaches to addressing wicked problems, based on Roberts (2000) and Rittel (1972)

APPROACH	COMMENTS
Authoritative	Place the responsibility for solving the problems in the hands of a few people. This reduces the number of stakeholders and hence eliminates some competing points of view, thereby reducing the complexity of the problem. The disadvantage, an appreciation of all the perspectives needed to tackle the problem may be lost.
Competitive	Pits opposing points of view against each other, requiring competing parties to come up with their preferred solutions. This approach weighs different solutions against each other and the best one is chosen. The disadvantage is that this adversarial approach creates a confrontational environment which can discourage sharing and hence there may not be an incentive to come up with the best possible solution.
Collaborative	Aims to engage all stakeholders to find the best possible solution for all stakeholders. Typically, this involves meetings in which issues and ideas are discussed and a common, agreed approach is formulated. Disadvantages of this approach are that it is time-consuming and, in some matters, at least one group of people may hold an absolute belief that contradicts absolute beliefs held by other groups; collaboration becomes impossible until one set of beliefs is relativized or abandoned entirely.

Within these approaches, various methods have been developed, such as problem structuring methods, that are normally used by a group of people in collaboration to create a consensus about, or at least to facilitate negotiations about, what needs to change (Rosenhead 2013) and shift the goal of action from "solution" to "intervention", which requires recognition that actions occur in an ongoing process, and further action will always be required (Knapp 2008). It is this recognition that actions occur in an ongoing process that is central to the study reported here, exploring how wicked problems associated with the environmental restoration of previously mined land might be addressed.

An understanding of ecological principles is seen as key to the transformative change required to achieve cities that are resilient to climate change and other stressors, which is an example of a wicked problem (Royal Society 2014). Cities are increasingly looking to green infrastructure to address some of the challenges they face. Benefits from green infrastructure include air temperature cooling, stormwater absorption, drinking water supply protection, and improved human health (Elmqvist *et al.* 2013). These, and other benefits, are the urban ecosystem services that residents and visitors to urban areas derive from local and regional ecosystems (Gómez-Baggethun *et al.* 2013; Larondelle *et al.* 2014). These benefits are not the products of the ecosystems alone, rather, they are the product of people and ecosystems in combination in complex social-ecological systems of industrial and post-industrial cities where choices made to address these challenges affect social, economic or environmental problems and trade-offs elsewhere (Andersson *et al.* 2015; McPhearson *et al.* 2016), that is to say, it is a wicked problem. Designing and managing urban ecosystems to ensure a resilient and equitable supply of ecosystem services requires expert, local knowledge (Faehnle *et al.* 2015) as well as inputs from engineers, architects, and designers (Ahern *et al.* 2014) as it is important that there is a match between the supply of urban ecosystem services to those who will benefit from them at a local scale as well as wider scales (McPhearson *et al.* 2015; Haase *et al.* 2014). There is a positive feedback loop that operates in this system, one that is usually expressed in negative language – urbanization and soil sealing provoke changes, predominantly a decline, in species diversity and human well-being in cities; both "are inextricably linked" (MEA 2005).

Urban ecosystem services are usually classified into four categories: (1) provisioning services; (2) regulating services; (3) habitat or supporting services; and (4) cultural services (MEA 2005; Cowling *et al.* 2008; TEEB 2011). Provisioning services include material outputs from ecosystems, for example, food, water, medicinal plants, and other resources. Regulating services maintain functions, for example, air and soil quality and flood prevention and control of stormwater and disease. Habitat and supporting services underpin the other services by providing living spaces for organisms. Supporting services also maintain plant and animal diversity. Cultural services include the non-material, socio-ecological benefits (such as psychological and cognitive benefits) people obtain from contact with the natural environment, such as recreation, aesthetic, spiritual, and psychological benefits and tourism (TEEB 2011). Recently the concept of ecosystem services has been modified and extended to that of natural capital.

The term "capital" is used to describe a stock or resource from which revenue or yield can be extracted. Human well-being is one such yield and arises from the use of a combination of types of capital: social capital, human capital, and built capital, all of which are based on natural capital. Four basic categories of natural capital are generally recognised: air, water (fresh, groundwater and marine), land (including soil, space and landscape), and habitats (including the ecosystems, and the flora and fauna which they support) (Ekins *et al.* 2003).

When natural capital assets are depleted in quantity or degraded in quality, which often occurs when natural resources have been extracted from the ground due to mining, the flows of beneficial services from such post-industrial land to people are affected. Policy decisions lead to actions that can impact natural capital via changes in ecosystem structure and function, which in turn alter the ecosystem services delivered. For example, recreational, amenity and other benefits to the population living around a parcel of post-industrial land will be affected by policies that determine the way that land is used, whether that is for legal or illegal (such as fly tipping, off-road biking and 4×4 vehicles, drug-taking, etc.) actions (Rands *et al.* 2010).

In 1978, Michael Morris MP asked the then Secretary of State for the Environment, The Right Honourable Peter Shore MP to set out the extent of derelict land in each of the London boroughs, the metropolitan districts, and the twenty largest non-metropolitan districts in England, expressed in acres and as a percentage of the total area. There were two areas with over 5% of the land listed as derelict: Wigan (2,612 acres (1,057 ha), 5.25%), in Greater Manchester, and Stoke-on-Trent (1,292 acres (523 ha), 5.57%) (Hansard 1978). In a similar question asked by Dr. Jack Cunningham MP on 6 July 1987, the Secretary of State's (at this time The Right Honourable Nicholas Ridley MP) reply included the information that dereliction in Wigan had risen to 6.4% of the borough (Hansard 1987). No figure was given for Stoke-on-Trent. The increase in derelict land in Wigan can be explained by the closure of collieries and other industrial sites in the nine years separating the two figures. Further closures continued until the early 1990s. Hence, Wigan is a borough in which there is much derelict land, and where changes in addressing this wicked problem can be examined.

METHODS

This is a case study of the legacy of mining in Wigan (a large town in Greater Manchester, England, 26 km northwest of Manchester) in general and of three distinct areas: Bickershaw and Garswood Hall collieries, and a number of collieries in the centre of Wigan in particular. Walk-over surveys, interviews with the land manager, user surveys, and an analysis of management plans for these post-industrial sites were conducted, and content analyses of themes revealed changes in the ecosystem service benefits flowing from the natural capital of these areas.

THE LEGACY OF MINING IN WIGAN

The geology around Wigan contains a number of faults that bring coal strata (locally called coal measures) towards the surface so that they can be mined. In his *History of Wigan*, Sinclair (1882) stated that the inhabitants of Wigan "were allowed to gather sticks from the forests of the Rector; then peat was discovered to be an excellent and more easily conveyed substitute, and finally cannel coal at the beginning of the fourteenth century was discovered in some neighbouring fault and found to give both great heat and excellent light". Cannel or Kennel was described in 1680 by R. Blome as "the choicest coal in England" and in 1933, Lord Crawford, the 27th Earl, addressed the Manchester Statistical Society and stated that "cannel is the greatest, indeed the noblest fuel in the world. It is noted for containing a large proportion of hydrogen and a low percentage of vegetable structure. The high calorific value, the rich yield of gas, the brilliance of its flame, the steadiness of its combustion, the cleanliness of the raw fuel and its ashlessness when burned earned it a worldwide reputation" (reported by Anderson & France (1994)). It is clear to see from this description why this natural resource was mined in the Wigan area. The coal measures in Wigan form part of the larger Lancashire coalfields.

McNeil & Nevell (2000) describe how maps from the late eighteenth century show the burgeoning of towns such as Wigan and the neighbouring cities of Salford and Manchester, as they became known for three industries: coal, engineering, and textiles. There was a steady increase in coal output throughout the eighteenth century, peaking in the middle of the nineteenth. The Northern Mine Research Society (n.d.) lists 68 mines in the Wigan area between the 1820s and 1992. There were 54 collieries operating in and around the town at the peak of this activity in 1854, with around 1,000 pit shafts within 8 km (5 miles) of the town centre. Locally, the mined coal was used to fuel engineering (including an iron works operating five blast furnaces at its peak), and the textile industry (Wigan was a centre for cotton spinning and weaving). Wigan's dominance as a cotton town began in the late nineteenth century and lasted until the mid-twentieth century. Other impacts of this industrialization were seen. For example, in the eighteenth century, roads were rough, so the Lancashire landowners, the mine owners of the time, developed the first canals in Britain to improve the transportation of coal from this coalfield to fuel the growth of cities such as Manchester, Salford, and Liverpool (Davies 2010).

From the mid-nineteenth century, railways began to take over from canals as the main means of transporting coal. In the 1830s, Wigan became one of the first towns in Britain to be served by a railway; the line had connections to Preston and to the Manchester and Liverpool Railway (McNeil & Nevell 2000).

Anderson & France (1994) report that coal was not the only natural resource extracted from the ground in Wigan. The many bands of ironstone, which generally has a content of 25–30%, were mined. Fireclay was extracted to make refectory bricks for use in connection with, initially five blast furnaces. Boulder clay was extracted and turned into bricks. Clay was also used for pottery. Glass-making was also an industry set up in the area, drawing on coal for fuel, clay for various types of brick, and sand, though it is thought that silica sand was not abundant in this region and soda would have been brought in. Building stone was extracted from several quarries.

In the twentieth century, manufacturing declined. By the end of the century, the collieries, mills, and ironworks had all closed. The last colliery in the area closed in 1993.

The ecosystems benefits following from the mining of coal included the provisioning services associated with the energy in the coal; the economic benefits to the mine owners, the miners, and those whose work was related to mining; and the social benefits associated with the close-knit mining communities, including the brass band musical heritage. Though life for the workers was tough and there were ecosystem dis-services resulting in poverty and ill-health, as described most famously by George Orwell (1937) in *The Road to Wigan Pier*. With the closure of the pits, there was a legacy of a high level of derelict land in the borough. Hackett (1977) identified six forms of dereliction (Tab. 2), all could be found to a greater or lesser extent within Wigan and which, in the mid-twentieth century, were addressed by a limited range of policies (Tab. 3).

Table 2. Types of dereliction and associated legacy based on Hackett (1977)

TYPE OF DERELICTION	LEGACY
Deep mining	Pit shafts, spoil heaps, derelict buildings and equipment, and subsidence on the surface.
Surface mining	Holes in the ground which may be either a single large excavation or a series of ridges and either filled with water or dry; overburden may be in mounds around the excavation site; spoil heads; derelict buildings and equipment.
Industry	Derelict buildings and equipment; soil contaminated to a greater or lesser extent; lagoons which may be an innocuous body of water or contain toxic elements or slurry.
Derelict housing	Housing for employees of the various industries which does not meet today's standards.
Transport systems	Roads, railways, canals, cuttings, bridges, and other structural works.
Refuse tips	Quarries and other excavations filled with refuse and illegally dumped materials by business or individuals (fly tipping).

BICKERSHAW, GARSWOOD HALL AND CENTRAL WIGAN

Detailed case studies at Bickershaw, Garswood Hall (now landscaped as Three Sisters Local Nature Reserve), and in the centre of Wigan (now the Wigan Flashes Local Nature Reserve) were conducted to identify changes in these policies (Tab. 3) in the second decade of the twenty-first century.

At Bickershaw, Leigh, Wigan, the first shaft was sunk in 1830 and the last colliery closed in 1992. The landscape legacy of the collieries was one that was bare and dominated by large spoil heaps, some of which still exist today. The substrata throughout the area are mining spoils, covered with clay and silt. The site is likely to be rich with iron deposits. The fields laid out as part of the restoration of post-mining land on the site are now abandoned and dominated by tall ruderal grass. Plans in 2010 to develop the site into a golf course around did not proceed, though some enabling work took place: a metalled road was constructed, and the hydrology of the site was changed, with concrete-lined ditches installed to facilitate the rapid flow of water off the site in order to dry it out. There is unmanaged access to the site and this has led to some anti-social behaviour on the site and specifically its use for off-road motorcycling and 4×4 motor vehicles. Along with this off-road activity other anti-social activities were also prevalent (Fig. 1).

Table 3. Mid-twentieth-century policies to address dereliction based on Hackett (1977)

FEATURE	MID-TWENTIETH-CENTURY POLICIES
Spoil heaps	Retain some spoil heaps as features in the landscape and plant with trees, either to add interest or provide a viewing platform. Grade out spoil heaps to produce new contours which are better related to the surrounding landscape. Apply a surface treatment that allows grass to grow with trees on the steeper slopes.
Subsidence and holes in the ground	Where a large area of subsidence fills with water, the resulting lake or "flash" may be developed as an amenity. Where subsidence is limited to small, isolated pockets, they were infilled to level off the landscape. Where the floor of the excavation is impermeable to water or where there is a high-water table, a lake may form in the excavation. Where the pit remains dry and is not too deep, it may be filled and used for agricultural purposes. Alternatively, the pit may be filled with waste as a land-fill site, eventually bringing the landscape back to the original level.
Pit shafts	Pit shafts across the landscape require some form of protection to avoid accidents with people falling into them.
Buildings and equipment	Renovate them for other uses. Buildings may be damaged or derelict and where repair is not possible or desirable, they may be demolished. Areas formerly occupied by the chemical industry or ironworks may be contaminated to a greater or lesser extent. There may also be a lagoon containing an innocuous body of water or toxic elements or slurry which can be difficult to deal with.
Contaminated land	May be treated to remove contamination or sealed over and often planted with grass and trees.
Lagoons	Drained and infilled.
Derelict housing	Renovated or demolished to be replaced by more modern housing. Some material or entire houses may be retained as a cultural link to the former housing as part of museums.
Transport systems	Retain to serve new development within the town or city. Fill in cuttings, removing bridges and other structural works is likely to be only economical where land values are high. Maintain structure that remain so that they are safe.
Refuse tips	Care needs to be exercised in removing these materials as their exact composition may not be known. Typically restored to agriculture or horticulture use or converted into housing, industry, recreation and nature reserves.

Bickershaw is at the head of the Common Lane Brook sub-catchment. A ditch leads from the northeast of the site to a group of small (*circa* 75 m^2) ponds, which became largely silted and dominated by *Glyseria maxima* (Great Manna Grass) with very little open water, though there is a small wetland with associated *Salix* spp. (willows) and fenland grasses. It is now acknowledged that by applying ecological principles, this area of the site can deliver flood mitigation ecosystem services and water-, land-, and habitat-related natural capital. The current ditch network within the wetland has to be converted back to streams with meanders, pools and riffles to allow hydrological processes to develop. It will also create a network of ponds in the existing area (Fig. 2).

Figure 1. The results of 4×4s and other behaviours at Bickershaw
(Photo: Mark Champion)

Figure 2. Before, during, and after work to slow the flow
(Photo: Mark Champion)

This work slows the flow of water draining from the site and further increases the availability of habitat for *Arvicola amphibious* (European water vole) and *Triturus cristatus* (northern crested newt). Both species were found in 2019 in areas of the site that have been restored and where they were once absent. The ditches that flow towards the marshy areas adjacent to Fir Tree Flash have been removed from their concrete-lined channels, profiled, and will be maintained. There is also the opportunity to create small ponds along its length. The area of small ponds has been re-contoured to make a larger body of open water with the potential to retain and store water during high flow regimes. This scheme will have a number of natural capital benefits (Tab. 4).

Table 4. Natural capital benefits from revised management of Bickershaw

RESOURCE/ENVIRONMENTAL RECEPTOR	CAUSE OF IMPACT	DESCRIPTION OF IMPACT	ACTION TO MANAGE IMPACT	TARGET OUTCOME
Population	Increased floodplain	Some reduction in flood risk	Wigan Council to approve scheme	One hectare of extra floodplain
Flora and fauna/ /biodiversity	Increased wetland area	Increased quality habitat	Ditches and bed lowering	One hectare of extra wetland
Air and climate	No change			
Water	Slowing of Upper Common Lane Brook	Reduced soluble iron	Bunds to slow water flow	Clearer water with less iron pollution
Soil, geology and hydro-geology	No change			
Landscape and visual amenity	Tidying and reducing derelict land	Increased naturalness	Careful landscaping	5 hectares of improved wetland and fen
Cultural heritage, archaeology and material assets	Tidying and reducing derelict land	Increased naturalness	Careful landscaping	5 hectares of improved wetland and fen
Traffic and transport	No change			
Use of natural Resources	No change			
Other including cumulative impacts	Improvements throughout Hey Brook	Better habitats for wildlife and people	Careful landscaping	Reduction in areas of post-industrial landscape

As a result of both work by the police and innovative conservation activities (the area is now too wet for off-road motorcycling and 4×4 vehicle (Fig. 3)), there has been a reduction in anti-social behaviour, there is a richer biodiversity, and measures have been taken to improve water retention on the site, providing flood mitigation to protect local communities. Today increasing numbers of people use the site for birdwatching, dog walking, and other more socially acceptable activities. The site is delivering both regulatory (flood mitigation) and cultural ecosystem services.

Three Sisters, Wigan, is on the site of the Garswood Hall Collieries, which closed in 1958. Above ground, women workers, who were called "the Pit Brow Lasses", screened, sorted, cleaned, and graded the coal. The dirt from the screens was carried to and dumped on three heaps which became known as the Three Sisters. The area of the old colliery was converted into the Three Sisters Recreation Area, which includes a motor sports arena (developed in 1974), a boating lake, and nature trails which cater to walkers, cyclists and horse riders in a nature reserve.

The nature reserve is now recognised as holding high biodiversity value, as a local community/visitor resource, an environmental education resource, and an initial "Gateway" into Wigan's Wetlands and Greenheart. In 1977, Ratcliffe set out what has now become known as the Ratcliffe Criteria (Ratcliffe 1977): these are criteria used for assessment of conservation status. The primary criteria defined were typicalness, fragility, size, diversity, naturalness, and rarity. The secondary criteria were recorded history, position in an ecological/geographical unit (ecological coherence), potential value, and intrinsic appeal. The application of these criteria is complex, and different criteria, combinations or emphases are needed for evaluation and selection between the use of the exemplary site and critical standard principles, between habitats and taxonomic groups, and within individual habitats and taxonomic groups.

Figure 3. Area of wetland created at Bickershaw as a result of innovative conservation activities to slow the flow of water off the site (Photo: Mark Champion)

In revisions to these criteria (Bainbridge *et al.* 2013), it was suggested that there should no longer be a distinction between primary and secondary criteria and that the effects of a range of environmental pressures, particularly climate change, and of the ways in which we can mitigate for these effects, demand that ecological coherence and potential value are now given greater attention in the site evaluation and selection process. And, finally, that consideration of ecological coherence and potential value should form part of the assessment of criteria in the selection of new sites for designation. An evaluation of Three Sisters against these criteria is presented in Table 5.

Table 5. Ratcliffe criteria assessment of Three Sisters Local Nature Reserve (based on the reserve management plan, Lancashire Wildlife Trust (2005))

CRITERIA	COMMENTS
Size	The site covers approximately 30.5 hectares.
Diversity	The area is rich botanically and has a good and varied range of habitats.
Fragility	The heath and mossland habitats and ecosystems are intrinsically fragile.
Rarity	The site is home to the nationally scarce and protected *Arvicola terrestris* (water vole). The wet woodland and fen are important for *Poecile montanus* (willow tit), *Locustella naevia* (grasshopper warbler), *Perdix perdix (*grey partridge), *Turdus philomelos* (song thrush), and *Emberiza schoeniclus* (reed bunting) are all Red-listed species of conservation concern. Amber-listed species are *Alcedo atthis (*kingfisher) (which nest nearby and use the site to feed), *Motacila cinereal* (grey wagtail), *Prunella modularis* (dunnock), *Turdus viscivorus* (mistlethrush), *Phylloscpus trochillus* (willow warbler), *Regulus* (goldcrest), and *Carduelis*

CRITERIA	COMMENTS
	cabaret (lesser redpoll). Sphagnum mosslands are developing but are simultaneously being colonized by birch woodland.
Typicalness	The fen, wetland, woodland, mossland, and heathland habitats are valuable as they provide a good example of how an ex-industrial site can be restored to habitats with potential for colonization by predominantly native local species.
Naturalness	The site has developed due to landscaping of the three large colliery spoil tips.
Position in an ecological unit	This reserve lies within the Urban Mersey Basin Natural Area (2,360 square kilometres) which includes some of the most densely populated urban areas in Britain. A range of habitats has been created by industry, with reservoirs, canals, mining subsidence lakes and even some kinds of industrial spoil adding to the variety of habitats and species in this natural area.
Potential value	Further management prescriptions will add to the site's value and definite prescriptions will benefit some of the more locally rare species and those with a high appeal to visitors. There are also opportunities for highlighting habitats and species that will raise awareness of wildlife and countryside issues among the local community.

Not addressed in this summary is the extensive swards of *Danthonia decumbens* (Heath Grass) (Fig. 4), which occur on the remains of the spoil heads. This is a grass of acid pastures and heathland, on sandy or peat soils, which are also often damp. In the UK, the nearest site where swards of heath grass such as those at Three Sisters is on the Great Orme, on the coast of north Wales. *D. decumbens* is not a grass that would be expected in Wigan. This illustrates how the unusual soil types, a legacy of industrial activity, can lead to unexpected floral components of the ecological communities. Those who champion "natural" environments may see this as an ecosystem disservice, while those who embrace the social-ecological model may be more accepting of this example of recombinant ecology resulting from the disturbance and alteration of soil types associated, in this case, with mining. Elsewhere across Wigan there are unusual communities of orchids, these novel habitats being one of the only locations where the hybrid of *Dactylorhiza purpurella* and *D. praetermissa* occurs in the zone of species overlap (Champion 2018). Both the *D. decumbens* and the orchids raise questions about the importance of naturalness within post-industrial landscapes. Their presence challenges the applicability of the Ratcliffe criteria for such landscapes and more widely in urban areas where recombinant communities which have not previously occurred now exist (James 2018). In the revised Ratcliffe criteria set out by Bainbridge *et al.* (2013), there is a presumption against species which are not native and have been introduced accidentally or deliberately to a site or to its biotic communities, stating that they should generally be regarded as undesirable and negative features. There are some exceptions noted, that is to say, where planned reintroductions have taken place and populations have become self-sustaining. Does this go far enough in the post-industrial environment?

Wigan Flashes, a 246-ha Local Nature Reserve in the north-west of England, 1 kilometre to the south of Wigan town centre (Ordnance Survey map reference SD585035), comprise distinct but closely associated areas of open water and wetland shallow wetlands (Westwood Flash, Pearson's Flash, Turner's Flash, Scotsman's Flash, Ochre Flash, Horrock's Flash and Bryn Marsh). The flashes were formed when subsidence occurred following the extraction of coal by both Worsley Mesnes and Pearson and Knowles Collieries from under previously agricultural fields. The main habitats are set out in Table 6.

Management of the site aims to enhance it to become one of the most important wildlife sites in the north-west of England. An important element of this management has been the creation and improvement of nearly 10 km of paths across the site. The ecosystem services at Wigan Flashes are joining once desperate mining communities and, as part of public health initiatives, are addressing preventable twenty-first century lifestyle related health issues, such as obesity, diabetes, cardiovascular disease, cancer, depression, dementia and orthopaedic injury important to health and wellbeing. Today nearly 100,000 visitors use Wigan Flashes annually: an order of magnitude higher than before the paths were instigated.

Figure 4. *Danthonia decumbens* (heath grass)
(Photo: Mark Champion)

Table 6. Habitats within Wigan Flashes

	HABITAT COVER								
	Reedbed (ha)	Fen (ha)	Wet woodland (ha)	Woodland (ha)	Mossland (ha)	Heathland (ha)	Open water (ha)	Streams and ditches (ha)	Grassland (ha)
Bryn Marsh	11.20	1.50	1.90	-	-	-	2.39	14.25	-
Hawkley Hall	7.25	-	-	-	-	-	0.20	-	-
Horrock's Ditch	-	-	-	-	-	-	-	19.83	-
Horrock's Flash	18.00	4.10	3.80	3.70	7.00	-	12.69	-	-
Ince Brook	-	-	-	-	-	-	-	0.69	-
Landfill	-	-	-	-	-	4.90	-	-	34.21
Ochre Flash	0.11	1.80	-	6.40	-	-	2.40	-	-
Pearson's Flash	7.50	2.20		0.60	0.02	-	7.58	7.53	-
Reed Brook	-	-	-	-	-	-	-	2.55	-
Scotman's Flash	1.50	4.97	-	2.00	-	-	26.74	-	-
Turner's Flash	5.38	-	-	-	-	0.17	4.85	1.50	-
Westwood Flash	0.46	-	1.00	8.10	-	0.16	1.89	-	0.30
Total	46.15	14.57	6.70	20.80	2.02	5.23	58.74	46.35	34.51

Casual observation of people at Wigan Flashes suggests that these visitors are engaging with the site in different ways. Barton & Pretty (2010) describe three levels of engagement with nature – viewing nature, functional engagement and active participation – leading to potentially different outcomes. Keniger *et al.* (2013), in their review, recognize the same three categories but use different names. Keniger *et al.* (2013) called the first category "indirect", which they described as viewing nature through a window or in a photograph – experiencing nature while not being physically present in it. This, they stated, becomes important when direct access to nature is restricted. The second type of interaction was called "incidental" and relates to activities where being in the presence of nature is secondary, for example, dog walking and commuting. The final category, called "intentional", implies a positive decision to visit nature and green spaces and actively participate in activities such as bird watching or fungal forays. Understanding the balance between the last two of these categories within a given site will provide insights into site usage and brings into sharp relief questions concerning the value of sites, such as this one, for purposes other than those that are purely focused on conservation, and wildlife protection and enhancement, for example, the Dragon Boat activities of Wigan Ladies' Breast Cancer Dragon Boat Group, which is for those living with breast cancer and their supporters (Fig. 5).

Figure 5. Wigan Ladies' Breast Cancer Dragon Boat Group
(Photo: Mark Champion)

Respondents in an unpublished survey conducted in 2012 for the site manager reported that around two-thirds of the activities of those visiting Wigan Flashes were categorized as being incidental, compared to just over one-third where the interaction with the site was intentional. This ratio was roughly the same for females and males. The proportion of intentional activities by those under 16 was relatively high (around 40%), but the level of this type of interaction fell to between 20–25% in the age groups covering ages 16–45, before climbing again to over 30% in the age range 46–55, to over 40% in the 56–65 age group and to 50% in the over-65s (Tab. 7).

Table 7. Level of engagement (a participant may use a site in different ways at different times and hence may be classed as both incidentally and intentionally interacting with nature in Wigan Flashes)

AGE (YEARS)	LEVEL OF ENGAGEMENT											
	INCIDENTAL						INTENTIONAL					
	Female		Male		Total		Female		Male		Total	
	No.	(%)	No.	(%)	No.	(%)	No.	(%)	No.	(%)	No.	(%)
<16	8	36.3	14	63.6	22	61.1	4	28.6	10	71.4	14	38.9
16–25	9	34.6	17	65.4	26	76.5	2	25.0	6	75.0	8	23.5
26–35	16	42.1	22	57.9	38	77.6	3	27.3	8	72.7	11	22.4
36–45	9	17.0	30	76.9	39	73.6	7	50.0	7	50.0	14	26.4
46–55	14	38.9	22	61.1	36	66.7	8	44.4	10	55.6	18	33.3
56–65	19	37.2	32	62.7	51	56.7	17	43.6	22	56.4	39	43.3
>65	10	30.3	23	69.7	33	50.0	7	10.6	26	39.4	33	50.0
Total	85	63.9	160	64.0	245	64.0	48	36.1	90	36.0	138	36.0

The most incidental reason for visiting Wigan Flashes was to walk the dog (67.7%) while nature-oriented activities were stated as a reason to visit by slightly fewer than one-third of the visitors (a respondent may have indicated more than one reason) (Tab. 8).

Table 8. Reasons for visiting Wigan Flashes (an individual could state that they came for more than one reason, e.g., for walking and bird watching)

REASON FOR VISIT	VISITORS IN EACH REASON CATEGORY					
	Female		Male		Total	
	No.	(%)	No.	(%)	No.	(%)
Walking	53	39.8	106	42.2	159	41.4
Birdwatching/wildlife interest	46	34.6	84	33.5	130	33.8
Cycling	14	10.5	44	17.5	58	15.1
Travelling to work/commuting	7	5.3	23	9.2	30	7.8
Dog walking	65	48.9	105	41.8	170	44.2
Other	24	18.0	82	32.7	106	27.6

The respondents were asked about their perception of the contribution of their visit to their health and well-being. Overall, 95.3% perceived that their visit was beneficial to their health and wellbeing. Every one of the 197 people in the age range 36–65 years perceived their visit in this way. The major benefits to well-being were seen as physical (66.7%) with fewer respondents citing psychological (39.1%) and social well-being benefits from visiting the site (5.2%) (Tab. 9).

Table 9. Perceived well-being benefits of those visiting Wigan Flashes

	NO.	TOTAL FOR ALL TYPES OF WELL-BEING (%)	PHYSICAL WELL-BEING (%)	PSYCHOLOGICAL WELL-BEING (%)	SOCIAL WELL-BEING (%)
Females and males	384	95.3	66.7	39.1	5.2
Females	133	97.0	65.1	34.9	9.3
Males	251	94.4	67.5	41.4	3.0

DISCUSSION

The use, value, and social equity associated with the re-purposing of derelict sites to provide local benefits for human health and well-being are open to examination. The UN Sustainable Development Goal (SDG) 11, the urban SDG, focuses attention on making cities and human settlements inclusive, safe, resilient, and sustainable. Repurposing derelict land has a significant role, as illustrated in this paper, in addressing that goal.

Where once the Lancashire Coalfield was at the heart of the Industrial Revolution and mining of the natural resources meant that provisioning ecosystem services dominated the landscape, now it is regulating and predominantly cultural ecosystem services that are to the fore. The reversal in management policy at Bickershaw from one which was predicated on drying the land to facilitate construction to one which "slows the flow", thereby, retaining water in the site to mitigate flooding indicates how thinking on the natural capital of the site has changed. The move towards restoring ecological functionality, in this case, reconnecting a stream with the wider landscape and creating a larger wetland has had other benefits in reducing anti-social behaviour and facilitating the use of the site by more people undertaking more acceptable activities. A regulating service has had cultural benefits.

The unexpected flora at Three Sisters, a direct result of the industrial substrata of the area, raises questions on the applicability of the Ratcliffe criteria in urban landscapes where the soil has been modified by industrial use. These synanthropic conditions are exploited by species that are not typically of the area. They are not "natural" but still of value.

Several theories provide perspectives on the person-environment relationship, and there are theories and models available to understand health outcomes: attention restoration theory; sense of place; therapeutic landscapes; psychophysiological theory; the bio-psychosocial model and Irvine *et al.*'s (2013) framework of five dimensions (physical (bio), cognitive, affective (psycho), social and spiritual). In this study, respondents were asked about three of these domains – physical (bio), affective (psycho), and social. Of those interviewed, 66.7% considered that the visit to the park was beneficial to their physical health, 39.1% to their psychological health, and 5.2% to their social well-being. There was little difference between women and men with regards to their perception of the benefits to their physical health, but more males than females perceived benefits to their psychological health, a situation that was reversed with regards to social well-being. The UK Government's advice on the health benefits of walking is that these accrue from moderate-intensity aerobic activity – walking fast enough to raise the heart rate and break a sweat – it needs to be faster than a stroll (NHS Choices n.d.). So, while people believe that they are benefiting their physical health, it may be that their level of activity is not sufficient to elicit the benefits they desire. A further study, recording the level of activity, would be required. A suggestion that people are not achieving the health benefits that they perceive they are achieving comes from the twinned observation that 44.2% of those interviewed were in Wigan Flashes in order to walk a dog (Table 8), and that this activity is generally carried out at a strolling pace rather than at a moderately-intense level. Dog walking is among the world's most popular recreational activities and much dog walking takes place in natural areas; it is a primary motivation for many to engage in physical activity. The health and well-being benefits associated with dog walking are considerable and the accrued cost savings associated with disease prevention and healthcare costs associated with this activity run into hundreds of millions of pounds per year (Bauman *et al.* 2001; Cutt *et al.* 2007). Yet a study in Australia has shown that dog walking in woodland leads to a 35% reduction in bird diversity and a 41% reduction in abundance both in areas where dog walking is common and where dogs are prohibited (Banks & Bryant 2007).

Coombes *et al.* (2010) reported that people living closest to formal parks are more likely to achieve the physical activity recommendations and provides evidence that access to urban green space is associated with raised levels of physical activity and that the likelihood of being overweight or obese is reduced. It is thought that this is due to the provision of good path networks found in formal open spaces. In the study reported in this paper, nearly 80% of those visiting Wigan Flashes travelled less than 4 km to the park and over 30% travelled less than 0.5 km. These people were using the park very frequently, with over 75% of those interviewed visiting daily (40.4%) or weekly (35.7%). These data confirm the importance of Wigan Flashes to the local community and the suggestions of other researchers.

Wigan Flashes is managed predominantly for nature and for the provision of supporting cultural ecosystem services (Tab. 3). The data obtained from those interviewed as part of the study reported here were predominantly from local people, using the park very frequently for incidental reasons (such as dog walking). Over 95% of visitors perceived that the visit was beneficial to their health (Tab. 8). These data raise questions around the future direction of the management for Wigan Flashes. That is to say, how can the site be managed to facilitate incidental use by the growing number of people over 56 years of age and at the same time maintain and enhance the experience for the 45% of those visitors in this age group who engage in activities such as bird watching (the intentional engagement)? It is clear from the data collected in this study that provision should be aimed at this growing sector of society who derive much from visiting areas such as Wigan Flashes, and that the management plan should place greater emphasis on the provision of ecosystem services associated with this age group than it currently does.

There were, in 2019, 19% of the UK population 65 years of age and over. It is projected that by 2043, 24% of the population (17.4 million people) will be in that age group (Lewis 2021). Hence, the data obtained in this study suggest that sites such as Wigan Flashes will become increasingly important in the society of tomorrow. Conversely, though, there is research examining the growing disconnect between children and their local, natural environments. This divide has been labelled with terms such as "biophobia" (Wilson 1984), "extinction of experience" (Pyle 1993) and "nature deficit disorder" (Louv 2008). Hence, it is possible that as young people grow older, they will not visit areas such as Wigan Flashes to the same degree as older people currently do. To establish this would require cohort studies following groups and mapping their changing attitudes and behaviours. Wigan Flashes is not unique, it is a typical urban nature area, and, as such, the implications from this study have relevance to all such sites.

These three cases have demonstrated how flood mitigation sits alongside biodiverse-rich habitats, and the provision of a safe and welcoming outdoor environment in which people can take responsibility for their own health. Such is the diversity of ecosystem services in the urban and post-industrial area of the north-west of England.

Natural areas within the urban environment, such as the three case study sites discussed in this chapter, have the potential to improve air and water quality, mitigate flooding, enhance physical and mental health, and promote social and cultural well-being (Keeler *et al.* 2019). There is mounting evidence that interacting with nature delivers measurable health and well-being benefits to people (Keniger *et al.* 2013). However, only a certain proportion of a population will interact intentionally (for example, by watching wildlife, or volunteering) with nature (Lepczyk *et al.* 2004; Gaston *et al.* 2007). The data obtained in the study of Wigan Flashes indicate that this is around 36% and that there was no overall difference when females were compared to males (Tab. 6).

With the advent of methods of enquiry using Big Data comes opportunities for investigations that have not previously been possible. The use of social media data opens up opportunities beyond the limited spatial and highly costly social surveys that have been conducted to date (McPhearson *et al.* 2016). Taking advantage of these advances in data collection and analysis it ought to be possible to develop more extensive and dynamic insights in to the ecological, social and health functioning of urban green spaces such as those described here.

ACCOMPANYING VIDEO

A video of the two authors of this chapter exploring the sites described in this chapter is available at: https://salford.figshare.com/articles/media/Ecosystem_Services_mp4/7981736

REFERENCES

Ahern J., Cilliers S. & Niemelä J. 2014: The concept of ecosystem services in adaptive urban planning and design: A framework for supporting innovation. Landscape and Urban Planning, 125, 254–259.

Anderson D. & France A.A. 1994: Wigan Coal and Iron. Smiths of Wigan, Smiths Books, Wigan.

Andersson M., McPhearson T., Kremer P., Gomez-Baggethun E., Haase D., Tuvendal M. & Wuster D. 2015: Scale and context dependence of ecosystem service providing units. Ecosystem Services, 12, 157–164.

Bainbridge I., Brown A., Burnett N., Corbett P., Cork C., Ferris R., Howe M., Maddock A., Mountford E. & Pritchard S. (eds.) 2013: Guidelines for the selection of biological SSSIs. Part 1: Rationale, Operational Approach and Criteria for Site Selection. JNCC, Peterborough.

Banks P.B. & Bryant J.V. 2007: Four-legged friend or foe? Dog walking displaces native birds from natural areas. Biology Letters, 3 (6), 611–613.

Barton J. & Pretty J. 2010: Urban ecology and human health and wellbeing. [In:] Gaston K.J. (ed.): Urban Ecology. Cambridge University Press, Cambridge, 202–229.

Bauman A.E., Russell S.J., Furber S.E. & Dobson A.J. 2001: The epidemiology of dog walking: an unmet need for human and canine health. The Medical Journal of Australia, 175 (11–12), 632–634.

Champion M. 2018: Biodiversity impacts at various scales following conservation management of synanthropic sites in North West England. PhD thesis, Edge Hill University.

Coombes E., Jones P.A. & Hillsdon M. 2010: The relationship of physical activity and overweight to objectively measured green space accessibility and use. Social Science & Medicine, 70 (6), 816–822.

Cowling R.M., Egoh B., Knight A.T., O'Farrell P.J., Reyers B., Rouget M., Roux D.J., Welz A. & Wilhelm-Rechman A. 2008: An operational model for mainstreaming ecosystem services for implementation. Proceedings of the National Academy of Sciences of the United States of America, 105 (28), 9483–9488.

Cutt H., Giles-Corti B., Knuiman M. & Burke V. 2007: Dog ownership, health and physical activity: A critical review of the literature. Health & Place, 13 (1), 261–272.

Davies A. 2010: Coal Mining in Lancashire & Cheshire. Amberly Publishing, Stroud.

Ekins P., Simon S., Deutsch L., Folke C. & De Groot R. 2003: A framework for the practical application of the concepts of critical natural capital and strong sustainability. Ecological Economics, 44 (2–3), 165–185.

Elmqvist T., Fragkias M., Goodness J., Güneralp B., Marcotullio P.J., McDonald R.I., Parnell S., Schewenius M., Sendstad M., Seto K.C. & Wilkinson C. (eds.) 2013: Urbanization, Biodiversity and Ecosystem Services: Challenges and Opportunities. A Global Assessment. Springer, New York.

Faehnle M., Söderman T., Schulman H. & Lehvävirta S. 2015: Scale-sensitive integration of ecosystem services in urban planning. GeoJournal, 80 (3), 411–425.

Gaston K.J., Fuller R.A., Loram A., MacDonald C., Power S. & Dempsey N. 2007: Urban domestic gardens (XI): variation in urban wildlife gardening in the United Kingdom. Biodiversity and Conservation, 16, 3227–3238.

Gómez-Baggethun E., Gren Å., Barton D.N., Langemeyer J., McPhearson T., O'Farrell P., Andersson E., Hamstead Z. & Kremer P. 2013: Urban ecosystem services. [In:] Elmqvist T., Fragkias M., Goodness J., Güneralp B., Marcotullio P.J., McDonald R.I., Parnell S., Schewenius M., Sendstad M., Seto K.C. & Wilkinson C. (eds.): Urbanization, Biodiversity and Ecosystem Services: Challenges and Opportunities. A Global Assessment. Springer, New York, 175–251.

Haase D. 2014: The nature of urban land use and why it is a special case. [In:] Seto K.C. & Reenberg A. (eds.): Rethinking Global Land Use in an Urban Era. Strüngmann Forum Reports. Volume 14. Cambridge, MIT Press.

Haase D., Larondelle N., Andersson E., Artmann M., Borgström S., Breuste J., Gomez-Baggethun E., Gren Å., Hamstead Z., Hansen R., Kabisch N., Kremer P., Langemeyer J., Rall E.L., McPhearson T., Pauleit S., Qureshi S., Schwarz N., Voigt A., Wurster D. & Elmqvist T. 2014: A quantitative review of urban ecosystem services assessments: concepts, models, and implementation. Ambio, 43, 413–433.

Hackett B. 1977: Landscape Reclamation Practice. IPC Science and Technology Press, Guildford.

Hansard 1978: HC Deb 27 July 1978, Vol. 954, cc896-7W. https://api.parliament.uk/historic-hansard/written-answers/1978/jul/27/derelict-land, access: 30.11.2021.
Hansard 1987: HC Deb 06 July 1987, Vol. 119, cc54-5W. https://api.parliament.uk/historic-hansard/written-answers/1987/jul/06/derelict-land, access: 30.11.2021.
Irvine K.N., Warber S.L., Devine-Wright P. & Gaston K.J. 2013: Understanding urban green space as a health resource: a qualitative comparison of visit motivation and derived effects among park users in Sheffield, UK. International Journal of Environmental Research and Public Health, 10 (1), 417–442.
James P. 2018: The Biology of Urban Environments. Oxford University Press, Oxford.
Keeler B.L., Hamel R., McPhearson T., Hamann M.H., Donahue M.L, Meza Prado K.A., Arkema K.K., Bratman G.N., Brauman K.A., Finlay J.C., Guerry A.D., Hobbie S.E., Johnson J.A., MacDonald G.K., McDonald R.I., Neverisky N. & Wood S.A. 2019: Social-ecological and technological factors moderate the value of urban nature. Nature Sustainability, 2, 29–38.
Keniger L.E., Gaston K.J., Irvine K.N. & Fuller R.A. 2013: What are the benefits of interacting with nature? International Journal of Environmental Research and Public Health, 10, 913–935.
Knapp R. 2008: Wholesome Design for Wicked Problems. Public Sphere Project. http://www.publicsphereproject.org/content/wholesome-design-wicked-problems, access: 28.04.2019.
Lancashire Wildlife Trust 2005: Wigan Flashes: 2nd LNR Management Plan 2006–2011. Lancashire Wildlife Trust, Preston.
Larondelle N., Hamstead Z.A., Kremer P., Haase D. & McPhearson T. 2014: Applying a novel urban structure classification to compare the relationships of urban structure and surface temperature in Berlin and New York City. Applied Geography, 53, 427–437.
Lepczyk C.A., Mertig A.G. & Liu J. 2004: Assessing landowner activities related to birds across rural-to-urban landscapes. Environment Management, 33, 110–125.
Lewis A. 2021: Housing an ageing population: a reading list. House of Commons Library, Briefing Paper 09239, 3 June 2021.
Louv R. 2008: Last Child in the Woods. Saving Our Children from Nature-Deficit Disorder. Algonquin Books, Chapel Hill.
McNeil R. & Nevell M. 2000: A Guide to the Industrial Archaeology of Greater Manchester. Association for Industrial Archaeology Annual Conference, Manchester.
McPhearson T., Andersson E., Elmqvist T. & Frantzeskaki N. 2015: Resilience of and through urban ecosystem services. Ecosystem Services, 12, 152–156.
McPhearson T., Pickett S.T.A., Grimm N.B., Niemelä J., Alberti M., Elmqvist T., Weber C., Hasse D., Breuste J. & Qureshi S. 2016: Advancing urban ecology toward a science of cities. BioScience, 66 (3), 198–212.
MEA – Millennium Ecosystem Assessment. 2005: Ecosystems and Human Well-being: Biodiversity Synthesis. World Resources Institute, Washington DC.
NHS Choices (n.d.): Walking for health. http://www.nhs.uk/Livewell/getting-started-guides/Pages/getting-started-walking.aspx, access: 20.04.2013.
Northern Mine Research Society (n.d.): Wigan Coalfield. https://www.nmrs.org.uk/mines-map/coal-mining-in-the-british-isles/lancashire-coalfield/wigan-coalfield/, access: 21.04.2019.
Orwell G. 1937: The Road to Wigan Pier. Victor Gollancz, London.
Pyle R.M. 1993: The Thunder Tree: Lessons from an Urban Wildland. Houghton Mifflin, Boston.
Rands M.R.W., Adams W.M., Bennun L., Butchart S.H.M., Clements A., Coomes D., Entwistle A., Hodge I., Kapos V., Scharlemann J.P.W., Sutherland W.J. & Vira B. 2010: Biodiversity conservation: challenges beyond 2010. Science, 329, 1298–1303.
Ratcliffe D.A. (ed.) 1977: A Nature Conservation Review. Cambridge University Press, Cambridge, 7–10.
Rittel H. 1972: On the planning crisis: systems analysis of the 'First and Second Generations'. Bedriftsøkonomen, 8, 390–396.
Roberts N. 2000: Wicked problems and network approaches to resolution. International Public Management Review, 1 (1), 1–19.
Rosenhead J. 2013: Problem structuring methods. [In:] Gass I.S. & Fu M.C. (eds.): Encyclopedia of Operations Research and Management Science. 3rd ed. Springer Verlag, New York, 1162–1172.
Royal Society 2014: Resilience to extreme weather. The Royal Society Science Policy Centre report 02/14. Report DES3400. The Royal Society Science Policy Centre, London.
Sinclair D. 1882: The History of Wigan. Volume 1. Wall, Wigan and Kent and Co., London.
TEEB 2011: TEEB Manual for Cities: Ecosystem Services in Urban Management. The Economics of Ecosystems and Biodiversity, Geneva.

United Nations 2012: World Urbanization Prospects: The 2011 Revision. UN Department of Economic and Social Affairs, New York.
United Nations 2014: World Urbanization Prospects: The 2014 Revision Highlights. UN Department of Economic and Social Affairs, New York.
Wilson E. 1984: Biophilia. Harvard University Press, Cambridge.

4.3. Role of Post-industrial Sites in Maintaining Species Diversity of Rare, Endangered and Protected Vascular Plant Species on the Example of the Urban-Industrial Landscapes

Barbara Bacler-Żbikowska
Department of Pharmaceutical Botany,
Faculty of Pharmaceutical Sciences in Sosnowiec,
Medical University of Silesia in Katowice;
30 Ostrogórska Str., 41-200 Sosnowiec, Poland;
bbacler@sum.edu.pl; ORCID: 0000-0003-4139-6307

Teresa Nowak
Institute of Biology, Biotechnology and Environmental Protection,
Faculty of Natural Sciences, University of Silesia in Katowice;
28 Jagiellońska Str., 40-032 Katowice, Poland;
teresa.nowak@us.edu.pl; ORCID: 0000-0001-7519-6309

ABSTRACT: Long-term industrial development has changed the natural environment significantly. The exploitation of coal, sand, gravel, limestone, and dolomites caused changes in the surface and landscape. Some of the side products of excavation of mineral resources provide specific mineral oligotrophic habitats in the urban-industrial ecological mosaic. This study aimed to present the systematic and ecological diversity of protected and endangered species spontaneously colonizing and growing in post-industrial habitats.

The conducted study has revealed that spontaneous flora and vegetation of habitats like quarries, sand and gravel pits, heaps, and subsidence reservoirs are outstanding due to their floristic richness and the participation of rare protected and endangered species. The occurrence of protected and endangered species enriches the ecosystem function in densely populated urban-industrial landscape (habitat and ecosystem) mosaics by maintaining biodiversity in ecosystem services for humans.

The group of endangered and protected vascular plant species recorded in post-industrial areas of the Silesian Upland covers 7.4% of species in the whole flora of the Silesian Province. The analyzed species belong to 59 botanical families, consisting of 45% of the recorded flora of the Silesian Province. The 23 species represent the *Orchidaceae* family (50% of all species occurring in Poland, all members of the *Orchidaceae* family are protected). The presence of 66 endangered species in the Silesian Voivodeship and 21 in Poland proves that the ecological habitat conditions of post-industrial areas are suitable as secondary habitats for these species. Hence, post-industrial sites help to essentially support the survival of the endangered species. As a result of spontaneous colonization and succession, post-industrial areas have become refugees for species disappearing in natural habitats.

KEYWORDS: Biodiversity and ecosystem functioning, post-industrial habitats, spontaneous colonization, ecosystem services

INTRODUCTION

Human activity in the Upper Silesian Industrial District, like in many European industrial centres, has been growing over the last centuries. As a result of the long industrial development, the natural environment has changed significantly. The exploitation of gravel, limestone, coal, sand, and dolomites caused changes in the surface and landscape. Mineral resources exploitation (open cast and deep mining) changes the hydrology and influences surface waters. Besides the intense environmental transformation, some of the side products of mineral resources excavation provide additional, specific, mineral oligotrophic habitats in the urban-industrial habitat mosaic, such as water and wetland habitats. An example of an extending area of water body habitats is where the continuous collapse of the mine sidewalks caused land subsidence, up to 10 m deep, with the subsidence areas becoming filled with groundwater over time.

The largest industrial centres in Poland are in the Silesian Upland – a physical and geographical macroregion in southern Poland. Industrial activity in the Upper Silesian Industrial District has been growing over many centuries. However, during the last 200 years, it has been connected with intense industrialization. The main branches of industry developed in this area, which exerted the most significant impact on the landscape and the natural landscape environment, are the mining and metallurgy industry in the broad sense. In the Tarnowskie Góry and Olkusz area, zinc and lead are extracted, while in the area near Częstochowa, iron ore was also a vital natural resource. The development of mining entailed the need to extract filling sand (Malon 2010; Dulias 2018). An enormous amount of filling sands occurs in the area from Kuźnica Warężyńska near Dąbrowa Górnicza and the Błędowska Desert in the vicinity of Jaworzno. Sandy deposits of glacial and partly aeolian accumulation occur here. The second most prosperous area is the Odra River valley in the Racibórz Basin and the western part of the Silesian Upland (Malon 2010). Other mineral resources such as sedimentary rocks of various ages were also exploited (Dulias 2018).

Due to such use, the natural environment of Silesia has changed significantly. The exploitation of mineral resources by mining disturbs hydrology and influences surface waters. Due to their floristic richness, including the participation of rare and endangered species, the flora of quarries, sand and gravel pits and heaps, sinking reservoirs, and post-coal mining heaps have been the subject of interest for many botanists, biologists, ecologists, and environmentalists. As a result of spontaneous colonization and succession, post-mineral excavation areas have become refugia for species disappearing in natural habitats (Tokarska-Guzik & Rostański 2001; Woźniak *et al.* 2018).

This study aimed to present the taxonomic and ecological diversity of a selected group of protected and endangered species which are spontaneously colonizing and growing on post-mineral excavation habitats. A second aim is to show the importance of maintaining functional biodiversity by presenting the spectrum of ecological indexes represented by the analysed group of protected and endangered species affecting the ecosystem services provided by post-mineral excavation habitats in densely populated urban-industrial landscapes.

MATERIALS AND METHODS

Characteristics of the study area

Geographically, the largest urban and industrial agglomeration in Poland is located in the Katowice Upland (341.13), which is the central part of the Silesian Upland (341.1) and the eastern part of the Tarnogórski Hump (341.12; Kondracki 2011). The second-largest in size is the Rybnickie Plateau (341.15) – the Rybnik Industrial District bordering on the north with the Katowice Upland. The coal mining industry in this area has been developing for over 200 years (Kondracki 2011).

Characteristics of the studied post-mineral excavation sites

Quarries

Carbonate rocks are essential natural resources mined in the Silesian Upland. Most of the resources have been identified in the northern part – on the Tarnogórski Region, in the north part of the Katowice Upland, and within the Jaworznickie Hills (Dulias & Hibszer 2004). In the vicinity of Strzelce Opolskie (Górażdże and Strzelce Opolskie), limestone is mined for the cement industry. The closed quarries in Żychcice, Rogoźnik and Sadowa Góra in Jaworzno (Dulias 2018) were also important for this industry. Many smaller quarries supplied limestone and marl for the limestone industry such as Sosnowiec, Czeladź, Mikołów-Mokre and Ligota Dolna.

The walls and bottoms of the excavations are an ideal habitat for colonization by limestone rock and xerothermic species. Due to the varied exposure of the excavation walls, they are inhabited by both shade-loving and photophilous species. Sometimes water is accumulated on the bottom of the excavation, creating reservoirs or periodically existing ponds. At that time, in the absence of treatments, such flooding was colonized by hammer plant species and peat bog vegetation. The development of excavations for recreational and educational purposes is also noted.

Opencast sandpits

Sand excavations cover about 50 km^2 in the Śląskie Voivodeship (Czylok 2004). The largest in terms of area include: Szczakowa, Dziećkowice, Kuźnica Warężyńska, Dzierżno Duże, Maczki-Bór Zachód and Maczki-Bór Wschód. The pits are 15–20 m deep (Dulias 2018). Some mines are currently closed, and some have been closed for several dozen years (Czylok 2004). They have been reclaimed and managed to establish forests or are subject to secondary vegetation succession, and others are filled with water. In the case of open excavations, the level of groundwater at the bottom is always changing. An alkaline reaction and high mineralization characterize them. The soil formed in these excavations has the form of initial peat with changeable pH depending on groundwater level (Czylok & Szymczyk 2009). Opencast sand pits often provide substitute habitats for numerous scarce plant species.

Heaps

In the Silesian Upland, especially in its central and southwestern parts, post-black coal mine heaps are very common. There are also heaps, much less frequent, associated with zinc and lead ores (Dulias 2018). Heaps and post-mining sites are diverse in petrographic, mineral, chemical, granulometric, and pH composition. They also have very different shapes (Żmuda 1973). These areas are mineral soil substrates. Due to spontaneous processes in many places, formation of plant communities both common and rare species and even species under protection (Rostański 2006; Woźniak 2010; Holeksa *et al.* 2015; Nowak *et al.* 2015).

Subsidence reservoirs

The presence of several thousand water reservoirs in the Silesian Upland, which arose as a result of direct or indirect human influence, is sometimes called the "anthropogenic lake district" (Rzętała 2008; Rzętała & Jaguś 2012). Some of them are associated with subsidence (Rzętała 2008). These reservoirs are created due to underground mining of coal, zinc, and lead ores (Machowski 2003; 2010). Two hundred fifty-one subsidence reservoirs are present (Jankowski & Wach 1980; Machowski 2003). Subsidence reservoirs perform an essential natural function, contributing to creating wetland habitats for plant and animal species not previously found in the area. They are critical as breeding habitats for birds (Machowski 2003).

DATA ACQUISITION

The analyses of species and subspecies growing spontaneously on post-mineral excavation sites included those recorded in the Red List of vascular plants of the Śląskie Voivodeship (Parusel & Urbisz 2012) and protected by law (Journal of Laws 2014, item 1409).

Data on the occurrence of the examined flora elements of the Silesian Upland were obtained from scientific and popular science publications containing detailed data (Celiński *et al.* 1974-1975; 1976; Duda 1992; Orczewska *et al.* 1993; Stebel 1993; Michalska 1994; Domański *et al.* 1995; Hereźniak 1995; Holak & Pawełko 1995; Czylok & Rahmonow 1996; Babczyńska-Sendek & Andrzejczuk 1997; Babczyńska-Sendek *et al.* 1997; Bernacki *et al.* 1997; Tokarska-Guzik 1997; Kołodziejek 1999; Madowicz 2000; Szeląg 2000; Woźniak & Kompała 2000; Woźniak & Rostański 2000; Hereźniak *et al.* 2001; Pasierbiński & Rostański 2001; Czylok & Baryła 2003; Spałek 2003; Szymczyk *et al.* 2003; 2011; Bzdęga *et al.* 2004; Absalon *et al.* 2005; Rostański 2006; Bula 2007; Chmura & Molenda 2007; Szczepańczyk 2007; Zalewska-Gałosz 2008; Cabała *et al.* 2009; Czylok & Szymczyk 2009; Czylok *et al.* 2009; Kompała-Bąba & Bąba 2009; Krajewski 2009; 2011; 2012; 2016a; 2016b; Błońska 2010; Błońska *et al.* 2011; Parusel & Parusel 2011; Tokarska-Guzik *et al.* 2011; Gębicki *et al.* 2013; Bula *et al.* 2014; Osiadacz & Kręciała 2014; Śliwińska-Wyrzychowska *et al.* 2014; Krajewski & Płachno 2015; Krajewski *et al.* 2015; Sieka *et al.* 2015).

Unpublished data collected and made available by the Center for Nature Heritage of Upper Silesia (CDPGŚ database) was also used. An analysis of the hazard category of the compiled species was carried out in the Śląskie Voivodeship, Poland, Europe, and the World using available studies (Parusel & Urbisz 2012; Kaźmierczakowa *et al.* 2016).

The types of post-industrial habitats were also listed, where the species occurrence was recorded: quarries, sand/gravel pits, heaps, and subsidence reservoirs. The ecological characteristics of species include the following habitat parameters: light indicator (L), thermal index (T), soil moisture index (W), trophic index (T), soil acidity index (R), soil resistance index NaCl or/and in water (S) and an indicator of resistance to increased content of heavy metals (M). The values of the listed indicators are based on available sources (Zarzycki *et al.* 2002; Pladias 2014-2020). Moreover, the syntaxonomic affiliation of individual species was determined (Zarzycki *et al.* 2002; Matuszkiewicz 2012). Thanks to syntaxonomic characteristics, information on the importance of post-mineral excavation habitats as refugee habitats was presented.

RESULTS

The group of endangered and protected vascular plant species recorded in post-industrial areas of the Silesian Upland consists of 7.4% of the total number of species in the flora of the Silesian Province. Precious is the occurrence of 23 species from the *Orchidaceae* family, about 50% of all species occurring in Poland, and the botanical family, whose representatives are all legally protected in our country. As many as 66 endangered species in the Silesian Voivodeship and 21 in Poland confirm that the ecological conditions in post-industrial areas are suitable for creating substitute habitats for these species.

The characteristics of ecological features of the analysed species present a broad spectrum, reflected by the diversity of habitat conditions characterizing post-industrial areas. The colonization by the species in question is not supported. It occurs spontaneously, which means that the post-industrial sites help the endangered species to survive. Such areas can contribute to the enrichment of biodiversity. In the future, such results may also be the basis for further comparative analysis with other post-industrial, post-excavation regions in Poland and abroad.

Results of the analysis of the examined protected and endangered species flora

In the post-mineral excavation sites of the Silesian Upland, 154 protected and endangered species, belonging to 59 botanical families, make up 45% of the recorded flora of the Silesian Province. The most represented were the following families: *Orchidaceae* (23 species), *Cyperaceae* (15), *Rosaceae* (8), *Fabaceae* and *Asteraceae* (7 species), *Poaceae* and *Potamogetonaceae* (6 species). Five species are spore plants – ferns, horsetails, and club mosses, and the rest are seed plants.

Thirty species under strict and 33 under partial protection were recorded. The participation of species representing individual threat categories is presented in Table 1. Many of the recorded species are also endangered in Poland.

Table 1. Participation of protected and endangered plant species recorded on the post-mineral excavation sites in terms of threat categories and different spatial scales

THREAT CATEGORIES	SILESIA VOIVODESHIP	POLAND	EUROPE	WORLD
CR – Critically Endangered	3	3	--	--
EN – Endangered	26	4	--	--
VU – Vulnerable	66	21	--	--
NT – Near Threatened	20	38	5	1
DD – Data Deficient	8	3	1	1
LC – Least Concern	29	--	52	20
R (Re) – Regionally Extinct	1	1	--	--
E – Extinct	--	--	1	--

The most significant number of protected and endangered species was recorded in quarries (80), in a similar number on sand and gravel pits (58) and on heaps (55), while the least in sinkhole reservoirs (35). As many as 14 species were present on three different types of objects: *Centaurium erythraea* subsp. *erythraea*, *Epipactis atrorubens*, *E. palustris*, *Eriophorum latifolium*, *Malaxis monophyllos*, *Melampyrum sylvaticum*, *Myricaria germanica*, *Najas marina*, *Neottia nidus-avis*, *Polypodium vulgare*, *Potentilla collina*, *Pyrola chlorantha*, *P. minor*, and *Viola rupestris*. Two species – *Schoenoplectus tabernaemontani* and *Utricularia australis* – occurred on four types of post-industrial facilities.

The analysis of the group of protection and endangered species regarding the participation of particular light index values shows the most of the analysed species growing on the post-mineral excavation sites represent the habitats of moderate light and full light (Fig. 1). Most of the analysed species represent the habitats of moderately cool and moderately warm conditions (Fig. 2). The number of protected and endangered species growing on the post-mineral excavation sites regarding particular soil moisture indexes, shows that species of fresh soil habitats clearly dominate. Nearly half of the species represent aquatic habitats and the number of dry soil species is smaller (Fig. 3). Among the analysed spectrum of the trophy index, the mesotrophic species are the largest group (Fig. 4). The presence of species occurring on alkaline soils is justified by the presence of the largest number of protected and threatened species in quarries. The second largest group is made up of soil species from alkaline to neutral (Fig. 5).

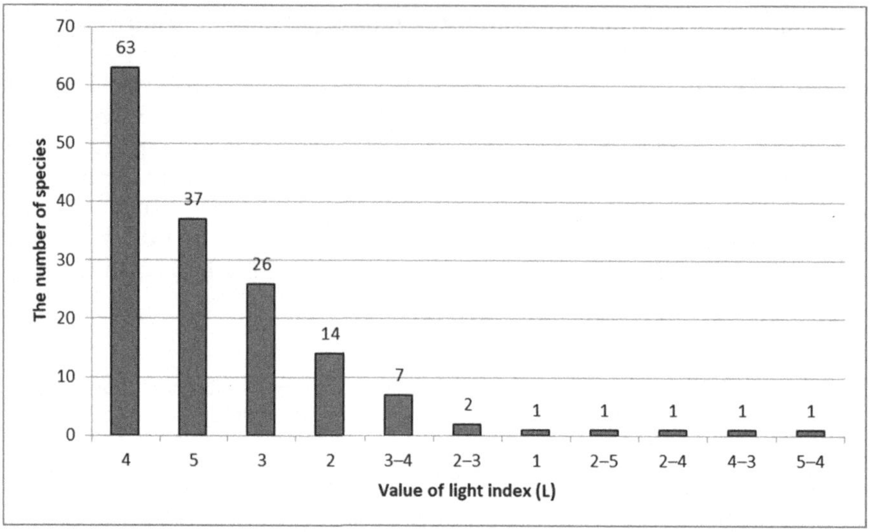

Figure 1. The number of species of the group of protected and endangered species growing on the post-mineral excavation sites representing particular light indexes: 1 – deep shade, 2 – moderate shade, 3 – half shade, 4 – moderate light, 5 – full light

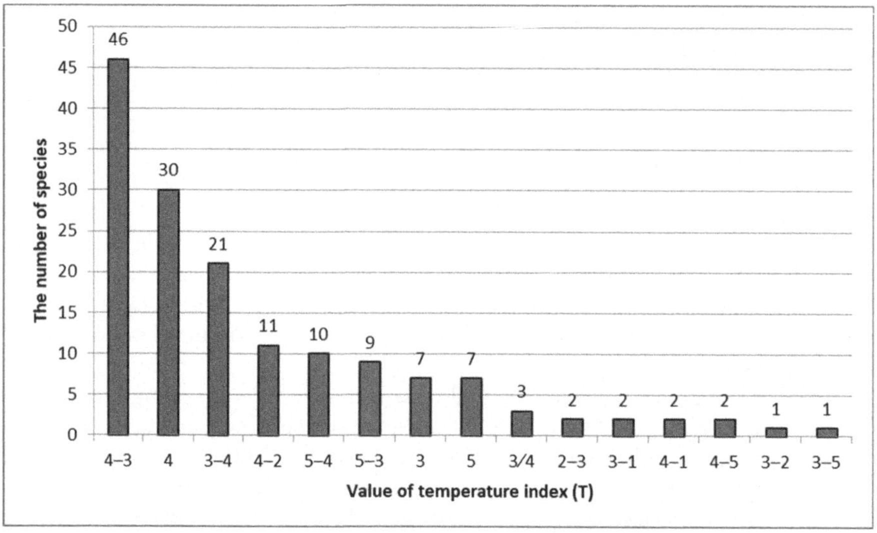

Figure 2. Comparison of number of protected and endangered species growing on post-mineral excavation sites representing particular temperature indexes: 1 – the coldest areas of the country, mainly alpine and subalpine floors, 2 – moderately cold areas, mainly the subalpine and upper regions, 3 – moderately cold climatic conditions, lower regions, northern lowlands and special micro-habitats – raised bogs, 4 – moderately warm climatic conditions, mainly lowlands, 5 – the warmest regions and micro-habitats, thermally advantaged areas

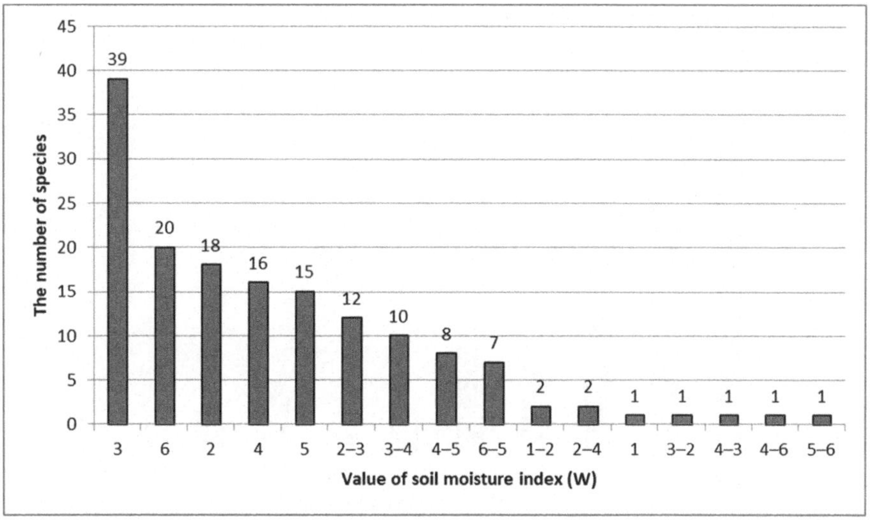

Figure 3. Comparison of number of protected and endangered species growing on the post-mineral excavation sites representing particular soil moisture indexes: 1 – very dry habitats, 2 – dry habitats, 3 – moderately moist habitats, 4 – very moist habitats, 5 – wet habitats, 6 – aquatic habitats

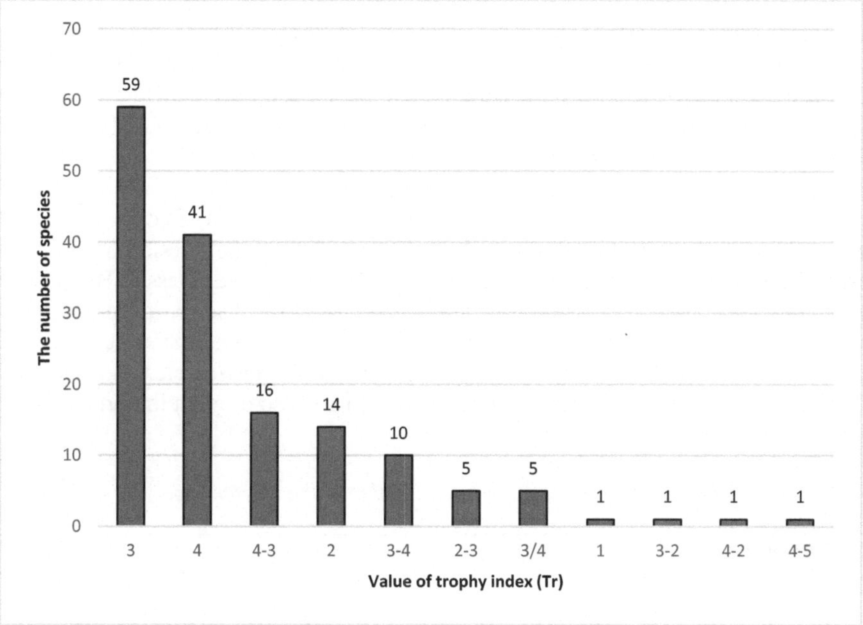

Figure 4. Comparison of number of protected and endangered species growing on the post-mineral excavation sites representing particular trophy indexes: 1 – extremely poor (extremely oligotrophic) soils (water) – raised bogs, loose sands, dry coniferous forests, 2 – poor (oligotrophic) soils (water) – fresh pine coniferous forests, 3 – moderately rich (mesotrophic) soils (water) – mixed forest, tall, acidophilous oak and beech forests, 4 – rich (eutrophic) soils (water) – lowland, fertile beech forests, 5 – very rich (extremely fertile, over-fertilised) soils (water)

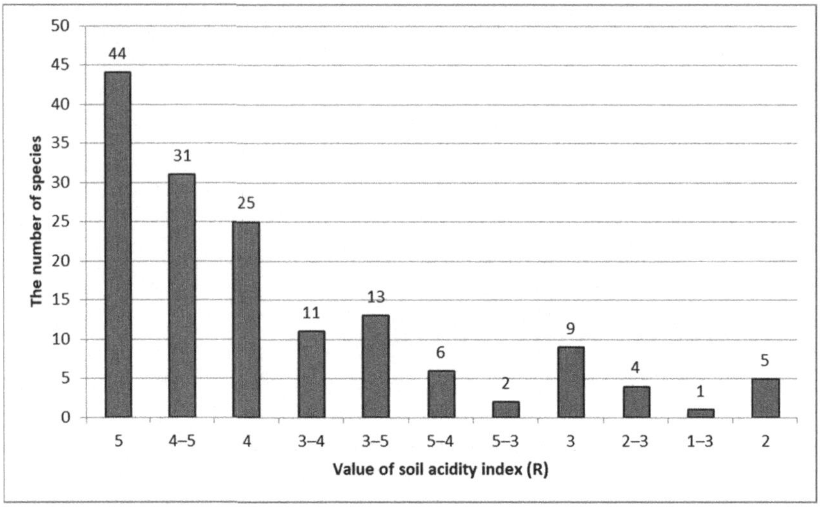

Figure 5. Comparison of number of protected and endangered species growing on the post-mineral excavation sites representing particular soil acidity indexes: 1 – highly acidic soils (pH < 4), 2 – acidic soils (4 ≤ pH < 5), 3 – moderately acidic (5 ≤ pH < 6), 4 – subneutral to neutral (6 ≤ pH < 7), 5 – alkaline (pH ≥ 7)

Among the analysed species, twelve – *Centaurium pulchellum*, *Dactylorhiza incarnata*, *Ophioglossum vulgatum*, *Batrachium aquatile*, *B. circinatum*, *B. trichophyllum*, *Callitriche autumnalis*, *Cladium mariscus*, *Nuphar lutea*, *Nymphaea alba*, *Najas marina*, *Zannichellia palustris* were observed as optional halophytes. In turn *Bulboschoenus maritimus* occurred mainly on soils with increased NaCl content (obligatory halophyte).

According to the criteria used in this study, three species (*Botrychium lunaria*, *Carlina acaulis*, *Alchemilla glacescens*) were classified as tolerant of increased contents of heavy metals.

Endangered and protected species of post-industrial habitats of the Silesian Upland represented 23 phytosociological classes. Some species are classified in two or three classes. However, the most probable one based on local conditions was selected for analysis. The most represented are deciduous forest species from the *Querco-Fagetea* class (33 species) and xerothermic grassland species from the *Festuco-Brometea* class (28). Water species of the *Potametea* class (20) and low-marsh meadows of the *Scheuchzerio-Caricetea nigrae* class (18) are also of great importance (Tab. 2).

Table 2. The phytosociological affiliation of the recorded species

PHYTOSOCIOLOGICAL CLASS	CHARACTERISTICS OF SYNTAXON	NUMBER OF SPECIES
Alnetea glutinosae	Forests with dominant black alder *Alnus glutinosa* or thickets of broad-leaved willows with alder, in depressions with complex drainage during periodically high water levels, on wet peat or peat-mineral soils	2
Asplenietea rupestria	Vegetation of this unit represents natural communities of rock crevices in rocky parts of the mountains; in the lowlands more often as anthropogenic communities in secondary habitats (walls, ruins)	2
Bidentetea tripartiti	Natural and semi-natural therophyte communities on silty shores of water and periodically flooded depressions	1
Chenopodietea	The previously distinguished vegetation unit which was gathering vegetation patches occurring among root crops, e.g., potato, beetroot, carrots	1

PHYTOSOCIOLOGICAL CLASS	CHARACTERISTICS OF SYNTAXON	NUMBER OF SPECIES
Epilobietea angustifolii	Unit of nitrophilic communities occurring on strata frequently trampled, and habitats of ruderal origin	1
Festuco-Brometea	The communities of thermophilic grasslands of a steppe character, on a substrate with an alkaline reaction, rich in calcium	28
Isoëto-Nanojuncetea	Vegetation patches of small summer and autumn therophytes, appearing ephemerally on moist and mineral wet habitats	1
Koelerio glaucae- -Corynephoretea canescentis	This unit gathers vegetation patches of psammophilus grasslands occurring in dry and relatively poor sandy or gravelly habitats	1
Littorelletea uniflorae	The vegetation communities which are composed of small aquatic or semi-aquatic perennials occurring in inland soft, oligo- or mesotrophic water bodies built up mostly by perennial plants	1
Molinio-Arrhenatheretea	Semi-natural and anthropogenic meadow and pasture communities on meso- and eutrophic undisturbed mineral and organic-mineral soils, possibly mineralized and dried muck made of low peat	11
Nardo-Callunetea	Communities of moors and poor *Nardus* grasslands. Semi-natural and anthropogenic communities, which in the original landscape occupied a small area under specific environmental conditions, spread spontaneously and developed in their present form as a result of human activity	4
Oxycocco-Sphagnetea	Shrub-sphagnum, sometimes with woody plants, communities of wet heaths, and raised bogs in acidic oligo- and dystrophic habitats fed exclusively or mainly by rainwater	1
Phragmitetea	Communities of rushes and salt meadows. The societies of grassland rushes, large sludge rushes, and other rushes with the participation of magnificent dicotyledonous perennials occurring in the coastal and coastal zones of inland stagnant and flowing water reservoirs	9
Potametea	The unit gather (Eurosiberian) communities of freshwater macrophytes in mesotrophic and eutrophic inland water reservoirs	20
Querco-Fagetea	This unit gathers Meso- and eutrophic deciduous forests that shed their leaves for winter, growing on mineral soils with varying degrees of moisture and representing in the humid parts of temperate continents a zonal (zonal) type of plants formation	33
Ruppietea maritimae	Vegetation patches of mainly rooted aquatic plant communities; extremely specialized communities of open so-called submarine meadows in the sublittoral zone, covering the seabed to a depth of 4 m	1
Scheuchzerio-Caricetea nigrae	The vegetation of low sage communities of marsh, and transitional fens, rich in bryophytes, create habitats of Nature 2000 identified under the number 7140	18
Stellarietea mediae	Communities of arable fields that arise spontaneously in the conditions of human agricultural extensive activity. For a long time, these vegetation patches were considered as undesirable vegetation and plants competing with crops	1
Thlaspietea rotundifolii	Pioneering vegetation communities of movable or poorly fixed screes in rocky parts of mountains and stonework in mountain sections of river valleys	1
Trifolio-Geranietea sanguinei	Thermophilic fringe vegetation communities in the contact zone of some forest or shrub communities with grass and grassland communities	4
Utricularietea intermedio-minoris	A specialized ecological group of vegetation communities occurring in shallow dystrophic water reservoirs on a peat substrate, in valleys and depressions in the peat bog complex, and in oligo- and dystrophic reservoirs	3
Vaccinio-Piceetea	Holarctic-boreal acidophilic, oligo- and mesotrophic forest vegetation communities with a predominance of coniferous woody species, shrubs, and mesophilic bryophytes	9

RESEARCH RESULTS AND MANAGEMENT DIRECTIONS

Regardless of the above-presented results, the post-industrial, post-mineral excavation of large areas in the Silesian Upland is still analysed primarily due to its economic importance (Henzel *et al.* 2009; Zagórska 2013). According to the current law prescription, the analysed sites undergo reclamation and development management plans (Ostręga & Uberman 2010; Pałasz 2012 and literature cited therein). Unfortunately, the law does not follow the knowledge concerning the dynamic biological, ecological, and ecosystem functioning processes. In some cases, the lack of application of the newest environmental experiences and scientific expertise in environmental management decisions and restoration actions, is causing more harm to the ecosystem's functioning than the mineral resources excavations themselves. Hopefully, global change challenges will force the law to follow the enhancement of ecosystem functioning of the *de novo* established human-created habitats. The issues of effective reclamation and its directions are the subject of numerous studies (e.g., Różkowski *et al.* 2010; Poros & Sobczyk 2013; 2014; Pietrzyk-Sokulska & Kulczycka 2014; Sobala & Pukowiec 2014; Bednarczyk *et al.* 2015). Most of them are agricultural and engineering-focused, ignoring the ecosystem functioning parameters, neglecting that all human life aspects depends on ecosystem functioning – ecosystem services. Unfortunately, it is common practice that the site is more likely to be restored in an old fashioned, wrong way, harmful for ecosystem function. The restored area is reused for economic purposes while preserving cultural heritage, and biodiversity enhancement centres are scarce. From a nature perspective, attempts of reclamation treatments seldom bring the expected results and sometimes harm the natural environment more than the industry activity alone (Woźniak 2010; Rostański *et al.* 2015; Woźniak *et al.* 2018). The same was observed and recorded in other countries a long time ago (Box 1999; Tropek *et al.* 2012). The high environmental value of many of the post-mineral excavation sites is neglected and ignored because of a lack of knowledge, partially because law and regulations do not follow the current scientific achievements.

Plants associated with soils rich in heavy metals – metallophytes' natural habitats – are the calamine grasslands. The calamine grasslands are an excellent example of a specific group of species established on post-mineral excavation habitats. The reclamation activities are destroying the valuable metallophytes communities (Rostański *et al.* 2015).

Other critical examples are peat bogs, wetlands, and water ecosystems. Generally, the natural potential of these ecosystems can be lost during the wrong so-called reclamation actions.

The described biodiversity of post-industrial areas contributed to paying attention to the natural approach to the reclamation process (Rostański 2000; 2001; 2006; Tokarska-Guzik 2000; Woźniak & Kompała 2000; Rostański & Tokarska-Guzik 2001). Research on organism strategies, intense adaptation processes, and mechanisms of spontaneous natural development occurring during the colonization process of post-industrial areas, provides vast possibilities for practitioners (Woźniak 2006; 2010; Woźniak & Cohn 2007; Błońska *et al.* 2019a; 2019b; 2020; Kompała-Bąba *et al.* 2020). The natural processes and organism's adaptations are of great economic importance and contribute to fully implementing the well-understood sustainable development principles in the undertaken reclamation activities. In the Anthropocene, each action which affects environmental ecosystem functioning should proceed, applying the Natural Capital and Novel Ecosystems concepts.

Biodiversity and natural values of post-mineral exploitation sites

Post-industrial, particularly post-mineral exploitation sites subjected to spontaneous succession, are characterized by specific habitat conditions, and thus unique biocoenoses often arise (Tokarska-Guzik & Rostański 2001; Woźniak 2010; Woźniak *et al.* 2018). These habitats often provide conditions for high levels of biodiversity to establish and develop. 581 plant species were found on the dumps after hard coal mining (Rostański 2006), while in the Olkusz Ore District, there were more than 800 species of vascular plants (Nowak *et al.* 2011), which is about 20% of Poland's flora. The described areas can be enclaves of rare and disappearing plant species in a given area (Tokarska-Guzik 1996; Rostański 1997; 2000; Tokarska-Guzik & Rostański 2001; Woźniak 2010; Woźniak *et al.* 2018). Metalliferous areas are habitats where calcareous grasslands develop (Kapusta *et al.* 2010) – a natural habitat protected under European Union law (Council Directive 1992). Specific forms of nature protection like SSSI are available to preserve precious wildlife in post-industrial areas in the UK (Rostański *et al.* 2015). Polish legislation, however, lacks analogous arrangements (Rostański 2006). Many ecosystem functions of the post-industrial, post-mineral excavation sites could provide ecosystem services for people in the urban and industrialized lands. The best-developed galman grasslands in the Olkusz Ore District are preserved in ecological reserves in the Natura 2000 area with calamine grasslands (Kowolik *et al.* 2010; Rostański *et al.* 2015; Regional Directorate for Environment Protection in Katowice 2019). Some sites have also been included in the UNESCO World Heritage List. The surface form of protection as a natural and landscape complex called "Żabie Doły" covered former flotation settlers on the border of Bytom, Chorzów and Piekary Śląskie towns. The primary biological value of this area is its richness in bird species. The following example is the open cast sandpit, which exhibits high natural values (Kompała-Bąba & Bąba 2009; Krajewski 2009). The habitat conditions enable the colonization of one representative of the orchid family – *Liparis loeselii*. This plant species is present in Annex II of the Habitats Directive (Council Directive 1992). Due to the presence of the protected species *Liparis loeselii*, a Natura 2000 area has been established – "Lipienniki in Dąbrowa Górnicza" on the "Kuźnica Warężyńska" and "Torfowisko Sosnowiec-Bory" excavations in the area of the Maczki-Bór mine – and partly covers the ecological use of "Torfowisko Bory" (Regional Directorate for Environment Protection in Katowice 2019). In addition, natural habitats are protected under the Habitats Directive (1992) on post-mineral excavation sites: transitional peat bogs and swamps ("Lipienniki in Dąbrowa Górnicza", "Torfowisko Sosnowiec-Bory"), shores and drained bottoms of water reservoirs with communities from *Littorelletea*, hard-water oligos and mesotrophic reservoirs with underwater meadows of *Charetea* shores ("Lipienniki in Dąbrowa Górnicza"). The Tarnowskie Góry area is related to lead and silver mining. The system of underground corridors has become an ideal place for bats (Annex II of the Habitats Directive). Therefore, to protect these animals, the Natura 2000 "Underground Tarnogórsko-Bytomskie Underground" area was established. These examples confirm the importance of secondary habitats for preserving biodiversity in an area of increasing human impact.

In industrial, post-mineral excavation space, spontaneous development and the constant adaptation and evolution of new ecological units and systems in post-industrial areas indirectly contribute to improving the functioning and quality of human life. For all types of such regions, unexpectedly large species diversity has been recorded. The organisms' diversity results from specific, varied micro-habitats present on a relatively small area. Such variety of micro-habitats contributes to increasing the potential for colonization by a broad group of species. In addition to plants, the biocenosis function for various animal groups is emphasized (Tropek & Konvicka 2008; Tropek *et al.* 2010; 2013; Rostański *et al.* 2015). Thus, a specific "network of life" is created, increasing the surface of green wildlife areas and supporting the flow of genes in regions with significant fragmentation of habitats.

Biodiversity as the basis of ecosystem functions

The close relationship between biodiversity (including rare and protected species) and ecosystem functioning and ecosystem services has been proven. However, it is still poorly understood (Balvanera *et al.* 2016). Many authors accept that biodiversity is an ecosystem service itself, and many field studies provide supporting data (Balvanera *et al.* 2016; de Groot *et al.* 2016). It is very often difficult to estimate the role of individual species in providing specific ecosystem services (de Groot *et al.* 2016). Scientists undertake many projects regarding this problem. Species diversity plays an essential role in increasing resistance to invasive species and other disturbances (Wilsey & Polley

2002), which is extremely important in areas transformed by the human economy. Although seemingly most species seem superfluous, long-term studies show that high species diversity gives more excellent stability to the ecosystem, which leads to more significant and more stable provision of ecosystem services (Yachi & Loreau 1999).

Large species diversity is, above all a precious source of difficult to estimate non-material benefits. It is underestimated and even ignored. However, it predominantly affects human life, including economic, and physical, mental, spiritual, and aesthetic health, not to mention nutrition and shelter. The richness of vascular plant species and the related above and below ground diversity of organisms also shows a relationship between, among others, services such as atmosphere regulation or pollinators, prevention of erosion, and stimulation of soil-forming processes (Balvanera *et al.* 2016). Identified donors of ecosystem services in the studied area are rare and endangered species on the scale of the Silesian Voivodeship, Poland, Europe, and the world. In the case of plants, there is a clear beneficial relationship between the richness of plant species and ecosystem services related to the supply of usable raw materials (e.g., medicinal, cosmetic, nutritional), genetic resources, erosion control, or soil-forming processes (Balvanera *et al.* 2016; de Groot *et al.* 2016).

Rare, endangered, and protected species biodiversity enhancement and the concept of ecosystem services in terms of post-mineral excavation sites

Natural resources are of interest to various disciplines, social, economic, and above all-natural sciences. The concept of ecosystem services allows combining research results from multiple fields and determining the functions of selected ecosystems beneficial for humans (Solon 2008).

Currently, the Millennium Ecosystem Assessment (MEA 2005) divides ecosystems services into four main groups:

1. Supply services related to:
   - food production (breeding of plants and animals) and obtaining food from the wild (collecting, hunting, and fishing);
   - obtaining raw materials of biological origin, including wood or fibrous raw materials;
   - genetic, medical, or decorative resources related to biological diversity;
   - water supply.
2. Regulatory services related to:
   - regulation of air composition;
   - climate regulation (temperature, rainfall) and atmospheric composition;
   - regulation of extreme phenomena, i.e., floods, storms, hurricanes, and fires, and prevention of erosion;
   - biological regulation, i.e., regulation of pest numbers, pollination, and their spread;
   - regulation of soil processes (soil-forming processes and improving the usable quality of soils);
   - regulation of the absorption of dust, gas, and water pollution.
3. Supporting services, among others elemental circulation, primary production, or habitat functions.
4. Cultural functions such as aesthetic, tourist, spiritual, scientific or educational functions.

All services mutually support each other and affect the integrity of the entire ecosystem. For example, procurement services are directly related to acquiring specific ecosystem goods, which may impoverish other benefits.

It is necessary to identify the ecosystem component directly responsible for providing the service to identify and evaluate a particular ecosystem service. Pollinators or flowering plants considered rare and endangered might be an example of an ecosystem component available to identify and assess a functional ecosystem service unit (Kremen 2005).

There are various connections, both positive and negative, between individual functions. Changing the quality of one ecosystem function causes changes in the related ecosystem services (Solon 2008). At the same time, the growth and protection of biodiversity causes an increase in all ecosystem services (Hockings 2003; Perrings *et al.* 2011).

In turn, obtaining natural resources is associated with a temporary decrease in some ecosystem services. Therefore, it is necessary to reconstruct the nature of the transformed exploitation areas to restore and enhance the natural regeneration of the biological ecosystem services and, consequently, economic value (DEFRA 2011).

We live in an age called the Anthropocene. The number of natural habitats decreases year after year in favour of areas subject to substantial human influence. In recent years, there have been increasing numbers of publications regarding the valuation of ecosystem services in natural areas and areas used by man (e.g., Rosin *et al.* 2011; Szkaradkiewicz *et al.* 2014; Wieliczko 2016), and protected areas (Boćkowski & Rogowski 2018).

Traditionally, human changed areas were considered to be more impoverished areas. However, scientific research is often different. Thinking about the natural environment with the concept of natural capital, novel ecosystems, and ecosystem services, gives a new perspective for practitioners. Spatial planning and management of urbanized areas and implementation of sustainable development in such areas require detailed analyses of the factors influencing the ecosystem functioning to enhance the rebuilding and reestablishment of the type of ecosystem which best adapted to new habitat conditions. All natural resources and natural conditions considered from the perspective of their suitability for humans, meeting their broadly understood needs, represent the environment's potential (Sudra 2015). It depends on the structure (biodiversity) and functioning of a given ecosystem and processes occurring in neighbouring ecosystems (Solon 2008).

Human activities sometimes contribute to the reduction of the potential of ecosystems. It is essential to identify when to restore environmental benefits and when natural processes are the most beneficial for the ecosystem and biological and ecological approaches to re-establish.

Investments in the scientific understanding of the mechanisms of reconstruction and sustainable use of ecosystems generate beneficial natural, economic, and social results (de Groot *et al.* 2010). The recreation of ecosystem function of the terrain's ability to provide ecosystem services, even of the sites most distorted in an irreversible manner, e.g., by opencast mines, are possible through natural processes (Zipper *et al.* 2011; Larondelle & Haase 2012).

The assessment of the costs necessary to preserve ecosystem services is met by Landscape Equivalency Analysis making it possible to balance economic activities with ecosystem protection goals.

SUMMARY

The group of endangered and protected vascular plant species recorded in post-industrial, post-mineral excavation areas of the Silesian Upland consists of 7.4% of the total number of species in the flora of the Silesian Province. It is very diverse both systematically and in variety of habitats. There are 23 species from the *Orchidaceae* family, about 50% of all species occurring in Poland, and the botanical family, whose representatives are all legally protected in our country. As many as 66 endangered species in the Silesian Voivodeship and 21 in Poland confirm that the ecological conditions in post-industrial areas are suitable for creating substitute habitats for these species.

The ecological features of the analysed group of endangered and protected vascular plant species present a broad spectrum, representing the diversity of habitat conditions characterizing the post-mineral excavation areas. Humans, e.g., through reclamation treatments, do not support settlement by plants. It occurs spontaneously during the colonization and establishment process. The post-industrial sites enhance the survival of endangered species. Such habitats contribute to the enrichment of biodiversity. At the same time, biodiversity is the prerequisite condition for ecosystem functioning and consequently for ecosystem services. In the future, the post-mineral excavation sites may be the basis for further comparative analysis with other post-industrial areas in Poland and abroad.

REFERENCES

Absalon D., Leśniok M., Nita J., Parusel J.B. & Wower A. 2005: Będzin – przyroda i krajobrazy. Urząd Miejski w Będzinie, Będzin.

Babczyńska-Sendek B. & Andrzejczuk I. 1997: Goryczka krzyżowa *Gentiana cruciata* L. w okolicach Tarnowskich Gór. Natura Silesiae Superioris, 1, 33–42.

Babczyńska-Sendek B., Kompała A. & Maćkowiak G. 1997: Nowe stanowisko *Euphorbia epithymoides* (*Euphorbiaceae*) na Wyżynie Śląskiej. Fragmenta Floristica et Geobotanica, Series Polonica, 4, 371–374.

Balvanera P., Quijas S., Martín-López B., Barrios E., Dee L., Isbell F., Durance I., White P., Blanchard R. & de Groot R. 2016: The links between biodiversity and ecosystem services. [In:] Potschin M., Haines-Young R., Fish R. & Turner R.K. (eds.): Routledge Handbook od Ecosystem Services. 1st ed. Routledge, Taylor & Francis Group, London–New York, 45–59.

Bednarczyk S., Galiniak G., Różkowski K. & Kaznowska-Opala K. 2015: Charakterystyka rekultywacji terenów poeksploatacyjnych w Kopalni Piasku „Szczakowa" w Jaworznie. Przegląd Górniczy, 71 (9), 85–91.

Bernacki L., Butwiłowska B. & Bernacka M. 1997: Występowanie *Iris sibirica* L. i *Gladiolus imbricatus* L. w zachodniej części Płaskowyżu Rybnickiego (Wyżyna Śląska). Acta Biologica Silesiana, 30 (47), 153–160.

Błońska A. 2010: Siedliska antropogeniczne na Wyżynie Śląskiej jako miejsca występowania rzadkich i zagrożonych gatunków torfowiskowych klasy *Scheuchzerio-Caricetea nigrae* (Nordh. 1937) R. Tx 1937. Woda-Środowisko-Obszary Wiejskie, 10 (1), 7–19.

Błońska A., Babczyńska-Sendek B. & Kołtuniak A. 2011: Nowe stanowisko *Orchis militaris* (*Orchidaceae*) na Garbie Tarnogórskim (Wyżyna Śląska). Fragmenta Floristica et Geobotanica Polonica, 1 (18), 177–180.

Błońska A., Kidawa J., Molenda T. & Chmura D. 2020: Hydrogeochemical conditions of the development of anthropogenic carbonate swamps: A case study of an abandoned Polish sandpit. Polish Journal of Environmental Studies, 29 (1), 561–569.

Błońska A., Kompała-Bąba A., Sierka E., Besenyei L., Magurno F., Frydecka K., Bierza W. & Woźniak G. 2019a: Impact of selected plant species on enzymatic activity of soil substratum on post-mining heaps. Journal of Ecological Engineering, 20 (1), 138–144.

Błońska A., Kompała-Bąba A., Sierka E., Bierza W., Magurno F., Besenyei L., Ryś K. & Woźniak G. 2019b: Diversity of vegetation dominated by selected grass species on coal-mine spoil heaps in terms of reclamation of post-industrial areas. Journal of Ecological Engineering, 20 (2), 209–217.

Boćkowski M. & Rogowski W. 2018: Wycena usług ekosystemowych oraz ich zastosowanie w rachunku ekonomicznym – praktyczne przykłady w zarządzaniu zasobami przyrodniczymi. Studia i Prace Kolegium Zarządzania i Finansów, 167, 37–64.

Box J. 1999: Nature conservation and post-industrial landscapes. Industrial Archaeology Review, 21 (2), 137–146.

Bula R. 2007: Materiały do atlasu rozmieszczenia oraz stanu zasobów roślin chronionych i zagrożonych rejonu górnośląskiego – PRESS. Część 11. *Iris sibirica* L. (*Iridaceae*). Natura Silesiae Superioris, 10, 5–18.

Bula R., Kloczkowska A., Parusel J.B., Romańczyk M. & Sokół K. 2014: Monitoring przyrodniczy Programu aktywizacji gospodarczej oraz zachowania dziedzictwa kulturowego Beskidów i Jury Krakowsko-Częstochowskiej Owca Plus na lata 2010–2014. Raport 2014. Centrum Dziedzictwa Przyrody Górnego Śląska.

Bzdęga K., Pasierbiński A. & Chmura D. 2004: Rola hałd pogórniczych w utrzymaniu regionalnej różnorodności florystycznej Górnego Śląska. Chrońmy Przyrodę Ojczystą, 60 (2), 17–33.

Cabała S., Gębicki C., Pierzgalski K. & Zygmunt J. 2009: Przyroda Częstochowy – strefy ochronne i stanowiska cenne przyrodniczo. Częstochowa.

Celiński F., Ludera F., Rostański K., Sendek A. & Wika S. 1974-1975: Nowe stanowiska rzadkich roślin naczyniowych na Górnym Śląsku i terenach przyległych. Cz. I i II. Zeszyty Przyrodnicze OTPN, 14–15, 11–31.

Celiński F., Rostański K., Sendek A., Wika S. & Cabała S. 1976: Nowe stanowiska rzadkich roślin naczyniowych na Górnym Śląsku i terenach przyległych. Cz. III. Zeszyty Przyrodnicze OTPN, 16, 15–31.

Chmura D. & Molenda T. 2007: Nowe stanowisko omiegu górskiego *Doronicum austriacum* Jacg. na Górnym Śląsku. Chrońmy Przyrodę Ojczystą, 63 (1), 20–24.

Council Directive 1992: Council Directive 92/43/EEC of 21 May 1992 on the conservation of natural habitats and of wild fauna and flora.

Czylok A. & Baryła J. 2003: Notatki florystyczne i ekologiczne z okolic Dąbrowy Górniczej i Sławkowa (Wyżyna Śląska). Natura Silesiae Superioris, 7, 11–17.

Czylok A. & Rahmonow O. 1996: Unikatowe układy fitocenotyczne w wyrobiskach wschodniej części województwa katowickiego. Kształtowanie środowiska geograficznego i ochrona przyrody na obszarach uprzemysłowionych i zurbanizowanych, 23, 27–31.

Czylok A. & Szymczyk A. 2009: Sand quarries as biotopes of rare and critically endangered plant species. [In:] Mirek Z. & Nikel A. (eds.): Rare, relict and endangered plants and fungi in Poland. W. Szafer Institute of Botany, Polish Academy of Sciences, Kraków, 187–192.

Czylok A. 2004: Wyrobiska po eksploatacji piasku na Wyżynie Śląskiej i ich roślinność. [In:] Partyka J. (ed.): Zróżnicowanie i przemiany środowiska przyrodniczo-kulturowego Wyżyny Krakowsko-Częstochowskiej. Tom 1. Przyroda. Wyd. Ojcowski Park Narodowy, Ojców, 205–212.

Czylok A., Tyc A. & Kubajak A. 2009: Przewodnik przyrodniczy po Dąbrowie Górniczej. Wydawnictwo Kubajak, Krzeszowice.

de Groot R., Jax K. & Harrison P. 2016: Links between biodiversity and ecosystem services. [In:] Potschin M. & Jax K. (eds.): OpenNESS Ecosystem Services Reference Book. EC FP7 Grant Agreement no. 308428.

de Groot R.S., Alkemade R., Braat L., Hein L. & Willemen L. 2010: Challenges in integrating the concept of ecosystem services and values in landscape planning, management and decision making. Ecological Complexity, 7 (3), 260–272.

DEFRA 2011: Biodiversity 2020: A strategy for England's wildlife and ecosystem services. Department for Environment, Food and Rural Affairs, London.

Domański R., Stebel A.M. & Stebel A. 1995: Godne ochrony obiekty przyrodnicze w południowej części województwa katowickiego. Część II. Kształtowanie środowiska geograficznego i ochrona przyrody na obszarach uprzemysłowionych i zurbanizowanych, 18, 11–17.

Duda W. 1992: Flora naczyniowa gminy Poraj. Ziemia Częstochowska, 18, 23–50.

Dulias R. & Hibszer A. 2004: Województwo śląskie: przyroda, gospodarka, dziedzictwo kulturowe. Wydawnictwo Kubajak, Kraków.

Dulias R. 2018: Geografia fizyczna Wyżyny Śląskiej. Wydawnictwo Uniwersytetu Śląskiego, Katowice.

Gębicki C., Śliwińska-Wyrzychowska A. & Karlikowska J. 2013: Skalne Jeziorko (stanowisko 2). [In:] Śliwińska-Wyrzychowska A. (ed.): Kopalnia przywrócona naturze – Przewodnik po przyrodniczej ścieżce dydaktyczno-edukacyjnej na obszarze nieczynnej kopalni odkrywkowej "Lipówka" w Rudnikach koło Częstochowy. Agencja Wydawnicza "ARGI", Częstochowa, 29–34.

Henzel H., Śmietana K., Zagórska E. & Bolek T. 2009: Klasyfikacja terenów poprzemysłowych w województwie śląskim. Świat Nieruchomości, 1 (67), 16–21.

Hereźniak J. 1995: Regionalna lista zagrożonych gatunków roślin naczyniowych w północnej części Wyżyny Śląsko-Krakowskiej. [In:] Materiały z Konferencji i Sympozjów 50 Zjazdu PTB. 26 June–01 July 1995, Poland, Kraków, 1–139.

Hereźniak J., Grzyl A., Kołodziejek J. & Lawrynowicz M. 2001: Materiały do flory północnej części Wyżyny Śląsko-Krakowskiej – rzadkie i interesujące gatunki roślin naczyniowych. Cz. 2. Fragmenta Floristica et Geobotanica Polonica, 8, 35–41.

Hockings M. 2003: Systems for assessing the effectiveness of management in protected areas. BioScience, 53 (9), 823–832.

Holak E. & Pawełko K. 1995: Nowe stanowisko łączenia baldaszkowatego (*Butomus umbellatus* L.) na terenie Górnośląskiego Okręgu Przemysłowego. Kształtowanie środowiska geograficznego i ochrona przyrody na obszarach uprzemysłowionych i zurbanizowanych, 19, 12–15.

Holeksa J., Błońska A., Kompała-Bąba A., Woźniak G., Kurek P., Szarek-Łukaszewska G., Grodzińska K. & Żywiec M. 2015: The vegetation of the Olkusz Ore-bearing Region. [In:] Godzik B. (ed.): Natural and historical values of the Olkusz Ore-Bearing Region. W. Szafer Institute of Botany, Polish Academy of Sciences, Kraków, 105–127.

Jankowski A.T. & Wach J. 1980: Uwagi o zbiornikach antropogenicznych na terenie GOP i jego obrzeżach. [In:] Materiały VII Sympozjum Polsko-Czechosłowackiego „Przeobrażenia środowiska geograficznego

w obszarach uprzemysłowionych i zurbanizowanych". UŚ, Oddział Katowicki PTG, Sosnowiec–Kozubnik, 65–76.
Journal of Laws (2014) item 1409 – Regulation of the Minister of the Environment on the Protection of the Species of Plants, 9 October 2014.
Kapusta P., Szarek-Łukaszewska G., Grodzińska K. & Godzik B. 2010: Murawy galmanowe okolic Olkusza (południowa Polska) i problemy ich ochrony. Chrońmy Przyrodę Ojczystą, 66 (1), 27–34.
Kaźmierczakowa R., Bloch-Orłowska J., Celka Z., Cwener A., Dajdok Z., Michalska-Hejduk D., Pawlikowski P., Szczęśniak E. & Ziarnek K. 2016: Polish red list of pteridophytes and flowering plants. Institute of Nature Conservation, Polish Academy of Sciences, Kraków.
Kołodziejek J. 1999: Rzadkie i interesujące gatunki roślin naczyniowych na górniczo zniekształconych obszarach środkowej części Częstochowskiego Okręgu Rudonośnego. Fragmenta Floristica et Geobotanica Polonica, 6, 39–43.
Kompała-Bąba A. & Bąba W. 2009: Threatened and protected species in the Kuźnica Warężyńska sandpit (Wyżyna Śląska Upland, S Poland). [In:] Mirek Z. & Nikel A. (eds.): Rare, relict and endangered plants and fungi in Poland. W. Szafer Institute of Botany, Polish Academy of Sciences, Kraków, 259–268.
Kompała-Bąba A., Sierka E., Dyderski M.K., Bierza W., Magurno F., Besenyei L., Błońska A., Ryś K., Jagodziński A.M. & Woźniak G. 2020: Do the dominant plant species impact the substrate and vegetation composition of post-coal mining spoil heaps? Ecological Engineering, 143, 105685.
Kondracki J. 2011: Geografia regionalna Polski. Wydawnictwo Naukowe PWN, Warszawa.
Kowolik M., Szarek-Łukaszewska G. & Jędrzejczyk-Korycińska M. 2010: Użytek ekologiczny „Pleszczotka górska" w cynkowo-ołowiowym terenie pogórniczym – potrzeba aktywnej ochrony. Chrońmy Przyrodę Ojczystą, 66 (1), 35–38.
Krajewski Ł. & Płachno B.J. 2015: *Utricularia bremii* (*Lentibulariaceae*) in Poland. Polish Botanical Journal, 60 (1), 105–109.
Krajewski Ł. 2009: Przyroda zbiornika Kuźnica Warężyńska. Cz. 3. Flora. Przyroda Górnego Śląska, 57, 8–9.
Krajewski Ł. 2011: Zespół *Cladietum marisci* w piaskowni w Dąbrowie Górniczej na tle rozmieszczenia kłoci wiechowatej *Cladium mariscus* w Polsce. Chrońmy Przyrodę Ojczystą, 67 (3), 276–283.
Krajewski Ł. 2012: Ramienice (*Characeae*) Zagłębia Dąbrowskiego (S Poland). Natura Silesiae Superioris, 13, 13–56.
Krajewski Ł. 2016a: Cibora żółta w Dąbrowie Górniczej. Przyroda Górnego Śląska, 83, 3–15.
Krajewski Ł. 2016b: Ramienice na hałdzie KWK „Porąbka-Klimontów" w Sosnowcu. Przyroda Górnego Śląska, 84 (3), 6–7.
Krajewski Ł., Pawlikowski P., Gutowska E., Jarzombkowski F., Kauzal P., Kotowska K., Kowalska M., Brzezińska K. & Dzierża P. 2015: Nowe dane o rozmieszczeniu i warunkach siedliskowych ramienic (*Characeae*) Polski (2010–2012) z uwzględnieniem terenów chronionych i objętych programem rolnośrodowiskowym. Woda-Środowisko-Obszary Wiejskie, 15 (2), 65–85.
Kremen C. 2005: Managing ecosystem services: what do we need to know about their ecology? Ecology Letters, 8 (5), 468–479.
Larondelle N. & Haase D. 2012: Valuing post-mining landscapes using an ecosystem services approach – An example from Germany. Ecological Indicators, 18, 567–574.
Machowski R. 2003: Zbiorniki w nieckach osiadania i zapadliskach w krajobrazie Wyżyny Śląskiej. Woda w przestrzeni przyrodniczej i kulturowej. Prace Komisji Krajobrazu Kulturowego, 2, 115–123.
Machowski R. 2010: Przemiany geosystemów zbiorników wodnych powstałych w nieckach osiadania na Wyżynie Katowickiej. Wydawnictwo Uniwersytetu Śląskiego, Katowice.
Madowicz A. 2000: Problem zagospodarowania terenów poprzemysłowych na przykładzie kamieniołomu G. Kijowa w Będzinie. Kształtowanie środowiska geograficznego i ochrona przyrody na obszarach uprzemysłowionych i zurbanizowanych, 30, 19–28.
Malon A. 2010: Piaski podsadzkowe. Państwowy Instytut Geologiczny – Państwowy Instytut Badawczy. http://geoportal.pgi.gov.pl/css/surowce/images/2009/pdf/09_42_piaski_podsadzkowe.pdf, access: 02.07.2021.
Matuszkiewicz W. 2012: Przewodnik do oznaczania zbiorowisk roślinnych Polski. Wydawnictwo Naukowe PWN, Warszawa.
MEA – Millennium Ecosystem Assessment. 2005: Ecosystems and Human Well-being: Biodiversity Synthesis. World Resources Institute, Washington DC.
Michalska D. 1994: Zmiany we florze i roślinności w rezerwacie „Góra Zborów" w ostatnich dwudziestu latach. Fragmenta Floristica et Geobotanica Series Polonica, 1, 181–207.

Nowak T., Jędrzejczyk-Korycińska M., Kapusta P. & Szarek-Łukaszewska G. 2015: Characteristics of the vascular plant flora in the Olkusz Ore-bearing Region. [In:] Godzik B. (ed.): Natural and historical values of the Olkusz Ore-bearing Region. W. Szafer Institute of Botany, Polish Academy of Sciences, Kraków, 147–166.

Nowak T., Kapusta P., Jędrzejczyk-Korycińska M., Szarek-Łukaszewska G. & Godzik B. 2011: The vascular plants of the Olkusz Ore-bearing Region [Rośliny naczyniowe Olkuskiego Okręgu Rudnego]. W. Szafer Institute of Botany, Polish Academy of Sciences, Kraków.

Orczewska A., Czylok A. & Kimsa S. 1993: Charakterystyka florystyczno-faunistyczna projektowanego rezerwatu przyrody na wyrobisku kopalni dolomitu „Blachówka" w Bytomiu. Kształtowanie środowiska geograficznego i ochrona przyrody na obszarach uprzemysłowionych i zurbanizowanych, 9, 13–22.

Osiadacz B. & Kręciała M. 2014: *Ophrys apifera* Huds. (Orchidaceae), a new orchid species to the flora of Poland. Biodiversity: Research and Conservation, 36 (1), 11–16.

Ostręga A. & Uberman R. 2010: Kierunki rekultywacji i zagospodarowania – sposób wyboru, klasyfikacja i przykłady. Górnictwo i Geoinżynieria, 34 (4), 445–461.

Pałasz J. 2012: Uwarunkowania organizacyjno-prawne procesów rekultywacji i rewitalizacji terenów poprzemysłowych w Polsce. Zeszyty Naukowe Politechniki Śląskiej. Seria Organizacja i Zarządzanie, 62 (1875), 5–23.

Parusel J.B. & Urbisz A. 2012: Czerwona lista roślin naczyniowych województwa śląskiego. [In:] Parusel J.B. (ed.): Czerwone listy wybranych grup grzybów i roślin województwa śląskiego. Centrum Dziedzictwa Przyrody Górnego Śląska, Katowice, Raporty Opinie, 6 (2), 105–177.

Parusel T. & Parusel J. 2011: Ekologiczna analiza flory naczyniowej kamieniołomu Penczków w Mysłowicach-Dziećkowicach. Kształtowanie środowiska geograficznego i ochrona przyrody na obszarach uprzemysłowionych i zurbanizowanych, 43, 58–69.

Pasierbiński A. & Rostański A. 2001: Zróżnicowanie flory naczyniowej zwałowisk pogórniczych zlokalizowanych na terenach leśnych aglomeracji katowickiej. Natura Silesiae Superioris, Suplement, 5, 19–31.

Perrings C., Duraiappah A., Larigauderie A. & Mooney H. 2011: The Biodiversity and Ecosystem Services Science–Policy Interface. Science, 331 (6021), 1139–1140.

Pietrzyk-Sokulska E. & Kulczycka J. 2014: Rekultywacja i zagospodarowanie wyrobisk poeksploatacyjnych – – problemy środowiskowe, ekonomiczne i społeczne. Technika Poszukiwań Geologicznych, Geotermia, Zrównoważony Rozwój, 2, 29–40.

Pladias (2014–2020): Database of the Czech Flora and Vegetation. www.pladias.cz, access: 18.11.2019.

Poros M. & Sobczyk W. 2013: Rewitalizacja terenu pogórniczego po kopalni surowców skalnych na przykładzie kamieniołomu Wietrznia w Kielcach. Rocznik Ochrona Środowiska, 15 (3), 2369–2380.

Poros M. & Sobczyk W. 2014: Kierunki rekultywacji terenów pogórniczych obszaru chęcińsko-kieleckiego w kontekście ich wykorzystania w aktywnej edukacji geologicznej. Rocznik Ochrona Środowiska, 16 (1), 386–403.

Regional Directorate for Environment Protection in Katowice 2019: Obszary Natura 2000. Regionalna Dyrekcja Ochrony Środowiska w Katowicach. http://natura2000.gdos.gov.pl/wyszukiwarka-n2k, access: 18.11.2019.

Rosin Z.M., Takacs V., Báldi A., Banaszak-Cibicka W., Dajdok Z., Dolata P.T., Kwieciński Z., Łangowska A., Moroń D., Skórka P., Tobółka M., Tryjanowski P. & Wuczyński A. 2011: Koncepcja świadczeń ekosystemowych i jej znaczenie w ochronie przyrody krajobrazu rolniczego. Chrońmy Przyrodę Ojczystą, 67 (1), 3–20.

Rostański A. & Tokarska-Guzik B. 2001: Możliwości i ograniczenia przyrodniczego zagospodarowania terenów poprzemysłowych. Natura Silesiae Superioris, Suplement, 5, 5–17.

Rostański A. 1997: Flora spontaniczna hałd Górnego Śląska. Archiwum Ochrony Środowiska, 23 (3–4), 159–165.

Rostański A. 2000: Rekultywacja i zagospodarowanie nieużytków poprzemysłowych – rozwiązania alternatywne. Inżynieria Ekologiczna, 1, 81–86.

Rostański A. 2001: Rola lokalnych zasobów genowych w zagospodarowaniu nieużytków poprzemysłowych. [In:] Materiały Sympozjum – Warsztaty 2001 nt. Przywracanie wartości użytkowych terenom górniczym, 163–172.

Rostański A. 2006: Spontaniczne kształtowanie się pokrywy roślinnej na zwałowiskach po górnictwie węgla kamiennego na Górnym Śląsku [Spontaneous plant cover on colliery spoil heaps in Upper Silesia (Southern Poland)]. University of Silesia, Katowice.

Rostański A., Jędrzejczyk-Korycińska M. & Nowak T. 2015: Ochrona przyrody obszarów metalonośnych. [In:] Wierzbicka M. (ed.): Ekotoksykologia: rośliny, gleby, metale. Wydawnictwo Uniwersytetu Warszawskiego, Warszawa, 507–521.
Różkowski K., Polak K. & Cała M. 2010: Wybrane problemy związane z rekultywacją wyrobisk w kierunku wodnym. Górnictwo i Geoinżynieria, 34 (4), 517–525.
Rzętała M. & Jaguś A. 2012: New lake district in Europe: origin and hydrochemical characteristic. Water and Environment Journal, 26 (1), 108–117.
Rzętała M. 2008: Funkcjonowanie zbiorników wodnych oraz przebieg procesów limnicznych w warunkach zróżnicowanej antropopresji na przykładzie regionu górnośląskiego. Wydawnictwo Uniwersytetu Śląskiego, Katowice.
Sieka P., Urbisz A. & Babczyńska-Sendek B. 2015: Godne ochrony stanowisko flory oraz roślinności kserotermicznej na wzgórzu Golcówka w Imielinie (Wyżyna Śląska). Chrońmy Przyrodę Ojczystą, 71 (5), 380–387.
Sobala M. & Pukowiec K. 2014: Stan zagospodarowania nieczynnych kamieniołomów na terenie Beskidu Śląskiego i Żywieckiego a polityka przestrzenna gmin. Prace Komisji Krajobrazu Kulturowego, 26, 127–138.
Solon J. 2008: Koncepcja "Ecosystem Services" i jej zastosowania w badaniach ekologiczno-krajobrazowych. [In:] Chmielewski T.J. (ed.): Struktura i funkcjonowanie systemów krajobrazowych: meta-analizy, modele, teorie i ich zastosowania. Problemy Ekologii Krajobrazu, 21, 25–44.
Spałek K. 2003: Materiały do rozmieszczenia rzadkich i interesujących gatunków paprotników (*Pteridophyta*) na Równinie Opolskiej. Fragmenta Floristica et Geobotanica Polonica, 10, 209–220.
Stebel A. 1993: Nowe stanowiska widłaka goździstego *Lycopodium clavatum* L. na Płaskowyżu Rybnickim i w południowo-wschodniej części Kotliny Raciborskiej. Kształtowanie środowiska geograficznego i ochrona przyrody na obszarach uprzemysłowionych i zurbanizowanych, 10, 43–45.
Sudra P. 2015: Usługi ekosystemowe na tle wybranych koncepcji ekologii miasta. Człowiek i Środowisko, 39 (1), 61–73.
Szczepańczyk A. 2007: Powyrobiskowe tereny w zachodniej części miasta Piekary Śląskie jako potencjalny użytek ekologiczny. Kształtowanie środowiska geograficznego i ochrona przyrody na obszarach uprzemysłowionych i zurbanizowanych, 38, 46–54.
Szeląg Z. 2000: Materiały do flory Wyżyny Krakowsko-Częstochowskiej. Fragmenta Floristica et Geobotanica Polonica, 7, 93–103.
Szkaradkiewicz M., Kazak J., Sabura K. & Oleszek J. 2014: Usługi ekosystemów jeziornych – klasyfikacja, waloryzacja, zastosowania. Acta Scientiarum Polonorum, Formatio Circumiectus, 13 (3), 91–99.
Szymczyk A., Rahmonov O. & Parusel T. 2011: Zróżnicowanie ekologiczne flory i roślinności wyrobiska po eksploatacji piasków „Siemonia". Acta Geographica Silesiana, 9, 63–74.
Szymczyk A., Stebel A. & Czylok A. 2003: Wroniec widlasty *Huperzia selago* na Wyżynie Śląskiej. Chrońmy Przyrodę Ojczystą, 59 (3), 75–78.
Śliwińska-Wyrzychowska A., Bogdanowicz M., Musielińska R., Bąbelewska A. & Witkowska E. 2014: Krajobrazowe i botaniczne walory nieczynnego kamieniołomu Lipówka w Rudnikach koło Częstochowy. Prace Komisji Krajobrazu Kulturowego, 26, 45–55.
Tokarska-Guzik B. & Rostański A. 2001: Możliwości i ograniczenia przyrodniczego zagospodarowania terenów poprzemysłowych. [In:] Rostański K. (ed.): Hałdy poprzemysłowe – obiekty badań botanicznych. Natura Silesiae Superioris, Suplement, 5–17.
Tokarska-Guzik B. 1996: Rola hałd zasadowych w utrzymaniu lokalnej bioróżnorodności. Przegląd Przyrodniczy, 7 (3–4), 261–266.
Tokarska-Guzik B. 1997: Rozmieszczenie i zasoby roślin chronionych na terenie miasta Jaworzno. Acta Biologica Silesiana, 30, 106–124.
Tokarska-Guzik B. 2000: Przyrodnicze zagospodarowanie nieużytków miejsko-przemysłowych na przykładzie centrów górniczych Europy. Inżynieria Ekologiczna, 1, 72–80.
Tokarska-Guzik B., Rostański A., Gorczyca J., Herczek A. & Dulias R. 2011: Waloryzacja przyrodnicza miasta Jaworzna. Zakład Badawczo-Usługowy "Ekos". https://www.um.jaworzno.pl/pl/natura/przyroda_i_ekologia/39/waloryzacja_przyrodnicza_miasta.html, access: 20.06.2021.
Tropek R. & Konvicka M. 2008: Can quarries supplement rare xeric habitats in a piedmont region? Spiders of the Blansky les Mts, Czech Republic. Land Degradation & Development, 19 (1), 104–114.
Tropek R., Cerna I., Straka J., Cizek O. & Konvicka M. 2013: Is coal combustion the last chance for vanishing insects of inland drift sand dunes in Europe? Biological Conservation, 162, 60–64.

Tropek R., Kadlec T., Hejda M., Kocarek P., Skuhrovec J., Malenovsky I., Vodka S., Spitzer L., Banar P. & Konvicka M. 2012: Technical reclamations are wasting the conservation potential of post-mining sites. A case study of black coal spoil dumps. Ecological Engineering, 43, 13–18.
Tropek R., Kadlec T., Karesova P., Spitzer L., Kocarek P., Malenovsky I., Banar P., Tuf I.H., Hejda M. & Konvicka M. 2010: Spontaneous succession in limestone quarries as an effective restoration tool for endangered arthropods and plants. Journal of Applied Ecology, 47 (1), 139–147.
Wieliczko B. 2016: Wykorzystanie usług ekosystemów w zarządzaniu zasobami naturalnymi w rolnictwie. Studia i Prace WNEiZ US, 46 (2), 135–144.
Wilsey B.J. & Polley H.W. 2002: Reductions in grassland species evenness increase dicot seedling invasion and spittle bug infestation. Ecology Letters, 5 (5), 676–684.
Woźniak G. & Cohn E. 2007: Monitoring of spontaneous vegetation dynamics on post coal mining waste sites in Upper Silesia, Poland. [In:] Sarsby R.W. & Felton A.J. (eds.): Geotechnical and Environmental Aspects of Waste Disposal Sites – Proceedings of the 4th International Symposium on Geotechnics Related to the Environment. Taylor and Francis Group, London, 289–294.
Woźniak G. & Kompała A. 2000: Rola procesów naturalnych w rekultywacji nieużytków poprzemysłowych. Inżynieria Ekologiczna, 1, 87–93.
Woźniak G. & Rostański A. 2000: Rola traw w spontanicznej sukcesji roślinnej na osadnikach ziemnych wód kopalnianych na Górnym Śląsku. Łąkarstwo w Polsce, 3, 159–167.
Woźniak G. 2006: Colonization process on coal mine sedimentation pools (Upper Silesia, Poland). Polish Botanical Studies, 22, 561–568.
Woźniak G. 2010: Zróżnicowanie roślinności na zwałach pogórniczych Górnego Śląska [Diversity of vegetation on coal-mine heaps of the Upper Silesia (Poland)]. W. Szafer Institute of Botany, Polish Academy of Sciences, Kraków.
Woźniak G., Sierka E. & Wheeler A. 2018: Urban and industrial habitats: How important they are for ecosystem services. [In:] Hufnagel L. (ed.): Ecosystem Services and Global Ecology. IntechOpen, 169–194.
Yachi S. & Loreau M. 1999: Biodiversity and ecosystem productivity in a fluctuating environment: the insurance hypothesis. Proceedings of the National Academy of Sciences of the United States of America, 96 (4), 1463–1468.
Zagórska E. 2013: Wojewódzkie bazy danych o terenach poprzemysłowych i zdegradowanych jako przykład podejmowanych działań w dziedzinie rewitalizacji terenów poprzemysłowych i zdegradowanych. Studia Ekonomiczne, 155, 108–121.
Zalewska-Gałosz J. 2008: Rodzaj *Potamogeton* L. w Polsce – taksonomia i rozmieszczenie. Instytut Botaniki Uniwersytetu Jagiellońskiego, Kraków.
Zarzycki K., Trzcińska-Tacik H., Różański W., Szelag Z., Wołek J. & Korzeniak U. 2002: Ecological indicator values of vascular plants of Poland. W. Szafer Institute of Botany, Polish Academy of Sciences, Kraków.
Zipper C.E., Burger J.A., Skousen J.G., Angel P.N., Barton C.D., Davis V. & Franklin J.A. 2011: Restoring forests and associated ecosystem services on Appalachian coal surface mines. Environmental Management, 47, 751–765.
Żmuda S. 1973: Antropogeniczne przeobrażenia środowiska przyrodniczego konurbacji górnośląskiej. Śląski Instytut Naukowy, PWN, Warszawa–Kraków.

4.4. Ecosystem Services and Post-industrial Areas

Edyta Sierka
Institute of Biology, Biotechnology and Environmental Protection,
Faculty of Natural Sciences, University of Silesia in Katowice;
28 Jagiellońska Str., 40-032 Katowice, Poland;
edyta.sierka@us.edu.pl; ORCID: 0000-0003-3317-4552

Łukasz Radosz
Institute of Biology, Biotechnology and Environmental Protection,
Faculty of Natural Sciences, University of Silesia in Katowice;
28 Jagiellońska Str., 40-032 Katowice, Poland;
lukasz.radosz@onet.eu; ORCID: 0000-0002-0270-7291

Karolina Ryś
Institute of Biology, Biotechnology and Environmental Protection,
Faculty of Natural Sciences, University of Silesia in Katowice;
28 Jagiellońska Str., 40-032 Katowice, Poland;
karorys@o2.pl; ORCID: 0000-0002-3391-5501

Gabriela Woźniak
Institute of Biology, Biotechnology and Environmental Protection,
Faculty of Natural Sciences, University of Silesia in Katowice;
28 Jagiellońska Str., 40-032 Katowice, Poland;
gabriela.wozniak@us.edu.pl; ORCID: 0000-0003-1936-2880

ABSTRACT: Resource management in a sustainable way is based on a thorough understanding of the complexity of Earth's natural ecosystems and how they can be managed without compromising future generations. Sustainable management of natural resources becomes much more complicated when there is a severe and persistent anthropogenic impact. The interdisciplinary approach should improve understanding, assessment, and maintenance of natural capital and related ecosystem services in urban-industrial areas. In ecological restoration, the biggest challenge is to find a consensus of suitable biodiversity indicators and economically possible measures, which will produce multiple socially and ecologically guided environmental benefits. There is difficulty reaching such a consensus because of the complexity and different ways of understanding the biodiversity concept. Restoration projects should involve ecologically-based methods and approaches, fulfilling many stakeholders' expectations for sustainable development and human well-being. Integrated natural and human models for sustainable management can help to understand the dynamics of ecosystems, including biodiversity and trophic levels, to simulate and evaluate different management scenarios concerning biodiversity and ecosystem services. There is still a need to increase understanding of the role of biodiversity and ecosystem service identification as essential factors influencing the dynamics of the ecosystem and sustainable management scenarios.

KEYWORDS: Ecosystem functioning, ecosystems of urban-industrial areas, biodiversity, natural capital

INTRODUCTION

Humanity is increasingly urban, depending on industrial and agricultural products, but continues to depend on nature, i.e. natural capital, for its survival. People are dependent on natural and anthropogenic functioning of ecosystems, and sustainable management of natural resources based on understanding the functioning of the Earth's natural ecosystems can enable their survival (Natural Resources Wales 2013) in the future (Haughton & Hunter 1994).

These mechanisms change when there is a solid and constant anthropogenic impact. Therefore, a multi-level approach should be taken to improve understanding, assessment, and maintenance of ecosystem services. An ecosystem is "a set of interacting species and their local, non-biological environment functioning together to sustain life" (Moll & Petit 1994). Ecosystem services are "the benefits human populations derive, directly or indirectly, from ecosystem functions" according to Costanza *et al.* (1997). Many of these ecological services are not consumed by humans directly but are needed to sustain the ecosystems themselves (Bolund & Hunhammar 1999). It is essential information for considering that, in the twentieth century the Earth has entered the Anthropocene epoch (Crutzen 2002). The main characteristic of the Anthropocene epoch is that human influences are shifting natural conditions beyond their limits, beyond the biological requirements which humans need for their existence (Steffen *et al.* 2007).

Large areas of the Earth's land surface are transformed through intensive agriculture, expanding urbanization, and industrialization. Often such human activities are changing the world's ecosystems and landscapes in drastic ways. Intensive research has revealed that the pressure of land use throughout the globe has influenced the environment, ranging from modification in the composition of atmospheric gases to the extensive transformation of the Earth's ecosystems (Wackernagel *et al.* 2002). The Millennium Ecosystem Assessment (MEA) revealed that 60% of ecosystem services are at risk because natural resources have been affected by exploitation and unsustainable use (MEA 2005).

According to the Millennium Ecosystem Assessment (MEA 2005), "Ecosystem services are indispensable to the living conditions of all people in all places". Ecosystem services can only be provided by ecosystems, which are functioning effectively. However, there is a good evidence base that outlines the importance of biodiversity to ecosystem functioning. Still, less research is focused on the direct relationship between biodiversity and ecosystem services (Binner *et al.* 2017), e.g., concerning urban areas, particularly the understanding of biodiversity in urban woodlands. Many world ecosystems have been damaged or disturbed by human activity. Those changed ecosystems need to be restored and managed accordingly, including the concept of Novel ecosystems when necessary (Hoekstra *et al.* 2005; Hobbs 2007).

The fact that human quality of life quality could increase while environmental quality had been deteriorating is considered as the Environmentalist's Paradox (Raudsepp-Hearne *et al.* 2010). Because humans as well as the ecosystem services they need depend on nature (Pascual *et al.* 2017), the approach that natural capital could be decoupled from human needs seems paradoxical.

Natural capital is the key provider of natural assets, among which ecosystem services are fundamental. Frequently the terminology regarding natural capital and ecosystem services is used interchangeably, complicating the understanding of this complex subject (Binner *et al.* 2017). Natural capital can be considered the stock, or natural assets, within an ecosystem or an area. It is a part of total capital, which is natural, human, social, and manufactured capital combined. These forms of capital can be partly substitutable for one another (Sen 1999). Furthermore, substitution may be a rational choice for humans, but there may be no substitutes for some ecosystem services. States have to co-create a symbiotic future of natural forces (soil, water, air, and living organisms) with human forces (innovations, development, and human well-being).

An essential element in the delivery of ecosystem services is biodiversity, understood as all living organisms. Still, in many cases, it is not the diversity of the microorganisms that is important, but the presence of a viable population of at least one species representing a particular functional group. Biological diversity, numbers of species, is also a service in itself. Moreover, there are the fundamental inter-connected relationships between biodiversity and ecosystem functions, including diversity (Fig. 1).

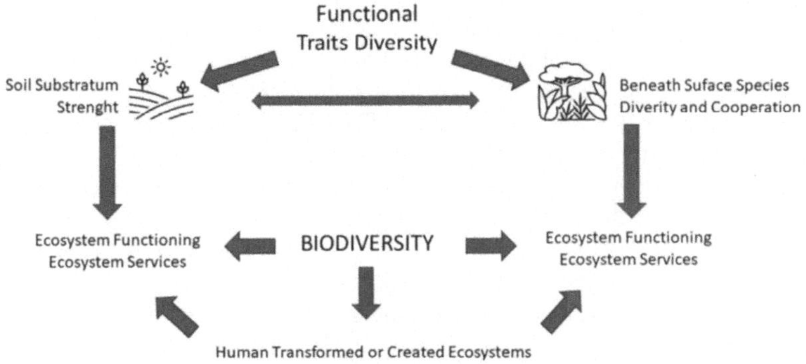

Figure 1. The basic inter-connected relationships between biodiversity, ecosystem functioning and ecosystem services (Woźniak *et al.* (2018) – changed)

Many of the mechanisms ruling ecosystem functioning are not fully recognized and understood. The relationships among the ecosystem elements are complex. The models created have to be simplified and translated into more informative formats for stakeholders and decision-makers. In this way, the decision-makers should be able to incorporate the ecosystem principles into management practices. Management practices may facilitate the enhancement of ecosystem services for human quality of life in urban-industrial sites.

Many of the urban, industrial and post-industrial areas represent complex land-cover mosaics. Some habitats of the mosaic represent sites where the "novel ecosystems" assemblage *sensu* Hobbs *et al.* (2013) develop in terms of their composition of ecological components and interactions (Wu 2014). The community composition in the city or industrial areas, i.e., organism composition and relationships developing on soil/soil substratum, is different from non-industrial habitats. In such new environmental situations, in habitats under constant human pressure in urban areas or created by human activity, some organisms perform various functions. In industrial or post-industrial areas, understanding which features of particular organisms, communities, vegetation type, or habitat characteristics are most important is crucial (Piekarska-Stachowiak *et al.* 2014). The most vital point for understanding the ecosystem function (ecosystem services) of industrial areas is the biodiversity-ecosystem, function-ecosystem service relationship. In the environment of urban and industrial areas, it might be expected that various aspects of the urban biodiversity-ecosystem service relationship are unique. Many industrial areas are deficient in nutrients (oligotrophic) and are at the initial developmental stage. These sites are valuable in terms of their potential for biodiversity enhancement (Fig. 1). This uniqueness implies an urgent need for the study of the biodiversity-ecosystem, function-ecosystem service relationship on the one hand, and the need for the decision makers and stakeholders to take this uniqueness into account in policy and management recommendations on the other hand. This uniqueness also implies that there is a high potential for the enhancement of those habitats. However, ecosystem dynamics in industrial landscapes are poorly understood (Pataki *et al.* 2011; Gómez-Baggethun & Barton 2013), especially when it comes to designing, creating, and restoring ecological processes, functions, and services in those areas (Benayas *et al.* 2009; Pataki *et al.* 2011).

There are severe contemporary constraints in understanding the biodiversity-ecosystem service relationship, which is a concept of the Blue-Green City. This concept supports and enhances the natural potential, primarily due to composition of primary producers using them, e.g., to reduce flood risk or to help improve the air, soil, and water quality. When people use nature (green plants or blue water) to help manage and enhance urban-industrial environments, e.g., in managing stormwater, it is often referred to as blue-green infrastructure (Fig. 2). Green infrastructure as an ecological-environmental entity is a larger concept associated with the services provided through ecological functioning for the social, economic, and environmental health of the surrounding urban-industrial environment. The aim of the concept of the Blue-Green City approach is to recreate a water-based, hydro-biological system based on biogeochemical processes by joining water management with the green infrastructure in urban areas. The Blue-Green project includes, for example, the idea of managing flood risk by combining the hydrological and ecological, e.g., specific vegetation type potential of the urban-industrial landscape. The interaction between blue and green plants and water systems can enrich the urban-industrial environment. An example of such solutions is the Blue-Green City project in Newcastle, UK (Newcastle 2016). In terms of ecology and hydrology, the aims of the Newcastle project are the following:

- the creation of an urban flood model to simulate the movement of water and sediment through blue-green features;
- improve water quality, habitat, and biodiversity using a system of blue-green natural components using the power and potential of biological processes (*Blue-Green Cities Research Project: http://www.bluegreencities.ac.uk*).

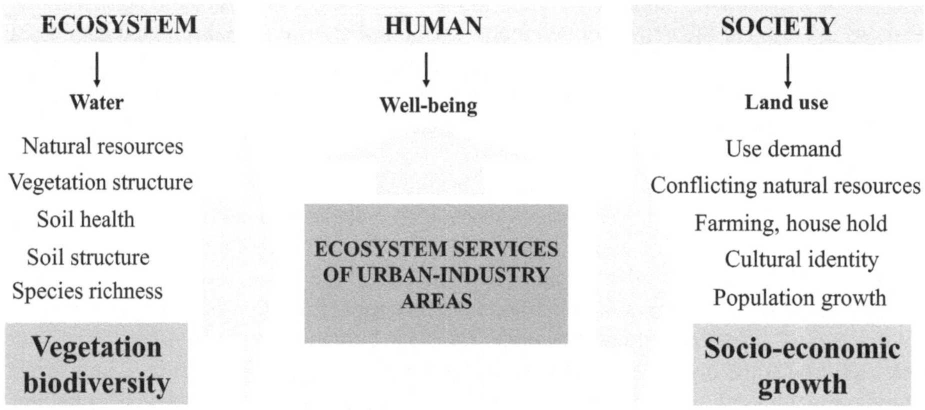

Figure 2. The complexity of the Blue-Green City concept in relation to the special mosaic of urban--industrial sites, land management, land requirements, and demand (Woźniak *et al.* 2018 – changed)

The Newcastle project is an example of an enterprise that considers both the ecological and hydrological elements, which are equally crucial for urban-industrial ecosystem functioning. The Anthropocene epoch is forcing the humans to change their way of thinking about urban-industry environmental management and introduce a different way of acting. Humans need to focus on nature-based approaches. All the possibilities of enhancing biodiversity have to be used in management practices. The development and enhancement of varied microhabitats support increasing biodiversity. The monitoring of a biodiversity index might be an approach to a nature based solution.

Restoration and regeneration of areas transformed and changed by industry can be a long and complicated process (Fig. 3). Post-industrial sites generally represent heavily altered and affected ecosystems, which have lost their primary biodiversity and most of their ecosystem functions and services (Bradshaw 2000). The wide range of biodiversity restoration and ecosystem services in post-industrial (particularly post-mining) sites has received wide attention among restoration scientists (Prach & Pyšek 2001; Rahmonov 2009; Walker *et al.* 2014; Tarvainen & Tolvanen 2015). Although the scientific attention to ecosystem services has been growing, there has been a solid tendency to conduct short-term experimental studies in which biodiversity was experimentally manipulated (in the laboratory or much less frequently in the field; Cardinale *et al.* 2012). However, some studies on vegetation development and spontaneous succession on urban and post-coal-mine sites were performed over ten years. Long-term research conducted in the field provided exciting results about the mechanisms concerning spontaneous vegetation, ecosystem development, and biodiversity enrichment in a broad spatiotemporal context (Woźniak 2010; Kompała-Bąba 2013).

Increasing the biodiversity (as the prerequisite) of all ecosystem services, which are dependent on ecosystem functions, is the main aim of ecological restoration (Whisenant 1999; Farber *et al.* 2006). In post-mining and post-industrial sites, the restoration and enhancement of biodiversity and ecosystem function should be related to the broader landscape (Fig. 4) and various local micro-habitats in a broad spatiotemporal context. Limiting the repair and management planning exclusively to the post-industrial and post-mining sites is the most common mistake of decision-makers. The essential prerequisites of soil/soil substratum physical features like erosion control, water infiltration, recognition, assessment, and the biotic spoil (spoil substratum) parameters, including bacteria and arbuscular mycorrhiza fungi (AMF), require wise, scientifically based decisions.

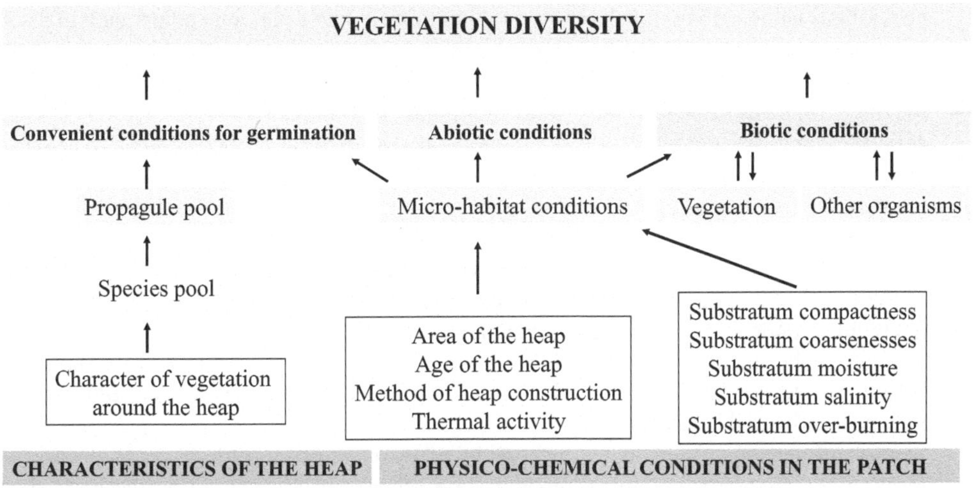

Figure 3. Scheme presenting the relationships between chosen abiotic parameters influencing vegetation diversity during spontaneous succession on post-coal mine heaps (Woźniak 2010 – changed)

The decision process must include the actions supporting the spontaneous vegetation development and introducing specific plant species possessing the most necessary traits for the particular circumstances, all of which are prerequisites for establishment of the best-adapted vegetation (Chmura *et al.* 2013; Markowicz *et al.* 2015; Stefanowicz *et al.* 2015; Woźniak *et al.* 2015; Bąba *et al.* 2016). Plants and vegetation as the primary producers are the fundamental elements of each developing ecosystem. Their restoration or enhancement is the basis for re-establishment of primary productivity of urban post-industrial sites of diverse organisms, carbon sequestration, and increasing the regulation and aesthetic values of sites and the landscape. Ecologists (SER 2004; Tarvainen & Tolvanen 2015) prefer to emphasize the reestablishment or the increase of biodiversity as a goal of restoration.

Post-industrial sites need to be managed in an area-specific way precisely adjusted to each object and microhabitat variety. The site-specific approach is necessary for choosing the most appropriate restoration method, which is the optimal effect in terms of environmental/ecosystem function recovery. The ecosystem reestablishment approach presents a type of gradient, or a continuum, of ecological restoration depending on the intensity and level of spontaneous processes. The intervention levels may have a broad range, and technical reclamation may be necessary in cases of, e.g., chemical industry toxic contamination. Sometimes it involves heavy interventions, such as the restructuring of landforms, importing soil and/or removal of deposited material on the one hand, and the spontaneous colonization by organisms and succession, so that the ecosystem is expected to recover principally through natural processes, on the other hand (Woźniak 2010; Holl & Aide 2011; Walker *et al.* 2014).

For post-industrial ecosystem development and reestablishment of functioning, and the ecosystem services that may be accrued, primary or secondary succession through natural, spontaneous processes is the most appropriate for several reasons:

- the biotic and abiotic site conditions of post-industrial sites are so different from the natural and semi-natural ones that it is inappropriate to use the experience from natural habitats for reclamation practice; the agricultural and forest bridging engineering practices are unacceptable;
- many of the post-industrial, particularly the post-mineral excavation sites, can be classified as novel ecosystems *sensu* Hobbs *et al.* (2013). According to novel ecosystems theory, the habitats of novel ecosystems cannot recover to the previous state. The only way for the novel ecosystems habitats to establish ecosystem functioning is to develop in adjustment to the new environmental conditions, with previously unknown nonanalogous plant species vegetation and ecosystem composition;
- the high microsite heterogeneity and variety on post-industrial sites requires low-scale actions which are not economically beneficial; the natural, spontaneous processes can achieve it;
- recognition and increasing understanding of natural, spontaneous succession enables the facilitation and enhancement of biological processes by assisted restoration of ecosystem function reestablishment, to speed up when necessary natural regeneration and reestablishment of the ecosystem under adverse environmental conditions (Woźniak *et al.* 2015);
- it should be understood and accepted that the target ecosystem mosaic may not always be an accurate replacement of the original ecosystem that was lost by open cast or deep mining or industrial activities, but a system of synusia of living organisms that are best adjusted to the new post-industrial conditions;
- factors influencing spontaneous succession of post-industrial sites have to be assessed, through studies of various measures and approaches, and this should be the basis for the planning of effective ecological restoration (Prach & Pyšek 2001; Alday *et al.* 2016; Horáčková *et al.* 2016; Nikolic *et al.* 2016);
- at the beginning of spontaneous succession, the early successional stages create a mosaic of species group composition traits that is of high conservation value, e.g., annual and early successional species (Chmura *et al.* 2011; Řehounková *et al.* 2016; Tropek *et al.* 2016);
- the maintenance of early successional stages should be a goal of restoration projects; the open habitats of early succession are appropriate for wildlife, e.g., habitats for butterflies in the urban-industrial landscape are scarce;
- technical reclamation, in comparison to spontaneous succession, can negatively influence the local biodiversity since it decreases the number of habitats that support specialized threatened species (Prach & Pyšek 2001; Řehounková *et al.* 2016). It may even enhance and maintain the pool of seeded alien species that may spread to the surrounding landscape and additionally threaten the urban-industrial environment (Rydgren *et al.* 2016);

- spontaneous primary succession in post-industrial and urban areas frequently leads to establishing a self-sustaining, well-functioning ecosystem. However, they may be novel ecosystems *sensu* Hobbs *et al.* (2013) which are different ecosystems compared to those that occur in natural and semi-natural habitats;
- habitat and microhabitat differences caused by adverse oligotrophic mineral environmental conditions, and the omission of and ignoring the rules and mechanisms of biological processes, are very often the reasons why reclamation fails;
- some post-industrial sites present the requirements for development of extreme endemism. Such sites are environmental islands, and processes of microevolution could be expected to occur. This issue still needs to be investigated.

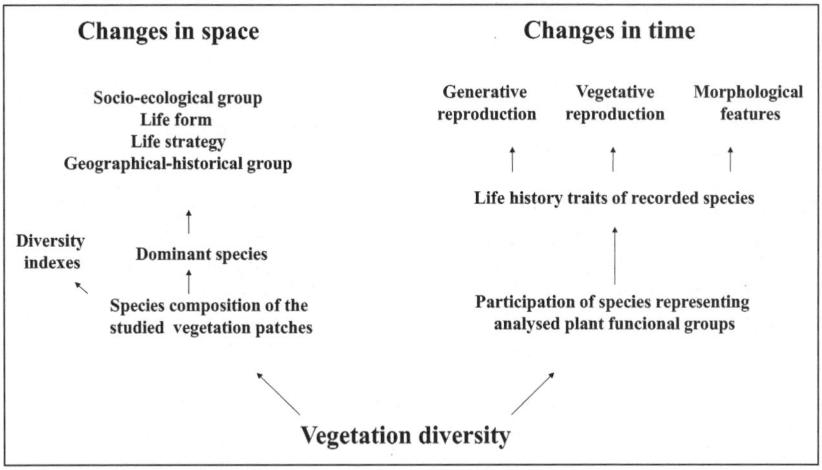

Figure 4. Some aspects of plant functional diversity of vegetation development on post-coal-mining heaps (Woźniak 2010 – changed)

CONCLUSIONS AND PERSPECTIVES

At present, we are facing a significant challenge concerning the proper ecosystem and environmental management of urban and industrial habitats: How important are some of the novel ecosystems for ecosystem services? The answer to this question requires understanding that in the ecological restoration – ecosystem functioning reestablishment, the biggest challenge is to find a consensus of suitable biodiversity enhancement, which will produce multiple socially and ecologically guided environmental benefits. The biodiversity changes can be monitored by using multiscale biodiversity indicators followed by economically possible measurements. Our contemporary "pyramids" (as Żelaziński J. said), like their Egyptian counterparts, also contain treasures (in particular, environmental treasures). In time, we will find that the landscape we have known so well since childhood has changed, and we miss the "hills of anthropogenic origin" that are so memorable.

Many years of development of city and industrial activity and frequent failure to undertake adequate environmental protection projects have led to upsetting of the natural ability of the environment to self-regulate and balance. However, there are also entirely degraded ecosystems whose conditions require de novo restoration to preserve Nature-Based Solutions to fulfil the broadest range of ecosystem services, following the principles of sustainable development. Sustainable development is understood as re-establishing ecosystem functioning for the efficient provision of ecosystem services for human society which can be based only on supporting and enhancing natural processes. The self-re-established ecosystem will be best adapted to the variety of post-mineral mining habitat conditions. Attention should be paid above all to the need to maintain the comprehensive nature of the transformation process, including social, technical, and environmental issues, based on the newest scientific knowledge.

The relationships between ecosystem services and multidimensional human needs are complex. The oversimplification of these relationships through an omission of locally relevant ecosystem functional elements in environmental management can lead to the loss of valuable novel ecosystems, which in turn can significantly affect biodiversity. Therefore, it is important to make appropriate decisions in the field of environmental management, taking into account social, technical and environmental issues, based on the latest scientific knowledge (Lapointe *et al.* 2021).

REFERENCES

Alday J.G., Zaldívar P., Torroba-Balmori P., Fernández-Santos B. & Martínez-Ruiz C. 2016: Natural forest expansion on reclaimed coal mines in Northern Spain: the role of native shrubs as suitable microsites. Environmental Science and Pollution Research, 23 (14), 13606–13616.

Bąba W., Błońska A., Kompala-Bąba A., Małkowski Ł., Ziemer B., Sierka E., Nowak T., Woźniak G. & Besenyei L. 2016: Arbuscular mycorrhizal fungi (AMF) root colonization dynamics of *Molinia caerulea* (L.) Moench. in grasslands and post-industrial sites. Ecological Engineering, 95, 817–827.

Benayas J.M.R., Newton A.C., Diaz A. & Bullock J.M. 2009: Enhancement of biodiversity and ecosystem services by ecological restoration: A meta-analysis. Science, 325 (5944), 1121–1124.

Binner A.R., Smith G., Bateman I.J., Day B.H., Agarwala M. & Harwood A. 2017: Valuing the social and environmental contribution of woodlands and trees in England, Scotland and Wales. Forestry Commission Research Report, Forestry Commission, Edinburgh.

Bolund P. & Hunhammar S. 1999: Ecosystem services in urban areas. Ecological Economics, 29 (2), 293–301.

Bradshaw A. 2000: The use of natural processes in reclamation – advantages and difficulties. Landscape and Urban Planning, 51 (2–4), 89–100.

Cardinale B.J., Duffy J.E., Gonzalez A., Hooper D.U., Perrings C., Venail P., Narwani A., Mace G.M., Tilman D., Wardle D.A., Kinzig A.P., Daily G.C., Loreau M., Grace J.B., Larigauderie A., Srivastava D.S. & Naeem S. 2012: Biodiversity loss and its impact on humanity. Nature, 486 (7401), 59–67.

Chmura D., Molenda T., Błońska A. & Woźniak G. 2011: Sites of leachate inflows on coalmine heaps as refuges of rare mountainous species. Polish Journal of Environmental Studies, 20 (3), 551–557.

Chmura D., Nejfeld P., Borowska M., Woźniak G., Nowak T. & Tokarska-Guzik B. 2013: The importance of land use type in *Fallopia* (*Reynoutria*) *japonica* invasion in the suburban environment. Polish Journal of Ecology, 61 (2), 379–384.

Costanza R., d'Arge R., de Groot R., Farber S., Grasso M., Hannon B., Limburg K., Naeem S., O'Neill R.V, Paruelo J., Raskin R.G., Sutton P. & van den Belt M. 1997: The value of the world's ecosystem services and natural capital. Nature, 387 (15), 253–260.

Crutzen P.J. 2002: Geology of mankind. Nature, 415, 23.

Farber S., Costanza R., Childers D.L., Erickson J., Gross K., Grove M., Hopkinson C.S., Kahn J., Pincetl S., Troy A., Warren P. & Wilson M. 2006: Linking ecology and economics for ecosystem management. BioScience, 56 (2), 121–133.

Gómez-Baggethun E. & Barton D.N. 2013: Classifying and valuing ecosystem services for urban planning. Ecological Economics, 86, 235–245.

Haughton G. & Hunter C. 1994: Sustainable Cities. Regional Policy and Development. Jessica Kingsley Publishers Ltd, Regional Studies Association, London.

Hobbs R.J. 2007: Setting effective and realistic restoration goals: key directions for research. Restoration Ecology, 15 (2), 354–357.

Hobbs R.J., Higgs E.S. & Hall C.M. 2013: Defining novel ecosystems. [In:] Hobbs R.J., Higgs E.S. & Hall C.M. (eds.): Novel Ecosystems: Intervening in the New Ecological World Order. John Wiley & Sons Ltd., Chichester, 58–60.

Hoekstra J.M., Boucher T.M., Ricketts T.H. & Roberts C. 2005: Confronting a biome crisis: global disparities of habitat loss and protection. Ecology Letters, 8 (1), 23–29.
Holl K.D. & Aide T.M. 2011: When and where to actively restore ecosystems? Forest Ecology and Management, 261 (10), 1558–1563.
Horáčková M., Řehounková K. & Prach K. 2016: Are seed and dispersal characteristics of plants capable of predicting colonization of post-mining sites? Environmental Science and Pollution Research, 23 (14), 13617–13625.
Kompała-Bąba A. 2013: Abiotic and biotic factors affecting the diversity of ruderal vegetation (Silesian Uplands Poland). Sorus, Poznań.
Lapointe M., Gurney G.G., Coulthard S. & Cumming G.S. 2021: Ecosystem services, well-being benefits and urbanization associations in a Small Island Developing State. People and Nature, 3 (2), 391–404.
Markowicz A., Woźniak G., Borymski S., Piotrowska-Seget Z. & Chmura D. 2015: Links in the functional diversity between soil microorganisms and plant communities during natural succession in coal mine spoil heaps. Ecological Research, 30, 1005–1014.
MEA – Millennium Ecosystem Assessment. 2005: Ecosystems and Human Well-being: Biodiversity Synthesis. World Resources Institute, Washington DC.
Moll G. & Petit J. 1994: The urban ecosystem: putting nature back in the picture. Urban Forests, 14 (5), 8–15.
Natural Resources Wales 2013: Introducing Sustainable Management of Natural Resources. Cyfoeth Naturiol Cymru, Natural Resources Wales. https://naturalresources.wales, access: 02.06.2021.
Newcastle 2016: http://www.bluegreencities.ac.uk/about/blue-greencitiesdefinition.aspx, access: 02.06.2021.
Nikolic N., Böcker R. & Nikolic M. 2016: Long-term passive restoration following fluvial deposition of sulphidic copper tailings: nature filters out the solutions. Environmental Science and Pollution Research, 23 (14), 13672–13680.
Pascual U., Balvanera P., Díaz S., Pataki G., Roth E., Stenseke M., Watson R.T., Dessane E.B., Islar M., Kelemen E., Maris V., Quaas M., Subramanian S.M., Wittmer H., Adlan A., Ahn S., Al-Hafedh Y.S., Amankwah E., Asah S.T., Berry P., Bilgin A., Breslow S.J., Bullock C., Cáceres D., Daly-Hassen H., Figueroa E., Golden C.D., Gómez-Baggethun E., González-Jiménez D., Houdet J., Keune H., Kumar R., Ma K., May P.H., Mead A., O'Farrell P., Pandit R., Pengue W., Pichis-Madruga R., Popa F., Preston S., Pacheco-Balanza D., Saarikoski H., Strassburg B.B., van den Belt M., Verma M., Wickson F. & Yagi N. 2017: Valuing nature's contributions to people: the IPBES approach. Current Opinion in Environmental Sustainability, 26, 7–16.
Pataki D.E., Carreiro M.M., Cherrier J., Grulke N.E., Jennings V., Pincetl S., Pouyat R.V., Whitlow T.H. & Zipperer W.C. 2011: Coupling biogeochemical cycles in urban environments: ecosystem services, green solutions, and misconceptions. Frontiers in Ecology and the Environment, 9 (1), 27–36.
Piekarska-Stachowiak A., Szary M., Ziemer B., Besenyei L. & Woźniak G. 2014: An application of the plant functional group concept to restoration practice on coal mine spoil heaps. Ecological Research, 29, 843–853.
Prach K. & Pyšek P. 2001: Using spontaneous succession for restoration of human-disturbed habitats: Experience from Central Europe. Ecological Engineering, 17 (1), 55–62.
Rahmonov O. 2009: The chemical composition of plant litter of black locust (*Robinia pseudoacacia* L.) and its ecological role in sandy ecosystems. Acta Ecologica Sinica, 29 (4), 237–243.
Raudsepp-Hearne C., Peterson G.D., Tengö M., Bennett E.M., Holland T., Benessaiah K., MacDonald G.K. & Pfeifer L. 2010: Untangling the environmentalist's paradox: Why is human well-being increasing as ecosystem services degrade? BioScience, 60 (8), 576–589.
Řehounková K., Čížek L., Řehounek J., Šebelíková L., Tropek R., Lencová K., Bogusch P., Marhoul P. & Máca J. 2016: Additional disturbances as a beneficial tool for restoration of post-mining sites: a multi-taxa approach. Environmental Science and Pollution Research, 23 (14), 13745–13753.
Rydgren K., Auestad I., Hamre L.N., Hagen D., Rosef L. & Skjerdal G. 2016: Long-term persistence of seeded grass species: an unwanted side effect of ecological restoration. Environmental Science and Pollution Research, 23 (14), 13591–13597.
Sen A. 1999: Development as Freedom. Oxford University Press, Oxford.
SER – Society for Ecological Restoration International Science & Policy Working Group. 2004: The SER International Primer on Ecological Restoration. Society for Ecological Restoration International Science & Policy Working Group. www.ser.org & Tucson, Society for Ecological Restoration International.

Stefanowicz A.M., Kapusta P., Błońska A., Kompała-Bąba A. & Woźniak G. 2015: Effects of *Calamagrostis epigejos*, *Chamaenerion palustre* and *Tussilago farfara* on nutrient availability and microbial activity in the surface layer of spoil heaps after hard coal mining. Ecological Engineering, 83, 328–337.
Steffen W., Crutzen P.J. & McNeill J.R. 2007: The Anthropocene: are humans now overwhelming the great forces of nature. AMBIO: A Journal of the Human Environment, 36 (8), 614–621.
Tarvainen O. & Tolvanen A. 2015: Healing the wounds in the landscape – reclaiming gravel roads in conservation areas. Environmental Science and Pollution Research, 23 (14), 13732–13744.
Tropek R., Cerna I., Straka J., Kocarek P., Malenovsky I., Tichanek F. & Sebek P. 2016: In search for a compromise between biodiversity conservation and human health protection in restoration of fly ash deposits: effect of anti-dust treatments on five groups of arthropods. Environmental Science and Pollution Research, 23 (14), 13653–13660.
Wackernagel M., Schulz N.B., Deumling D., Linares A.C., Jenkins M., Kapos V., Monfreda C., Loh J., Myers N., Norgaard R. & Randers J. 2002: Tracking the ecological overshoot of the human economy. Proceedings of the National Academy of Sciences of the United States of America, 99 (14), 9266–9271.
Walker L.R., Hölzel N., Marrs R., del Moral R. & Prach K. 2014: Optimization of intervention levels in ecological restoration. Applied Vegetation Science, 17 (2), 187–192.
Whisenant S.G. 1999: Repairing Damaged Wildlands: A Process-Orientated, Landscape-Scale Approach. Cambridge University Press, Cambridge.
Woźniak G. 2010: Zróżnicowanie roślinności na zwałach pogórniczych Górnego Śląska [Diversity of vegetation on coal-mine heaps of the Upper Silesia (Poland)]. W. Szafer Institute of Botany, Polish Academy of Sciences, Kraków.
Woźniak G., Markowicz A., Borymski S., Piotrowska-Seget Z., Chmura D. & Besenyei L. 2015: The relationship between successional vascular plant assemblages and associated microbial communities on coal mine spoil heaps. Community Ecology, 16 (1), 23–32.
Woźniak G., Sierka E. & Wheeler A. 2018: Urban and industrial habitats: how important they are for ecosystem services. [In:] Hufnagel L. (ed.): Ecosystem Services and Global Ecology. IntechOpen, 169–194.
Wu J. 2014: Urban ecology and sustainability: The state-of-the-science and future directions. Landscape and Urban Planning, 125, 209–221.

4.5. Plant Responses to Harsh Conditions of Post-industrial Habitats

Agnieszka Szuba
Department of Genetics and Environmental Interactions,
Institute of Dendrology, Polish Academy of Sciences;
5 Parkowa Str., 62-035 Kórnik, Poland;
aszuba@man.poznan.pl; ORCID: 0000-0001-6278-3897

Ewelina Ratajczak
Department of Developmental Biology,
Institute of Dendrology, Polish Academy of Sciences;
5 Parkowa Str., 62-035 Kórnik, Poland;
eratajcz@man.poznan.pl; ORCID: 0000-0003-2710-4638

Anna Kasprowicz-Maluśki
Department of Molecular and Cellular Biology,
Institute of Molecular Biology and Biotechnology,
Faculty of Biology, Adam Mickiewicz University in Poznań;
6 Uniwersytetu Poznańskiego Str., 61-614 Poznań, Poland;
akas@amu.edu.pl; ORCID: 0000-0003-3699-9306

Emilia Pers-Kamczyc
Department of Genetics and Environmental Interactions,
Institute of Dendrology, Polish Academy of Sciences;
5 Parkowa Str., 62-035 Kórnik, Poland;
epk@man.poznan.pl; ORCID: 0000-0002-5610-2124

ABSTRACT: Post-industrial habitats are often burdened with various anthropogenic disturbances like contamination with heavy metals, polycyclic hydrocarbons and other xenobiotics, or are exposed to drought. As a consequence, these habitats usually represent a very harsh environment for plant life. Plants growing in such areas must cope with continuous exposure to stress. Stress negatively affects plants and triggers specific responses at the physiological, biochemical and molecular levels. In plant tissues, exposure to stress causes generation of reactive oxygen species (ROS), accumulation of toxic metabolites like malondialdehyde (MDA), and is followed by the increased production of antioxidant and other anti-stress compounds like glutathione (GSH), ascorbic acid (AsA), proline, polyamines, glycine or betaine. Plant cell exposition to stress conditions results in damage to lipids, proteins and DNA, chromosomal aberration, and disruption of spindle activity, which in turn causes cell injury and results in plant growth dysfunctions: lower growth rates or even plant cell death. The plants which can survive acute exposure to stress are forced to live in prolonged or even chronic stress conditions. During this time various responses to stress are observed, some of them are common to all stress durations, and some are characteristic for particular stress phases or particular stressors. Plants that survived these events become acclimatized to particular harsh habitats and, in the end, adapted to them. These processes of acquisition and maintenance of stress tolerance are utilized during phytoremediation – an eco-friendly method of remediation of polluted post-industrial areas. This chapter highlights the major molecular background of observed stress-triggered damages and the plant (cell) stress responses.

KEYWORDS: Post-industrial areas, abiotic stress, acclimatization and adaptation, growth reduction, oxidative stress, proteomics, metabolomics, transcriptomics, epigenetics, phytoremediation

INTRODUCTION

Post-industrial areas are usually heavily modified environments (Fig. 1), most frequently burdened with abiotic stress conditions. Anthropogenic activity frequently results in the modification of natural soil layers and subsequent disturbances in availability of nutrients, disturbances in natural watercourses together with strong light exposure commonly observed on various post-mining heaps or post-industrial landfills, what results in drought stress. Pollution with various salts, heavy metals, or organic compounds like polycyclic aromatic hydrocarbons, pesticides, petroleum, etc. results with strong environmental contamination (Foresman 1986; Szuba & Lorenc-Plucińska 2018; https://www.eea.europa.eu/publications/a-decade-of-industrial-pollution-data). Moreover, in post-industrial areas, combination of various abiotic stresses are usually present, thus all living organisms, including plants, are exposed to the combined abiotic stress conditions known to be significantly more dangerous to plants functioning compared to simple stress scenarios (Li *et al.* 2017; Szuba & Lorenc-Plucińska 2018). Post-industrial areas, especially directly after the cessation of industrial activity, are also characterized by very low biodiversity (Fig. 1). For all these reasons plants growing in post-industrial areas are exposed to various stressors and may be more sensitive to subsequent biotic stress impacts (Schoeneweiss 1981) and again, more weakened.

Because of their immobility, the successful response of plants to stress conditions of post-industrial areas is crucial for their survival. The majority of published research concerned biochemical studies with analyses of plant response to imposed, short-term single stress laboratory scenarios, with only a few molecular reports from environmental studies. The lack of molecular studies of plants growing in real environmental conditions can be related to the limited number of variable parameters that can be controlled and with a majority of them, like the presence of diseases, pathogens, and pests or weather conditions (one of the most important environmental factors; Bălan *et al.* 2009), can only be monitored. Such lack of control may be problematic during sensitive experiments at the molecular OMIC level (level of metabolome, proteome, transcriptome, or epigenome) as these dynamic "structures" react to every change of plant habitats (Cramer *et al.* 2011; Poorter *et al.* 2012; Alexandersson *et al.* 2014). In addition, the variability among individuals used in the field experiments can be very large as individuals are exposed to different micro-environments, which further complicates conclusions of *in situ* biochemical studies (Szuba *et al.* 2017; Szuba & Lorenc-Plucińska 2018). Nevertheless, some molecular studies have been carried out on plants growing *in situ* conditions (Zhao *et al.* 2005; Kerner *et al.* 2014), and even comparing geographically separated plots (Hoehenwarter *et al.* 2008; Ma *et al.* 2015; Szuba & Lorenc-Plucińska 2018). Still, the plant response to abiotic stress is very well known, including at the molecular, OMIC level (Mosa *et al.* 2017) largely thanks to the numerous available laboratory reports (Fig. 2). The most frequently analyzed abiotic stresses include drought and heavy metal stresses, the same ones that are most often observed in post-industrial areas (Szuba & Lorenc-Plucińska 2018). These facts, combined with the knowledge gained from various *in situ* experiments, help us understand the molecular background of plant functioning in anthropogenically modified harsh habitats (Fig. 2), and predict the plant's successful long-term survival.

Figure 1. Examples of harsh post-industrial areas.
From the top: 1) the post-mining heap in South Poland; 2) the buffer zone of the copper smelter (KGHM, Głogów, Poland) and 3) the former tannery landfill surrounded by a vast forest (Warcino, Poland)
(Photo: Agnieszka Szuba)

ABIOTIC STRESS: PLANT CELL DAMAGE AND PLANT (CELL) RESPONSE

It is well known that abiotic stress causes various direct and indirect physiological and biochemical effects like decrease in plant biomass, seed quality and germination rates, as well as root elongation, loss of turgor pressure and osmotic adjustments, reduced internal CO_2 concentration, decreased stomatal conductance to CO_2, decreased pigment contents, reduced leaf water potential, and slowdown of net photosynthesis and respiration (Fig. 2) (Bah *et al.* 2010; Vannini *et al.* 2011; Sharmin *et al.* 2012). These effects are caused by negative changes at the molecular level that cause plant cell dysfunctions such as damage of photosynthetic apparatus, membrane breakdown, damage of cytoskeleton, loss of spindle activity etc., and cause cell injury or even plant cell death, but also trigger numerous cell responses (Fig. 2) (Shukla *et al.* 2011; Shahid *et al.* 2012; Huang *et al.* 2014; Trinh *et al.* 2014; Onaga & Wydra 2016). The most common result of the energy-consuming anti-stress processes is reduced plant growth.

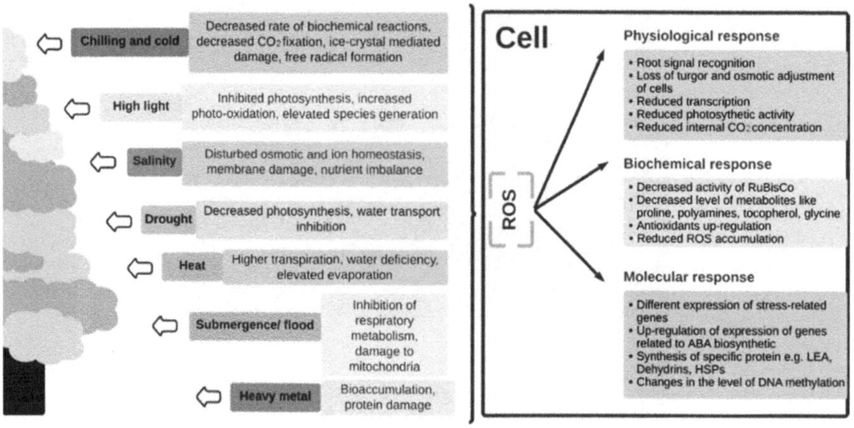

Figure 2. Effects of various abiotic stresses on plants and major plant cell stress responses. Simplified scheme presents selected, major negative effects characteristic for various abiotic stresses and major consequences of the stress-triggered ROS burst common to various stress conditions

In all cases, the plant stress response may be divided into several phases (Baena-González 2010; Harb *et al.* 2010). The early stress response (the so-called "shock" for cells) usually involves activating signaling pathways – mitogen-activated protein kinase (MAPK) activation, Ca^{2+} signatures, and phytohormone shifts are common early symptoms of stress (Leone *et al.* 1994; Yang *et al.* 2014; Szuba & Lorenc-Plucińska 2018). Plants react with transcriptome adjustments, usually resulting in up-regulation of translation stress-related genes as well as up-regulation of genes related with metal transporters and with increased abundance of proteins involved, e.g., in phenolic compounds, lipids, amino acids, or hormone metabolism (Li *et al.* 2016).

Subsequently, in most cases, significant oxidative bursts are detected and consequently, the activation of various antioxidant systems is often observed as plant responses common to a wide range of stresses (Fig. 2) (e.g. Zhang & Sonnewald 2017). A common consequence of various abiotic stresses is damage to macromolecules like DNA mutations, lipid peroxidation or incorrect protein folding and aggregation. As the result, abiotic stress disrupts key metabolic processes in plant cells (Michalak 2006; Jiang & Liu 2010; Fahr *et al.* 2013; Luo *et al.* 2014; Hasan *et al.* 2017). Stress negatively affects cell cycle control, and central metabolisms: nitrogen metabolism, the Krebs cycle, or glycolysis influence sugar and energy metabolism, or photosynthesis is downregulated (e.g. Bah *et al.* 2010; Sharmin *et al.* 2012; Evlard *et al.* 2014; Huang *et al.* 2014; Lomaglio *et al.* 2015; Zemleduch-Barylska & Lorenc-Plucińska 2015; 2016; Yıldız & Terzi 2016). The common effect is an alerted, usually reduced amount of proteins involved in the production of adenosine triphosphate (ATP) and nicotinamide adenine dinucleotide (NAD+) – the most important cellular energy carriers necessary for plant growth (Ahsan *et al.* 2009; Vannini *et al.* 2011; Bhargava & Sawant 2013; Zeng *et al.* 2014).

In the long term, plants "try to" protect macromolecules (Pandey *et al.* 2015; Zhang & Sonnewald 2017; Szuba & Lorenc-Plucińska 2018), including by overproduction of heat shock proteins (HSPs) (for details please see below), protection of the photosynthetic apparatus by sequestration and stabilization of carotenoids (Langenkämper *et al.* 2001) or by remodeling of the cell wall (Kasprowicz *et al.* 2009; Rodakowska *et al.* 2009; Zemleduch-Barylska & Lorenc-Plucińska 2015). During these processes, various groups of molecules are involved that are over-produced, like osmoprotectants (carbohydrates, amino acids, etc.), antioxidants (carbohydrates, flavonoids, carotenoids, xanthophylls, tocopherols, glutathione, ascorbic acid (AsA) etc.), and signaling molecules (carotenoids, jasmonates, salicylates or polyamines). Interestingly, some chemical compounds may play multiple roles, for example, sugars or proline (e.g. Arbona *et al.* 2017). Sugars are one of the most important regulators that facilitate many physiological processes, such as photosynthesis, seed germination, flowering, senescence, and many more under various abiotic stresses (Ciereszko 2018). Sugars can also play a protective role against stress factors, for example as osmoprotectants or donors of carbon skeletons, in the regulation of many defensive reactions or signaling molecules. They can be transported over long distances and are often storage substances (Ciereszko 2009; Sami *et al.* 2016). The role of carbohydrates, e.g. nonreducing sugars, is crucial for plant cell protection during stress. Sugars such as sucrose and the raffinose family of oligosaccharides (RFO: raffinose or stachyose) are used by plants as storage material (e.g. accumulated in seeds during development) but also are key plant cell osmoprotectants (Pukacka *et al.* 2009). Proline, a proteinogenic amino acid, is the next important multifunctional stress-related compound, accumulation of which is connected with water loss prevention mechanisms and other adaptive responses in higher plants (Hayat *et al.* 2012). This intracellular signal compound is accumulated during conditions of drought, high salinity, heavy metals, and biotic stresses (Liang *et al.* 2013). Proline reduces oxidative damage to the lipid membranes, stabilizes the structures of the proteins, and plays an important role in ROS removal (Kumar *et al.* 2015). Under stress the level of proline rises, which guarantees higher plant resistance to environmental stress (Kijowska-Oberc *et al.* 2020).

Apart from common responses to various stressors, it is believed that different stresses trigger specific molecular responses characteristic for a particular stress (Fig. 2) (Hu *et al.* 2015; Pandey *et al.* 2015; Zhang & Sonnewald 2017). For example, under acute osmotic stress, probably the most important environmental disturbance, activation of the Salt Overly Sensitive (SOS) pathway is one of the characteristic early symptoms (Leone *et al.* 1994; Droillard *et al.* 2000; Shanker & Venkateswarlu 2011; Yang *et al.* 2014; Che-Othman *et al.* 2017; Szuba & Lorenc-Plucińska 2018). Drought, or more generally water stress, causes an additional strong misbalance of exo- and endocytosis to preserve membrane integrity during cell shrinking (Zwiewka *et al.* 2015). The ion channels, like potassium transporters, are therefore regulated, frequently under abscisic acid (ABA) control, to maintain cellular turgor and prevent direct ion toxicity (under salt stress) (Verslues *et al.* 2006; Shanker & Venkateswarlu 2011; Osakabe *et al.* 2014; Hu *et al.* 2015; Pandey *et al.* 2015; Zhang & Sonnewald 2017). The common denominator of the osmotic stress response is also overproduction of various, usually small, water-soluble osmoprotective molecules like small proteins, compounds like amino acids, amines betaines (glycine betaine), organic acid sugars, or pyrols (Shanker & Venkateswarlu 2011). The specific response, one of the best recognized, had also been developed by plants exposed to the heavy metal (HM) stress (Szuba & Lorenc-Plucińska 2018; Szuba *et al.* 2020). The defensive mechanisms against heavy metal stresses have a common denominator which is a decreasing concentration of free heavy metal ions in the plant cell cytosol/protoplast – the plant cell interior where the most "precious" plant cell macromolecules are located. However, the detoxification of HMs mechanisms begins outside the plant cell and is connected with the exudation of various organic ligands into the rhizosphere. These compounds are characterized by a strong affinity to HMs and increase the retention of HMs in the soil, decreasing the level of harmful elements that affected the plant itself (Shahid *et al.* 2012; Reddy *et al.* 2016; Fernández-Fuego *et al.* 2017). The next barrier for heavy metals is the plant cell wall (CW), a well-known first line of plant cell defense against harmful conditions (Kasprowicz *et al.* 2009; Rodakowska *et al.* 2009). In this complex structure, HMs bound to cellulose, hemicellulose, lignin, callose, and pectin (Williams *et al.* 2000; Krzesłowska 2011). For this reason, one of the characteristic plant cell responses to HMs stress is CW remodelling, such as an increasing amount of pectines etc. to increase HMs binding sites (Krzesłowska 2011; Luo *et al.* 2014; Li *et al.* 2016; Shi *et al.* 2019). Nevertheless, even modified CW has its maximal capacity, and when all HMs binding sites are occupied by harmful ions there is a significant danger of plasma membrane (PM) damage. As the integrity of PM is one of the most crucial elements of survival of all living cells, the excess of HMs ions has to be removed from the direct vicinity of the PM. Heavy metals, but also other compounds like salts or xenobiotics are transported via the PM, usually actively, with the use of various ATP-related transporters (Williams *et al.* 2000; Colpaert *et al.* 2011). Because the proton gradient is necessary for active transport of numerous compounds, various ATPases, like P-type ATPases or PPi-ATPases, play a crucial role during the stress response and are often detected among the over-produced molecules during various OMIC studies of stressed plants (e.g. Klychnikov *et al.* 2007; Zhang *et al.* 2017a; Szuba *et al.* 2020).

Once heavy metals enter the symplast, they are bound to various chelates, such as phytochelatins (PCs), phenolic compounds, or metallothionines (MTs) (Reis *et al.* 2015; Reddy *et al.* 2016; Hasan *et al.* 2017; Jalmi *et al.* 2018). This process is crucial to maintain the minimal possible concentrations of free metal ions, especially ions of non-essential heavy metals like Pb or Cd in the cytoplasm (Hasan *et al.* 2017). Ion homeostasis is finally obtained when PCs-HMs or MTs-HMs complexes are redirected to the vacuole, the final and main place of HMs sequestration in the plant protoplast (Sharma *et al.* 2016; Hasan *et al.* 2017). For this reason, increased levels of Vacuolar-H^+-ATPases are often observed in HMs-stressed plants, as protons are similarly important for the proper functioning of tonoplast transporters as during the trespass of plasma membranes (Luo *et al.* 2014; Sharma *et al.* 2016; Zhang & Sonnewald 2017; Zhang *et al.* 2017a; 2017b). If all mentioned processes are sufficient, the negative phenotypic changes caused by HMs exposition will be not observed (Szuba *et al.* 2020). Only when HMs sequestration processes are inefficient, mostly when the concentrations of heavy metals exceed the tolerance threshold of a particular plant, harmful ions in the cytosol trigger the ROS burst. Reactive oxygen or nitrogen species (ROS and RNS) are the main reason for the molecular damages caused by heavy metals, but also by numerous other abiotic stressors (Jalmi *et al.* 2018).

The most common negative effect of abiotic stresses is therefore the ROS burst and subsequent damages of key macromolecules by over-produced radicals (Fig. 2) (Mourato *et al.* 2012). ROS are over-produced even during exposition to low levels of stress and endogenous generation of ROS belongs to one of the first stress effects (Sinha *et al.* 1997; Panda & Choudhry 2005; Ahsan *et al.* 2009; Hänsch & Mendel 2009; Puig & Peñarrubia 2009; Wang *et al.* 2013; Trinh *et al.* 2014). ROS include superoxide anion radicals, hydroxyl radicals, singlet oxygen, and hydrogen peroxide in plant cells, and activate signaling pathways leading to numerous changes in physiological, biochemical, and molecular mechanisms in cellular metabolism (Das & Roychoudhury 2014). The excess of radicals leads to the subsequent production of various toxic compounds – products of damaged cell compartments (like MDA being a product of lipid membrane peroxidation; Skórzyńska-Polit 2007; Labudda 2013). Therefore, a common response of plants cells exposed to a wide range of various abiotic stress conditions, is to reduce the level of the oxidative stress (Fig. 2). To avoid negative effects of ROS, enzymatic (e.g., peroxidases, catalase, superoxide dismutase (SOD)) or non-enzymatic (e.g., GSH or AsA) antioxidants are activated in plant cells (Vannini *et al.* 2011; Mourato *et al.* 2012; Pourrut *et al.* 2013; Huang *et al.* 2014; Zeng *et al.* 2014; Zemleduch-Barylska & Lorenc-Plucińska 2015; Ashraf *et al.* 2017; Hasan *et al.* 2017; Szuba & Lorenc-Plucińska 2018). However, during stress exposition, not all antioxidative enzymes are necessarily up-regulated, thus the up-regulation and the down-regulation of other anti-ROS enzymes was frequently observed in stressed plants. For example, in leaves of poplars exposed to chromium, significant up-regulation of Cu-Zn SOD and downregulation of ascorbate peroxidase (APX) was observed (Szuba & Lorenc-Plucińska 2018). Such diverse responses may be connected with the various roles of antioxidative enzymes, including numerous physiological functions, for example during redox signaling or photosynthesis. The complex interactions of ROS and the antioxidant system play an important role in many stresses, including responses to environmental conditions and regulation of plant growth (Foyer & Noctor 2005). This interaction is controlled by the redox systems in plant cells. All living organisms are oxidation-reduction (redox) systems, and therefore the redox state is considered to be a key determinant of cell function (Foyer & Noctor 2005; Dietz *et al.* 2016). Many metabolic processes in cells induce oxidation and reduction reactions, so it is believed that all biological responses are controlled by redox systems. The balance of redox state is regulated by a system of non-enzymatic and enzymatic antioxidants. This system is crucial for the resistance of plant cells to environmental stress and rapid adaptation to the environmental conditions. Thiol group containing proteins, mainly peroxiredoxins (Prxs) and thioredoxins (Trxs), are one of its key elements. Changes in thiol status of proteins probably play a role in redox signaling under biotic and abiotic stress (König *et al.* 2002; Foyer & Noctor 2013; Sevilla *et al.* 2015). Thioredoxins and peroxiredoxins are small proteins (around 12kDa) regulating the function of target proteins through oxidoreductase activity. Trxs couples with Trx-dependent peroxidases (Prxs) to scavenge hydrogen peroxide (H_2O_2) and peroxinitrite (Barranco-Medina *et al.* 2007; Sevilla *et al.* 2015). Trx/Prx system plays a key role in the control of the redox state of proteins containing disulfide bonds, and hence also in the regulation of the process of seed germination, as well as subsequent plant development (Rouhier *et al.* 2015), and most important, during an environmental stress adaptation (Spoel & Loake 2011; Liu *et al.* 2013). Prxs play a role in functional change from peroxidase and chaperone activity. Switching between the roles of peroxidases and chaperones is possibly regulated by stress or by prevailing environmental conditions (Kim *et al.* 2012b; König *et al.* 2013).

The regulatory redox signaling network in plant cells detects every metabolic imbalance and modulates a rapid response at the level of transcription and translation (Foyer & Noctor 2016; Mangano *et al.* 2016). As a result, plants usually can quickly adapt to changing environmental conditions and produce high-quality seeds. At the same time, when the intensity of stress is too high, the negative phenotypic changes are accompanied by decreased amounts of antioxidants, both enzymatic and non-enzymatic. Such effects belong to commonly observed symptoms of the insufficient antioxidative response – the plant stops coping with stress. As a consequence, the proper interpretation of anti-ROS response is a challenging task, and drawing a quick conclusion may lead to misinterpretations, as the low level of antioxidative enzyme activity will not always mean a low level of oxidative stress. Therefore, the analysis of oxidative stress should be more complex and should be based on simultaneous analyses of the level of antioxidants and radicals (Szuba *et al.* 2020).

Apart from the over-production of direct enzymatic ROS scavengers, like SOD involved in dismutation of the most harmful radical, superoxide ($O_2^{-\cdot}$), activation of enzymes indirectly involved in ROS scavenging is frequently observed under stress conditions. Among these "non-obvious" enzymes involved in response to oxidative stress are key enzymes involved in the biosynthesis of various compounds characterized with antioxidative properties, like chalcone synthase – a key enzyme of flavonoid biosynthesis or polyphenol oxidase – an enzyme involved in pigment biosynthesis (Kim *et al.* 2012a; Mourato *et al.* 2012; Szuba & Lorenc-Plucińska 2018) as well as enzymes involved in detoxification of highly reactive cytotoxic byproducts like glyoxalases and enzymes detoxifying methylglyoxal (Yadav *et al.* 2008; Szuba & Lorenc-Plucińska 2018). Also, among the metabolites, apart from the most characteristic antioxidants such as GSH or AsA, very often various compounds with a different "main" biological activity may exhibit ROS scavenging activity, such as proline mentioned above (Shanker & Venkateswarlu 2011).

The antioxidative response is crucial to protect cell integrity and to protect the proper activity of various macromolecules, with proteins among the most important ones (Wang *et al.* 2004; Hasan *et al.* 2017).

Proteins, especially the biggest and most abundant ones, for example the large subunit of RuBisCo, are usually significantly exposed to stressors (Fig. 2) (Hendry & Price 1993; Vajpayee *et al.* 2000; Hänsch & Mendel 2009; Bah *et al.* 2010; Nagajyoti *et al.* 2010; Zeng *et al.* 2014; Lomaglio *et al.* 2015). As a consequence, proteins are very strongly influenced by the action of stress-triggered ROS, but in some cases (e.g. during heavy metal stress) also directly via binding Cr or Pb ions to the particular amino acid residues (Vajpayee *et al.* 2000; Salnikow & Zhitkovich 2008; Ahsan *et al.* 2009; Todeschini *et al.* 2011; Kandziora-Ciupa *et al.* 2013). Signals caused by protein dysfunctions, breakdown, or crosslinking are commonly observed, especially during high-intensity abiotic stress (exposure to strong abiotic stress frequently results in significant decrease of overall protein level).

One of the stress-related damages of proteins is also various protein post-transcriptional modifications (PTMs). Protein PTMs like nitration or carboxylation are a known negative consequence of oxidative stress (action of ROS or RNS), even called stress footprints (Szuba & Wojtaszek 2010; Szuba *et al.* 2015; Yang *et al.* 2015; Falfushynska *et al.* 2016). Such PTMs inhibit protein activity (Harb *et al.* 2010; Gietler *et al.* 2017), especially when they occur in the catalytic center of enzymes. Protein ubiquitination is a PTM related to marking the ROS-damaged proteins to be disassembled. An increase in the level of ubiquitination together with increased activity of proteasomes is frequently observed under stress conditions because the misfolded proteins may be toxic to the plant protoplast and have to be quickly removed (Vannini *et al.* 2011; Huang *et al.* 2014). PTMs are an interesting subject of stress-related plant study as there is some data suggesting differential roles of PTMs during short- and chronic stress (Tanou *et al.* 2012; Szuba *et al.* 2015), but also because for stress-related PTMs being previously considered exclusively as stress footprints the physiological roles were recently discovered (Harb *et al.* 2010; Mata-Pérez *et al.* 2016).

The major part of protecting proteomes from these negative effects of stress is the massive overproduction of various heat shock proteins (HSPs) (Wang *et al.* 2004; Goupil *et al.* 2009; Sharmin *et al.* 2012; Hasan *et al.* 2017). HSPs prevent protein aggregation, assist in refolding, and play a role in protein translocation in signal transduction and transcriptional activation. HSP70 is typically the most common chaperonin family (Ireland *et al.* 2004; Wang *et al.* 2004), but overexpression of other groups like HSP90s or HSP80s has been also reported when abiotic stress levels are high (Ahsan *et al.* 2009; Goupil *et al.* 2009). Proteins may also be protected in other ways, such as photosynthetic proteins which are protected by an increase in the amount of proteins associated with plastids and lipids. These structural proteins are involved in indirect plant stress responses as their main role is the storage of carotenoids (Rey *et al.* 2000) which are responsible (among others) for quenching photodynamic reactions and, generally, are known antioxidants with the key protective role during stress (Halliwell 1987). Thylakoids and stromal elements are also protected by ascorbate peroxidases detoxifying the harmful photosynthesis products (Davletova *et al.* 2005), which links these proteins directly to photosynthesis protection.

One of the most important macromolecules damaged by ROS action is DNA (Machado *et al.* 1997; Møller *et al.* 2007; Nagajyoti *et al.* 2010; Huang *et al.* 2014). Heavy metals like chromium may create direct Cr-DNA adducts whereas Pb is known to induce DNA breaks. DNA is also damaged by the action of stress-induced radicals or via DNA-protein crosslinking (Salnikow & Zhitkovich 2008). As DNA is a crucial molecule for plant cell survival, signals of DNA protection and repair are commonly observed under abiotic stress (Panda & Choudhry 2005; Mourato *et al.* 2012; Wang *et al.* 2013; Huang *et al.* 2014). The efficient DNA repair systems are connected with abiotic stress tolerance in plants and are, generally, similar to mechanisms involved in the protection of other key components of plant cells. They include activation of enzymatic antioxidative response like up-regulation of catalase or cytosolic ascorbate peroxidase (Vanderauwera *et al.* 2011). Other enzymes involved in the DNA damage response under various abiotic stresses are, for example, thiazole biosynthesis enzyme, an enzyme involved in the biosynthesis of thiamine (Machado *et al.* 1997) or phosphoglycolate phosphatase, involved among others in the metabolism of 2-phosphoglycolate – a product of ROS-induced DNA lesions (Pellicer *et al.* 2003). Also, proteins indirectly associated with stress responses may act protectively on DNA or processes of transcription. Among them, the glycine-rich RNA-binding protein involved in mRNA splicing plays important role in abiotic stress adaptation (Kim *et al.* 2010).

It is generally believed that plants may implement two major strategies of life in a harsh environment: to tolerate or to avoid stress. The second one is important because the final effect of stress action strongly depends on the dose of stress and the stress effect may vary depending on both the exposure time and the concentrations of xenobiotics, salts, or heavy metals (Vajpayee *et al.* 2000; Ireland *et al.* 2004; Bah *et al.* 2010; Shukla *et al.* 2011; Wang *et al.* 2013; Huang *et al.* 2014; Zeng *et al.* 2014). As the stress duration is rather independent of plants, the bioavailability of negative factors may be, within some limits, modulated by plant reactions. These aspects are probably best recognized in the case of heavy metals – one of the most common pollutants of anthropogenically-modified habitats. The critical concentrations of numerous xenobiotics and heavy metals are defined by generally accepted standards applicable e.g. in agriculture. Some examples include (mg kg^{-1} dry weight of soil): Fe 20–2500, Mn 1500–3000, but only Cr 75–100 or Cu 60–125 (Kabata-Pendias & Pendias 2001; Becker & Asch 2005). However, numerous pollutants especially dangerous to living organisms, e.g. non-essential heavy metals such as lead or cadmium, have a strong potential to precipitate with most anions (Kopittke *et al.* 2008; Tangahu *et al.* 2011). This high potential to precipitate, or high affinity to absorb various organic compounds like phenolics or low molecular mass organic acids are "utilized" by plants to decrease the level of bioavailable ions, which constitute a real danger for leaving cells (Fig. 3).

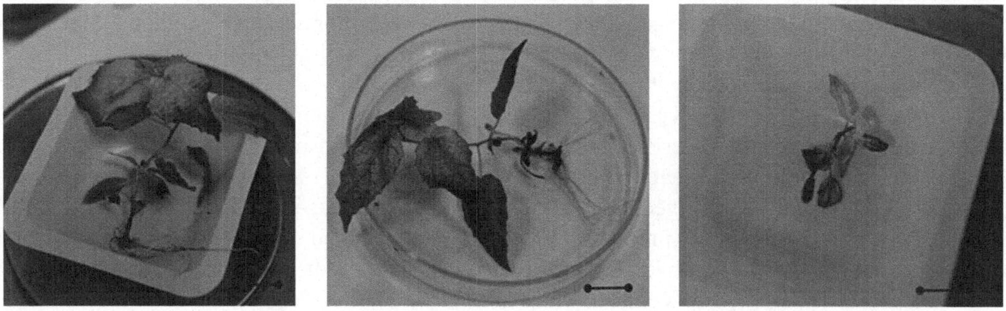

Figure 3. *Populus* × *canescens* grown for 6 weeks *in vitro* exposed to various heavy metal stresses. From left: 1) representative image of poplar microcuttings grown in normal conditions (MS agar medium); 2) poplar microcuttings grown in MS medium supplemented with 1.5 mM $Pb(NO_3)_2$ and 3) poplar microcuttings grown in MS medium supplemented with 0.75 mM $Cu(NO_3)_2$. Lead ions form insoluble salts with numerous anions present in the culture medium in contrast to copper ions. The participation of Pb^{2+} significantly decreased the bioavailability of lead. It is clearly visible that despite the use of doubled concentrations as compared to copper, the presence of high concentrations of lead ions did not cause significant phenotypic changes. Decreased growth and chlorosis caused by heavy metal stress is clearly visible in Cu^{2+} - exposed plants. Scale: 1 cm (Photo: Agnieszka Szuba)

Thus, the bioavailability of hazardous ions depends not only on the general soil solution composition (Tangahu *et al.* 2011), but also on the concentrations of various active organic compounds exuded by roots into the rhizosphere (Jiang & Liu 2010; Luo *et al.* 2014; Ashraf *et al.* 2017; Szuba *et al.* 2017; Szuba *et al.* 2020). Alleviating stress effects by decreasing HMs bioavailability also belongs to the major element of another mechanism of increased plant stress tolerance – symbiosis (Jentschke & Godbold 2000; Smith & Read 2008; Bhargava *et al.* 2012; Ali *et al.* 2013; Szuba *et al.* 2017; Zhang *et al.* 2017b). The positive effect of symbionts, especially mycorrhizal fungi, in the alleviation of stress is so significant that mycorrhizae are generally believed to be one of the major plant tolerance mechanisms (Szuba *et al.* 2020). The fungal hyphae not only create a bio-filter that decreases access of pollutants to plant cells (e.g. in the form of mantle created by ectomycorrhizal fungal species), but also exude various chelating compounds much more intensively than plant roots themselves (Marschner *et al.* 1998; Jentschke & Godbold 2000; Bellion *et al.* 2006; Wang *et al.* 2010; Colpaert *et al.* 2011; Luo *et al.* 2014; Fernández-Fuego *et al.* 2017; Szuba *et al.* 2017). Moreover, by increasing nutrient, water, and antioxidative status of plants such symbionts positively influence plant vitality and resistance to subsequent stress (Cicatelli *et al.* 2010; Colpaert *et al.* 2011; Luo *et al.* 2014; Yamaji *et al.* 2016; Wang *et al.* 2018; Shi *et al.* 2019). Ectomycorrhizal fungi also influence plant cell responses by modulation of the level of transcription of plant genes related to defense processes, energy metabolism, or CW biosynthesis (Luo *et al.* 2014; Wang *et al.* 2018; Shi *et al.* 2019). The network of interactions is complex and includes various other microorganisms that could significantly influence plant host tolerance (Grover *et al.* 2011) by sequestration of HMs, but also by induction of HSPs or synthesis of ACC-deaminase (reviewed in Grover *et al.* 2011). Even though those microorganisms are also negatively affected by stress (Kandeler *et al.* 1996; Stępniewska & Wolińska 2005; Chaperon & Sauvé 2008; Pan & Yu 2011; Santos *et al.* 2011) they still provide a significant increase in plant tolerance. For this reason, it is necessary to have a holistic view of the symbiotically-enhanced mechanisms of adaptation in field conditions.

PROLONGED STRESS→CHRONIC STRESS→ACCLIMATIZATION→ADAPTATION

Available reports suggest that the molecular status of plants exposed to prolonged stress conditions is generally very similar to that observed in plants exposed to shorter stress durations. Significant disturbances of major cellular pathways such as down-regulation of glycolysis and up-regulation of ATP production, flavonoid biosynthesis and protein turnover (Simova-Stoilova *et al.* 2015), disorders of energy conversion, the Krebs cycle, glycolysis or photosystem II protein components (Król & Weidner 2017) were still observed in plants during persistent osmotic stress lasting more than a week. Similarly, during persistent abiotic stress, increased photorespiration, enhanced glycolysis, and tricarboxylic acid cycle changed cell wall functioning or proteasome activity (Bah *et al.* 2010). Prolonged osmotic stress leads to premature senescence and reduced energy production e.g. via cell swelling caused by harmful ions. Ion homeostasis and activation of ion pumps play a key role in cellular homeostasis (Shanker & Venkateswarlu 2011).

Reports referring to persistent stresses (lasting longer than a week) showed that in oaks a 20-day long drought causes significant down-regulation of glycolysis and up-regulation of ATP production, flavonoid biosynthesis, and protein turn-over (Pandey *et al.* 2015). In grapevines, persistent drought (2 weeks) caused disorders of the main cellular cycles like energy conversion, the Krebs cycle, glycolysis or photosystem II protein components (Rodakowska *et al.* 2009). Similarly, poplar cuttings grown for 22 weeks in pots with solid tannery waste revealed increased photorespiration, enhanced glycolysis and tricarboxylic acid cycle, changed cell wall functioning or proteasome activity (Verslues *et al.* 2006).

Although exposure to chronic harsh conditions is the reality of plant life in post-industrial habitats, there is a lack of molecular data about plants able to live when exposed to much longer, chronic stress conditions. Some of them can be obtained from time-course experiments in which plant responses to shorter and longer-term stress duration are compared. These reports consider weeks, days, or even hours (Huang *et al.* 2014; Meisrimler *et al.* 2016) as a long-term study, but have nevertheless observed increased mitigation of negative effects under longer stress duration, e.g. via stabilization of protein metabolic processing and energy use, maintenance of ROS production at a low level (Meisrimler *et al.* 2016), activation of the ubiquitin-proteasome system pathway or by DNA repair (Huang *et al.* 2014).

The proteome status of *Populus alba* 'Villafranca' plants grown for years (4y) on a contaminated area of a tannery landfill (Szuba & Lorenc-Plucińska 2018) significantly differed from plants with short-term stress duration. Poplars, which were able to survive in the environment of the tannery landfill coped with continuous harsh conditions (mainly HMs stress, but also triggered by other negative factors like decreased mineral and water access caused by bio-concurrence (Fig. 1 and 4) and unfavorable soil structure (Szuba & Lorenc-Plucińska 2018), which caused a continuous ROS burst. The plant response at the proteome level suggested that chronically stressed plants invest in macromolecule protection to prevent molecule devastation and the high costs associated with increased protein or DNA degradation and re-synthesis. This is quite different from responses to shorter-term stress durations (Huang *et al.* 2014), where for example, strong signals of increased proteasome activity are commonly observed. Interestingly, available reports suggest, that during exposition to chronic stress, the protein protection was dominated not by HSP70s but rather by other HSPs subfamilies like HSP80s and HSP90s (Szuba & Lorenc-Plucińska 2018) which are rarely reported under shorter stress duration. This shift is connected with various "specializations" of HSPs subfamilies. Although all heat shock proteins are involved in the stabilization of proper protein structures (Wang *et al.* 2004), the HSP80 subfamily also plays roles in signal transduction or cell cycle control, whereas HSP90s play a role in stress adaptation and morphological evolution (Richter & Buchner 2001; Wang *et al.* 2004). HSP90 affects morphogenetic responses, and promotes optimal functioning of plants via reducing the effects of destabilizing environmental stimuli (Queitsch *et al.* 2002) and as such is an important player during plant growth in harsh post-industrial areas.

Figure 4. Examples of the ongoing phytoremediation activity on the harsh post-industrial area. From the top: 1) the former tannery landfill (Warcino, Poland) with 8-year-old poplars – the stage of the selection of the *Populus* spp. cultivar the most suitable for the harsh conditions of the landfill; 2) the same tannery landfill a few years later: forest workers planting the one of the cultivars selected during the initial study (*Populus maximowiczii* × *P. trichocarpa* Torr. et Gray clone 'Hybrid 275'). (Photo: Agnieszka Szuba)

But most importantly, despite intensive antioxidant protection, symptoms of central metabolism dysregulation were not observed in chronically stressed plants (Szuba & Lorenc-Plucińska 2018; Kasprowicz-Maluśki, unpublished data). This result was in stark contrast to shorter-stress durations (even persistent ones; see above) and highlighted a possible mechanism of life in chronic, long-term stress (years): a postulated, but still unknown at the molecular level, mechanism of establishing a "new homeostasis" of plant cells able to survive to the harsh conditions (Gaspar *et al.* 2002; Gupta & Huang 2014). Due to limited information on acclimatized/adapted plants at the OMIC level, it is tempting to learn from animal research. Changes found during analysis of *Corbicula fluminea*, a freshwater clam, revealed that adaptation to chronically warmed waters (Falfushynska *et al.* 2016) seemingly triggered a very similar response to that observed in aforementioned poplars (Szuba & Lorenc-Plucińska 2018), mainly the activation of antioxidants and HSP systems. However, in contrast to the above mentioned "plant results", these responses were antagonistic. Thus, because of the different mechanisms of ROS action etc., mechanisms of the chronic stress response cannot, unfortunately, be directly transferred from animal systems. Thus OMIC analyses of the plant response to chronic stress are still in progress, and some unknown molecular mechanisms may be discovered. For example, Li *et al.* (2017) suggested that adaptation to stress conditions (complex desert stress) in grasses may increase protein adhesion to membranes (even of RuBisCo) and in consequence decrease protein solubility.

The single available experimental reports in this field may support the thesis of increased molecular stability during chronic stress (Szuba & Lorenc-Plucińska 2018). For example, Medrano *et al.* (1997) analyzed drought adaptations in 2 consecutive years – in the second year, the results of the stressed variant were more similar to controls than those from the first year. Comparisons between sensitive and tolerant ecotypes found in harsh environments are also interesting (e.g. Li *et al.* 2017). Such studies have generally analyzed subsequent generations and the processes of adaptation with their consequences, in response to short-term stress (Sairam *et al.* 2005; Hu *et al.* 2015). The majority of these reports support the thesis on the molecular stabilization of chronically stressed plants: adapted ecotypes revealed higher tolerance to short-term stress (Kovalchuk *et al.* 2004; Sairam *et al.* 2005), and molecular trials were less distributed compared to controls (sensitive ecotypes).

Epigenetic analysis of native *Arabidopsis* (seeds) in the Chernobyl zone in 1989-92 revealed that the frequencies of extrachromosomal homologue recombination of adapted and control *Arabidopsis* plants tended to become more similar over time (Kovalchuk *et al.* 2004). The authors postulated some unidentified mechanisms of plant protection, which still await discovery.

Without a doubt, most data on acclimatization and adaptation comes from epigenetic research, because it is believed that epigenetic changes are crucial to these processes (Mirouze & Paszkowski 2011; Bräutigam *et al.* 2013; Sahu *et al.* 2013; Latzel *et al.* 2016; Lämke & Bäurle 2017). Epigenetic analyses concern several precisely controlled processes like histone post-translational modifications, transposon mobility, siRNA mediated methylation, host methyltransferase activity, and the best recognized: methylation of DNAs' cytosines (5mC). In plants, 5mC frequency may affect from 6% up to even 60% (in poplars) of all cytosines (Raj *et al.* 2011; Sahu *et al.* 2013). DNA methylation occurs in symmetrical (CG and CHG, where H = A, T or C) and in nonsymetrical (CHH) sequence contexts, and the sequence context is important in adaptive processes. It is believed that symmetrical methylations (GC) are the most stable during cell division (Mirouze & Paszkowski 2011; Lämke & Bäurle 2017) and as such are most important during transgenerational inheritance of stress memory (González *et al.* 2016). Epigenetically triggered phenotypic plasticity (Gratani 2014) is especially important for plants with relatively long lifespans like trees (Raj *et al.* 2011) and most probably play a key role in plant survival in harsh post-industrial areas. It is known that the stability of epigenetic marks may decrease in subsequent sexual generations (Sahu *et al.* 2013). Interestingly, the site-specific offspring epigenetic "memory" has also been postulated for vegetatively propagated woody plants (Raj *et al.* 2011; Schönberger *et al.* 2016), which point to the possibilities of using the vegetative progeny of the best acclimatized/adapted genotypes selected during the initial targeted study for the forestation of particular post-industrial areas (Fig. 4) (Lorenc-Plucińska *et al.* 2017; Szuba & Lorenc-Plucińska 2018).

Stress conditions may result in a decrease (Sahu *et al.* 2013) or increase (Kovalchuk *et al.* 2004; Sahu *et al.* 2013; González *et al.* 2016) in global genome methylation, but significant shifts in the methylation of particular genes may also occur without changes in global DNA methylation level (Chwialkowska *et al.* 2016). Moreover, it is known that methylation differentially affects plant organs (Vining *et al.* 2012; Chwialkowska *et al.* 2016).

Usually, the epigenetic marks are strictly correlated with transcription level (Vining *et al.* 2012; Chwialkowska *et al.* 2016) because high 5mC levels prevent gene transcription. Interestingly, data are suggesting that epigenetic changes are also strictly connected with other OMIC levels – namely may be guided by metabolome shifts in plants, but the mechanisms involved are poorly studied. Still, it has been shown that accumulation of S-adenosyl methionine (SAM) may regulate DNA methylation, that iron-sulphur clusters may regulate DNA demethylation, and that GSH regulates histone glutathionylation (Shen *et al.* 2016). DNA methylation may also be controlled by phytohormonal fluxes, which are in turn influenced by various abiotic and biotic factors. Among the various phytohormones, abscisic acid (ABA) seems to be the most interesting, as ABA is a major abiotic stress-response hormone and plays a crucial role in facilitating plant survival and growth under a wide range of stress conditions, during stress alleviation and adaptation (Leone *et al.* 1994; Sahu *et al.* 2013; Wani *et al.* 2016). Most interestingly, ABA levels and their cross-talk with other phytohormones (Szuba & Wojtaszek 2010; Yang *et al.* 2014) differ between short- and long-term stress responses (Leone *et al.* 1994). It is believed that epigenetic markers may be especially influenced by ROS (Cyr & Domann 2011), and might be linked through redox intermediates like NAD+, SAM, or 2-oxoglutarat) with the central metabolism, but there is almost a lack of data about plants (Shen *et al.* 2016).

The transgenerational modification and memory of stress tolerance is probably the first step of adaptation – the evolution of new genotypes perfectly suitable to a particular local environment.

Usually, research on ecotypes has focussed on different genets with unknown genetic differences found in nature which experienced short-term stress in laboratory conditions that are different from their "native" environment (Kovalchuk *et al.* 2004; Paerek *et al.* 2010). This may lead to misinterpretation because of a lack of some of the components of complex stress conditions which occur in nature, as was suggested by Schönberger *et al.* (2016) during the analysis of P deficiency. The issues related to plant response to combined stresses are a topic that has recently been under constant interest because combined stress is closer to the reality of field conditions (Rivero *et al.* 2014; Pandey *et al.* 2015; Wani *et al.* 2016; Li *et al.* 2017; Zhang & Sonnewald 2017). Combined stress, due to its complexity, is considered more dangerous to plants and triggers more significant growth inhibition (Medrano *et al.* 1997; Szuba unpublished data). Some available data suggest distinct and common responses like oxidative damage, accumulation of glycine or trehalose (Rivero *et al.* 2014; Wani *et al.* 2016), and the suggested differences with response to single-stress scenarios has to be analyzed in more detail, especially *in situ*. As the use of OMIC methods in field conditions of a real biotope is rare (Szuba & Lorenc-Plucińska 2018) there is a clear necessity for intensive molecular study of plants grown in their native habitats, especially in the contest to the acquisition and maintenance of increased tolerance to harsh stress conditions.

THE ROLE OF ADAPTED GENOTYPES: PHYTOREMEDIATION

Molecular damages caused by acute and chronic stresses may be fatal and resulted in plant cell death (Mosa *et al.* 2017) as not all plant individuals can survive in harsh conditions of post-industrial habitats. Acclimatization and, finally, adaptation processes, can result in the generation of new genotypes better adapted to the local environment. For this reason, the genetic plasticity and the possibility to select suitable genotypes – those best adapted to a particular habitat – is one of the most important elements of the green management of post-industrial areas (Pajević *et al.* 2009; Szuba & Lorenc-Plucińska 2015; 2018). Such plants will be able to survive and function relatively normally in harsh post-industrial areas (Fig. 4).

Anthropogenically modified habitats are usually significantly polluted and may constitute a danger to humans and animals. Therefore, apart from the need to increase the biodiversity in post-industrial areas, there is an urgent need to eliminate the risk of potential poisoning – the need to remediate the area and create, above all, a safe ecosystem. The classical physical and chemical methods of remediation of polluted areas are not always socially acceptable and are frequently very costly (Avudainayagam *et al.* 2003; Liu *et al.* 2018; Szuba & Lorenc-Plucińska 2018). An interesting alternative method for soil remediation, also resulting in immediate biodiversity increase, is phytoremediation – use of green plants, in some cases with the associated fungi or other microorganisms and soil amendments for remediation of toxic environmental contaminants (Fig. 4) (Ashraf *et al.* 2019; Yan *et al.* 2020).

Plants used for phytoremediation recruit all previously described processes of stress tolerance with an "additional" effect in form of absorption of a broad spectrum of compounds from contaminated soil and its accumulation in various organs (Nagajyoti *et al.* 2010; Baldantoni *et al.* 2014; Pająk *et al.* 2017). The extraction of active chelating compounds directly immobilized harmful compounds whereas the activation of beneficial soil bacteria resulted in the decomposition of various contaminants (Shukla *et al.* 2011) and with a further decrease in the environmental danger (Marmiroli *et al.* 2011; Szuba & Lorenc-Plucińska 2018). For all these reasons phytoremediation is recently one of the most popular methods of remediation of polluted lands (Ashraf *et al.* 2019; Yan *et al.* 2020), with emphasis on various post-mining areas, buffer zones of smelters, former disposal areas like landfills (Fig. 1 and 4), radioactive waste or closed polygons polluted with airfare chemicals (Qi *et al.* 2015; Lorenc-Plucińska *et al.* 2017; Cary *et al.* 2021; Wu *et al.* 2021). Concerning usually very vast areas where remediation is needed (Fig. 1), phytoremediation is the only possible method because of the extremely high costs of utilization of physical or chemical "classical" techniques (Liu *et al.* 2018).

Among the phytoremediation approaches utilized in land environments, we distinguish the phytodegradation, a process of degradation of toxic compounds in the rhizosphere or in the plant organs, occurring thanks to the activity of various plants, but also associated microorganisms with their enzymes involved in "normal" metabolism. The metabolic processes resulted in degradation or inactivation (phytotransformation) of xenobiotics and are especially useful for remediation of habitats polluted with persistent organic compounds like herbicides or fuels (Chlebek & Hupert-Kocurek 2019). The next process, phytovolatilization involves uptake of harmful substances from the soil and release into the air, most frequently with a simultaneous phytotransformation to the more volatile and less toxic form. In this process, there is, however, a risk of environmental recontamination, as the compounds are released via transpiration into the atmosphere (Limmer & Burken 2016).

Phytostabilization is a process based on a decrease of bioavailability of toxic substances and their immobilization in the soil, making them less dangerous for plants, animals, or humans. The pollutant still is present in the particular area, but it is characterized by significantly reduced mobility. A very important role during the phytostabilization process is played by root exudates and symbiotic microorganisms, especially mycorrhizal fungi, modulating the bioavailability of toxic compounds during processes of so-called enhanced phytoremediation (Ullah *et al.* 2015; Meena *et al.* 2017; Manoj *et al.* 2020). Phytoremediation is a process involved in the remediation of environments polluted mainly with heavy metals but also organic compounds.

Finally, phytoremediation approaches include phytoextraction (Anderson *et al.* 1999; Chaney & Baklanov 2017; Suman *et al.* 2018; Szuba & Lorenc-Plucińska 2018; Ashraf *et al.* 2019; Yan *et al.* 2020), probably the most important process of phytoremediation. Phytoextraction results in a real environmental cleaning. It consists of the permanent removal of harmful substances from the polluted habitats after uptake by plants followed by harvest and removal of the plants. Phytoextraction is a process mainly related to the removal of heavy metals from the environment. Interestingly, phytoextraction of rare earth elements is highly efficient and may be profitable. The examples of phytoextraction of Ni or Au, with the use of *Brassica juncea* or *Alyssum bertolonii* are more and more frequent. So-called phytomining is the branch of industry with profits at the net-yield level of 500 up to 1000 USA dollars per ha (Anderson *et al.* 1999; Chaney & Baklanov 2017).

The successful uptake of contaminants into harvestable plant biomass depends on several important factors. One of the most important ones, the bioconcentration factor (BCF), determines the plant's ability to uptake and concentrate metals in plant tissues. Phytoextraction is an efficient process only when the concentrations that can be obtained in the tissues are much higher than those observed in the soil, i.e. when the bioconcentration factor is greater than 1 (Rascio & Navari-Izzo 2011):

$$\text{Bioconcentration Factor (BCF)} = \frac{C\ \text{harvested tissue}}{C\ \text{soil}}$$

where $C_{\text{harvested tissue}}$ is the concentration of the target metal in the tissues of plants harvested from the polluted area and C_{soil} is the concentration of the same metal in the remediated soil.

Since most of the heavy metals taken up by plants remain in their underground part (Krzesłowska 2011; Pourrut *et al.* 2013; Szuba & Lorenc-Plucińska 2018; Szuba *et al.* 2020), which is much harder to harvest and remove from the contaminated area, another important factor is the translocation factor (TF), which determines what part of the collected heavy metal pool is transported to the aboveground part of the plant:

$$\text{Translocation Factor (TF)} = \frac{C\ \text{shoot}}{C\ \text{root}}$$

where C_{shoot} is concentration (mg/kg dry weight) or content (mg/plant) of the element in the aboveground part of the plant and C_{root} is concentration or content of a particular metal in roots.

High values of BCF and TF are the important features of hyperaccumulators – plants that can function despite the very high concentration of HMs in their tissues (for example: 100 mg/kg of Cd or Se, 300 mg/kg of Cr or Cu, 1000 mg/kg of As or Ni, and even up to 10000 mg/kg of Mn in dried foliage; Suman *et al.* 2018) and survive in extremely polluted environments. Hyperaccumulators are most frequently selected for phytoremediation use (Rascio & Navari-Izzo 2011). Among the known hyperaccumulators, the most numerous are annual plants like *Brassica* spp. However, the best hyperaccumulators are also characterized by big and deep root systems (to penetrate the largest possible soil volume) and fast growth. They are resistant to pests and pathogens and easy to harvest (Tangahu *et al.* 2011; Lorenc-Plucińska *et al.* 2017; Yan *et al.* 2020). For all these reasons, among the plants, the most frequently chosen for phytoremediation are also representatives of genus *Salix* spp. or *Populus* spp. (Hrynkiewicz & Baum 2013; Szuba & Lorenc-Plucińska 2018; Pilipović *et al.* 2021). The biomass for energy production produced by fast-growing species (like poplars or willows) or the acquisition of metals with marked values (like Au or Ni by Indian mustard) is an additional advantage of using the proper hyperaccumulators along with the gradually improved soil quality and reduced environmental risk – the main goal of phytoremediation (Ashraf *et al.* 2019; Yan *et al.* 2020; Pilipović *et al.* 2021).

As the phytoremediation of post-industrial areas is in great demand, work is underway on improving this methodology and developing and utilization of new phytoremediation technologies (however, performed mainly in the controlled conditions of the laboratory or the greenhouse; Ruiz & Daniell 2009). In the recent decade, targeted microorganism-assisted phytoremediation seems to be one of the major research directions, as the understanding of the necessity of a holistic view of plant-soil interactions grows constantly (Pilipović *et al.* 2021; Tiodar *et al.* 2021). More and more examples of enhanced phytoremediation efficiency have come from genetic engineering studies – with a flagship example in the form of manipulation of the enzymes involved in mercury emission (Ruiz & Daniell 2009; Tiodar *et al.* 2021). The examples of successful use of genetic engineering also concern woody species, where overexpression of gamma-glutamylcysteine synthetase increased the detoxification of cadmium in poplars (He *et al.* 2015).

Nowadays, the latest laboratory techniques are more and more boldly used to support phytoremediation processes. For example, by CRISPS-cas9 technology, a genome editing system was used to enhance the phytoremediation of hazardous pollutants like metal(loid)s (Sarma *et al.* 2021). Also, nanoparticles like F-Fe^0, Ip-Ag^0 or Br-Ag^0 are starting to be used to enhance the decontamination processes and increase the efficiency of the phytoremediation (under the aegis of the nano-phytoremediation approach; Romeh & Saber 2020). Finally, attempts are underway to use machine learning processes to accelerate the selection of new hyperaccumulators (Mota-Merlo & Martos 2021). Still, these modern studies are in their infancy, and similarly, the number of research reports on broadly defined environmental phytoremediation at the molecular level, done with the use of research material taken directly from remediated habitats, is very few.

ACKNOWLEDGEMENTS

The authors thank Prof. G. Lorenc-Plucińska, Prof. A. M. Jagodziński and Mrs. K. Grewling, all of whom are from the Institute of Dendrology, Polish Academy of Sciences, Kórnik, Poland, for their valuable assistance and discussions. We thank Warcino Forest District for their practical support. This work was funded by the National Science Center, Poland (DEC-2020/39/B/NZ9/03336), and was supported by the Institute of Dendrology PAS.

REFERENCES

Ahsan N., Renaut J. & Komatsu S. 2009: Recent developments in the application of proteomics to the analysis of plant responses to heavy metals. Proteomics, 9 (10), 2602–2621.

Alexandersson E., Jacobson D., Vivier M.A., Weckwerth W. & Andreasson E. 2014: Field-omics – understanding large-scale molecular data from field crops. Frontiers in Plant Science, 5, 286.

Ali H., Khan E. & Sajad M.A. 2013: Phytoremediation of heavy metals – Concepts and applications. Chemosphere, 91 (7), 869–881.

Anderson C.W.N., Brooks R.R., Chiarucci A., LaCoste C.J., Leblanc M., Robinson B.H., Simcock R. & Stewart R.B. 1999: Phytomining for nickel, thallium and gold. Journal of Geochemical Exploration, 67 (1–3), 407–415.

Arbona V., Manzi M., Zandalinas S.I., Vives-Peris V., Pérez-Clemente R.M. & Gómez-Cadenas A. 2017: Physiological, metabolic, and molecular responses of plants to abiotic stress. [In:] Sarwat M., Ahmad A., Abdin M.Z. & Ibrahim M.M. (eds.): Stress Signaling in Plants: Genomics and Proteomics Perspective. Volume 2. Springer, Cham, 1–35.

Ashraf S., Ali Q., Zahir Z.A., Ashraf S. & Asghar H.N. 2019: Phytoremediation: Environmentally sustainable way for reclamation of heavy metal polluted soils. Ecotoxicology and Environmental Safety, 174, 714–727.

Ashraf U., Kanu A.S., Deng Q., Mo Z., Pan S., Tian H. & Tang X. 2017: Lead (Pb) toxicity; physio-biochemical mechanisms, grain yield, quality, and Pb distribution proportions in scented rice. Frontiers in Plant Science, 8, 259.

Avudainayagam S., Megharaj M., Owens G., Kookana R.S., Chittleborough D. & Naidu R. 2003: Chemistry of chromium in soils with emphasis on tannery waste sites. Reviews of Environmental Contamination and Toxicology, 178, 53–91.

Baena-González E. 2010: Energy signaling in the regulation of gene expression during stress. Molecular Plant, 3 (2), 300–313.

Bah A.M., Sun H., Chen F., Zhou J., Dai H., Zhang G. & Wu F. 2010: Comparative proteomic analysis of *Typha angustifolia* leaf under chromium, cadmium and lead stress. Journal of Hazardous Materials, 184 (1–3), 191–203.

Bălan M.C., Jäntschi L., Bolboacă S. & Ştefu M. 2009: Weather monitoring setup and analysis of air, soil and leaf parameters. Bulletin of the University of Agricultural Sciences and Veterinary Medicine, Horticulture, 66 (2), 672–679.

Baldantoni D., Cicatelli A., Bellino A. & Castiglione S. 2014: Different behaviours in phytoremediation capacity of two heavy metal tolerant poplar clones in relation to iron and other trace elements. Journal of Environmental Management, 146, 94–99.

Barranco-Medina S., Krell T., Finkemeier I., Sevilla F., Lázaro J.-J. & Dietz K.-J. 2007: Biochemical and molecular characterization of the mitochondrial peroxiredoxin PsPrxII F from *Pisum sativum*. Plant Physiology and Biochemistry, 45, 729–739.

Becker M. & Asch F. 2005: Iron toxicity in rice – conditions and management concepts. Journal of Plant Nutrition and Soil Science, 168, 558–573.

Bellion M., Courbot M., Jacob C., Blaudez D. & Chalot M. 2006: Extracellular and cellular mechanisms sustaining metal tolerance in ectomycorrhizal fungi. FEMS Microbiology Letters, 254 (2), 173–181.

Bhargava A., Carmona F.F., Bhargava M. & Srivastava S. 2012: Approaches for enhanced phytoextraction of heavy metals. Journal of Environmental Management, 105, 103–120.

Bhargava S. & Sawant K. 2013: Drought stress adaptation: metabolic adjustment and regulation of gene expression. Plant Breeding, 132 (1), 21–32.

Bräutigam K., Vining K.J., Lafon-Placette C., Fossdal C.G., Mirouze M., Marcos J.G., Fluch S., Fraga M.F., Guevara M.Á., Abarca D., Johnsen Ø., Maury S., Strauss S.H., Campbell M.M., Rohde A., Díaz-Sala C. & Cervera M.-T. 2013: Epigenetic regulation of adaptive responses of forest tree species to the environment. Ecology and Evolution, 3 (2), 399–415.

Cary T.J., Rylott E.L., Zhang L., Routsong R.M., Palazzo A.J., Strand S.E. & Bruce N.C. 2021: Field trial demonstrating phytoremediation of the military explosive RDX by XplA/XplB-expressing switchgrass. Nature Biotechnology, 39 (10), 1216–1219.

Chaney R.L. & Baklanov I.A. 2017: Phytoremediation and Phytomining: Status and Promise. Advances in Botanical Research, 83, 189–221.

Chaperon S. & Sauvé S. 2008: Toxicity interactions of cadmium, copper, and lead on soil urease and dehydrogenase activity in relation to chemical speciation. Ecotoxicology and Environmental Safety, 70 (1), 1–9.

Che-Othman M.H., Millar A.H. & Taylor N.L. 2017: Connecting salt stress signalling pathways with salinity-induced changes in mitochondrial metabolic processes in C3 plants. Plant, Cell & Environment, 40 (12), 2875–2905.

Chlebek D. & Hupert-Kocurek K. 2019: Endophytic bacteria in the phytodegradation of persistent organic pollutants. Advancements of Microbiology, 58 (1), 70–79.

Chwialkowska K., Nowakowska U., Mroziewicz A., Szarejko I. & Kwasniewski M. 2016: Water-deficiency conditions differently modulate the methylome of roots and leaves in barley (*Hordeum vulgare* L.). Journal of Experimental Botany, 67 (4), 1109–1121.

Cicatelli A., Lingua G., Todeschini V., Biondi S., Torrigiani P. & Castiglione S. 2010: Arbuscular mycorrhizal fungi restore normal growth in a white poplar clone grown on heavy metal-contaminated soil, and this is associated with upregulation of foliar metallothionein and polyamine biosynthetic gene expression. Annals of Botany, 106 (5), 791–802.

Ciereszko I. 2009: Sucrose metabolism in plant tissues under stress conditions: Key enzymes, localization and function. [In:] Maksymiec W. (ed.): Compartmentation of Responses to Stresses in Higher Plants, True or False. Transworld Research Network, Kerala, 193–218.

Ciereszko I. 2018: Regulatory roles of sugars in plant growth and development. Acta Societatis Botanicorum Poloniae, 87 (2), 3583.

Colpaert J.V., Wevers J.H.L., Krznaric E. & Adriaensen K. 2011: How metal-tolerant ecotypes of ectomycorrhizal fungi protect plants from heavy metal pollution. Annals of Forest Science, 68, 17–24.

Cramer G.R., Urano K., Delrot S., Pezzotti M. & Shinozaki K. 2011: Effects of abiotic stress on plants: a systems biology perspective. BMC Plant Biology, 11, 163.

Cyr A.R. & Domann F.E. 2011: The redox basis of epigenetic modifications: from mechanisms to functional consequences. Antioxidants & Redox Signaling, 15 (2), 551–589.

Das K. & Roychoudhury A. 2014: Reactive oxygen species (ROS) and response of antioxidants as ROS-scavengers during environmental stress in plants. Frontiers in Environmental Science, 2, 53.

Davletova S., Rizhsky L., Liang H., Shengqiang Z., Oliver D.J., Coutu J., Shulaev V., Schlauch K. & Mittler R. 2005: Cytosolic ascorbate peroxidase 1 is a central component of the reactive oxygen gene network of Arabidopsis. The Plant Cell, 17 (1), 268–281.

Dietz K.-J., Mittler R. & Noctor G. 2016: Recent progress in understanding the role of reactive oxygen species in plant cell signaling. Plant Physiology, 171 (3), 1535–1539.

Droillard M.J., Thibivilliers S., Cazalé A.C., Barbier-Brygoo H. & Laurière C. 2000: Protein kinases induced by osmotic stresses and elicitor molecules in tobacco cell suspensions: two crossroad MAP kinases and on osmoregulation-specific protein kinase. FEBS Letters, 474, 217–222.

Evlard A., Sergeant K., Ferrandis S., Printz B., Renaut J., Guignard C., Paul R., Hausman J.-F. & Campanella B. 2014: Physiological and proteomic responses of different willow clones (*Salix fragilis* × *alba*) exposed to dredged sediment contaminated by heavy metals. International Journal of Phytoremediation, 16 (11), 1148–1169.

Fahr M., Laplaze L., Bendaou N., Hocher V., Mzibri M.E., Bogusz D. & Smouni A. 2013: Effect of lead on root growth. Frontiers in Plant Science, 4, 175.

Falfushynska H., Phan T. & Sokolova I.M. 2016: Long-term acclimation to different thermal regimes affects molecular responses to heat stress in a freshwater clam *Corbicula fluminea*. Scientific Reports, 6, 39476.

Fernández-Fuego D., Bertrand A. & González A. 2017: Metal accumulation and detoxification mechanisms in mycorrhizal *Betula pubescens*. Environmental Pollution, 231, 1153–1162.

Foresman T.W. 1986: Mapping, monitoring, and modelling of hazardous waste sites. Science of the Total Environment 56, 255–263.

Foyer C. & Noctor G. 2005: Redox homeostasis and antioxidant signaling: a metabolic interface between stress perception and physiological responses. The Plant Cell, 17 (7), 1866–1875.

Foyer C.H. & Noctor G. 2013: Redox signaling in plants. Antioxidants & Redox Signaling, 18 (16), 2087–2090.

Foyer C.H. & Noctor G. 2016: Stress-triggered redox signalling: what's in pROSpect? Plant Cell & Environment, 39 (5), 951–964.

Gaspar T., Franck T., Bisbis B., Kevers C., Jouve L., Hausman J.F. & Dommes J. 2002: Concepts in plant stress physiology. Application to plant tissue cultures. Plant Growth Regulation, 37, 263–285.

Gietler M., Nykiel M., Orzechowski S., Fettke J. & Zagdańska B. 2017: Protein carbonylation linked to the wheat seedlings tolerance to the water deficiency. Environmental and Experimental Botany, 137, 84–95.

González A.P.R., Chrtek J., Dobrev P.I., Dumalasová V., Fehrer J., Mráz P. & Latzer V. 2016: Stress-induced memory alters growth of clonal offspring of white clover (*Trifolium repens*). American Journal of Botany, 103 (9), 1567–1574.

Goupil P., Souguir D., Ferjani E., Faure O., Hitmi A. & Ledoigt G. 2009: Expression of stress-related genes in tomato plants exposed to arsenic and chromium in nutrient solution. Journal of Plant Physiology, 166 (13), 1446–1452.

Gratani L. 2014: Plant phenotypic plasticity in response to environmental factors. Advances in Botany, 2014, 208747.

Grover M., Ali S.Z., Sandhya V., Rasul A. & Venkateswarlu B. 2011: Role of microorganisms in adaptation of agriculture crops to abiotic stresses. World Journal of Microbiology & Biotechnology, 27, 1231–1240.

Gupta B. & Huang B. 2014: Mechanism of salinity tolerance in plants: physiological, biochemical, and molecular characterization. International Journal of Genomics, 2014, 701596.

Halliwell B. 1987: Oxidative damage, lipid peroxidation and antioxidant protection in chloroplasts. Chemistry and Physics of Lipids, 44 (2–4), 327–340.

Hänsch R. & Mendel R.R. 2009: Physiological functions of mineral micronutrients (Cu, Zn, Mn, Fe, Ni, Mo, B, Cl). Current Opinion in Plant Biology, 12 (3), 259–266.

Harb A., Krishnan A., Ambavaram M.M.R. & Pereira A. 2010: Molecular and physiological analysis of drought stress in Arabidopsis reveals early responses leading to acclimation in plant growth. Plant Physiology, 154 (3), 1254–1271.

Hasan M.K., Cheng Y., Kanwar M.K., Chu X.-Y., Ahammed G.J. & Qi Z.-Y. 2017: Responses of plant proteins to heavy metal stress – A review. Frontiers in Plant Science, 8, 1492.

Hayat S., Hayat Q., Alyemeni M.N., Wan A.S., Pichtel J. & Ahmad A. 2012: Role of proline under changing environments. Plant Signaling & Behavior, 7 (11), 1456–1466.

He J., Li H., Ma C., Zhang Y., Polle A., Rennenberg H., Cheng X. & Luo Z.-B. 2015: Overexpression of bacterial γ-glutamylcysteine synthetase mediates changes in cadmium influx, allocation and detoxification in poplar. New Phytologist, 205 (1), 240–254.

Hendry G.A.F. & Price A.H. 1993: Stress indicators: Chlorophylls and carotenoids. [In:] Hendry G.A.F. & Grime J.P. (eds.): Methods in Comparative Plant Ecology. Chapman & Hall, London, 148–152.

Hoehenwarter W., van Dongen J.T., Wienkoop S., Steinfath M., Hummel J., Erban A., Sulpice R., Regierer B., Kopka J., Geigenberger P. & Weckwerth W. 2008: A rapid approach for phenotype-screening and database independent detection of cSNP/protein polymorphism using mass accuracy precursor alignment. Proteomics, 8, 4214–4225.

Hrynkiewicz K. & Baum C. 2013: Selection of ectomycorrhizal willow genotype in phytoextraction of heavy metals. Environmental Technology, 34 (2), 225–230.

Hu J., Rampitsch C. & Bykova N.V. 2015: Advances in plant proteomics toward improvement of crop productivity and stress resistance. Frontiers in Plant Science, 6, 209.

Huang T.-L., Huang L.-Y., Fu S.-F., Trinh N.-N. & Huang H.-J. 2014: Genomic profiling of rice roots with short- and long-term chromium stress. Plant Molecular Biology, 86 (1–2), 157–170.

Ireland E.H., Harding S.J., Bonwick G.A., Jones M., Smith C.J. & Williams J.H.H. 2004: Evaluation of heat shock protein 70 as a biomarker of environmental stress in *Fucus serratus* and *Lemna minor*. Biomarkers, 9 (2), 139–155.

Jalmi S.K., Bhagat P.K., Verma D., Noryang S., Tayyeba S., Singh K., Sharma D. & Sinha A.K. 2018: Traversing the links between heavy metal stress and plant signaling. Frontiers in Plant Science, 9, 12.

Jentschke G. & Godbold D.L. 2000: Metal toxicity and ectomycorrhizas. Physiologia Plantarum, 109 (2), 107–116.

Jiang W. & Liu D. 2010: Pb-induced cellular defense system in the root meristematic cells of *Allium sativum* L. BMC Plant Biology, 10, 40.

Kabata-Pendias A. & Pendias H. 2001: Trace elements in soils and plants. CRC Press, Boca Raton.

Kandeler F., Kampichler C. & Horak O. 1996: Influence of heavy metals on the functional diversity of soil microbial communities. Biology and Fertility of Soils, 23, 299–306.

Kandziora-Ciupa M., Ciepał R., Nadgórska-Socha A. & Barczyk G. 2013: A comparative study of heavy metal accumulation and antioxidant responses in *Vaccinium myrtillus* L. leaves in polluted and non-polluted areas. Environmental Science and Pollution Research International, 20 (7), 4920–4932.

Kasprowicz A., Kierzkowski D., Maruniewicz M., Derba-Maceluch M., Rodakowska E., Zawadzki P., Szuba A. & Wojtaszek P. 2009: Mechanical integration of plant cell. [In:] Baluška F. (ed.): Plant-Environment Interactions. Signaling and Communication in Plants. Springer-Verlag, Berlin Heidelberg, 1–21.

Kerner R., Delgado-Eckert E., Ernst D., Dupuy J.-W., Grams T.E.E., Winkler J.B., Lindermayr C. & Müller-Starck G. 2014: Large-scale protein analysis of European beech trees following four vegetation periods of twice ambient ozone exposure. Journal of Proteomics, 109, 417–435.

Kijowska-Oberc J., Staszak A.M., Wawrzyniak M.K. & Ratajczak E. 2020. Changes in proline levels during seed development of orthodox and recalcitrant seeds of genus *Acer* in a climate change scenario. Forests, 11, 1362.

Kim B.-G., Lee E.-R. & Ahn J.-H. 2012a: Analysis of flavonoid contents and expression of flavonoid biosynthetic genes in *Populus euramericana* Guinier in response to abiotic stress. Journal of the Korean Society for Applied Biological Chemistry, 55, 141−145.

Kim J.Y., Kim W.Y., Kwak K.J., Oh S.H., Han Y.S. & Kang H. 2010: Glycine-rich RNA-binding proteins are functionally conserved in *Arabidopsis thaliana* and *Oryza sativa* during cold adaptation process. Journal of Experimental Botany, 61 (9), 2317–2325.

Kim S.Y., Jung Y.J., Shin M.R., Park J.H., Nawkar G.M., Maibam P., Lee E.S., Kim K.-S., Paeng S.K., Kim W.Y., Lee K.O., Yun D.-J., Kang C.H. & Lee S.Y. 2012b: Molecular and functional properties of three different peroxiredoxin isotypes in Chinese cabbage. Molecules and Cells, 33, 27–33.

Klychnikov O.I., Li K.W., Lill H. & de Boer A.H. 2007: The V-ATPase from etiolated barley (*Hordeum vulgare* L.) shoots is activated by blue light and interacts with 14-3-3 proteins. Journal of Experimental Botany, 58 (5), 1013–1023.

König J., Baier M., Horling F., Kahmann U., Harris G., Schürmann P. & Dietz K.-J. 2002: The plant-specific function of 2-Cys peroxiredoxin-mediated detoxification of peroxides in the redox-hierachy of photosynthetic electron flux. Proceedings of the National Academy of Sciences of the United States of America, 99 (8), 5738–5743.

König J., Galliardt H., Jütte P., Schäper S., Dittmann L. & Dietz K.-J. 2013: The conformational bases for the two functionalities of 2-cysteine peroxiredoxins as peroxidase and chaperone. Journal of Experimental Botany, 64 (11), 3483–3497.

Kopittke P.M., Asher C.J. & Menzies N.W. 2008: Prediction of Pb speciation in concentrated and dilute nutrient solutions. Environmental Pollution, 153 (3), 548–554.

Kovalchuk I., Abramov V., Pogribny I. & Kovalchuk O. 2004: Molecular aspects of plant adaptation to life in the Chernobyl zone. Plant Physiology, 135 (1), 357–363.

Król A. & Weidner S. 2017: Changes in the proteome of grapevine leaves (*Vitis vinifera* L.) during long-term drought stress. Journal of Plant Physiology, 211, 114–126.

Krzesłowska M. 2011: The cell wall in plant cell response to trace metals: polysaccharide remodeling and its role in defense strategy. Acta Physiologiae Plantarum, 33, 35–51.

Kumar V., Shriram V., Hossai M.A. & Kishor P.B.K. 2015: Engineering proline metabolism for enhanced plant salt stress tolerance. [In:] Wani S.H. & Hossain M.A. (eds.): Managing Salt Tolerance in Plants: Molecular and Genomic Perspectives. CRC Press, Taylor & Francis Group Book, Boca Raton, 353–372.

Labudda M. 2013: Lipid peroxidation as a biochemical marker for oxidative stress during drought. An effective tool for plant breeding. E-wydawnictwo, Poland. http://www.e-wydawnictwo.eu/Document/DocumentPreview/3342, access: 11.12.2021.

Lämke J. & Bäurle I. 2017: Epigenetic and chromatin-based mechanisms in environmental stress adaptation and stress memory in plants. Genome Biology, 18, 124.

Langenkämper G., Manac'h N., Broin M., Cuiné S., Becuwe N., Kuntz M. & Rey P. 2001: Accumulation of plastid lipid-associated proteins (fibrillin/CDSP34) upon oxidative stress, ageing and biotic stress in Solanaceae and in response to drought in other species. Journal of Experimental Botany, 52 (360), 1545–1554.

Latzel V., González A.P.R. & Rosenthal J. 2016: Epigenetic memory as a basis for intelligent behavior in clonal plants. Frontiers in Plant Science, 7, 1354.

Leone A., Costa A., Tucci M. & Grillo S. 1994: Comparative analysis of short- and long-term changes in gene expression caused by low water potential in potato (*Solanum tuberosum*) cell-suspension cultures. Plant Physiology, 106 (2), 703–712.

Li G.K., Gao J., Peng H., Shen Y.O., Ding H.P., Zhang Z.M., Pan G.T. & Lin H.J. 2016: Proteomic changes in maize as a response to heavy metal (lead) stress revealed by iTRAQ quantitative proteomics. Genetics and Molecular Research, 15 (1), gmr.15017254.

Li L., Chen X., Shi L., Wang C., Fu B., Qiu T. & Cui S. 2017: A proteome translocation response to complex desert stress environments in perennial *Phragmites* sympatric ecotypes with contrasting water availability. Frontiers in Plant Science, 8, 511.

Liang X., Zhang L., Natarajan S.K. & Becker D.F. 2013: Proline mechanisms of stress survival. Antioxidants & Redox Signaling, 19 (9), 998–1011.

Limmer M. & Burken J. 2016: Phytovolatilization of organic contaminants. Environmental Science & Technology, 50 (13), 6632–6643.

Liu L., Li W., Song W. & Guo M. 2018: Remediation techniques for heavy metal-contaminated soils: Principles and applicability. Science of the Total Environment, 633, 206–219.

Liu Y., Zhang C., Chen J., Guo L., Li X., Li W., Yu Z., Deng J., Zhang P., Zhang K. & Zhang L. 2013. Arabidopsis heat shock factor HsfA1a directly senses heat stress, pH changes, and hydrogen peroxide via the engagement of redox state. Plant Physiology and Biochemistry, 64, 92–98.

Lomaglio T., Rocco M., Trupiano D., De Zio E., Grosso A., Marra M., Delfine S., Chiatante D., Morabito D. & Scippa G.S. 2015: Effect of short-term cadmium stress on *Populus nigra* L. detached leaves. Journal of Plant Physiology, 182, 40–48.

Lorenc-Plucińska G., Walentynowicz M. & Lewandowski A. 2017: Poplar growth and wood production on a grassland irrigated for decades with potato starch wastewater. Agroforestry Systems, 91, 307–324.

Luo Z.-B., Wu C., Zhang C., Li H., Lipka U. & Polle A. 2014: The role of ectomycorrhizas in heavy metal stress tolerance of host plants. Environmental and Experimental Botany, 108, 47–62.

Ma L., Sun X., Kong X., Galvan J.V., Li X., Yang S., Yang Y., Yang Y. & Hu X. 2015: Physiological, biochemical and proteomics analysis reveals the adaptation strategies of the alpine plant *Potentilla saundersiana* at altitude gradient of the Northwestern Tibetan Plateau. Journal of Proteomics, 112, 63–82.

Machado C.R., Praekelt U.M., de Oliveira R.C., Barbosa A.C.C., Byrne K.L., Meacock P.A. & Menck C.F.M. 1997: Dual role for the yeast *THI4* gene in thiamine biosynthesis and DNA damage tolerance. Journal of Molecular Biology, 273 (1), 114–121.

Mangano S., Juárez S.P.D. & Estevez J.M. 2016: ROS regulation of polar growth in plant cells. Plant Physiology, 171 (3), 1593–1605.

Manoj S.R., Karthik C., Kadirvelu K., Arulselvi P.I., Shanmugasundaram T., Bruno B. & Rajkumar M. 2020: Understanding the molecular mechanisms for the enhanced phytoremediation of heavy metals through plant growth promoting rhizobacteria: A review. Journal of Environmental Management, 254, 109779.

Marmiroli M., Pietrini F., Maestri E., Zacchini M., Marmiroli N. & Massacci A. 2011: Growth, physiological and molecular traits in Salicaceae trees investigated for phytoremediation of heavy metals and organics. Tree Physiology, 31 (12), 1319–1334.

Marschner P., Jentschke G. & Godbold D.L. 1998: Cation exchange capacity and lead sorption in ectomycorrhizal fungi. Plant and Soil, 205, 93–98.

Mata-Pérez C., Begara-Morales J.C., Chaki M., Sánchez-Calvo B., Valderrama R., Padilla M.N., Corpas F.J. & Barroso J.B. 2016: Protein tyrosine nitration during development and abiotic stress response in plants. Frontiers in Plant Science, 7, 1699.

Medrano H., Parry M.A.J., Socias X. & Lavlor D.W. 1997: Long term water stress inactivates Rubisco in subterranean clover. Annals of Applied Biology, 131 (3), 491–501.

Meena K.K., Sorty A.M., Bitla U.M., Choudhary K., Gupta P., Pareek A., Singh D.P., Prabha R., Sahu P.K., Gupta V.K., Singh H.B., Krishanani K.K. & Minhas P.S. 2017: Abiotic stress responses and microbe-mediated mitigation in plants: The omics strategies. Frontiers in Plant Science, 8, 172.
Meisrimler C.-N., Wienkoop S., Lyon D., Geilfus C.-M. & Lüthje S. 2016: Long-term iron deficiency: Tracing changes in the proteome of different pea (*Pisum sativum* L.) cultivars. Journal of Proteomics, 140, 12–23.
Michalak A. 2006: Phenolic compounds and their antioxidant activity in plants growing under heavy metal stress. Polish Journal of Environmental Studies, 15 (4), 523–530.
Mirouze M. & Paszkowski J. 2011: Epigenetic contribution to stress adaptation in plants. Current Opinion in Plant Biology, 14 (3), 267–274.
Møller I.M., Jensen P.E. & Hansson A. 2007: Oxidative modifications to cellular components in plants. Annual Review of Plant Biology, 58, 459–481.
Mosa K.A., Ismail A. & Helmy M. 2017: Plant Stress Tolerance. An Integrated Omics Approach. Springer Briefs in Systems Biology. Springer, Cham.
Mota-Merlo M. & Martos V. 2021: Use of machine learning to establish limits in the classification of hyperaccumulator plants growing on serpentine, gypsum and dolomite soils. Mediterranean Botany, 42, e67609.
Mourato M., Reis R. & Martins L.L. 2012: Characterization of plant antioxidative system in response to abiotic stresses: A focus on heavy metal toxicity. [In:] Montanaro G. (ed.): Advances in Selected Plant Physiology Aspects. Intech, Rijeka, 23–44.
Nagajyoti P.C., Lee K.D. & Sreekanth T.V.M. 2010: Heavy metals, occurrence and toxicity for plants: a review. Environmental Chemistry Letters, 8, 199–216.
Onaga G. & Wydra K. 2016: Advances in Plant Tolerance to Abiotic Stresses. [In:] Abdurakhmonov I.Y. (ed.): Plant Genomics. IntechOpen. https://www.intechopen.com/chapters/51235, access: 11.12.2021.
Osakabe Y., Osakabe K., Shinozaki K. & Tran L.-S.P. 2014: Response of plants to water stress. Frontiers in Plant Science, 5, 86.
Paerek A., Sopory S.K., Bohnert H.J. & Govindjee (eds.) 2010: Abiotic Stress Adaptation in Plants. Physiological, Molecular and Genomic Foundation. Springer, Dordrecht.
Pająk M., Halecki W. & Gąsiorek M. 2017: Accumulative response of Scots pine (*Pinus sylvestris* L.) and silver birch (*Betula pendula* Roth) to heavy metals enhanced by Pb-Zn ore mining and processing plants: Explicitly spatial considerations of ordinary kriging based on a GIS approach. Chemosphere, 168, 851–859.
Pajević S., Borišev M., Nikolić N., Krstić B., Pilipović A. & Orlović S. 2009: Phytoremediation capacity of poplar (*Populus* spp.) and willow (*Salix* spp.) clones in relation to photosynthesis. Archives of Biological Sciences, 61 (2), 239–247.
Pan J. & Yu L. 2011: Effects of Cd or/and Pb on soil enzyme activities and microbial community structure. Ecological Engineering, 37, 1889–1894.
Panda S.K. & Choudhry S. 2005: Chromium stress in plants. Brazilian Journal of Plant Physiology, 17 (1), 95–102.
Pandey P., Ramegowda V. & Senthil-Kumar M. 2015: Shared and unique responses of plants to multiple individual stresses and stress combinations: physiological and molecular mechanisms. Frontiers in Plant Science, 6, 723.
Pellicer M.T., Nuñez M.F., Aguilar J., Badia J. & Baldoma L. 2003: Role of 2-phosphoglycolate phosphatase of *Escherichia coli* in metabolism of the 2-phosphoglucolate formed in DNA repair. Journal of Bacteriology, 185 (19), 5815–5821.
Pilipović A., Zalesny R.S. Jr., Rogers E.R., McMahon B.G., Nelson N.D., Burken J.G., Hallett R.A. & Lin C.-H. 2021: Establishment of regional phytoremediation buffer systems for ecological restoration in the Great Lakes Basin, USA. II. New clones show exceptional promise. Forests, 12, 474.
Poorter H., Fiorani F., Stitt M., Schurr U., Finck A., Gibon Y., Usadel B., Munns R., Atkin O.K., Tardieu F. & Pons T.L. 2012: The art of growing plants for experimental purposes: a practical guide for the plant biologist. Functional Plant Biology, 39 (11), 821–838.
Pourrut B., Shahid M., Douay F., Dumat C. & Pinelli E. 2013: Molecular mechanisms involved in lead uptake, toxicity and detoxification in higher plants. [In:] Gupta D.K., Corpas F.J. & Palma J.M. (eds.): Heavy Metal Stress in Plants. Springer-Verlag, Heidelberg, 121–147.
Puig S. & Peñarrubia L. 2009: Placing metal micronutrients in context: transport and distribution in plants. Current Opinion in Plant Biology, 12 (3), 299–306.
Pukacka S., Ratajczak E. & Kalemba E. 2009: Non-reducing sugar levels in beech (*Fagus sylvatica*) seeds as related to withstanding desiccation and storage. Journal of Plant Physiology, 166 (13), 1381–1390.

Qi L., Qin X., Li F.-M., Siddique K.H.M., Brandl H., Xu J.& Li X. 2015: Uptake and distribution of stable strontium in 26 cultivars of three crop species: oats, wheat, and barley for their potential use in phytoremediation. International Journal of Phytoremediation, 17 (3), 264–271.

Queitsch C., Sangster T.A. & Lindquist S. 2002: Hsp90 as a capacitor of phenotypic variation. Nature, 417, 618–624.

Raj S., Bräutigam K., Hamanishi E.T., Wilkins O., Thomas B.R., Schroeder W., Mansfield S.D., Plant A.L. & Campbell M.M. 2011: Clone history shapes *Populus* drought responses. Proceedings of the National Academy of Sciences of the United States of America, 108 (30), 12521–12526.

Rascio N. & Navari-Izzo F. 2011: Heavy metal hyperaccumulating plants: How and why do they do it? And what makes them so interesting? Plant Science, 180, 169–181.

Reddy M.S., Kour M., Aggarwal S., Ahuja S., Marmeisse R. & Fraissinet-Tachet L. 2016: Metal induction of a *Pisolithus albus* metallothionein and its potential involvement in heavy metal tolerance during mycorrhizal symbiosis. Environmental Microbiology, 18 (8), 2446–2454.

Reis G.S.M., de Almeida A.-A.F., de Almeida N.M., de Castro A.V., Mangabeira P.A.O. & Pirovani C.P. 2015: Molecular, biochemical and ultrastructural changes induced by Pb toxicity in seedlings of *Theobroma cacao* L. PLoS ONE, 10 (7), e0129696.

Rey P., Gillet B., Römer S., Eymery F., Massimino J., Peltier G. & Kuntz M. 2000: Over-expression of a pepper plastid lipid-associated protein in tobacco leads to changes in plastid ultrastructure and plant development upon stress. The Plant Journal, 21 (5), 483–494.

Richter K. & Buchner J. 2001: Hsp90: chaperoning signal transduction. Journal of Cellular Physiology, 188 (3), 281–290.

Rivero R.M., Mestre T.C., Mittler R., Rubio F., Garcia-Sanchez F. & Martinez V. 2014: The combined effect of salinity and heat reveals a specific physiological, biochemical and molecular response in tomato plants. Plant, Cell & Environment, 37 (5), 1059–1073.

Rodakowska E., Derba-Maceluch M., Kasprowicz A., Zawadzki P., Szuba A., Kierzkowski D. & Wojtaszek P. 2009: Signaling and Cell Walls. [In:] Baluška F. & Mancuso S. (eds.): Signaling in Plants. Springer Verlag, Berlin Heidelberg, 173–193.

Romeh A.A. & Saber R.A.I. 2020: Green nano-phytoremediation and solubility improving agents for the remediation of chlorfenapyr contaminated soil and water. Journal of Environmental Management, 260, 110104.

Rouhier N., Cerveau D., Couturier J., Reichheld J.-P. & Rey P. 2015: Involvement of thiol-based mechanisms in plant development. Biochimica et Biophysica Acta, 1850 (8), 1479–1496.

Ruiz O.N. & Daniell H. 2009: Genetic engineering to enhance mercury phytoremediation. Current Opinion in Biotechnology, 20 (2), 213–219.

Sahu P.P., Pandey G., Sharma N., Puranik S., Muthamilarasan M. & Prasad M. 2013: Epigenetic mechanisms of plant stress responses and adaptation. Plant Cell Reports, 32, 1151–1159.

Sairam R.K., Srivastava G.C., Agarwal S. & Meena R.C. 2005: Differences in antioxidant activity in response to salinity stress in tolerant and susceptible wheat genotypes. Biologia Plantarum, 49 (1), 85–91.

Salnikow K. & Zhitkovich A. 2008: Genetic and epigenetic mechanisms in metal carcinogenesis and cocarcinogenesis: nickel, arsenic, and chromium. Chemical Research in Toxicology, 21 (2), 28–44.

Sami F., Yusuf M., Faizan M., Faraz A. & Hayat S. 2016: Role of sugars under abiotic stress. Plant Physiology and Biochemistry, 109, 54–61.

Santos J.A., Nunes L.A.P.L., Melo W.J. & Araújo A.S.F. 2011: Tannery sludge compost amendment rates on soil microbial biomass of two different soils. European Journal of Soil Biology, 47, 146–151.

Sarma H., Islam N.F., Prasad R., Prasad M.N.V., Ma L.Q. & Rinklebe J. 2021: Enhancing phytoremediation of hazardous metal(loid)s using genome engineering CRISPR–Cas9 technology. Journal of Hazardous Materials, 414, 125493.

Schoeneweiss D.F 1981: The role of environmental stress in diseases of woody plants. Plant Disease, 65 (4), 308–314.

Schönberger B., Chen X., Mager S. & Ludewig U. 2016: Site-dependent differences in DNA methylation and their impact on plant establishment and phosphorus nutrition in *Populus trichocarpa*. PLoS ONE, 11 (12), e0168623.

Sevilla F., Camejo D., Ortiz-Espín A., Calderón A., Lázaro J.J. & Jiménez A. 2015: The thioredoxin/peroxiredoxin/sulfiredoxin system: current overview on its redox function in plants and regulation by reactive oxygen and nitrogen species. Journal of Experimental Botany, 66 (10), 2945–2955.

Shahid M., Pinelli E. & Dumat C. 2012: Review of Pb availability and toxicity to plants in relation with metal speciation; role of synthetic and natural organic ligands. Journal of Hazardous Materials, 219–220, 1–12.

Shanker A. & Venkateswarlu B. 2011: Abiotic stress in plants. Mechanisms and adaptations. IntechOpen. https://www.intechopen.com/books/371, access: 11.12.2021.

Sharma S.S., Dietz K.-J. & Mimura T. 2016: Vacuolar compartmentalization as indispensable component of heavy metal detoxification in plants. Plant, Cell & Environment, 39 (5), 1112–1126.

Sharmin S.A., Alam I., Kim K.-H., Kim Y.-G., Kim P.J., Bahk J.D. & Lee B.-H. 2012: Chromium-induced physiological and proteomic alterations in roots of *Miscanthus sinensis*. Plant Science, 187, 113–126.

Shen Y., Issakidis-Bourguet E. & Zhou D.-X. 2016: Perspectives on the interactions between metabolism, redox, and epigenetics in plants. Journal of Experimental Botany, 67 (18), 5291–5300.

Shi W., Zhang Y., Chen S., Polle A., Rennenberg H. & Luo Z.-B. 2019: Physiological and molecular mechanisms of heavy metal accumulation in nonmycorrhizal versus mycorrhizal plants. Plant, Cell & Environment, 42 (4), 1087–1103.

Shukla O.P., Juwarkar A.A., Singh S.K., Khan S. & Rai U.N. 2011: Growth responses and metal accumulation capabilities of woody plants during the phytoremediation of tannery sludge. Waste Management, 31 (1), 115–123.

Simova-Stoilova L.P., Romero-Rodríguez M.C., Sánchez-Lucas R., Navarro-Cerrillo R.M., Medina-Aunon J.A. & Jorrín-Novo J.V. 2015: 2-DE proteomics analysis of drought treated seedlings of *Quercus ilex* supports a root active strategy for metabolic adaptation in response to water shortage. Frontiers in Plant Science, 6, 627.

Sinha S., Gupta M. & Chandra P. 1997: Oxidative stress induced by iron in *Hydrilla verticillata* (l.f.) Royle: Response of antioxidants. Ecotoxicology and Environmental Safety, 38 (3), 286–291.

Skórzyńska-Polit E. 2007: Lipid peroxidation in plant cells, its physiological role and changes under heavy metal stress. Acta Societatis Botanicorum Poloniae, 76 (1), 49–54.

Smith S.E. & Read D.J. 2008: Mycorrhizal Symbiosis. 3rd ed. Academic Press, London.

Spoel S.H. & Loake G.J. 2011: Redox-based protein modifications: the missing link in plant immune signalling. Current Opinion in Plant Biology, 14 (4), 358–364.

Stępniewska Z. & Wolińska A. 2005: Soil dehydrogenase activity in the presence of chromium (III) and (VI). International Agrophysics, 19, 79–83.

Suman J., Uhlik O., Viktorova J. & Macek T. 2018: Phytoextraction of heavy metals: A promising tool for clean-up of polluted environment? Frontiers in Plant Science, 9, 1476.

Szuba A. & Lorenc-Plucińska G. 2015: Utilization of proteomics in experimental field conditions – A case study of poplars growing on grassland affected by long-term starch wastewater irrigation. Journal of Proteomics, 126, 200–217.

Szuba A. & Lorenc-Plucińska G. 2018: Field proteomics of *Populus alba* grown in a heavily modified environment – An example of a tannery waste landfill. Science of the Total Environment, 610–611, 1557–1571.

Szuba A. & Wojtaszek P. 2010: Structural modifications of proteins caused by nitric oxide. Postępy Biochemii, 56 (2), 107–114.

Szuba A., Karliński L., Krzesłowska M. & Hazubska-Przybył T. 2017: Inoculation with a Pb-tolerant strain of *Paxillus involutus* improves growth and Pb tolerance of *Populus* × *canescens* under *in vitro* conditions. Plant and Soil, 412, 253–266.

Szuba A., Kasprowicz-Maluśki A. & Wojtaszek P. 2015: Nitration of plant apoplastic proteins from cell suspension cultures. Journal of Proteomics, 120, 158–168.

Szuba A., Marczak Ł. & Kozłowski R. 2020: Role of the proteome in providing phenotypic stability in control and ectomycorrhizal poplar plants exposed to chronic mild Pb stress. Environmental Pollution, 264, 114585.

Tangahu B.V., Abdullah S.R.S., Basri H., Idris M., Anuar N. & Mukhlisin M. 2011: A review on heavy metals (As, Pb, and Hg) uptake by plants through phytoremediation. International Journal of Chemical Engineering, 2011, 939161.

Tanou G., Filippou P., Belghazi M., Job D., Diamantidis G., Fotopoulos V. & Molassiotis A. 2012: Oxidative and nitrosative-based signaling and associated post-translational modifications orchestrate the acclimation of citrus plants to salinity stress. The Plant Journal, 72 (4), 585–599.

Tiodar E.D., Văcar C.L. & Podar D. 2021: Phytoremediation and microorganisms-assisted phytoremediation of mercury-contaminated soils: challenges and perspectives. International Journal of Environmental Research and Public Health, 18 (5), 2435.

Todeschini V., Linqua G., D'Agostino G., Carniato F., Roccotiello E. & Berta G. 2011: Effects of high zinc concentration on poplar leaves: A morphological and biochemical study. Environmental and Experimental Botany, 71, 50–56.
Trinh N.-N., Huang T.-L., Chi W.-C., Fu S.-F., Chen C.-C. & Huang H.-J. 2014: Chromium stress response effect on signal transduction and expression of signaling genes in rice. Physiologia Plantarum, 150 (2), 205–224.
Ullah A., Mushtaq H., Ali H., Munis M.F.H., Javed M.T. & Chaudhary H.J. 2015: Diazotrophs-assisted phytoremediation of heavy metals: a novel approach. Environmental Science and Pollution Research, 22, 2505–2514.
Vajpayee P., Tripathi R.D., Rai U.N., Ali M.B. & Singh S.N. 2000: Chromium (VI) accumulation reduces chlorophyll biosynthesis, nitrate reductase activity and protein content in *Nymphaea alba* L. Chemosphere, 41 (7), 1075–1082.
Vanderauwera S., Suzuki N., Miller G., van de Cotte B., Morsa S., Ravanat J.-L., Hegie A., Triantaphylidès C., Shulaev V., Van Montagu M.C.E., Van Breusegem F. & Mittler R. 2011: Extranuclear protection of chromosomal DNA from oxidative stress. Proceedings of the National Academy of Sciences of the United States of America, 108 (4), 1711–1716.
Vannini C., Domingo G., Marsoni M., Bracale M., Sestili S., Ficcadenti N., Speranza A., Crinelli R., Carloni E. & Scoccianti V. 2011: Proteomic changes and molecular effects associated with Cr(III) and Cr(VI) treatments on germinating kiwifruit pollen. Phytochemistry, 72 (14–15), 1786–1795.
Verslues P.E., Agarwal M., Katiyar-Agarwal S., Zhu J. & Zhu J.-K. 2006: Methods and concepts in quantifying resistance to drought, salt and freezing, abiotic stresses that affect plant water status. The Plant Journal, 45 (4), 523–539.
Vining K.J., Pomraning K.R., Wilhelm L.J., Preist H.D., Pellegrini M., Mockler T.C., Freitag M. & Strauss S.H. 2012: Dynamic DNA cytosine methylation in the *Populus trichocarpa* genome: tissue-level variation and relationship to gene expression. BMC Genomics, 13, 27.
Wang C.-R., Tian Y., Wang X.-R., Yu H.-X., Lu X.-W., Wang C. & Wang H. 2010: Hormesis effects and implicative application in assessment of lead-contaminated soils in roots of *Vicia faba* seedlings. Chemosphere, 80 (9), 965–971.
Wang R., Gao F., Guo B.-Q., Huang J.-C., Wang L. & Zhou Y.-J. 2013: Short-term chromium-stress-induced alterations in the maize leaf proteome. International Journal of Molecular Sciences, 14, 11125–11144.
Wang W., Vinocur B., Shoseyov O. & Altman A. 2004: Role of plant heat-shock proteins and molecular chaperones in the abiotic stress response. Trends in Plant Science, 9 (5), 244–252.
Wang X.-H., Wang Q., Nie Z.-W., He L.-Y. & Sheng X.-F. 2018: *Ralstonia eutropha* Q2-8 reduces wheat plant above-ground tissue cadmium and arsenic uptake and increases the expression of the plant root cell wall organization and biosynthesis-related proteins. Environmental Pollution, 242 (B), 1488–1499.
Wang Y., Xu L., Chen Y., Shen H., Gong Y., Limera C. & Liu L. 2013: Transcriptome profiling of radish (*Raphanus sativus* L.) root and identification of genes involved in response to lead (Pb) stress with next generation sequencing. PLoS ONE, 8 (6), e66539.
Wani S.H., Kumar V., Shriram V. & Sah S.K. 2016: Phytohormones and their metabolic engineering for abiotic stress tolerance in crop plants. The Crop Journal, 4, 162–176.
Williams L.E., Pittman J.K. & Hall J.L. 2000: Emerging mechanisms for heavy metal transport in plants. Biochimica et Biophysica Acta, 1465 (1–2), 104–126.
Wu B., Peng H., Sheng M., Luo H., Wang X., Zhang R., Xu F. & Xu H. 2021: Evaluation of phytoremediation potential of native dominant plants and spatial distribution of heavy metals in abandoned mining area in Southwest China. Ecotoxicology and Environmental Safety, 220, 112368.
Yadav S.K., Singla-Pareek S.L. & Sopory S.K. 2008: An overview on the role of methylglyoxal and glyoxalases in plants. Drug Metabolism and Drug Interactions, 23, 51–68.
Yamaji K., Watanabe Y., Masuya H., Shigeto A., Yui H. & Haruma T. 2016: Root fungal endophytes enhance heavy-metal stress tolerance of *Clethra barbinervis* growing naturally at mining sites via growth enhancement, promotion of nutrient uptake and decrease of heavy-metal concentration. PLoS ONE, 11 (12), e0169089.
Yan A., Wang Y., Tan S.N., Yusof M.L.M., Ghosh S. & Chen Z. 2020: Phytoremediation: A promising approach for revegetation of heavy metal-polluted land. Frontiers in Plant Science, 11, 359.
Yang C., Liu J., Dong X., Cai Z., Tian W. & Wang X. 2014: Short-term and continuing stresses differentially interplay with multiple hormones to regulate plant survival and growth. Molecular Plant, 7 (5), 841–855.

Yang Y., Dong C., Yang S., Li X., Sun X. & Yang Y. 2015: Physiological and proteomic adaptation of the alpine grass *Stipa purpurea* to a drought gradient. PLoS ONE, 10 (2), e0117475.

Yıldız M. & Terzi H. 2016: Proteomic analysis of chromium stress and sulfur deficiency responses in leaves of two canola (*Brassica napus* L.) cultivars differing in Cr(VI) tolerance. Ecotoxicology and Environmental Safety, 124, 255–266.

Zemleduch-Barylska A. & Lorenc-Plucińska G. 2015: *Populus* × *canescens* grown on Cr-rich tannery waste: Comparison of leaf and root biochemical and proteomic responses. Plant Physiology and Biochemistry, 90, 1–13.

Zemleduch-Barylska A. & Lorenc-Plucińska G. 2016: Response of leaf and fine roots proteomes of *Salix viminalis* L. to growth on Cr-rich tannery waste. Environmental Science and Pollution Research, 23, 18394–18406.

Zeng F., Wu X., Qiu B., Wu F., Jiang L. & Zhang G. 2014: Physiological and proteomic alterations in rice (*Oryza sativa* L.) seedlings under hexavalent chromium stress. Planta, 240 (2), 291–308.

Zhang H. & Sonnewald U. 2017: Differences and commonalities of plant responses to single and combined stresses. The Plant Journal, 90 (5), 839–855.

Zhang J., Wei J., Li D., Kong X., Rengel Z., Chen L., Yang Y., Cui X. & Chen Q. 2017a: The role of the plasma membrane H^+-ATPase in plant responses to aluminum toxicity. Frontiers in Plant Science, 8, 1757.

Zhang Y., Sa G., Zhang Y., Zhu Z., Deng S., Sun J., Li N., Li J., Yao J., Zhao N., Zhao R., Ma X., Polle A. & Chen S. 2017b: *Paxillus involutus*-facilitated Cd^{2+} influx through plasma membrane Ca^{2+}-permeable channels is stimulated by H_2O_2 and H^+-ATPase in ectomycorrhizal *Populus* × *canescens* under cadmium stress. Frontiers in Plant Science, 7, 1975.

Zhao C., Wang J., Cao M., Zhao K., Shao J., Lei T., Yin J., Hill G.G., Xu N. & Liu S. 2005: Proteomic changes in rice leaves during development of field-grown rice plants. Proteomics, 5 (4), 961–972.

Zwiewka M., Nodzyński T., Robert S., Vanneste S. & Friml J. 2015: Osmotic Stress Modulates the Balance between Exocytosis and Clathrin-Mediated Endocytosis in *Arabidopsis thaliana*. Molecular Plant, 8, 1175–1187.

Part Five

Educational and Social Aspects of Post-mineral Exploitation Habitats

5.1. Environmental Knowledge and Understanding: an Important and Necessary Aspect of Corporate Social Responsibility

Andrzej M. Jagodziński
Department of Ecology,
Institute of Dendrology, Polish Academy of Sciences;
5 Parkowa Str., 62-035 Kórnik, Poland;
amj@man.poznan.pl; ORCID: 0000-0001-6899-0985

Gabriela Woźniak
Institute of Biology, Biotechnology and Environmental Protection,
Faculty of Natural Sciences, University of Silesia in Katowice;
28 Jagiellońska Str., 40-032 Katowice, Poland;
gabriela.wozniak@us.edu.pl; ORCID: 0000-0003-1936-2880

ABSTRACT: Human activities impact the environment for a long time. The environment is transformed due to agriculture, different types of industry, and exploitation of natural mineral resources. The abiotic and biotic elements and their relationships with the environment have been altered in many places. Knowledge and understanding of the inseparable links between human health and environmental health are increasing. However, environmental protection activities conducted up to now are insufficient on a global scale (global change). There is a need to analyze the reasons for inefficiency and list possible erroneous actions in various human activity on mining areas. As a result, a change of the current paradigm is necessary. The modern approach must include the current state of knowledge regarding mechanisms of ecosystem functioning, particularly the Novel Ecosystems and the natural environment, and their roles in providing ecosystem services for the urban-industrial landscape and societies. In post-mining areas, the innate potential of poor oligotrophic habitats determines their functioning and environmental value for the urban-industrial landscape mosaic. The species composition of spontaneously established vegetation and developing ecosystems are optimally adapted to the specific abiotic conditions of post-mining mineral habitats. A new type of internal relationships is developing with regard to the flows of matter and energy in novel ecosystems. The essential functions of ecosystems, including ecosystems created in mining areas, include the demand for regulation potential of natural processes (e.g., regulation of the water balance, retention and increases of biodiversity, carbon sequestration). Strengthening of the opportunities to adapt to newly established ecosystem functioning in a comprehensive environmental approach should be one of the aims of corporate social responsibility (CSR) in the face of global changes. The potential of the natural environment is defined as natural capital, on which ecosystem functioning (ecosystem services) and all aspects of the economy are dependent. We aimed to present the current interdisciplinary environmental knowledge which should be included in CSR with respect to all aspects of economic activity, to limit and counteract global change.

KEYWORDS: CSR, environmental and socio-ecological resistance, novel ecosystems function, natural capital of the post-mining areas

INTRODUCTION

In recent centuries, the environment has been transformed due to the excessive exploitation of natural resources (firstly agriculture and different types of industry (chemical, textile), and secondly the exploitation of natural mineral resources). Human activities have an increasing impact on the existing vegetation and ecosystems. The abiotic and biotic elements of the soil or soil substrate have been changed or ultimately destroyed in many places. Knowledge and understanding of the inseparable links between human health and environmental health are increasing (Hartig *et al.* 2014). There is also growing awareness of the fundamental importance of primary producers – plant species composition and vegetation (photosynthetic organisms) for the functioning of ecosystems and the environment.

Environmental protection activities conducted for many years are insufficient or inappropriate and ineffective considering the deteriorating condition of the environment on a global scale (global change). Therefore, there is a need to analyze the reason for inefficiency and list possible erroneous actions in various mining areas. Transformation of the paradigm is necessary. The modern approach must include the current state of knowledge regarding the mechanisms of functioning of ecosystems and the natural environment with respect to their roles in providing ecosystem services for the urban-industrial landscape and societies. Environmental managers should consider the post-mining areas as ecosystems in which anthropogenic elements dominate, and their natural potential to determine ecosystem functioning. The various functions of the biological system in post-industrial regions depend on the relationships among its components: the type of substrate, groundwater and surface water, the availability of living organisms that can colonize the post-mining areas to create vegetation, and ultimately ecosystems. Spontaneously established ecosystems are optimally adapted to the specific abiotic conditions of post-mining mineral habitats and create a new type, the so-called novel ecosystems of mining areas. The essential functions of ecosystems, including ecosystems created in mining areas, include the regulation potential of natural processes (e.g., regulation of the water balance, retention and increases of biodiversity, carbon sequestration). The current corporate social responsibility (CSR) task is to strengthen the opportunities of newly established ecosystem functioning in a comprehensive environmental approach. The potential of the natural environment is defined as natural capital, on which ecosystem functioning (ecosystem services) are dependent (Costanza *et al.* 1997; Daily 1997; MEA 2005).

THE FUNCTIONING OF ECOSYSTEMS AND CORPORATE SOCIAL RESPONSIBILITY

Corporate social responsibility should be considered an integral and essential responsibility of all business and industrial activity. CSR is about companies contributing responsibly to society, as part of their fundamental workings, in their ethical, sustainable development, social and economic behaviors. The modern CSR concept of development in post-mineral mining areas must consider providing ecosystem services of the post-excavation areas in terms of the natural capital and novel ecosystems concepts (Fig. 1).

Regardless of the severe disturbances, e.g., in post-industrial areas, new living conditions are created which support a previously unknown composition of various species – a unique ecosystem (*sensu* Hobbs *et al.* 2013) – significantly different from the formerly known plant communities in the surrounding ecosystems (e.g., Rawlik *et al.* 2018a; 2018b; 2021). At the end of the 20th Century, the concept of ecosystem services provided by ecosystem functions was introduced. Ecosystem services enable humankind to survive and develop. Ecosystem services are built, self-reconstructing, and only available thanks to properly functioning ecosystems. These are possible due to the natural capital that has become the subject of scientific research and analysis in ecology, nature protection, economics, and social sciences. Suppose one considers what is necessary for a human's life, from the air he/she breathes, the water consumed, food and shelter, to recreation and natural beauty. In that case, it turns out that human health depends on environmental health, ecosystem functioning, and the high-quality services it provides.

The existing relationships among environmental health, social aspects, human health, and economy, although complex, often indirect, displaced in space and time, and dependent on some modifying forces, are evident and inevitable (Fig. 1). The Millennium Ecosystem Assessment (MEA 2005) indicated that the primary ecosystem services on which human health directly depends are the following: freshwater, food, raw materials, medicines, nutrient cycling, water purification, control of infectious diseases, climate regulation, and recreation. It has been shown that direct contact with elements of wildlife and ecosystem services helps to improve the functioning of the immune system, mood, and concentration while reducing stress and increasing the benefits of physical exercise. As a consequence, this leads to the expected reduction in the occurrence of diseases in large urban-industrial centers of Europe. For millennia, ecosystem services seemed to be inexhaustible and available for free. However, with the growing global population and consumerist lifestyle, human pressure on ecosystems has increased (and is still growing). Deforestation, agricultural expansion, urbanization, extraction and combustion of fossil fuels, car and air travel, or the spread of invasive species, disrupt the stability of natural ecosystems, threatening environmental health, and as a consequence, the availability and quality of ecosystem services on which the condition of human life depends. If disturbed ecosystems cannot meet the needs of society's ecosystem services, this will, directly and indirectly, affect livelihoods, income, local migration, and even social or political conflicts.

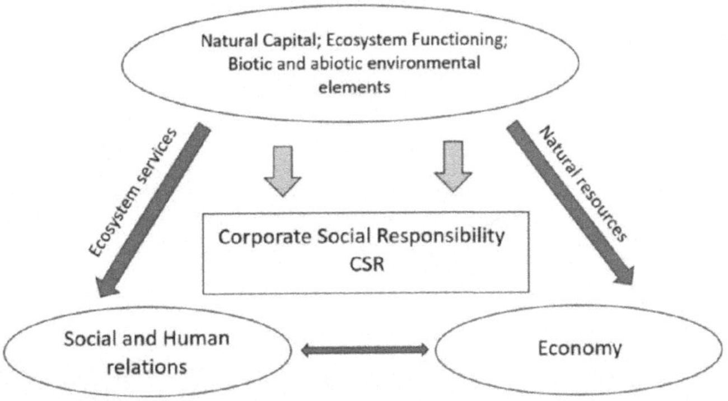

Figure 1. The relationships among Corporate Social Responsibility (CSR), environment (Natural Capital; Ecosystem Functioning; biotic and abiotic environmental elements), social-human links and the economy

Disturbed and non-optimally functioning ecosystems impact the physical and mental health of humans. There seems to be a stereotype in the human consciousness. Since the natural processes that drive ecosystems and provide ecosystem services are available for free, they have no economic value, so they can be considered worthless. This pernicious way of thinking is currently being changed by pointing out the importance, from a financial point of view, of facing the possibility of losing, e.g., the natural capital of pollination services. Natural capital is the only type of self-reproducible capital. Human well-being is a condition that arises as a result of adequate access to primary materials necessary for a good life, ensuring freedom of choice and action as well as health, good social relationships, and security. Well-being depends on the aggregate value of ecosystem goods and benefits provided (Solon *et al.* 2017).

We are not paid directly for services provided by the ecosystem (Society). We bear much higher costs for their loss, paying (Economy) for the reconstruction and construction of complex ecosystem functions and biogeo-chemical feedback relationships on which society is dependent (Fig. 2). The loss of soil fertility and erosion reduces yields due to the eradication of soil organisms, e.g., herbicides and insecticides. Biodiversity is considered the basis for natural ecosystems that provide countless services (ecosystem services). In this respect, properly functioning and diverse ecosystems (based on biogeochemical processes), adaptable and resistant to disturbances, should be maintained. Thanks to constant natural selection, living organisms, including plants, adapt to fulfill various functions in ecosystems and thus the natural environment. Living organisms (primarily plants), through adaptation processes, have adjusted to environmental conditions, including abiotic factors (e.g., precipitation, nutrient gradients) and biotic interactions (e.g., competition or herbivory, relationships between different groups of organisms). Assessment of patterns of plant functioning is necessary to understand nature (biosphere, atmosphere, and the dynamics of ecosystems, such as the formation of plant communities or the course and dynamics of nutrient cycles such as carbon and nitrogen). It is equally important to create conditions for developing and maintaining a mosaic of ecosystems, each shaped by natural processes in response to available habitat and natural environment conditions. It is also necessary to use spontaneous biological processes.

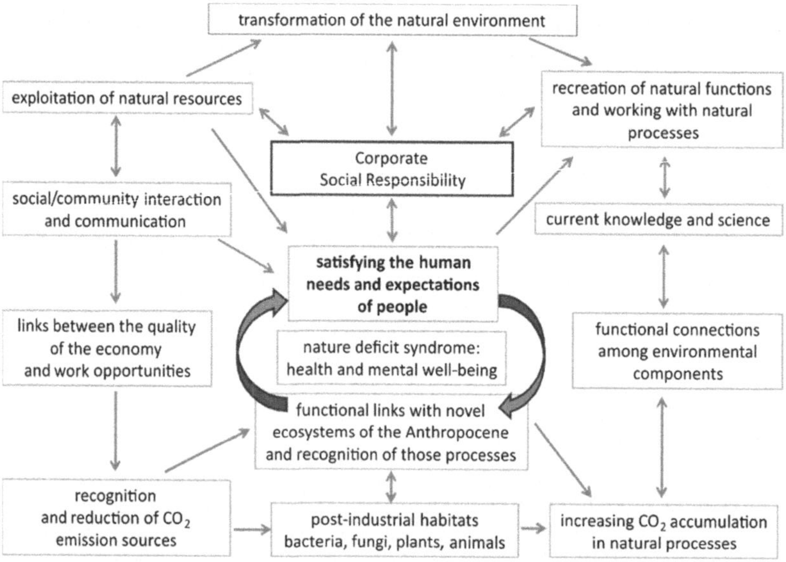

Figure 2. Environmental and socio-ecological factors may influence and enhance the ecological resistance due to proper management of spatial planning; CSR concerning the transition of the natural environment

Wherever there is a possibility, natural processes should be introduced into greenery arranged in urban and industrial areas. Today, most ecosystems have been transformed, directly or indirectly, by humans, by the unintentional or intentional introduction of new species that cause changes in the functioning of ecosystems, and degradation of abiotic elements, e.g., by excessive soil fertilization or industrial activity. As a result of these activities, novel ecosystems have been created, carrying – on the one hand – a new potential of ecosystem services and, on the other hand – requiring new paradigms indicating how to manage urban and industrial landscapes to optimize ecosystem functioning. It is essential to be aware that an even, regularly trimmed, single-species lawn provides negligible ecosystem services, compared to the same-sized varied meadow mimicking semi-natural flower meadows, mowed two (maximum three) times a season. Such a meadow creates the living conditions for countless insects, including pollinating insects. Understanding the functional mechanisms of developing ecosystems and establishing habitats resulting from mining activities (broadly understood as exploitation of mineral resources) and the use of comprehensive, interdisciplinary knowledge about the possibilities of giving new functions to the emerging ecosystems, should be the priority subject of all institutions involved. For effective management, taking into account ecosystem services, researchers suggest the use of proven solutions "based on nature" and its unsurpassed adaptability, allowing the adaptation of individual organisms, as well as entire biological systems, to adjust to changes in the environment.

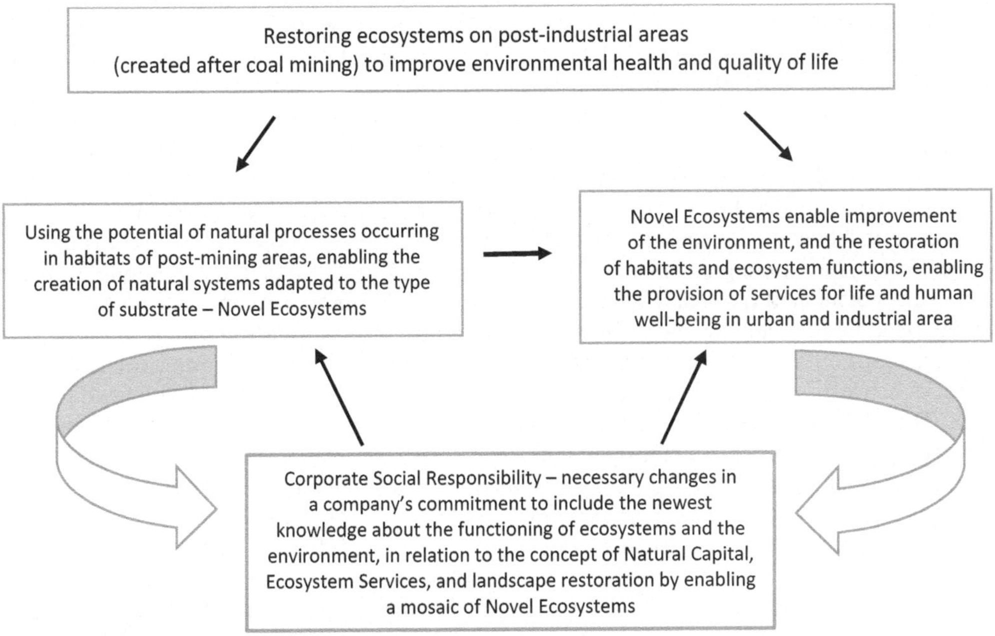

Figure 3. Characteristics of the necessary changes in the future management of post-mining areas to effectively restore natural habitats, strengthen natural capital, the function of ecosystems, and ecosystem services

The approach to adaptation of organisms within the meaning of "recombinant ecosystem – hybrid future" (*sensu* Rotherham 2017) can be applied "based on the ecosystem". In addition to their reconstruction, the recombinant ecosystem considers the social, economic, and cultural benefits. The benefits of ecosystem services and the risk of their loss are not always clearly visible and fully realized. Therefore, constant care for the proper functioning of ecosystems, to which the health and resilience of human survival are inseparably linked, is critical. The heaps occurring in Silesia, water reservoirs in depressions, sedimentation pools, or urban ecosystems, meet the criteria set by the conceptual and theoretical assumptions. They can undoubtedly be considered elements of novel ecosystems. Such elements significantly enrich the landscape of the urban-industrial agglomeration, as is the case in other urban industrial centers of Europe (Rotherham 2017). Nature, and its natural processes, can play a fundamental and essential role in restoring the biological and environmental value of post-mining landscapes. In Europe, many positive examples are enabling the processes of natural succession, which can be called spontaneous regeneration, of mining areas. Adopting a new approach that will consider the significant contribution and importance of biological processes is necessary for emergence of independently functioning, new types of novel ecosystems (Fig. 3).

CURRENT ENVIRONMENT MANAGEMENT VS. NECESSARY FUTURE ACTIVITIES

Novel ecosystems can emerge in the unusual conditions of mineral soil substrates which can prevail in post-mining sites. The new approach requires a change in the current paradigm concerning post-industrial areas. Post-industrial sites, particularly the post-mineral extraction areas, should be considered in association with the natural environment and ecosystem functioning, not only with respect to the economic benefits for human utility. This recommended approach will require essential updates in national legislation regarding the legal regulations about environmental protection. Novel ecosystems are established in special mineral site conditions of post-mining habitats. They create an unknown natural resource that can significantly improve ecosystem services, mainly those services, on a landscape scale, providing social, recreational, and health opportunities for local communities.

The change in emphasis from a solely economic approach to a socio-environmental approach can provide an additional platform for informing stakeholders and establishing relationships with communities in the area. Effective interaction and communication between groups and individual stakeholders is also an essential aspect of developing a positive partnership when deciding on the most appropriate development scenarios for a site. Suppose there is a decision to leave a site undeveloped to allow natural processes to occur and natural succession (i.e., re-wilding of the area). In that case, it should not be considered solely as a cheap or low-cost option that is convenient and easy to carry out for mining companies. Entrepreneurs need to understand the possibilities and environmental potential of the area to later manage it effectively for optimal emerging ecosystem functioning opportunities, and the most effective provision of ecosystem services (Fig. 2).

The natural environment is a complicated network of dynamic connections and often synergistic feedbacks between numerous biotic and abiotic elements operating at various spatial and temporal scales. The environment will not improve until it is accepted that the mosaic of ecosystems, with their inter-related functions, provide natural capital and ecosystem services in addition to economic benefits (Woźniak *et al.* 2018). Humanity has understood a small part of the dependencies driving ecosystem and environmental functioning, and there is still much more to do. Having access only to limited knowledge about the mechanisms of the functional and dynamic processes of novel ecosystems, ignoring or having an incomplete understanding of natural biogeochemical processes, one cannot feel self-confident in making decisions about the environment. Because of a lack of knowledge, some environmental management actions can cause serious, often irreversible, damage to the complex processes of ecosystems and their ecological functioning. How post-mining sites are managed in the future should be very different, depending on the post-mineral excavation functions envisaged for them or the ecosystem services the area can provide.

In some cases, the combination of natural processes and spontaneous succession can play a crucial role in self-multiplication of the natural capital and resulting ecosystem services at a site, and consequently benefit human health and well-being. Society should appreciate the power of nature itself in restoring best-adjusted ecosystems (landscape elements) depending on the prevailing habitat conditions or landscape (the mosaic of ecosystems) without additional human intervention (habitat and wildlife development). Schopf & Foster (2014) recorded the growth of a "natural wilderness" with a rich ecology on an area of post-industrial urban spoil in Toronto (Canada). In some circumstances, management decisions can support natural processes through appropriate environmental management interventions. It will not be possible to restore all areas to their former use (see Hobbs *et al.* 2006; 2013). However, utterly new habitat conditions can arise in some instances, i.e., completely new ecosystems are being established, the so-called "novel ecosystems" (Box 1999; Doley *et al.* 2012; Schopf & Foster 2014). Box (1999) cited an example of the range of vegetation types following natural regeneration on post-coal mining spoil heaps in Telford, UK, with one mound now classified as a Site of Special Scientific Interest. The research results exemplify why post-industrial site management must not be guided solely by engineering, agricultural, and practical approaches. When deciding on the functions to be performed by a given site, the increase of economic returns for investors should not be the only aim. Naturalist areas that can be identified as a valuable must, where possible, be left to regenerate to restore the environmental ecosystem functions lost by industrially transformed (including by mineral exploitation) areas. Therefore, in the proper management of post-industrial and post-exploitation regions, it will be essential to understand the natural processes in these areas (De Sousa 2014).

The managers responsible for Corporate Social Responsibility have to acknowledge the environmental value of oligotrophic terrestrial, wetland, and water habitats. Post-exploitation areas often consist of a mosaic of various microhabitats, which can be inhabited by an adequate range of species of plants and animal organisms that have become established in the course of ecological succession. The colonizing species represent different habitat preferences concerning light, water, nutrients, litter quality or pH (e.g., Skorupski *et al.* 2013; Urbanowski *et al.* 2018; 2021a, 2021b). Knowledge about the system of spontaneous successional processes is still not well understood. Therefore, research should include a better understanding of successional processes occurring on the surface of a given area and below the surface, within the soil (soil substrate), particularly on such unusual habitats as the post-industrial and post-mining sites that may develop novel ecosystems. Feedback relationships and processes related to the role of soil microorganisms in the formation of soil substrate, CO_2 sequestration in soil/soil substrates, and the relationships among soil organisms (bacteria, fungi, Enchytraeidae, Nematoda) and plants still require further research (e.g., Horodecki & Jagodziński 2017; 2019; Horodecki *et al.* 2019). It also seems essential to understand the role of mycorrhiza in the processes of plant colonization on post-industrial areas (often contaminated land; e.g., Kałucka *et al.* 2016; Kałucka & Jagodziński 2016; 2017). Understanding the diversity (both the taxonomic and functional diversity) of organisms has been proven to be extremely important (Hatfield *et al.* 2018), and future post-industrial activities must be based upon a complete understanding of the above processes (De Sousa 2014).

Environmental management activities can create positive opportunities for greater industrial and academic cooperation to aid effective post-industrial reclamation and restoration. One of the critical issues for the establishment of novel ecosystems on post-industrial landscapes seems to be the role of various organisms (plants, microorganisms) and the flexibility of living organisms to adapt and participate in the process of cleaning contaminated areas, e.g., those contaminated with heavy metals, petroleum substances, aromatic hydrocarbons, etc., as part of their life processes. The living organisms can have the biological potential to prevent the pollutants from further contaminating the water, soil, or air systems. In turn, natural processes, and biological potential can help to manage the sites, plants, and microorganisms, providing the abilities for remediation. This biological potential of organisms to adapt to the changing environmental habitat conditions is used in the remediation processes. The contaminants which may be bound in the soil, may be broken down into smaller constituent parts that other living organisms can metabolize (Ptaszek *et al.* 2020). However, it is crucial to understand the mechanisms of plant response to various stress conditions (drought, temperature, salinity, heavy metals, etc.) that the plants encounter in post-mining areas, to support or carefully mimic the spontaneous natural processes.

Some post-industrial sites are of little economic value and can become places of natural, environmental, social, and cultural significance. The earlier history of the area, industrial processes, and their current use appear to be crucial for proposing appropriate scenarios for the land use for a particular place. Where buildings of the former industrial infrastructure have survived, they may be left in the landscape as part of the area's industrial heritage (i.e., incorporated into the landscape and used to strengthen the historical function). The buildings can be adapted, for example, as museum buildings where visitors can learn about mining activities, the mining process, applied technologies, and post-industrial environmental restoration, or they can be used as potential ecological research and development centers. Such places can also be used for tourism, recreation, education, the organization of significant sports events, or as a meeting place for people during organized festivities, thereby contributing to the social and community enhancement of the area. It is worth mentioning that technological tools, particularly those in existing research geoinformatics, may facilitate the restoration and recovery of post-industrial sites. These jointly shared experiences may, in the future, promote cooperation and enable the identification of the best scenarios for the repair and redevelopment of various places which have been covered by industrial activity in the past. Some of these post-industrial landscapes could be wrongly called wastelands. Still, it has been shown that they could become fully functioning places that will provide sites for (novel) ecosystem establishment with a range of ecosystem services (regulatory, cultural, social, and economic) that serve society.

CONCLUSION

Management of the post-mining sites should be based on the newest scientific knowledge. Novel ecosystems enable improvement of the environment, the restoration of habitats and ecosystem functions, enabling the provision of services for life and human well-being in urban and industrial areas. Therefore, Corporate Social Responsibility is necessary to change a company's commitment to include the newest knowledge about the functioning of ecosystems and the environment concerning the concept of Natural Capital, Ecosystem Services, and landscape restoration. The opportunity given by the mosaic of Novel Ecosystems is used by restoring ecosystems on post-industrial areas (created after coal mining) to improve environmental health. The potential of natural processes occurring in the habitats of post-mining regions enables the creation of biological systems adapted to the type of substrate – Novel Ecosystems.

REFERENCES

Box J. 1999: Nature conservation and post-industrial landscapes. Industrial Archaeology Review, 21 (2), 137–146.

Costanza R., d'Arge R., de Groot R., Farber S., Grasso M., Hannon B., Limburg K., Naeem S., O'Neill R.V, Paruelo J., Raskin R.G., Sutton P. & van den Belt M. 1997: The value of the world's ecosystem services and natural capital. Nature, 387 (15), 253–260.

Daily G.C. (ed.) 1997: Nature's Services: Societal Dependence on Natural Ecosystems. Island Press, Washington, DC.

De Sousa C. 2014: The greening of urban post-industrial landscapes: past practices and emerging trends. Local Environment, 19 (10), 1049–1067.

Doley D., Audet P. & Mulligan D.R. 2012: Examining the Australian context for post-mined land rehabilitation: Reconciling a paradigm for the development of *natural* and *novel* ecosystems among post-disturbance landscapes. Agriculture, Ecosystems & Environment, 163, 85–93.

Hartig T., Mitchell R., de Vries S. & Frumkin H. 2014: Nature and Health. Annual Review of Public Health, 35, 207–228.

Hatfield J.H., Harrison M.L.K. & Banks-Leite C. 2018: Functional diversity metrics: how they are affected by landscape change and how they represent ecosystem functioning in the tropics. Current Landscape Ecology Reports, 3, 35–42.

Hobbs R.J., Arico S., Aronson J., Baron J.S., Bridgewater P., Cramer V.A., Epstein P.R., Ewel J.J., Klink C.A., Lugo A.E., Norton D., Ojima D., Richardson D.M., Sanderson E.W., Valladares F., Vilà M., Zamora R. & Zobel M. 2006: Novel ecosystems: theoretical and management aspects of the new ecological world order. Global Ecology and Biogeography, 15 (1), 1–7.

Hobbs R.J., Higgs E.S. & Hall C.M. (eds.) 2013: Novel Ecosystems: Intervening in the New Ecological World Order. John Wiley & Sons Ltd., Chichester.

Horodecki P. & Jagodziński A.M. 2017: Tree species effects on litter decomposition in pure stands on afforested post-mining sites. Forest Ecology and Management, 406, 1–11.

Horodecki P. & Jagodziński A.M. 2019: Site type effect on litter decomposition rates: A three-year comparison of decomposition process between spoil heap and forest sites. Forests, 10 (4), 353.

Horodecki P., Nowiński M. & Jagodziński A.M. 2019: Advantages of mixed tree stands in restoration of upper soil layers on postmining sites: A five-year leaf litter decomposition experiment. Land Degradation & Development, 30, 3–13.

Kałucka I.L. & Jagodziński A.M. 2016: Successional traits of ectomycorrhizal fungi in forest reclamation after surface mining and agricultural disturbances: A review. Dendrobiology, 76, 91–104.

Kałucka I.L. & Jagodziński A.M. 2017: Ectomycorrhizal fungi: a major player in early succession. [In:] Varma A., Prasad R. & Tuteja N. (eds.): Mycorrhiza – Function, Diversity, State of the Art. 4th ed. Springer International Publishing, 187–229.

Kałucka I.L., Jagodziński A.M. & Nowiński M. 2016: Biodiversity of ectomycorrhizal fungi in surface mine spoil restoration stands in Poland – first time recorded, rare and red-listed species. Acta Mycologica, 51 (2), 1080.

MEA – Millennium Ecosystem Assessment. 2005: Ecosystems and Human Well-being: Biodiversity Synthesis. World Resources Institute, Washington DC.

Ptaszek N., Pacwa-Płociniczak M., Noszczyńska M. & Płociniczak T. 2020: Comparative study on multiway enhanced bio- and phytoremediation of aged petroleum-contaminated soil. Agronomy, 10 (7), 947.

Rawlik M., Kasprowicz M. & Jagodziński A.M. 2018a: Differentiation of herb layer vascular flora in reclaimed areas depends on the species composition of forest stands. Forest Ecology and Management, 409, 541–551.

Rawlik M., Kasprowicz M., Jagodziński A.M., Kaźmierowski C., Łukowiak R. & Grzebisz W. 2021: Tree species have a greater influence on species composition of the herb layer than soil texture on a forested post-mining area. Land Degradation & Development, 32 (5), 2013–2024.

Rawlik M., Kasprowicz M., Jagodziński A.M., Kaźmierowski C., Łukowiak R. & Grzebisz W. 2018b: Canopy tree species determine herb layer biomass and species composition on a reclaimed mine spoil heap. Science of the Total Environment, 635, 1205–1214.

Rotherham I.D. 2017: Recombinant Ecology – A Hybrid Future? Springer International Publishing.

Schopf H. & Foster J. 2014: Buried localities: archaeological exploration of a Toronto dump and wilderness refuge. Local Environment, 19 (10), 1086–1109.

Skorupski M., Horodecki P. & Jagodziński A.M. 2013: Roztocze z rzędu Mesostigmata (Arachnida, Acari) na terenach przemysłowych i poprzemysłowych w Polsce [Mite species of Mesostigmata order (Arachnida, Acari) in industrial and postindustrial areas of Poland]. Nauka Przyroda Technologie, 7 (1), 11.

Solon J., Roo-Zielińska E., Affek A., Kowalska A., Kruczkowska B., Wolski J., Degórski M., Grabińska B., Kołaczkowska E., Regulska E. & Zawiska I. 2017: Świadczenia ekosystemowe w krajobrazie młodoglacjalnym. Ocena potencjału i wykorzystania. Wydawnictwo Akademickie SEDNO, Polska Akademia Nauk, Instytut Geografii i Przestrzennego Zagospodarowania im. Stanisława Leszczyckiego, Warszawa.

Urbanowski C.K., Horodecki P., Kamczyc J., Skorupski M. & Jagodziński A.M. 2018: Succession of mite assemblages (Acari, Mesostigmata) during decomposition of tree leaves in forest stands growing on reclaimed post-mining spoil heap and adjacent forest habitats. Forests, 9 (11), 718.

Urbanowski C.K., Horodecki P., Kamczyc J., Skorupski M. & Jagodziński A.M. 2021a: Does litter decomposition affect mite communities (Acari, Mesostigmata)? A five-year litterbag experiment with 14 tree species in mixed forest stands growing on post-industrial area. Geoderma, 391, 114963.

Urbanowski C.K., Horodecki P., Kamczyc J., Skorupski M. & Jagodziński A.M. 2021b: Predatory mite instars (Acari, Mesostigmata) and decomposing tree leaves in mixed and monoculture stands growing on a spoil heap and surrounding forests. Experimental and Applied Acarology, 84, 703–731.

Woźniak G., Sierka E. & Wheeler A. 2018: Urban and industrial habitats: how important they are for ecosystem services. [In:] Hufnagel L. (ed.): Ecosystem Services and Global Ecology. IntechOpen, 169–194.

5.2. Activities of Mining Enterprises for Repair of the Natural Environment in the Light of Applicable Legal Regulations

Edyta Sierka
Institute of Biology, Biotechnology and Environmental Protection, Faculty of Natural Sciences, University of Silesia in Katowice; 28 Jagiellońska Str., 40-032 Katowice, Poland; edyta.sierka@us.edu.pl; ORCID: 0000-0003-3317-4552

Beata Pabiszczak
Department of Reclamation and Environmental Protection, Restructuring of Mines Company S.A.; 10 Perla Str., 41-300 Dąbrowa Górnicza, Poland; beatapabiszczak@srk.com.pl

ABSTRACT: The paper aimed to summarize the reclamation practices applied in areas transformed by hard coal mining and identify the use of nature-based solutions. The direct impacts, such as post-coal mine heaps, and indirect impacts, such as land deformation due to subsidence, are observed. The results showed that the applied reclamation measures allow for their implementation following the provisions of the Spatial Development Plans and legal acts, which require restoring the reclamation areas to their proper condition. Entrepreneurs declare that they approach all issues related to changes and transformations of the environment caused by economic activity very carefully and rationally, which is included in the rehabilitation programs carried out. The methods used make little use of natural capital ecosystems created as a result of coal mining. The concepts of increasing the scope of ecosystem services to improve environmental quality are also not taken into account.

KEYWORDS: Hard coal exploitation, transformation of the environment, reclamation, nature based solutions

INTRODUCTION

"Our civilization … is founded on coal [...]. The machines that keep us alive, and the machines that make machines, are all directly or indirectly dependent upon coal. Practically everything we do, from eating an ice to crossing the Atlantic, and from baking a loaf to writing a novel, involves the use of coal, directly or indirectly" (Orwell 1937). Nowadays, the multiple effects of many years of exploitation of this raw material are felt socially, economically, and environmentally (Wang *et al.* 2009).

The mining industry's impact on the natural environment is generally perceived as an unfavorable activity. It manifests itself in a negative impact on humans (Blanco *et al.* 2019), referred to as ecosystem disservices. In industrial regions, adverse impacts are co-created both by humans (changes in the environment related to, e.g., the exploitation of raw materials) and ecosystems (impoverished biodiversity of ecosystems) (Cox *et al.* 2018).

However, the impacts of hard coal mining and processing can, directly and indirectly, impact air, water, soil, ecosystems, and animal and human health during every phase of the coal "life cycle", from resource extraction to final by-product disposal (Fig. 1).

Further treatment of the areas created by mining or processing is limited by the different legal regulations in each country, e.g., Polish Geological and Mining Law (2011).

Figure 1. General view of the waste rock heap in the area of Zabrze
Upper Silesia, South Poland (Photo: Edyta Sierka)

In addition, the managers apply good practices, taking into account the provisions of reclamation programs for post-mining areas and their intended purpose. Land reclamation has played a significant role in the urban development process of post-mining areas. The reclamation should be carried out as quickly as possible after the end of waste disposal The reclamation should be carried out as quickly as possible after the end of waste storage:

This study aimed to:

- present the reclamation practices conducted on post-coal mine sites by the administrators of these areas on the example activity of Restructuring of Mines Company S.A. (SRK);
- identify the possibilities for practitioners to introduce the newest scientific achievements into their applications.

MINING AND THE NATURAL ENVIRONMENT

The environmental impacts of natural resources mining can occur on a local, regional and global scale through direct and indirect effects of mining practices. The effects of mining activity can lead to erosion of the Earth's surface, the formation of sinkholes (sedimentation pools), loss of biodiversity, or contamination of soil, groundwater (Marschalko *et al.* 2012), and surface water with substances from mining processes (Huang *et al.* 2018). Industrial activity leads directly to the exclusion of agricultural and forest areas from their current use.

The mining areas become investment areas for construction of mining infrastructure, i.e., shafts, processing plants, railway sidings, transport routes, storage yards, and other facilities necessary for the mining operation. In close connection with the operation of mining sites, the areas are occupied by underground water sedimentation pools and post-mining by-product heaps. These so-called heaps are formed from the excavated material or arise due to processing and treatment of minerals.

Indirect effects of mining activities in the environment are manifested by changes in geomorphology, water relations, deformation of the land surface, or impacts on buildings, roads, and forest areas (Giam *et al.* 2018). The effects of mining activity also influence the functioning of local communities and the natural environment (Sierka *et al.* 2012).

POST-COAL MINE HEAPS, MINE WATER SEDIMENTATION POOLS, AND POST-MINING SUBSIDENCE RESERVOIRS

The most common form of impact on elements of the natural environment is the presence of post-coal mine heaps (Fig. 2).

The mineral material deposited on post-coal mine heaps is represented by the mineral forms in the Carboniferous geological period. It consists of, among other, sandstones of various grain sizes and clay (carbonate) binder, amorphous mudstones, or clay shales, which are clay fraction rock with an admixture of mudstone and coal shale (Ribeiro *et al.* 2011).

If it meets the required criteria, the material collected on post-coal mine heaps can be used to fill transformed areas (Finkelman *et al.* 2021), including road construction by builders.

The areas transformed by deep hard coal processing activities include sedimentation pools (Fig. 3), designed to purify the water, used in the technological process, by allowing gravity to remove from the coal dust suspension of water, prior to introducing it into surface waters.

The composition of the dust in sedimentation pools depends mainly on the size of the particles. The research showed that the coal dust masses collected in the sedimentation pools, depends mainly on the size and the content of carbon particles. Coal dust has been documented in the region's waters and on the surface of the ocean and the seabed, causing a severe impact on aquatic fauna and flora (Johnson & Bustin 2006), including ocean acidification (Brown & Spiegel 2019).

Figure 2. General view of the post-coal mine heaps in the area of Katowice, Upper Silesia, South Poland (Photo: Edyta Sierka)

As an indirect result of underground coal mining, the appearance of subsidence basins is observed (Fig. 4). With the emergence of a reservoir, large land areas are often occupied, and the existing vegetation is destroyed, which leads to a decrease in biodiversity. In the subsequent stages, the processes of spontaneous succession most often lead to the formation of a water ecosystem similar in nature to those formed in natural reservoirs (Pierzchała & Sierka 2020). The size, extent, and course of land subsidence depend on the depth of exploitation and the sequence of layers with different mechanical properties, the nature of tectonic dislocations, and the extraction system.

Figure 3. Coal dust sedimentation pool spontaneously covered by *Phragmites australis* (Photo: Gabriela Woźniak)

Generally land subsidence induced by underground coal mining is one of the adverse impacts of mining, which not only destroys soil structure and changes its properties, but also causes other eco-environmental problems (Marschalko *et al.* 2012), such as limitation of vegetation growth, reduction of crop production, plant death, acceleration of soil water erosion, degradation of the landscape, surface soil fracture, dropping of groundwater table, changes in topographic and hydrologic conditions, loss of agricultural land and topsoil, and so on (Kuter *et al.* 2014).

Figure 4. This reservoir was created in the subsidence trough above the exploited hard coal deposit at Sosnowiec, Upper Silesia, South Poland (Photo: Edyta Sierka)

Reservoirs in subsidence basins have been and are still of interest in terms of the impacts of underground hard coal mining on geomorphological changes (Marschalko *et al.* 2009), their dynamics and forecasting future surface transformations, and the effects of land depression in Europe (Doležalová *et al.* 2009). Various impacts of subsidence are indicated, which, apart from soil erosion, land surface changes, drainage, watering, and reduction of yields, also include removal of nitrogen availability for plants (Kuter *et al.* 2014). Due to the functioning of the reservoirs, an important issue is their role as reservoirs that receive or store water transferred from outside the catchment area, e.g., drainage water from a mine. These waters may be saline and contain coal dust, which is why ecosystems with particular habitat conditions are formed (Kašovská *et al.* 2014).

RECLAMATION AND DEVELOPMENT OF AREAS TRANSFORMED BY HARD COAL MINING IN THE LIGHT OF THE LAW

The exploitation of energy resources place for over 100 years. It has led to the transformation of land, which should be supported by remedial measures in the form of reclamation and development. According to provisions of the law, the other decisions about the so-called disused waste rock dumps depend on the purpose of the land based on Local Spatial Development Plans for individual municipalities.

Ecological restoration and mine reclamation have become important parts of the sustainable development strategy in many countries. Good planning and environmental management will minimize the impacts of mining on the environment and will help in preserving eco-diversity (Rosa *et al.* 2020). The steps are taken to reduce the nuisance of areas transformed by mining activities and restore their ecological functions. A reconstruction planning framework for such regions should include an integrated analysis of the ecosystem services and socio-ecological system, provided to facilitate community involvement and deliver social benefits (Woźniak *et al.* 2018). Ecological restoration and mine reclamation have become important parts of the sustainable development strategy in many countries.

From the point of view of Polish legislation, the rock is a type of waste[1]. According to the Act on waste and Directive 2008/98/EC (Directive), when managing it, the owner of the site is obliged to handle it following the established hierarchy. In the first place, the waste should be processed at the site of its origin. Waste that cannot be processed on site should be transferred to the nearest locations that allow such processing. The environmental cost caused by, for example, transport, should not outweigh the benefits of choosing a more pro-environmental technology. However, the concept of Natural Capital and ecosystem services also depending on the novel ecosystem functioning, is not introduced into the legislation. A deep understanding of all of these phenomena is crucial for efficient limitation of global change and unfortunately is not included in the Polish legislation system.

In the current legislation frame in Poland, the management of post-coal mine heaps depends on ownership of the land, types of collected material, and their thermal condition. The management or lack of control depends on the intended use of the areas based on the Local Spatial Development Plans for individual municipalities.

One of the methods is biological reclamation in the forest direction. Heaps located in areas belonging to the State Forests National Forest Holding undergo reclamation in the forest direction (tree plantations, to be precise). Generally, such reclamation consists of two stages. The first stage, technical reclamation, consists of shaping the heaped body to optimally fit into the surroundings and minimizing the possibility of aeration. In this stage, proper water management, i.e., retaining the necessary amount of water on the plateau and draining excess water into absorptive ditches or, if possible, into the natural environment. The body formed should be adequately compacted using spiked rollers. The next stage is biological reclamation consisting of fertilizing an appropriate layer of soil adapted to the composition of the planned planting, e.g., tree species (Fig. 5) or grasses (Fig. 6). Following the guidelines of the relevant forest districts, trees are planted in suitably dressed holes. There are also maintenance works that need to be performed, which last from 3 to 5 years. They consist of the annual use of cover protecting the seedlings from boars, mowing grass and weeds between seedlings, fertilizing, and watering.

---

[1] waste in Polish EU legislation is the Waste Act (2012 Dz.U.2021.0.779)

Figure 5. Planting areas of woody species (Photo: Beata Pabiszczak)

Figure 6. Heap after the Siersza mine (Photo: Beata Pabiszczak)

After the care and approval of the works carried out by the territorially competent Forest District (Fig. 7), the areas are transferred to the State Forests. Such investments are currently underway, for example in Siersza, the Trzebinia district in the Chrzanów district, in the Lesser Poland voivodship, after the Siersza Coal Mine. The task is 100% financed by the National Fund for Environmental Protection and Water Management in Warsaw.

A solution used in biological reclamation (Fig. 7) was the fertilization of soil substrate masses initially (Fig. 8) mixed with grasses.

Figure 7. Planting tree in the reclamation process of Siersza Coal Mine (SRK materials, 2019)

Figure 8. The area undergoing technical reclamation (SRK materials, 2019)

The applied method allowed for the maximum use of seeds, their growth in stages, preventing the seeds from being eaten, and most importantly, through layered germination and different layers of root system formation – maximum protection of the soil layers against erosion (Fig. 9).

Figure 9. The site after complete of biological reclamation (Photo: Edyta Sierka)

After shaping and compacting, a layer of fertile soil is frequently applied, and then a mixture of grasses was fed with artificial fertilizer. Then the so-called hydro-seeding was made.

Observations comparing the long-term effects of primary succession and the introduction of plants by planting or sowing are few (Fig. 10).

Figure 10. Natural succession in reclaimed areas (Photo: Edyta Sierka)

In the Bytom–Miechowice district, the Miechowice coal-mine started operating in 1902, and in 1996 it was connected with the Bobrek mine to create the Bobrek–Miechowice mine (Fig. 11). In 1999, the Bobrek–Miechowice mine was closed.

Figure 11. Miechowice – the area undergoing reclamation (SRK materials, 2019)

The area was subjected to technical reclamation consisting in the formation of the land (Fig. 12) and then biological reclamation (Fig. 13) used a mixture of grasses. The area in question is intended for sale for investment. Biological reclamation was financed by the National Fund for Environmental Protection and Water Management.

Figure 12. Miechowice – area under reclamation (SRK materials, 2019)

Reclamation aims to create conditions for vegetation and entire ecosystems, which will become an element integrated into the surroundings and fulfilling specific ecological functions. Unfortunately, regularly updated knowledge and the newest achievements of biological sciences about the natural processes in the changing environment are not shared among the practitioners. The legislation system is not enforcing the practitioners and decision-makers to follow the dynamic adaptation processes in nature. Human activities should support and enhance the natural adaptation process. Humans cannot act as effectively as the adaptation processes to the changing environment at various organizational levels, from the molecular to the landscape.

Figure 13. Miechowice heap – reclaimed area (SRK materials, 2019)

Restructuring of Mines Company S.A. (SRK) with the takeover of the "Kazimierz-Juliusz" (SRK) mine, Sosnowiec took over the area of many water basins in Sosnowiec. The reservoir has been filled and reclaimed towards forest. The remaining water bodies must be turned towards a recreational park based on arrangements with the Sosnowiec Commune. An essential element resulting from mining activities are continuous and discontinuous deformations of the landscape.

Discontinuous deformations are one of the effects of mining exploitation on the surface. They can take the form of cracks, crevices, faults, sinkholes, landslides, funnels, and uplift. In the course of the activities of SRK, we have seen many such deformities. Most of them were related to the so-called shallow exploitation, very often occurring due to shallow goaf "activation" by exploitation carried out at greater depths. This is the case in the forests of the Jaworzno Commune. As a result of the exploitation carried out in the 1970s, the post-exploitation voids of the 1920s became active.

In forest areas, discontinuous deformations in the form of valleys of varying depth and diameter were formed. Reclamation of these areas will consist of filling the funnels with material, fertilizing the appropriate type of soil, and planting trees following the guidelines of the Forest Inspectorate. It is worth noting that the reclamation works will have to be preceded by an engineer reconnaissance. An additional but significant difficulty in designing and carrying out reclamation works in forest areas is minimizing forest stand losses, i.e., limiting tree clearance to enable the task to be carried out.

As a result of the normalization of water relations, these reservoirs usually fill up spontaneously with water, and the natural succession of vegetation restores the dynamic ecological balance. For environmental reasons, the emerging water bodies are part of the so-called small retention. They are part of the concept of actions to mitigate climate change and support the development of biodiversity. It often turns out that reservoirs of such genesis are areas of the incredible richness of the organic world, refuges for flora and fauna, and abundant in rare and protected species. That may influence plants and invertebrates (Sierka *et al.* 2012).

However, from the legal point of view, it damages the environment, and measures must be taken to remedy it. These reservoirs were often covered with post-coal mine mineral rock material. Still, after a change in law, they are left without interference or strengthening the banks with mineral material as part of reclamation activities. These reservoirs are also places to collect water from the drainage of exploited deposits. They often show signs of salinity. In the areas taken over by SRK there are water reservoirs used by mines as sedimentation pools for underground water (and only they bear the hallmark of salinity) to pre-purify water of suspended solids before discharging it into the reservoir. The accumulated sediments are examined before the decision on the further treatment of the site is made. If the test results show that the collected sediments are of material value due to their carbon content, the deposits are selected and sold. Then the water is pumped out of the pools. The pools are filled with appropriate material due to changes in the legislation on the use of post-coal mine mineral rock material to fill unfavourably transformed areas. Material masses are required. The last stage is biological reclamation consisting of fertilizing the earth masses and then planting, bushing, or planting trees.

A separate issue is the water reservoirs formed in septic basins resulting from subsidence caused by mining exploitation. It is very often the case that an ecosystem was created in the resulting pool and in its surroundings, the violation of which would be a violation of nature. These pools remain intact, or cleaning works are carried out in their surroundings. Most often, these types of reservoirs are managed by the Angling Associations. Some pools were built in areas designated following the Local Spatial Development Plan for development other than water (Fig. 14).

Figure 14. Flooded areas require reclamation (the so-called "Montomet" Basin in Piekary Śląskie and Bytom) (SRK materials, 2019)

Prospective activities assume that corporate social responsibility should be carried out, e.g., due to the increasing pressure of society and local authorities in securing environmental problems for the areas where the exploitation and development of post-mining activities are carried out. These should be ways to continue a good, open dialogue with local communities.

Entrepreneurs declare that they approach all issues related to changes and transformations of the environment caused by economic activity very carefully and rationally, which is included in the rehabilitation programs carried out. Restructuring of Mines Company S.A. in Bytom, using both its funds from the sale and lease of property, and to a large extent from the co-financing from the National Fund for Environmental Protection and Water Management in Warsaw, fulfils its reclamation obligations imposed by administrative decisions. For the tasks to be financed from the National Fund for Environmental Protection and Water Management in Warsaw, the "Program for the implementation of projects in the field of environmental protection" was developed and approved by a Resolution of the Management Board of the SRK, which is being implemented successively.

Restoring functions to transformed regions is one of the most important and, simultaneously, more complex issues, mainly due to environmental, social, legal, technical, and economic aspects. Due to the necessity to operate, among others, extraction of mineral resources and their inevitable effects, it is also essential to carry out extensive activities and educational initiatives to familiarize stakeholders, local communities, contractors with this issue and the tasks carried out by entrepreneurs.

CONCLUSIONS

The presented practices for reclamation areas affected by direct or indirect industrial transformations include applications that effectively lead to diversified land development following applicable law.

As it seems, both the needs and the natural values of the existing objects are not considered in deciding on their future use. The analysis of the effects of hard coal mining and processing, and their impacts on the natural environment on a local and global scale, indicates that the latest scientific research results are used to a small extent in reclamation - restoration activities. There appeared a need to use the idea of natural capital and ecosystem services of functioning of novel ecosystems both in legal acts and in practice.

A deep understanding of the phenomena and processes is crucial for efficient limitation of global change and unfortunately is not included in the Polish and European legislation system.

The legislation system is not enforcing the practitioners and decision-makers to follow the dynamic adaptation processes in nature. Human activity should make use of nature-based solutions for better recondition of the environment.

REFERENCES

Blanco J., Dendoncker N., Barnaud C. & Sirami C. 2019: Ecosystem disservices matter: Towards their systematic integration within ecosystem service research and policy. Ecosystem Services, 36, 100913.

Brown B. & Spiegel S.J. 2019: Coal, climate justice, and the cultural politics of energy transition. Global Environmental Politics, 19 (2), 149–168.

Cox D.T.C., Hudson H.L., Plummer K.E., Siriwardena G.M., Anderson K., Hancock S., Devine-Wright P. & Gaston K.J. 2018: Covariation in urban birds providing cultural services or disservices and people. Journal of Applied Ecology, 55 (5), 2308–2319.

Doležalová H., Kajzar V., Souček K. & Staš L. 2009: Evaluation of mining subsidence using GPS data. Acta Geodynamica et Geomaterialia, 6 (155), 359–367.

Finkelman R.B., Wolfe A. & Hendryx M.S. 2021: The future environmental and health impacts of coal. Energy Geoscience, 2 (2), 99–112.

Giam X., Olden J.D. & Simberloff D. 2018: Impact of coal mining on stream biodiversity in the US and its regulatory implications. Nature Sustainability, 1 (4), 176–183.

Huang L., Krigsvoll G., Johansen F., Liu Y. & Zhang X. 2018: Carbon emission of global construction sector. Renewable and Sustainable Energy Reviews, 81 (2), 1906–1916.

Johnson R. & Bustin R.M. 2006: Coal dust dispersal around a marine coal terminal (1977–1999), British Columbia: The fate of coal dust in the marine environment. International Journal of Coal Geology, 68 (1–2), 57–69.

Kašovská K., Pierzchała Ł., Sierka E. & Stalmachová B. 2014: Impact of the salinity gradient on the mollusc fauna in flooded mine subsidences (Karvina, Czech Republic). Archives of Environmental Protection, 40 (1), 87–99.

Kuter N., Dilaver Z. & Gül E. 2014: Determination of suitable plant species for reclamation at an abandoned coal mine area. International Journal of Mining, Reclamation and Environment, 28 (5), 268–276.

Marschalko M., Třeslín L. & Fuka M. 2009: Possible impact of current and former mining activities on a research locality in the Ostrava-Karviná District. Górnictwo i Geoinżynieria, 33 (1), 429–442.

Marschalko M., Yilmaz I., Bednárik M. & Kubečka K. 2012: Deformation of slopes as a cause of underground mining activities: three case studies from Ostrava–Karviná coal field (Czech Republic). Environmental Monitoring and Assessment, 184, 6709–6733.

Orwell G. 1937: The Road to Wigan Pier. Part One, Chapter 2. Project Gutenberg of Australia, 18–31. https://libcom.org/files/wiganpier.pdf, access: 11.12.2021.

Pierzchała Ł. & Sierka E. 2020: Do submerged plants improve the water quality in mining subsidence reservoirs? Applied Ecology and Environmental Research, 18 (4), 5661–5672.

Ribeiro J., da Silva E.F., de Jesus A.P. & Flores D. 2011: Petrographic and geochemical characterization of coal waste piles from Douro Coalfield (NW Portugal). International Journal of Coal Geology, 87 (3–4), 226–236.

Rosa J.C.S., Morrison-Saunders A., Hughes M. & Sánchez L.E. 2020: Planning mine restoration through ecosystem services to enhance community engagement and deliver social benefits. Restoration Ecology, 28 (4), 937–946.

Sierka E., Stalmachová B., Molenda T., Chmura D. & Pierzchała Ł. 2012: Environmental and socio-economic importance of mining subsidence reservoirs. Ben – Technická Literatura, Praha.

Wang Y.-J., Zhang D.-C., Lian D.-J., Li Y.-F. & Wang X.-F. 2009: Environment cumulative effects of coal exploitation and its assessment. Procedia Earth and Planetary Science, 1 (1), 1072–1080.

Woźniak G., Sierka E. & Wheeler A. 2018: Urban and industrial habitats: how important they are for ecosystem services. [In:] Hufnagel L. (ed.): Ecosystem Services and Global Ecology. IntechOpen, 169–194.

Green Scenarios..., Dyczko, Jagodziński & Woźniak (eds.) © 2022 Taylor & Francis Group, London, UK

5.3. Values of Post-industrial Novel Ecosystems for Enhancement of Ecosystem Services in the Anthropocene Epoch

Agnieszka Błońska
Institute of Biology, Biotechnology and Environmental Protection,
Faculty of Natural Sciences, University of Silesia in Katowice;
28 Jagiellońska Str., 40-032 Katowice, Poland;
agnieszka.blonska@us.edu.pl; ORCID: 0000-0002-1696-0001

Barbara Bacler-Żbikowska
Department of Pharmaceutical Botany,
Faculty of Pharmaceutical Sciences in Sosnowiec, Medical University of Silesia in Katowice;
30 Ostrogórska Str., 41-200 Sosnowiec, Poland;
bbacler@sum.edu.pl; ORCID: 0000-0003-4139-6307

ABSTRACT: For the last two centuries, the natural environment has been severely altered or damaged. For a very long time, environmental protection has been part of the industrial strategy for restoration. Regardless of the frequent mention of ecological protection in official documents, environmental parameters are getting worse and worse – the global change crisis reflects the seriousness of the situation. The reason for that might be the focus exclusively on agriculture and forestry – to the exclusion of the enhancement of wildlife development and functioning of ecosystems best adapted to new human-created habitats. This enhancement has to be based on a scientific understanding of the dynamic changes in the functioning of novel adapted ecosystems. This approach is fundamental for urban areas and sites transformed by industry, particularly post-mineral excavations covering increasing areas of land at the global scale. At the same time, there are also very densely populated areas. Many studies show that biological, biogeochemical processes, serve as surprising solutions for re-establishing well-functioning ecosystems in new severe habitats.

Environmental management policies in the Anthropocene Epoch must include the importance of estimating the value of the post-excavation novel ecosystem habitats to enhance ecosystem services.

KEYWORDS: Post-mineral excavation sites, biodiversity enhancement, novel ecosystems functioning enhancement

INTRODUCTION

Anthropocene is the name of a new geological epoch initiated by human activity. The term was proposed by Paul Crutzen (the Nobel Prize winner in chemistry). At one point, human activity started to cause dangerous changes in the environment by introducing, e.g., hazardous chemicals, causing, among others, an ozone hole and climate change. Modifications in land management, deforestation, and fossil fuel extraction contribute to irreversible changes in geology and landscape on the Earth (Crutzen 2006). Most scientists consider the 18th Century industrial revolution to be the beginning of this epoch. An apparent acceleration occurred in the middle of the 20th Century when atomic energy was invented and first used. The artificial radioactive signal from the nuclear bomb explosion of 16 July 1945 was traceable in the environment (Zalasiewicz *et al.* 2015).

A steady increase in human economic activity has been recorded, which has an increasing impact on the environment and natural resources (Vitousek *et al.* 1997; Foley *et al.* 2005). Destroyed, simplified, fragmented ecosystems, in turn, are becoming more and more common around the world, covering a growing part of the Earth (Ellis *et al.* 2010). As a result, the functioning of natural ecosystems has been affected and is being transformed. Part of the altered ecosystems or established newly created habitats have become new ecosystems (novel ecosystems; Morse *et al.* 2014). Human activities, agriculture, urbanization and industry have a significant impact on many ecosystems. The era of the "new man" entails negative and often irreversible changes in nature. One of the most extreme manifestations of these changes is the loss of biodiversity associated with the massive extinction of species called the sixth catastrophe or sixth great extinction (Barnosky *et al.* 2011). In addition, human changes on the Earth's surface through rapid urbanization, are transforming the environment for agriculture and industry (Cardinale *et al.* 2012). The challenge is to support as much microhabitat diversity as possible to provide habitat conditions to establish biodiversity. Many post-mineral excavation poor oligotrophic sites offer refuge habitats for endangered species. Many of the species become endangered because of agricultural fertilizer, herbicide, and pesticide use.

In the 18th Century, industrialization, which began in Great Britain, spread around the world. Today, people live on a rapidly urbanizing and industrialized planet (McPhearson *et al.* 2016). At the current rate of development, the global population of cities will increase from about four billion, the level in the second decade of the 21st Century, to more than six billion by 2045. Urban and industrial landscapes are primarily associated with the everyday presence of humans in the environment. Unfortunately, there has been an increasing lack of space for wildlife (Haase 2014).

THE IDENTIFICATION OF *DE NOVO* HUMAN-CREATED HABITAT CONDITIONS

There is a wide range and immense diversity of living organisms and spontaneous biological phenomena in natural and semi-natural ecosystems, determining ecosystem processes and functioning (Irvine *et al.* 2013). The dynamics of human economic activity caused the transformation or diminishing of natural and semi-natural ecosystems and the emergence of new habitats. These sites are the direct result of deliberate or unintentional human action, which has led to the crossing of an ecological threshold. The unknown previous conditions facilitate the new trajectory of vegetation and ecosystem development and inhibit the return to the old system regardless of the presence or lack of additional human intervention. The new ecosystem is a unique set of *de novo* human-created habitat conditions, attracting colonization and establishment of the best-adapted flora, fauna, and many related above and below ground organisms. The resulting novel ecosystem is self-sufficient regarding its species composition, structure and biogeochemistry, and frequently can provide efficient ecosystem services (Woźniak *et al.* 2018). A characteristic feature of the new ecosystem is the change in species composition and relationships among species (e.g., plants and bacteria or fungi) compared with ecosystems recorded before the ecological threshold has been crossed (Morse *et al.* 2014).

The spontaneous processes in novel ecosystem habitats, such as post-mineral excavation pits and heaps connected with the natural mineral exploitation side-effect sites, have been studied for some time. Unexpectedly, such sites provide high microhabitat variety supporting the spontaneous colonization of different organisms, resulting in increased diversity of primary producers, the photosynthesizing organisms, mainly plant species. Biodiversity is an essential criterion for success of projects in restoring habitats in post-industrial, including post-mining areas (Fig. 1–4) (Cardinale *et al.* 2012; Morse *et al.* 2014; McPhearson *et al.* 2016; Woźniak *et al.* 2018).

Figure 1. *Epipactis palustris* often occurs in anthropogenic habitats (Photo: Agnieszka Błońska)

Figure 2. *Equisetum variegatum* – community in the sandpit in Dąbrowa Górnicza (Southern Poland) (Photo: Agnieszka Błońska)

Figure 3. *Liparis loeselii* in the Kuźnia Warężyńska sandpit (Photo: Agnieszka Błońska)

Figure 4. *Lycopodiella inundata* and *Drosera rotundifolia* in the sandpit (Photo: Agnieszka Błońska)

The UK Government, e.g., has identified ecosystem services in urban-industrial landscape enhancement as a critical tool for regional economic regeneration, habitat restoration, and the implementation of international commitments on biodiversity (Doick *et al.* 2009; Pediaditi *et al.* 2010). Urban and post-industrial areas, including post-mining sites that have been left behind and have become overgrown by spontaneous succession, are often valuable biodiversity and wildlife centres. They can compose a part of developing blue-green infrastructure for sustainable cities. Many of them form mosaics of various substrates on which different plant communities grow, creating an ecosystem together with the related animals (micro-, meso- and macrofauna) and microorganisms (Bonthoux *et al.* 2014; Woźniak *et al.* 2015). Post-industrial, particularly the post-mineral excavation areas, can be challenging environments for many reasons: toxicity, high daytime temperatures (in sheltered areas in direct sunlight), weak or non-existent soils, insufficient or excessive drainage due to compaction caused by heavy equipment or, in contrast, a complex, large pore substrate unable to hold water (Frouz *et al.* 2008; Woźniak 2010; Frouz 2014; Woźniak *et al.* 2015; Wiesmeier *et al.* 2019).

These substrates often consist of different mineral material types excavated from deep down. The soil-forming process on such habitats can take up to decades. This process involves living organisms, mainly plants, as a source of organic material, leading to a slow accumulation of organic matter. Such places do not have a seed bank and depend on plant species from other areas or brought in by animals (Frouz *et al.* 2008; Woźniak 2010; Woźniak *et al.* 2015). Apart from the environmental transformation conducted for 200 years, some surprisingly efficient self-naturalized processes have occurred. The industrial development of mining in Wigan, UK, is an example of how dynamic economic growth took place at the natural environment's expense. Due to spontaneous biological processes, the watercourse in the Wigan previous mining site changed into a system of ponds. In the small wetland areas, coastal vegetation has developed. According to natural hydrological processes, the current network of ditches and wetlands in the Wigan mining area will probably transform into streams with meanders and riverbeds. In some parts, the prolonged water flow has contributed to the colonization of these water bodies by rare animal species associated with humid habitats. In the UK, a particular program supports the Natural Process, which will bring several benefits in natural capital protection and contribute to better ecosystem services provision (Anderson & France 1994; Davies 2010; Rands *et al.* 2010).

POST-INDUSTRIAL NOVEL ECOSYSTEMS AS A SOURCE OF ECOSYSTEM SERVICES

Ecosystem services are typically classified into four categories (MEA 2005; Cowling *et al.* 2008; TEEB 2011). There are many classifications, but in natural and seminatural ecosystems, there are generally four main ecosystems services listed:

1. supply services – including starting material from ecosystems, for example, food, water, medicinal plants, and other resources;
2. regulatory services – maintain functions such as air and soil quality, flood prevention, and control of rainwater and disease;
3. habitat or ancillary services – provide a basis for other services, supporting living space for organisms and maintaining the diversity of plants and animals;
4. cultural services – including intangible, socioeconomic benefits (such as psychological and cognitive benefits) that people gain from contact with the environment, such as recreation, aesthetic, spiritual, psychological services, and tourism.

Wigan Flashes in the UK, a previous mining area, mainly supports natural, spontaneous processes to improve ecosystem functioning, including increasing biodiversity, carbon sequestration, and water retention. Apart from those essential services, the old mining area provides services related to cultural ecosystems.

A significant increase in biodiversity and improved retention in a previous coal mining area is highly valued and couldn't be provided due to anything other than the natural complex process. Nowadays, more and more people use this place to watch birds and walk. Thus, the site provides vital regulatory ecosystem services (flood mitigation) and recreational and cultural services.

Nature, including wildlife, has spontaneously developed in the vast majority of areas previously considered to be degraded due to industrial activity (Cohn *et al.* 2001). Thus, supporting processes that spontaneously operate at restoring and providing ecosystem services after almost a century have dominated the post-industrial landscape, now providing the essential regulation of ecosystem services on the urban-industrial landscape. However, the cultural ecosystem services are much more easily recognized by people.

Many organisms, including bees and other insects, are associated with spontaneously developing vegetation in open poor oligotrophic habitats. In the UK, the protection of such developing new systems exists in the British National Pollination Strategy (Defra 2018). The richness of plant species, and thus the diversity of flora, in the open, primarily oligotrophic post-industrial habitats, can help promote habitats for pollinators and key invertebrate species. Besides, if the post-mining, post-mineral excavation habitats are within urban boundaries, they may provide refuge habitats in urban areas (Buglife 2009; 2019). About 30 butterfly species are associated with post-industrial areas and protected in these habitats in the UK (Butterfly Conservation 2019). In particular, butterflies prefer open, diverse post-industrial habitats and can naturally colonize the highly dispersed fragments of the mosaic of vegetation remaining in the initial stages. Post-mineral excavation areas should also be managed for pollinators, as these post-industrial areas provide safe shelter from pesticides (Defra 2018).

To prevent the disappearance of rare vegetation and accompanying butterflies, it is sometimes necessary to introduce occasional disturbances and remove bushes (Butterfly Conservation 2019). In addition, ecosystems developing in post-mining areas, and located near urban settlements, often become places of recreation, walks, green gyms, wild playgrounds (bird watching), and some of them are also attractive viewpoints (Millward & Mostyn 1989; Icarus 2014). In the UK, city parks were established mainly in the 1970s, and they are not statutorily protected. These are primarily open public space areas on the edge of urban agglomerations with amenities such as toilets, a café, and parking. In the majority of cases, these were post-industrial areas, hence their proximity to urban areas. The largest park in Europe has a similar history – the Silesian Park, established in the 1970s and 1980s in the Upper Silesian Industrial District (Bacler-Żbikowska *et al.* 2015).

In the 1960s, many post-industrial, mostly post-coal-mine sites in the West Midlands, were identified as Sites of Conservation Interest (SINCs). In the Black Country and West Midlands, some post-industrial, particularly post-coal-mine sites, were recognized as high ecological and environmental value sites, due to the different species composition which has colonized those sites and the resulting ecological processes which have developed (Box *et al.* 1994). Box & Cossons (1988) found that post-industrial sites are almost exclusively home to rare species such as the *Lycopodiaceae.* They grow on the acidic clay substrate of an abandoned Telford opencast coal mine (Box & Cossons 1988). Such an example from Poland is the inactive quarry in Poland's most industrially transformed region – the Silesian Upland, which became the first Polish habitat for the bee orchid *Ophrys apifera* (Fig. 5; Osiadacz & Kręciała 2014).

In his book under the title "Review of England's wildlife sites in England", Sir John Lawton appealed to the public and the authorities to create a space for nature inside the urban-industrial landscape. The report suggests a targeted approach to conservation efforts to improve the condition and development of existing open areas, by connecting neighbouring sites to strengthen and expand spaces available to wildlife, for human well-being (Lawton *et al.* 2010).

Figure 5. *Ophrys apifera* on its first location in Poland near Imielin (Photo: Barbara Bacler-Żbikowska)

NATURAL ECOSYSTEM PROCESSES AND THE DYNAMIC ECONOMIC DEVELOPMENT

Long-term studies conducted in the Silesia industrial region (Poland) found that the first colonizers were algae. Apart from annual plant species, long-lived herbaceous plants become the first colonizers during spontaneous succession on the hard coal mine heaps. Some plant species will tolerate extraordinarily high or low pH levels or high salinity (Woźniak 2010). The natural processes often take a long time, creating a sustained system that does not need any additional investment or work load. Unfortunately, economically focused people looking for defined benefits do not want to dedicate time and wait for the solutions that nature can develop. The dynamic economic, so-called development strategy would instead cause additional harm to the environment and developing ecosystem functioning, rather than take enough time to understand the natural adaptation processes and supporting them when necessary.

These post-industrial places, both in the UK and Poland, can vary significantly in retaining water. Local water accumulations often take place in microhabitats, where the soil was heavily compacted. Interesting groups of aquatic plant species can quickly colonize the open water-filled hollows. In local depressions of various sizes and depths formed in post-mining areas, amphibians and aquatic invertebrates appear together with wetland vegetation. Observations conducted during the study in Poland and the Czech Republic were similar (Sierka *et al.* 2009; Woźniak 2010). The animals related to the particular wetland vegetation types significantly increase the values of biodiversity indicators. In the UK, where plants have naturally colonized and created communities, post-mining areas are being examined to create the conditions for their further development (Lunn 2000).

The valuable wetland vegetation and ecosystems are crucial elements of the natural values that spontaneously emerged, e.g., in the post-coal mine heaps in the UK's Wigan region and the Silesia Region in Poland. Apart from the wetland vegetation and ecosystems, *Danthonia decumbens* grasslands are a particular type of heathland vegetation in Great Britain. *D. decumbens* is not the grass that would be expected in the Wigan old mining area. However, a unique heritage of industrial activity can lead to new, unexpected habitats and associated plant communities of Novel ecosystems (Champion 2018; James 2018).

The slow initial colonization of post-mining areas in the UK often leads to many diverse assembled vegetation communities. This type of open mosaic habitat on previously managed land is a conservation priority. The habitat mosaic of open vegetation patches is protected by the UK Biodiversity Action Plan (BAP) and is a legally protected habitat (Rodwell 2000).

NOVEL ECOSYSTEMS ENHANCE ECOSYSTEMS SERVICES

Reversing the management policy based on dehumidification, to facilitate construction and move towards restoring functionality of aquatic and wetland ecosystems and creating a larger wetland area, slowing down the flow and retaining water to mitigate flooding, indicates how thinking has changed. In addition, the re-use of post-industrial and post-mining regions has changed in order to strengthen the significant role of environmental management in achieving sustainable water resources development. Canals and concrete embarkments are strictly avoided. Unfortunately, in Poland hydrotechnical canalization and regulation of water courses, is wrongly considered as the best solution for flooding. It is known worldwide that such an approach is the worst when water retention is the target.

The unexpected diversity (within and between sites) of the spontaneously developed vegetation on the heaps of deposited rock in Wigan and other European mining Regions, provides proof of the rightness of the distinction of Novel Ecosystems as a new system of biotic and abiotic elements adapted to habitats under significant human pressure in the Anthropocene. Preserving the diversity of vegetation and ecosystems in post-mining areas is essential for urban planners, as it helps international commitments to maintain and enhance biodiversity (Buglife 2009; 2019). There is a level of protection category lower than the Local Nature Reserve (LRP) and Country Park (CP) in the UK. Many places covered by this form of protection are post-industrial areas (particularly post-coal-mining).

The landscape approach would protect the ecosystem and environmental processes throughout the country. Therefore, the current national planning policy framework requires the authorities of different parts of the country to ensure efficient land use and create the necessary conditions to ensure sustainable use of previously converted land. The solutions explained above are an appropriate approach, particularly true for supporting wetland and aquatic habitats (Andersson *et al.* 2015). The proper recognition of the environmental potential of available land is a prerequisite to identify the most supportive management system for ecosystem functioning. The applied management depends on the characteristics of the site. Such an approach may include natural and recreational purposes and provide an opportunity for nature conservation. Enabling the preservation of the identified environmental value of land previously transformed by humans (industry or agriculture) should be considered. Thus, closed mine sites can constitute additional open spaces valuable to ecosystem recovery in the urban and industrial landscape (Bonthoux *et al.* 2014). Facilitating the restoration and deployment of open ecosystem services makes them available for spontaneous processes that help create strategic habitats in the face of challenges posed by global change. Increasing the area available for wildlife and strengthening ecological networks, primarily if they are close to existing natural sites, is the main point for environmental management to mitigate global change. The economy and short-term financial benefits cannot be the only prerequisite for decisions on the management of an area (Buglife 2019).

Understanding and perceiving the functioning of novel ecosystems must be the key to the successful transformation required for urban-industrial areas, to make them climate-resilient (Royal Society 2014). In urban-industrial areas, such ways of developing green infrastructure are increasingly leading to development of blue-green infrastructure to meet the challenges of climate change. Benefits resulting from the existence and development of blue-green infrastructure include: biodiversity increase, air temperature cooling, rainwater absorption, protection of drinking water supply (retention) and human health improvement (Elmqvist *et al.* 2013).

It is, therefore, necessary to use all available urban, industrial and post-industrial areas to develop blue-green infrastructure. These benefits are the urban-industrial ecosystem services that residents and visitors derive from local and regional ecosystems (Gómez-Baggethun *et al.* 2013; Larondelle *et al.* 2014). These benefits are not just products of the functioning of ecosystems, but rather the development of people and ecosystems combined in the complex social and environmental systems of industrial cities, in which the choices made to meet these challenges have an impact on the social, economic, and above all ecological situation (Andersson *et al.* 2015; McPhearson *et al.* 2016). The system of designing and managing urban ecosystems to ensure a resilient and equitable supply of ecosystem services requires local expertise (Faehnle *et al.* 2015). For some projects, the contributions from engineers, architects, and designers might be required (Ahern *et al.* 2014) because the supply of urban ecosystem services must be compatible with those who will use them locally and regionally (Haase *et al.* 2014; McPhearson *et al.* 2015). There is a positive feedback loop in this system, usually a negative feedback loop: urbanization and industrialization cause change. Hence, the decline in species diversity and, consequently, the ecosystem functioning efficiency and the availability of ecosystem services. However, "they are inextricably linked" (MEA 2005).

REFERENCES

Ahern J., Cilliers S. & Niemelä J. 2014: The concept of ecosystem services in adaptive urban planning and design: A framework for supporting innovation. Landscape and Urban Planning, 125, 254–259.

Anderson D. & France A.A. 1994: Wigan Coal and Iron. Smiths of Wigan, Smiths Books, Wigan.

Andersson M., McPhearson T., Kremer P., Gomez-Baggethun E., Haase D., Tuvendal M. & Wurster D. 2015: Scale and context dependence of ecosystem service providing units. Ecosystem Services, 12, 157–164.

Bacler-Żbikowska B., Grzędzicka E., Kowalik K., Poraj A., Drozd Ł., Frądczak M., Golasz K., Stodółka A. & Żbikowski J. 2015: Przyroda Parku Śląskiego. Fundacja Park Śląski, Chorzów.

Barnosky A.D., Matzke N., Tomiya S., Wogan G.O.U., Swartz B., Quental T.B., Marshall C., McGuire J.L., Lindsey E.L., Maguire K.C., Mersey B. & Ferrer E.A. 2011: Has the Earth's sixth mass extinction already arrived? Nature, 471 (7336), 51–57.

Bonthoux S., Brun M., Di Pietro F., Greulich S. & Bouché-Pillon S. 2014: How can wastelands promote biodiversity in cities? A review. Landscape and Urban Planning, 132, 79–88.

Box J., Douse A. & Kohler T. 1994: Non-statutory sites of importance for nature conservation in the West Midlands. Journal of Environmental Planning and Management, 37 (3), 361–367.

Box J.D. & Cossons V. 1988: Three species of clubmoss (Lycopodiaceae) at a lowland station in Shropshire. Watsonia, 17, 69–71.

Buglife 2009: Planning for Brownfield Biodiversity: A best practice guide. Buglife – The Invertebrate Conservation Trust, Peterborough.

Buglife 2019: Brownfield sites under threat. https://www.buglife.org.uk/resources/habitat-hub/brownfield-hub/brownfields-sites-under-threat, access: 27.05.2021.

Butterfly Conservation 2019: Brownfields for Butterflies. Available from: https://butterfly-conservation.org/sites/default/files/habitat-brownfield-leaflet.pdf, access: 27.05.2021.

Cardinale B.J., Duffy J.E., Gonzalez A., Hooper D.U., Perrings C., Venail P., Narwani A., Mace G.M., Tilman D., Wardle D.A., Kinzig A.P., Daily G.C., Loreau M., Grace J.B., Larigauderie A., Srivastava S.D. & Naeem S. 2012: Biodiversity loss and its impact on humanity. Nature, 486 (7401), 59–67.

Champion M. 2018: Biodiversity impacts at various scales following conservation management of synanthropic sites in North West England. PhD thesis, Edge Hill University.

Cohn E.V.J., Rostański A., Tokarska-Guzik B., Trueman I.C. & Woźniak G. 2001: The flora and vegetation of an old solvay process tip in Jaworzno (Upper Silesia, Poland). Acta Societatis Botanicorum Poloniae, 70 (1), 47–60.

Cowling R.M., Egoh B., Knight A.T., O'Farrell P.J., Reyers B., Rouget M., Roux D.J., Welz A. & Wilhelm-Rechman A. 2008: An operational model for mainstreaming ecosystem services for implementation. Proceedings of the National Academy of Sciences of the United States of America, 105 (28), 9483–9488.

Crutzen P.J. 2006: The "Anthropocene". [In:] Ehlers E. & Krafft T. (eds.): Earth System Science in the Anthropocene. Springer, Berlin, Heidelberg, 13–18.

Davies A. 2010: Coal Mining in Lancashire & Cheshire. Amberly Publishing, Stroud.

Defra 2018: National Pollinator Strategy: Implementation Plan, 2018–2021. Department for Environment, Food and Rural Affairs. https://assets.publishing.service.gov.uk/government/uploads/system/uploads/attachment_data/file/912826/nps-implementation-plan-2018-2021.pdf, access: 28.05.2021.

Doick K.J., Sellers G., Castan-Broto V. & Silverthorne T. 2009: Understanding success in the context of brownfield greening projects: The requirement for outcome evaluation of success assessment. Urban Forestry & Urban Greening, 8 (3), 163–178.

Ellis E.C., Goldewijk K.K., Siebert S., Lightman D. & Ramankutty N. 2010: Anthropogenic transformation of the biomes, 1700 to 2000. Global Ecology and Biogeography, 19 (5), 589–606.
Elmqvist T., Fragkias M., Goodness J., Güneralp B., Marcotullio P.J., McDonald R.I., Parnell S., Schewenius M., Sendstad M., Seto K.C. & Wilkinson C. 2013: Urbanization, Biodiversity and Ecosystem Services: Challenges and Opportunities. A Global Assessment. Springer, New York.
Faehnle M., Söderman T., Schulman H. & Lehvävirta S. 2015: Scale-sensitive integration of ecosystem services in urban planning. GeoJournal, 80, 411–425.
Foley J.A., DeFries R., Asner G.P., Barford C., Bonan G., Carpenter S.R., Chapin F.S., Coe M.T., Daily G.C., Gibbs H.K., Helkowski J.H., Holloway T., Howard E.A., Kucharik C.J., Monfreda C., Patz J.A., Prentice I.C., Ramankutty N. & Snyder P.K. 2005: Global consequences of land use. Science, 309 (5734), 570–574.
Frouz J. (ed.) 2014: Soil Biota and Ecosystem Development in Post Mining Sites. CRC Press, Taylor & Francis Group, Boca Raton.
Frouz J., Prach K., Pižl V., Háněl L., Starý J., Tajovský K., Materna J., Balík V., Kalčík J. & Řehounková K. 2008: Interactions between soil development, vegetation and soil fauna during spontaneous succession in post mining sites. European Journal of Soil Biology, 44 (1), 109–121.
Gómez-Baggethun E., Gren Å., Barton D.N., Langemeyer J., McPhearson T., O'Farrell P., Andersson E., Hamstead Z. & Kremer P. 2013: Urban ecosystem services. [In:] Elmqvist T., Fragkias M., Goodness J., Güneralp B., Marcotullio P.J., McDonald R.I., Parnell S., Schewenius M., Sendstad M., Seto K.C. & Wilkinson C. (eds.): Urbanization, Biodiversity and Ecosystem Services: Challenges and Opportunities. A Global Assessment. Springer, New York, 175–251.
Haase D. 2014: The nature of urban land use and why it is a special case. [In:] Seto K.C. & Reenberg A. (eds.): Rethinking Global Land Use in an Urban Era. Strüngmann Forum Reports. Volume 14. MIT Press, Cambridge.
Haase D., Larondelle N., Andersson E., Artmann M., Borgström S., Breuste J., Gomez-Baggethun E., Gren Å., Hamstead Z., Hansen R., Kabisch N., Kremer P., Langemeyer J., Rall E.L., McPhearson T., Pauleit S., Qureshi S., Schwarz N., Voigt A., Wurster D. & Elmqvist T. 2014: A quantitative review of urban ecosystem service assessments: concepts, models, and implementation. Ambio, 43, 413–433.
Icarus 2014: Access to Nature: inspiring people to engage with their natural environment. Final Evaluation Report. Part 2. Natural England.
Irvine K.N., Warber S.L., Devine-Wright P. & Gaston K.J. 2013: Understanding urban green space as a health resource: a qualitative comparison of visit motivation and derived effects among park users in Sheffield, UK. International Journal of Environmental Research and Public Health, 10 (1), 417–442.
James P. 2018: The Biology of Urban Environments. Oxford University Press, Oxford.
Larondelle N., Hamstead Z.A., Kremer P., Haase D. & McPhearson T. 2014: Applying a novel urban structure classification to compare the relationships of urban structure and surface temperature in Berlin and New York City. Applied Geography, 53, 427–437.
Lawton J.H., Brotherton P.N.M., Brown V.K., Elphick C., Fitter A.H., Forshaw J., Haddow R.W., Hilborne S., Leafe R.N., Mace G.M., Southgate M.P., Sutherland W.J., Tew T.E., Varley J. & Wynne G.R. 2010: Making Space for Nature: a review of England's wildlife sites and ecological network. Report to Department for Environment, Food and Rural Affairs (Defra).
Lunn J. 2000: Ecological and nature conservation aspects of land affected by mining in the Yorkshire Coalfield. Unpublished Master's thesis, Sheffield Hallam University.
McPhearson T., Andersson E., Elmqvist T. & Frantzeskaki N. 2015: Resilience of and through urban ecosystem services. Ecosystem Services, 12, 152–156.
McPhearson T., Pickett S.T.A., Grimm N.B., Niemelä J., Alberti M., Elmqvist T., Weber C., Hasse D., Breuste J. & Qureshi S. 2016: Advancing urban ecology toward a science of cities. BioScience, 66 (3), 198–212.
MEA – Millennium Ecosystem Assessment. 2005: Ecosystems and Human Well-being: Biodiversity Synthesis. World Resources Institute, Washington DC.
Millward A. & Mostyn B. 1989: People and nature in cities: the social aspects of planning and managing natural parks in urban areas. Urban Wildlife Now 2. Nature Conservancy Council, Peterborough.
Morse N.B., Pellissier P.A., Cianciola E.N., Brereton R.L., Sullivan M.M., Shonka N.K., Wheeler T.B. & Mc Dowell W.H. 2014: Novel ecosystems in the Anthropocene: a revision of the novel ecosystem concept for pragmatic applications. Ecology and Society, 19 (2), 12.
Osiadacz B. & Kręciała M. 2014: *Ophrys apifera* Huds. (Orchidaceae), a new orchid species to the flora of Poland. Biodiversity: Research and Conservation, 36 (1), 11–16.

Pediaditi K., Doick K.J. & Moffat A.J. 2010: Monitoring and evaluation practice for brownfield, regeneration to greenspace initiatives. A meta-evaluation of assessment and monitoring tools. Landscape and Urban Planning, 97 (1), 22–36.
Rands M.R.W., Adams W.M., Bennun L., Butchart S.H.M., Clements A., Coomes D., Entwistle A., Hodge I., Kapos V., Scharlemann J.P.W., Sutherland W.J. & Vira B. 2010: Biodiversity conservation: challenges beyond 2010. Science, 329 (5997), 1298–1302.
Rodwell J.S. (ed.) 2000: British Plant Communities. Volume 5: Maritime Communities and Vegetation of Open Habitats. Cambridge University Press, Cambridge.
Royal Society 2014: Resilience to Extreme Weather. Royal Society Science Policy Centre, Report DES3400. Report 02/14, Royal Society, London.
Sierka E., Molenda T. & Chmura D. 2009: Environmental repercussion of subsidence reservoirs reclamation. Journal of Water and Land Development, 13a (10), 41–52.
TEEB 2011: Manual for Cities: Ecosystem Services in Urban Management. http://www.teebweb.org/publication/teeb-manual-for-cities-ecosystem-services-in-urban-management, access: 28.05.2021.
Vitousek P.M., Mooney H.A., Lubchenco J. & Melillo J.M. 1997: Human domination of Earth's ecosystems. Science, 277 (5325), 494–499.
Wiesmeier M., Urbanski L., Hobley E., Lang B., von Lützow M., Marin-Spiotta E., van Wesemael B., Rabot E., Ließ M., Garcia-Franco N., Wollschläger U., Vogel H.-J. & Kögel-Knabner I. 2019: Soil organic carbon storage as a key function of soils – A review of drivers and indicators at various scales. Geoderma, 333 (1), 149–162.
Woźniak G. 2010: Zróżnicowanie roślinności na zwałach pogórniczych Górnego Śląska [Diversity of vegetation on coal-mine heaps of the Upper Silesia (Poland)]. W. Szafer Institute of Botany, Polish Academy of Sciences, Kraków.
Woźniak G., Markowicz A., Borymski S., Piotrowska-Seget Z., Chmura D. & Besenyei L. 2015: The relationship between successional vascular plant assemblages and associated microbial communities on coal mine spoil heaps. Community Ecology, 16 (1), 23–32.
Woźniak G., Sierka E. & Wheeler A. 2018: Urban and industrial habitats: how important they are for ecosystem services. [In:] Hufnagel L. (ed.): Ecosystem Services and Global Ecology. IntechOpen, 169–194.
Zalasiewicz J., Waters C.N., Williams M., Barnosky A.D., Cearreta A., Crutzen P., Ellis E., Ellis M.A., Fairchild I.J., Grinevald J., Haff P.K., Hajdas I., Leinfelder R., McNeill J., Odada E.O., Poirier C., Richter D., Steffen W., Summerhayes C., Syvitski J.P.M., Vidas D., Wagreich M., Wing S.L., Wolfe A.P., Zhisheng A. & Oreskes N. 2015: When did the Anthropocene begin? A mid-twentieth century boundary level is stratigraphically optimal. Quaternary International, 383, 196–203.

5.4. Educational Activities in the Areas Created as a Result of Mining Operations in the Silesian Province

Agnieszka Hutniczak
Institute of Biology, Biotechnology and Environmental Protection,
Faculty of Natural Sciences, University of Silesia in Katowice;
28 Jagiellońska Str., 40-032 Katowice, Poland;
agnieszka.hutniczak@us.edu.pl; ORCID: 0000-0002-6235-6139

Teresa Nowak
Institute of Biology, Biotechnology and Environmental Protection,
Faculty of Natural Sciences, University of Silesia in Katowice;
28 Jagiellońska Str., 40-032 Katowice, Poland;
teresa.nowak@us.edu.pl; ORCID: 0000-0001-7519-6309

Agnieszka Błońska
Institute of Biology, Biotechnology and Environmental Protection,
Faculty of Natural Sciences, University of Silesia in Katowice;
28 Jagiellońska Str., 40-032 Katowice, Poland;
agnieszka.blonska@us.edu.pl; ORCID: 0000-0002-1696-0001

ABSTRACT: Education is one of the most crucial areas of action in societies. Well-educated members of a community can make up their minds and make decisions based on the newest knowledge. There are some areas of knowledge that are developing particularly intensively. For education, it is essential to follow and adopt the development of scientific knowledge.

Environmental science is a complex interdisciplinary, intensively developing field of knowledge. Many feedback loops link the biotic and abiotic elements. The science is still ongoing to recognize and understand the complexity of the relationships between the biogeochemical parts of the environment and ecosystems. The so-called environmental protection activities used for a long time have not brought any improvements and human activities have caused the crisis of global changes. The reason for the failure of environmental protection activities might be an exclusive focus on agricultural and engineering approaches. The biological prerequisites of ecosystem functioning re-establishment have been neglected for decades.

In terms of the global change challenges, all efforts should focus on re-establishing natural capital, particularly the processes enhancing ecosystem functioning. Ecosystem functioning is fundamental for ecosystem services, which in turn determine human survival and life. Bearing this in mind, the educational system at all stages has to include the concept of biogeochemical feedback loops between the related processes that are still in dynamic balance with changing habitat conditions. On top of this, the adaptation processes are intense in sites modified, transformed, and created by humans (novel ecosystems) – the concept of self-renewing natural capital. All these concepts mentioned above have to be included in all levels of education.

Based on a query of the available educational and didactic materials, this chapter presents proposals for the use of selected didactic methods (both teaching and activating) during the organization of field workshops. Thanks to this, the educational goals will be fully achieved. These methods have been adapted to the individual target groups, the age of the participants, as well as the specific thematic scope. This work also contains a detailed catalog of proposed topics of nature workshops within thematic blocks, taking into account target groups of recipients.

KEYWORDS: Following the newest scientific achievements, post-mining and post-industrial areas

INTRODUCTION

Educational activities related to the natural environment are still the subject of current didactic studies (e.g., Łobocki 2000; Żytko 2002; Brown 2005; Budniak 2005; Będkowska 2006; Stawiński 2006; Danks & Schofield 2008; Bąk *et al.* 2014; Hutniczak 2014; Korwin-Szymanowska *et al.* 2015; Hopkins 2019). The extensive didactic material concerns the individual age groups of children and adolescents (e.g., Elbanowska 1983; 1994; Way 1997; Folwaczny 1999; Michalak 2004; Bouchardy 2008; Gąsecki 2008).

To realize the idea of "Green Scenarios" for the mining industry responding to environmental challenges of the Anthropocene epoch, knowledge and education are the fundamentals. Education activities should take into account issues that are the focus of key pro-environmental activities/scenarios. The contents of the presented article suggest taking up the following general issues using various teaching methods:

- biological diversity of areas resulting from mining exploitation;
- the potential of areas created as a result of exploitation of carbon storage in the form of biomass and soil organic matter (SOM);
- the potential of areas created as a result of exploitation to provide ecosystem services for the surrounding population and improve the environment on a local scale.

The project of exemplary educational activities envisages the use of many educational methods adapted to the transmitted content and target/age groups. It is very important that the workshops are conducted by qualified specialists.

The implementation of the proposed educational campaign will bring society closer to nature. In today's world, the deficit contact syndrome is becoming a challenge and calling for humans to return to nature. Direct human contact with nature is also possible in extreme habitats (heaps, brownfields, post-industrial areas) – in habitats where, contrary to belief, there are favourable physicochemical conditions for plant life.

Undoubtedly, these areas are interesting places in terms of nature, and their presence in the landscape gives the possibility of direct human contact with nature.

MATERIALS AND METHODS

The proposed educational activities are aimed at:

- raising the qualifications, awareness and commitment of employees for sustainable development;
- support and participation in promoting the idea of sustainable development in local and national social initiatives;
- developing a system of cooperation with local communities located within industrial and operational areas.

The basis for the creation of the presented catalog of classes was reading scientific studies allowing for the optimal use of methods and forms of work with selected target groups, taking into account different aspects of the natural environment. The following studies were used in the preparation of the review of selected didactic methods and types of activities: Way (1997), Okoń (1998), Gałązka (1999), Łobocki (2000), Stawiński (2006), Kaczmarzyk & Kopeć (2007), Małek (2007), Grabowski (2011), Halchynska & Navolska (2020), Ministry of Science and Higher Education (2021).

The project is supposed to include the following target groups:

- pre-schoolers (5–6 years old);
- pupils in grades I–III of primary school;
- pupils in grades IV–VIII of primary school;
- high school students (I–III);
- students of the University of the Third Age;
- people with intellectual disabilities (mild, moderate disability);
- specialists (e.g., employees of the Environmental Protection Department of City Hall, sanitary-epidemiological station, mine employees).

The collected material was used to develop an educational project, details of which are presented in tables 1–4 of this chapter. The presented catalog of classes can be used (slightly modifying the content (topics and issues)) for the implementation of field workshops in the areas created as a result of mining operations in the Silesian Province, but also in other post-industrial areas of Poland. The source materials used are quoted in the presented catalog of the proposed educational activities (Tab. 1, 3–4).

THE PROPOSED EDUCATIONAL CAMPAIGN – THE PROJECT OF EXEMPLARY EDUCATIONAL ACTIVITIES

As it results from the conducted review, the most effective didactic methods in the perception of the natural environment relate to the impact on all senses, through direct contact. Therefore, the following teaching methods are proposed for use during field workshops:

- Brainstorm (Osborne's method, brainstorming, ideas factory) – giving different associations or imaginative solutions. It is a way of solving problems. It is worth remembering that every idea is good, you should not criticize or evaluate ideas. Please give as many answers as possible.
- Drama – a technique that has no previously prepared script or cast. Its main essence is to empathize with participants in a given situation ("entering the role").
- Decision tree – a graphical record of a decision tree in which the order of levels that are completed is important, and so: the problem is recorded first, the second – solutions. At the third level, the positive and negative effects of individual solutions are recorded. Goals and values or the decision taken are at the top of the tree.
- Discussion – belongs to relatively difficult methods. It is a way of solving problems, expressing your own opinion and syntactic expression. Participants learn time discipline as well as respect for the beliefs of others. At the end of the discussion, it is worth remembering the summary, and highlighting the important aspects during the course of the classes.
- Method of didactic games (using the area) – this method teaches rational decision making based on the analysis of various situations. This method prepares participants for various social roles. It allows you to thoroughly understand natural phenomena.
- Land art – an artistic activity whose area is the natural environment. By using natural factors/elements, you create your own artistic object.

- Mental map – a method of visual representation of the problem. A problem, password or question is entered in the centre of the map. From that starting point, ideas associated with the main theme are entered.
- Mini-lecture – a short presentation of the issue by the lecturer.
- Observation – a learning method that allows you to acquire knowledge by seeing the surrounding reality.
- Heuristic talk (heureza) – a method based on guiding the thinking and activity of participants by means of subsequent questions or tasks.
- Work with text (source material) – this method is usually associated with reading a fragment of the text. By using this method, you can easily develop the participant's important skills such as: selection of information, preparation of abstract, summary or notes.
- Fun – performing certain activities for pleasure using your own imagination. It has a creative character.

During the classes, methods based on the practical activity of participants can also be used.

GUIDELINES FOR THE ORGANIZATION OF FIELD WORKSHOPS

- Topics and issues (content) may be modified depending on the workshop leader.
- Content and methods can be modified by the workshop leader on an ongoing basis depending on the intellectual level and commitment.
- A short break may be taken during the workshop.
- One workshop group consists of 10–15 people.
- For the following age groups (please provide a group supervisor!):
  - pre-schoolers (5–6 years old);
  - pupils in grades I–III of primary school;
  - pupils in grades IV–VIII of primary school;
  - high school students (I–III);
  - people with intellectual disabilities (mild, moderate disability);
- Off-road attire is required during field classes!
- Workshops will be cancelled if the weather conditions are bad.

Table 1. Raising the qualifications, awareness and commitment of employees for sustainable development – the proposed educational activities

Target group – – participants	Subject of classes	General objectives	Specific objectives (issues possible to develop depending on the teacher and the group's interest)	Learning methods	Means and teaching aids	Duration
Employees/specialists (e.g., employees of the Environmental Protection Departments of City Hall, sanitary and epidemiological station, mine employees)	Development of heaps in urban spaces Potential of areas created as a result of exploitation to provide ecosystem services Natural capital in post-industrial landscapes	– deepening cause and effect thinking – deepening social skills and developing the imagination of participants – ability to work in a group – cooperation – realizing the importance of elements of nature in everyday life of people – expressing your own feelings and thoughts – paying attention to the proper management of post-mining areas – paying attention to pro-ecological attitudes – raising awareness and commitment of participants/employees for sustainable development	– learning the etymology of the word "heap" – learning the specifics of the habitat (heap) as a place of occurrence of various plant species – post-industrial landscape as a colonization place for plants, including rare species – defining the concepts of "natural capital" and "critical natural capital" – defining the terms: "ecosystem services", "ecosystem benefits", "sustainable development" – defining the terms "reclamation", "revitalization" and indicating the differences between them – discussion of the reclamation goals – distinguishing reclamation methods – familiarizing with examples of post-mining land reclamation in Poland and abroad – indication of the need for sustainable development – identification of ecosystem services on the example of a heap	– lecture – observation – brainstorm – drama – discussion on the importance of ecosystem services for local people and improving the state of the environment on a local scale – decision tree method	– pens – photographs of selected vascular plant species – role cards about the characters played – keys for marking plants – gray paper (sheet)	1.5 h

Target group – – partici-pants	Subject of classes	General objectives	Specific objectives (issues possible to develop depending on the teacher and the group's interest)	Learning methods	Means and teaching aids	Duration
			– presenting the natural capital and ecosystem services/benefits – determining the positive and negative sides of the heap's presence in the landscape – discussion on the use of heaps in urban spaces – presenting the flow of benefits from natural capital – presenting the state of knowledge about ecosystem services – understanding the links between individual ecosystem services – showing areas related to the exploitation of mineral resources as unique habitats – ability to play the role (e.g., naturalist/ecologist, mine director, heapman, tree) – improving observation skills – shaping pro-ecological attitudes – enriching biological vocabulary			

Supplementary materials for participants, e.g., Szafer *et al.* (1988), Tokarska-Guzik (2000), Fisher *et al.* (2009), de Groot *et al.* (2010), Woźniak (2010), Tokarska-Guzik *et al.* (2012), Regulation of the Minister for Environment… (2014), Telega (2014; 2015), The General Directorate for Environmental Protection (2014; 2021), Hodgson & Winram (2017), Solon *et al.* (2017), Islam *et al.* (2019

Table 2. Support and participation in promoting the idea of sustainable development in local and national social initiatives – the proposed educational activities

Proposed activity	Goal	Description of the event	Target group
Ecological picnic: *"I'm not afraid of heaps..."*	– increasing the knowledge of society about the biological diversity of post-mining areas – showing the potential of the areas in question – promoting the idea of sustainable development in local social initiatives	– interdisciplinary nature event – various forms of activity are foreseen: workshops, lectures, games, thematic games, competitions with prizes – it is planned to organize a zone for exhibitors (e.g., related to ecological education or healthy food) – the opportunity to discuss natural topics with representatives of the world of science (e.g., sustainable development) – additional attraction – a brass band concert	All inhabitants of the province
The photo competition entitled *"The most beautiful view from the heap"*	– interest of participants in post-mining areas – developing pro-ecological attitudes – propagating the idea of sustainable development in local social initiatives	– task of the participants is the following: taking a photo of the city panorama/an interesting view from the top of any heap of Silesian Province and sending it to the organizers with a description (date of the photo, author, place of taking, information about the content of the photo) – participation in the competition – free of charge – prizes are foreseen for the winners!	Age group: primary school, classes IV–VIII

Table 3. Developing a system of cooperation with local communities located within industrial and operational areas – the proposed educational activities

Thematic block	Course contents	Objectives of the proposed educational and didactic activities
Biodiversity	– relationships between the diversity of spontaneous vegetation and the biological parameters of substrates resulting from the exploitation of mineral resources such as hard coal – carbon cycle in natural and semi-natural extremes – the basics of biogeohydrochemical processes – coal balance during spontaneous succession in post-mining and post-industrial areas – the role of microorganisms in the circulation of elements including carbon in post-mining and post-industrial substrates – what is metatranscript scripting and why is its use important in relation to climate change mitigation?	– explaining various aspects of the concept of biodiversity – presenting the importance of various dimensions of biodiversity for the functioning of ecosystems – presentation, explanation and comparison of the importance of biological, taxonomic and functional diversity – analysis and explanation of the importance of biodiversity for the intensity of the mineralization process in terrestrial and aquatic ecosystems – presentation of biological diversity found in spontaneous vegetation on heaps of gangue – presentation and explanation of the importance of functional diversity and its relationship with the amount of CO_2 released in patches of spontaneous vegetation on heaps of gangue – characteristics of vegetation (vascular plants) as the original producer of organic substances at individual stages of ecosystem development – presentation and explanation of the importance of species composition of spontaneous vegetation on heaps of gangue as a factor conditioning processes such as circulation of elements in the heap of gangue – characteristics and explanation of the importance of the structure of vegetation (root secretions) and their impact on properties of biomass (formed at the end of the season), and as a consequence, organic matter in soil/soil substrate – causes and effects of changes in the chemical composition, organic matter in soil/soil substrate and their impact on the rate and nature of the distribution of organic matter, and as a consequence, the rate of element circulation depending on the type and stage of terrestrial ecosystem development – presentation and explanation of the level of relationships and mutual feedbacks between plants, microorganisms (including: bacteria, ecto- and endo-mycorrhizal fungi) and soil fauna in natural and semi-natural ecosystems – characteristics of selected elements of ecosystems and explanation of the close dependencies, connections and their interactions allowing the flow of matter and energy – presentation and explanation of the significance of the rhizosphere as part of the soil environment/soil substrate in which biological and physicochemical processes are modified through directly adjacent plant roots – presentation and explanation of the importance of organic substances secreted by plant roots (organic acids, phenolic compounds, polysaccharides, amino acids, proteins, including enzymes) for changes occurring in the substrate (including carbon accumulation) – presentation and explanation of the relationship between the quantity and quality of substances secreted by roots on the effectiveness of life processes, physiological state (sensitivity to stress) of individual plants, as well as synergistic relationships between individuals of plants of the same and different species (inter- and intra-species competition) – presentation and explanation of the relationship between metabolic effort of plants during the formation of root secretions, and biological activity (including enzymes, total content of bacterial phospholipid fatty acids (PLFA)) of the post-industrial medium and the presence of, among others, arbuscular mycorrhizal fungi and the presence of glomalins (carbon accumulation in the soil, including gangue dumps)

Thematic block	Course contents	Objectives of the proposed educational and didactic activities
		– presenting and explaining the importance of soil/soil respiration processes in various types of ecosystems in relation to the effectiveness of activities related to limiting the effects of climate change – presenting and explaining the importance of aboveground functional diversity of plants and vegetation for the diversity of organisms present under the surface of the soil/soil substrate (above- and below-ground diversity relationships) – presenting and explaining the possibilities and usefulness of using the latest analytical methods such as: metatransgenomics (RNA extraction, cDNA synthesis, DSN (Duplex Nuclease Specific Normalization) enriched mRNA), single-stranded cDNA metatransgenomics using random hexamers to represent total soil RNA, followed by synthesis of double-stranded cDNA – presenting the characteristics and explaining the importance of functional and species biodiversity of soil microorganisms clusters using bioinformatics tools in choosing an effective method of limiting the effects of climate change – presentation and determination of the impact of selected biotic and abiotic factors on qualitative and quantitative parameters of carbon sequestration in soil/soil substrate
Conditions for carbon accumulation (biogeochemical processes on the habitats of "novel ecosystems")	– "novel ecosystems" as an example of specific habitat conditions for new, unique systems of diversity of organisms – Anthropocene epoch – an era requiring the use of all possible natural circumstances (on all scales) to limit the effects of climate change – abiotic parameters of substrates created in connection with the exploitation of mineral resources such as hard coal as a habitat meeting the criteria of "novel ecosystems" – areas after hard coal exploitation and the emergence of new, unheard of ecosystems, and systems of living organisms on various scales as an example of "novel ecosystems" – the potential of wetland, aquatic and terrestrial habitats arising in connection with the exploitation of mineral resources such	– explaining the nature and criteria for recognizing a habitat created in connection with human activities as "novel ecosystems" – presentation of the importance of identifying ecosystem functioning processes on new types of habitats arising in connection with the exploitation of mineral resources – presentation, explanation and comparison of the possibilities of carbon accumulation in the course of development of biological systems on the basis of new terrestrial and wetland habitats – analysis and explanation of the importance of carbon accumulation in wetland and aquatic terrestrial ecosystems (fresh and salt waters), in aerobic and anaerobic conditions, on natural and semi-natural habitats – presentation of the importance of identification (microevolution) of new organisms (new species of mycorrhizal fungi) found on gangue habitats, subject to the process of spontaneous succession (without remediation) – biological "novel ecosystems", as an introduction to the new order and the "Trojan horse" of nature protection and environmental protection – presentation and explanation of the meaning of "novel ecosystems" substrate parameters – the variety of granulometric structure of the substrate with the parameters of element circulation, compared to the substrates of natural and semi-natural ecosystems – characteristics and explanation of the importance of soil moisture in "novel ecosystems" in the element circulation process, compared to the soil of natural and semi-natural ecosystems, taking into account the role of montmorillonites in the structure of the soil (soil substrate) heaps of gangue – presentation and explanation of the significance of the diversity of the granulometric structure of the heaps of gangue as an example of a new habitat for the development of microbiology and biological activity of the post-mining habitat – presentation and explanation of processes occurring in the ground of areas transformed and shaped by human (mineral ground with a diversified granulometric structure, subject to physical, chemical and biological processes over time) in comparison with processes taking place in structured soil with a developed soil profile

Thematic block	Course contents	Objectives of the proposed educational and didactic activities
	as hard coal and the *de novo* ecosystems developed on them, for the accumulation of CO_2	
Strengthening ecosystem services that can be provided by systems created in connection with the exploitation of mineral resources	– "natural capital" – a new concept of perceiving the values that nature provides to all living organisms (including human) – relationships between the concepts of natural capital and ecosystem services in relation to areas created in connection with the exploitation of mineral resources – characteristics of ecosystem services depending on the adopted classification, comparison of the most commonly used classification – ecosystem services in areas after the exploitation of mineral resources, among others hard coal and the ecosystems developing on them	– the essence of the potential of phenomena characterized as "novel ecosystem", "natural capital" and "ecosystem services" in post-industrial areas in comparison with natural and semi-natural systems – determining how drainage of wetlands and peat bogs for agriculture impacts CO_2 emissions, whether fossil fuels are the main cause of greenhouse gas growth and climate change – presentation, characteristics and analysis of the biological basis for the production of ecosystem services – presentation and discussion of the possibilities of biological erosion prevention as an example of regulating ecosystem services – presenting and discussing the possibilities of mitigating the effects of widespread eutrophication by creating and protecting oligotrophic habitats in areas after the exploitation of mineral resources, as an example of regulating ecosystem services – presentation and possibilities of creating habitat protection for pollinating organisms, habitats with a clearly greater ability to accumulate CO_2, as an example of regulating ecosystem services – presentation and possibilities of creating protection of aquatic and wetland habitats – ecosystem systems, as an example of regulating ecosystem services in post-mining areas – presentation and development opportunities for living space services, habitats for migratory species and maintenance, supporting the diversity of the genetic pool – genetic variability of organisms generated by difficult habitat conditions, among other areas after the exploitation of mineral resources – presentation, characterization and analysis of cultural ecosystem services, aesthetic and landscape values, assessment of the possibility of indicating places for recreation and tourism, supporting cognitive development and spiritual experience of areas after exploitation of mineral resources – presentation of the possibilities to increase the value of ecosystem services through the skilful introduction of crops (honey, fiber, oil, medicinal) – presentation of the possibility of creating substitute substrates for rare and protected species on post-mining dumps, especially poor, oligotrophic mineral habitats – presentation of the possibilities of creating replacement habitats on post-mining dumps, endangered utilitarian, medicinal or/and cosmetic species, especially those related to habitats poor in nitrogen, phosphorus and potassium – presentation of the values and potential of post-mining dumps in relation to cleanliness (no industrial pollution of the 1960s, 1970s and 1980s) as a habitat for obtaining pure utility or herbal raw material

Supplementary materials for participants, e.g., Szafer *et al.* (1988), Costanza & Daly (1992), Costanza *et al.* (1997), Tokarska-Guzik (2000), Boyd & Banzhaf (2007), Fisher *et al.* (2009), Woźniak (2010), Reyers *et al.* (2012), Tokarska-Guzik *et al.* (2012), Regulation of the Minister for Environment (2014), Telega (2014; 2015), The General Directorate for Environmental Protection (2014; 2021), Hodgson & Winram (2017), Solon *et al.* (2017), Islam *et al.* (2019).

Table 4. A detailed catalog of selected proposed topics of nature workshops within thematic blocks, taking into account target groups of recipients (Elbanowska 1983; 1994; Szafer *et al.* 1988; Taylor 1991; Kielar-Turska 1992; Kayser 1993; Soida 1993; Durrell G. & Durrell L. 1994; Folwaczny 1999; Kujawiński 1999; Krzyżewska 2000; Dymara 2001; Nachtigall 2001; Gardner 2002; Krekeler & Rieper-Bastian 2002; Żytko 2002; Sternberg & Spear-Swerling 2003; Vopel 2003; Michalak 2004; Urbisz A. & Urbisz A. 2004; Walter 2004; Brown 2005; Budniak 2005; 2009; Saan 2005; 2006; Będkowska 2006; Muchacka 2006; Rusch 2006; Bouchardy 2008; Danks & Schofield 2008; Gąsecki 2008; Dobrosz-Teperek & Dasiewicz 2009; Sławińska 2010; Kamińska & Świderek 2011; Hutniczak 2014; Jelinek 2014; Korwin-Szymanowska *et al.* 2015; 2016; Hopkins 2019)

Age group	Subject of classes	General objectives	Specific objectives (issues possible to develop depending on the teacher and the group's interest)	Learning methods	Means and teaching aids	Duration
Pre-schoolers (5–6 years old)	What plant is it? Land art on the heap... Fun in the trailer – searching for nature letters	– shaping emotional bond with nature – awakening cognitive curiosity – developing teamwork skills – developing environmentally friendly attitudes	– learning about the specifics of the habitat (heap) as a place of occurrence of various plant species – learning about plants through many senses: sight, smell, touch – the indication of the parts the plant consists of (anatomy of plants) – using natural elements of the natural environment related to the heap to create their own artistic composition – paying attention to the heap as a viewpoint in the city – shaping the attentive nature observer attitude – improving the ability to listen and understand the teacher's questions – enriching natural vocabulary – shaping pro-ecological attitudes	– observation – heuristic talk – free expression of pupils – physical activity "herbaceous plant – shrub – tree" – sensors games – sensual games – elements of the field game	– colourful engravings of plants – magnifying glass – binoculars – nature riddles – sensory boxes (content, e.g., stones, leaves, pine cones, branches, soil, tree bark, dry pasta) – branches, leaves, stones and other elements of nature found by participants on the heap to create their own artistic composition – field mortars – nature lists	45 min

Age group	Subject of classes	General objectives	Specific objectives (issues possible to develop depending on the teacher and the group's interest)	Learning methods	Means and teaching aids	Duration
Pupils from in grades I–III of primary school	Searching for nature letters – we examine the recesses of the heap Heap – a strange garden? – diversity of the plant world	– shaping emotional bonds with nature – awakening cognitive curiosity – developing teamwork skills – developing environmentally friendly attitudes	– getting to know plants growing on heaps with the help of many senses: sight, smell, touch – the indication of threats to the areas in question – exploring the area through different senses – defining the terms "tree", "shrub", "herbaceous plant" and indicating the differences between them – getting to know individual parts in plant construction (plant anatomy) – paying attention to the heap as a viewpoint in the city – improving observation skills – improving the skills of listening and understanding the teacher's questions – shaping the attitude of a watchful nature observer – enriching natural vocabulary – shaping pro-ecological attitudes	– observation – practical action – examining plants with many senses – heuristic talk – free expressions of pupils – sensual games – elements of the field game – physical activity "herbaceous plant – shrub – tree" – fun "freeze frame" (optional) – creating your own postcard from a heap	– nature letters – magnifying glass – binoculars – field mortars – colourful engravings of plants – nature riddles – sheets of paper – pencils – crayons – adhesive tape – whistle – stopwatch	60 min
Pupils in grades IV–VIII of primary school	The diversity of the world of plants on heaps of the Upper Silesian region	– awakening cognitive curiosity – shaping cause and effect thinking – developing teamwork skills	– learning the etymology of the word "heap" – defining the concept of "biodiversity" – learning the specifics of the habitat (heap) as a place of occurrence of plant species – acquaint pupils with selected plants growing in heaps – learning about plant adaptations to occur in extreme habitats	– brainstorm – thematic exercises – discussion – "Nature Master" competition	– introductory exercise to understand biodiversity – photographs of selected species of vascular plants – sheets of paper	1.5 h

Age group	Subject of classes	General objectives	Specific objectives (issues possible to develop depending on the teacher and the group's interest)	Learning methods	Means and teaching aids	Duration
	We get to know common plant species in the Upper Silesian region Foreign plants on heaps, with particular emphasis on invasive species Trees and shrubs of post-industrial areas Mushrooms and mycorrhizal fungi on heaps Soil profile analysis	– enrichment of biological vocabulary – developing pro-ecological attitudes	– clarifying the concept of "extreme habitat" – examples of extreme habitats with particular emphasis on heaps – acquaintance with examples of plant adaptation to different living environments (e.g., xeromorphism) – paying attention to various life forms of plants – paying attention to the occurrence of rare and alien species (including invasive plants) in heaps – defining the terms "alien species", "invasive species" and "biological invasion" – getting to know some invasive plant species – discussion of the features of invasive plants and ways to prevent their spread – the indication of species protected by law (strict, partial) – explaining the need for plant protection in the environment – learning the healing and poisonous properties of selected plants – ability to recognize species of trees and shrubs appearing on the heap – noticing differences between a tree and a bush – acquiring the ability to distinguish individual leaves from composite leaves – introduction of the concept of "breast" – understanding the importance of trees in nature and in human life – discussion of the principles of plant collection and conservation for documentation purposes – familiarization with the specific aquatic environment of the settler and its function – identification of aquatic plants in the vicinity of the reservoir/settler	– mini-lecture – observation – heuristic talk – practical action: examining plants with many senses – free expression of pupils	– boxes with letters that make up the words "biodiversity" and "succession" – work cards (optional) – crossword with the slogan "heap" – magnifying glass – tailor meter – prizes for competition winners – pencils – pH meter – washers – soil profile – sample simple key for marking plants – jigsaw puzzles depicting selected plant species appearing in heaps	

Age group	Subject of classes	General objectives	Specific objectives (issues possible to develop depending on the teacher and the group's interest)	Learning methods	Means and teaching aids	Duration
	Calamine plants on heaps Spontaneous succession on the heap Forest on the heap Adaptation of plants to living in extreme habitats Underground water sedimentation tanks – a threat to groundwater?		– learning about how organisms adapt to life in water – defining the concepts of "soil" and "soil profile" – learning the function of soils – discussion of soil layers and the causes of their contamination – ability to provide elements of a simplified soil profile and perform soil pH measurement – *defining the terms* "*mycorrhizal fungi*" and "*mycorrhiza*" – learning about the features of mycorrhizal fungi – learning examples of mycorrhizal fungi species – discussion on the importance of mycorrhizal fungi and mycorrhiza in nature – clarifying the term "succession" – learning the stages of succession – defining the term "reclamation" – introduction to plant species appearing on the path of spontaneous succession – familiarizing with examples of post-mining land reclamation in Poland and abroad – discussion of the importance of post-mining areas in the landscape – clarification of the term "calamine plants" – learning about plants found on soils with high zinc and lead content – paying attention to the heap as a viewpoint in the city – shaping a positive attitude towards the role of plants in preserving life on earth – improving the skills of listening and understanding the teacher's questions – improving observation skills – enrichment of biological vocabulary			

Age group	Subject of classes	General objectives	Specific objectives (issues possible to develop depending on the teacher and the group's interest)	Learning methods	Means and teaching aids	Duration
High school students (I–III)	The diversity of the world of plants on heaps of the Upper Silesian region	– awakening cognitive curiosity – shaping cause and effect thinking – developing teamwork skills – enrichment of biological vocabulary – developing pro-ecological attitudes	– learning the etymology of the word "heap" – learning the specifics of the habitat (heap) as a place of occurrence of various plant species – to familiarize students with plants growing in heaps – defining the concept of "biodiversity" – examples of extreme habitats with particular emphasis on heaps – clarifying the concept of "extreme habitat" – clarifying the concept of "replacement habitats" and discussing their role for plant life – learning the healing and poisonous properties of selected plants – learning the principles of plant collection and conservation for documentation purposes – adapting plants to occur in extreme habitats – marking the occurrence of rare and alien species (including invasive plants) on heaps – indication of species protected by law (strict, partial) – defining the terms "alien species", "invasive species" and "biological invasion" – discussion of the features of invasive plants and ways to prevent their spread – paying attention to the categories of plant invasiveness in Poland – learning basic legal acts (national and international) regarding species protection in Poland – ability to recognize species of trees and shrubs appearing on the heap – acquiring the ability to distinguish individual leaves from composite leaves – introduction of the concept of "breast" – understanding the importance of trees in nature and in human life	– brainstorm – thematic exercises – discussion – "Nature Master" competition – mental map (with the slogan "heap") – mini-lecture – observation – heuristic talk – practical action: examining plants with many senses – free expression of participants	– introductory exercise to understand diversity – photographs of selected species of vascular plants – photographs of selected mycorrhizal fungi – work cards (optional) – sheets of paper – boxes with letters that make up the words "biodiversity" and "succession" – crossword with the slogan "heap" – magnifying glass – pencils – pH meter – washers – soil profile – examples of legal acts (national and international) regarding species protection in Poland – art accessories	1.5 h
	We get to know common plant species in the Upper Silesian region					
	Foreign plants on heaps, with particular emphasis on invasive species					
	Trees and shrubs of post-industrial areas					

Age group	Subject of classes	General objectives	Specific objectives (issues possible to develop depending on the teacher and the group's interest)	Learning methods	Means and teaching aids	Duration
	Mushrooms and mycorrhizal fungi on heaps		– defining the concepts of "soil" and "soil profile" – ability to provide elements of a simplified soil profile and perform soil pH measurement – discussion of soil layers and the causes of their contamination – learning the function of soils – defining the terms "mycorrhizal fungi" and "mycorrhiza" – discussion on the importance of mycorrhizal fungi and mycorrhiza in nature – learning about the features of mycorrhizal fungi – learning examples of mycorrhizal fungi species – clarification of the term "calamine plants" – learning about plants found on soils with high zinc and lead content – clarifying the term "succession" – learning the stages of succession – introduction to plant species appearing on the path of spontaneous succession – familiarizing with examples of post-mining land reclamation in Poland and abroad – defining the term "reclamation" – distinction of reclamation methods – knowledge of examples of plant adaptation to different living environments (e.g., xeromorphism) – familiarization with the specific aquatic environment of the settler and its function – identification of aquatic plants in the vicinity of the reservoir/settler – learning about how organisms adapt to life in water			
	Soil profile analysis					
	Calamine plants on heaps					
	Spontaneous succession on the heap					
	Forest on the heap					
	Adaptation of plants to living in extreme habitats					

Age group	Subject of classes	General objectives	Specific objectives (issues possible to develop depending on the teacher and the group's interest)	Learning methods	Means and teaching aids	Duration
	Underground water sedimentation tanks – a threat to ground water? Heap through the eyes of the artist...		– look at the heap through the prism of history, symbolism and art – learning the historical significance of heaps – tourist potential of heaps – determining the positive and negative sides of the heap's presence in the landscape – explaining the need for plant protection in the environment – enriching biological vocabulary – discussion of the importance of post-mining areas in the landscape – improving observation skills – shaping a positive attitude towards the role of plants in preserving life on earth – improving listening skills and understanding of the lecturer's questions – art workshop – preparing your own image – a heap in the Silesian landscape			
Students of the University of the Third Age	Heap – a unique feature of the Silesian landscape Heap – death and life of plants... Heap – potential for the environment	– awakening cognitive curiosity – deepening cause and effect thinking – enrichment of biological vocabulary and terminology related to post-mining areas – paying attention to the proper management of post-mining areas	– learning the specifics of the habitat (heap) as a place of occurrence of plant species – familiarizing with the term "biodiversity" – recognition of selected vascular plant species based on characteristic features – defining the terms "alien species", "invasive species" and "biological invasion" – drawing attention to the introduction of alien species into the natural environment – getting to know some invasive plant species – paying attention to the categories of plant invasiveness in Poland – discussion of the features of invasive plants and ways to prevent their spread – marking the presence of rare, alien (including invasive) species on the heap, as well as trees and shrubs – clarifying the concept of "extreme habitat"	– lecture – brainstorm – thematic exercises – discussion – mental map (with the slogan "heap") – observation – heuristic talk – practical action: examining plants with many senses – free expression of participants	– photographs of selected species of vascular plants – crossword with the slogan "heap" – magnifying glass – pencils – washers – example of a simple key for marking plants – art accessories	1.5 h

Age group	Subject of classes	General objectives	Specific objectives (issues possible to develop depending on the teacher and the group's interest)	Learning methods	Means and teaching aids	Duration
	Plants with medicinal properties on the heap Foreign plants on heaps, with particular emphasis on invasive species Heap through the eyes of the artist...	– paying attention to pro-ecological attitudes	– examples of extreme habitats with particular emphasis on heaps – knowledge of examples of plant adaptation to different living environments (e.g., xeromorphism) – paying attention to the healing and toxic properties of selected plants – learning the principles of plant collection and conservation for documentation purposes – explanation of the terms "herbal medicine", "aromatherapy" – look at the heap through the prism of history, symbolism and art – learning the historical significance of heaps and the etymology of the word "heaps" – discussion on the use of heaps in urban space – tourist potential of heaps – "Heap piles on heaps..." – determining the positive and negative sides of the heap's presence in the landscape – improving observation skills – improving listening skills and understanding of the lecturer's questions – shaping a positive attitude towards the role of plants in preserving life on earth – shaping pro-ecological attitudes – enriching biological vocabulary – art workshops – preparing your own image – a heap in the Silesian landscape			
People with intellectual	Land art on the heap	– developing and shaping cognitive processes and	– exploring the area through the senses – learning the specifics of the habitat (heap) as a place of occurrence of various plant species	– general mobility exercises – land art	– photographs of selected vascular plant species	45 min

Age group	Subject of classes	General objectives	Specific objectives (issues possible to develop depending on the teacher and the group's interest)	Learning methods	Means and teaching aids	Duration
disabilities (mild, moderate disability)	We create our own plant composition – a postcard from a heap	spatial orientation – developing the skills of observing the natural world	– participants learn about plants growing on heaps – using natural elements of the natural environment related to the heap to create their own artistic composition – improving small motor skills – improving fine motor skills – improving observation skills – improving listening skills and understanding of the lecturer's questions – shaping pro-ecological attitudes – enriching natural vocabulary	– observation – practical action: examining plants with many senses – free expression of participants, sensory games	– branches, leaves, stones and other elements of nature found by participants on the heap – sheets of paper – teaching materials of the Harmonia publishing house – sensory boxes (e.g., stones, leaves, pine cones, branches, soil, tree bark, dry pasta) – adhesive tape – plant specimens, pebbles, branches, soil collected by participants	
	I and nature – the meeting on a heap...					
	What plant is it?					

CONCLUSIONS

Conducting educational activities related to the natural environment is necessary so that all users of a given area are fully aware of the processes taking place in it and their interrelationships. This will enable the most effective land development while maintaining the idea of sustainable development.

Directing specific content (topics and issues) to different age groups while adapting appropriate teaching methods is an essential element of nature education. Such an approach will ensure continuity in the education of specialists, and at the same time, it will arouse the interest of the participants in the immediate environment.

It should be emphasized that the overriding goal of educational activities, especially in areas modified, transformed, and created by human, is to teach observation of the environment in which we live. Moreover, it is essential to develop social responsibility for the inhabited areas and develop creativity in solving the specific natural problems affecting the quality of human life and thus the development of activities in contact with nature.

REFERENCES

Bąk A., Leśny A. & Palamer-Kabacińska E. 2014: Przygoda w edukacji i edukacja w przygodzie. Outdoor i adventure education w Polsce. Fundacja Pracownia Nauki i Przygody, Warszawa.

Będkowska H. 2006: Gry i zabawy w lesie. Agencja Reklamowo-Wydawnicza Arkadiusz Grzegorczyk, Warszawa.

Bouchardy C. 2008: Przyjaciel Natury. Dla młodych odkrywców przyrody. Wydawnictwo Book House, Warszawa.

Boyd J. & Banzhaf S. 2007: What are ecosystem services? The need for standardized environmental accounting units. Ecological Economics, 63 (2–3), 616–626.

Brown S.E. 2005: Robimy eksperymenty. Wydawnictwo Liber, Poznań.

Budniak A. 2005: Aktywizacja uczniów klas początkowych poprzez doświadczenia przyrodnicze. [In:] Kuźma J. & Morbitzer J. (eds.): Edukacja – szkoła – nauczyciel. Promowanie rozwoju dziecka. Wydawnictwo Naukowe Akademii Pedagogicznej, Kraków, 416–420.

Budniak A. 2009: Doświadczenia przyrodnicze w poznawaniu środowiska przez uczniów klas początkowych. Wydawnictwo Deni-Press, Katowice.

Costanza R. & Daly H.E. 1992: Natural capital and sustainable development. Conservation Biology, 6 (1), 37–46.

Costanza R., d'Arge R., de Groot R., Farber S., Grasso M., Hannon B., Limburg K., Naeem S., O'Neill R.V, Paruelo J., Raskin R.G., Sutton P. & van den Belt M. 1997: The value of the world's ecosystem services and natural capital. Nature, 387 (15), 253–260.

Danks F. & Schofield J. 2008: Przyrodniczy plac zabaw. Gry i zabawy na świeżym powietrzu. Pomysły na cały rok. Wydawnictwo Klub Dla Ciebie, Warszawa.

de Groot R.S., Alkemade R., Braat L., Hein L. & Willemen L. 2010: Challenges in integrating the concept of ecosystem services and values in landscape planning, management and decision making. Ecological Complexity, 7 (3), 260–272.

Dobrosz-Teperek K. & Dasiewicz B. 2009: Edukacja poprzez zmysły i doświadczenia. Meritum, 2 (13), 48–52.

Durrell G. & Durell L. 1994: Poradnik przyrodnika. Wszechstronny i praktyczny przewodnik po świecie przyrody. Wydawnictwo Świat Książki, Warszawa.

Dymara B. (ed.) 2001: Dziecko w świecie współdziałania. Oficyna Wydawnicza Impuls, Kraków.

Elbanowska S. 1983: Przyroda nieożywiona w wychowaniu przedszkolnym. Wydawnictwo Szkolne i Pedagogiczne, Warszawa.

Elbanowska S. 1994: Jak zadziwić przedszkolaka tym, co świeci, pływa, lata. Seria wydawnicza 4 książek: Jesień, Zima, Wiosna, Lato. Wydawnictwo Medium, Warszawa.

Fisher B., Turner R.K. & Morling P. 2009: Defining and classifying ecosystem services for decision making. Ecological Economics, 68 (3), 643–653.

Folwaczny A. 1999: Propozycje zajęć w lesie dla różnych grup wiekowych. [In:] Wika S. (ed.): Lasy województwa śląskiego wczoraj – dziś – jutro. Wydawnictwo Kubajak, Krzeszowice, 139–144.
Gałązka A. 1999: Drama – metodą twórczej edukacji. Chowanna, XLII (LIII), 1–2 (12–13), 54–68.
Gardner H. 2002: Inteligencje wielorakie. Teoria w praktyce. Wydawnictwo Media Rodzina, Poznań.
Gąsecki K. 2008: Wychowanie do dialogu z przyrodą w edukacji wczesnoszkolnej. [In:] Komorowska-Zielony A. (ed.): Twórcze działania przyrodnicze i matematyczne w edukacji wczesnoszkolnej. Wydawnictwo Uniwersytetu Gdańskiego, Gdańsk, 30–31.
Grabowski M. 2011: Zabawa jako podstawowa forma aktywizacji dziecka w wieku wczesnoszkolnym. Edukacja Elementarna w Teorii i Praktyce: kwartalnik dla nauczycieli, 2–3, 7–16.
Halchynska O. & Navolska L. 2020: Land art obiekty z metalu jako trend sztuki współczesnej. Knowledge, Education, Law, Management, 5 (33), 1, 31–38.
Hodgson S. & Winram M. 2017: Natural Capital Account 2016–2017. Forestry Commission, England.
Hopkins A. 2019: Jaka to roślina? Ćwiczenia rozwijające i kształtujące procesy poznawcze. Grupa Wydawnicza Harmonia, Gdańsk.
Hutniczak A. 2014: Zajęcia w terenie ważnym elementem edukacji ekologicznej. [In:] Terlecka M.K. (ed.): Edukacja ekologiczna – wybrane problemy. Wydawnictwo Armagraf, Krosno, 83–94.
Islam M., Yamaguchi R., Sugiawan Y. & Managi S. 2019: Valuing natural capital and ecosystem services: a literature review. Sustainability Science, 14, 159–174.
Jelinek J.A. 2014: Edukacja przyrodnicza starszych przedszkolaków w domu i w przedszkolu. [In:] Gruszczyk-Kolczyńska E. (ed.): Starsze przedszkolaki. Jak skutecznie je wychowywać i kształcić w przedszkolu i w domu. Wydawnictwo Bliżej Przedszkola, Kraków, 276–305.
Kaczmarzyk M. & Kopeć D. 2007: Dydaktyka zdrowego rozsądku – podręcznik dla studentów przygotowujących się do zawodu nauczyciela, nauczycieli stażystów, nauczycieli kontraktowych i opiekunów stażu. Wydawnictwo Edukacyjne Wiking, Wrocław.
Kamińska E. & Świderek G. 2011: Nie znikaj – materiały dla nauczycieli. Wydawnictwo Źródła, Łódź.
Kayser R. 1993: Na wycieczce na biwaku. Poradnik młodego turysty. Wydawnictwo Arkady, Warszawa.
Kielar-Turska M. 1992: Jak pomagać dziecku w poznawaniu świata. Wydawnictwo Szkolne i Pedagogiczne, Warszawa.
Korwin-Szymanowska A., Lewandowska E. & Tuszyńska L. 2015: Edukacja środowiskowa w kształceniu nauczycieli w perspektywie praktycznej. Wydawnictwo Akademii Pedagogiki Specjalnej, Warszawa.
Korwin-Szymanowska A., Lewandowska E. & Witkowska-Tomaszewska A. 2016: Edukacja dla zrównoważonego rozwoju w perspektywie wyzwań społeczeństwa wiedzy. Wydawnictwo Akademii Pedagogiki Specjalnej, Warszawa.
Krekeler H. & Rieper-Bastian M. 2002: Fascynujące eksperymenty. Łatwe, odkrywcze, zaskakujące. Wydawnictwo Hubert, Warszawa.
Krzyżewska J. 2000: Aktywizujące metody i techniki w edukacji wczesnoszkolnej. Suwałki, Wydawnictwo AU OMEGA.
Kujawiński J. (ed.) 1999: Rozwijanie twórczej aktywności uczniów klas początkowych. Wydawnictwo Szkolne i Pedagogiczne, Warszawa.
Łobocki M. 2000: Metody i techniki badań pedagogicznych. Oficyna Wydawnicza Impuls, Kraków.
Małek E. 2007: Stosowanie zasad i metod nauczania 514[02].Z4.02. Poradnik dla ucznia. Wydawca Instytut Technologii Eksploatacji – Państwowy Instytut Badawczy, Radom. https://zsm2krakow.pl/images/dokumenty/dydaktyka/tuf/technik.uslug.fryzjerskich_514[02]_z4.02_u.pdf, access: 24.07.2021.
Michalak R. 2004: Aktywizowanie ucznia w edukacji wczesnoszkolnej. Oficyna Wydawnicza Impuls, Kraków.
Ministry of Science and Higher Education. 2021: Land art – sztuka ziemi. Interwencja w świat przyrody. https://zpe.gov.pl/a/land-art---sztuka-ziemi-interwencja-w-swiat-przyrody/D3YpTu5cS, access: 24.07.2021.
Muchacka B. 2006: Zabawy badawcze w edukacji przedszkolnej. Wydawnictwo Naukowe Akademii Pedagogicznej, Kraków.
Nachtigall W. 2001: Natura matką wynalazków. Wielka księga bioniki. Wydawnictwo Debit, Bielsko-Biała.
Okoń W. 1998: Nowy słownik pedagogiczny. Wydawnictwo Akademickie Żak, Warszawa.
Regulation of the Minister for Environment of 9 October 2014 on plant species protection (Dz. U. 2014, poz. 1409).
Reyers B., Polasky S., Tallis H., Mooney H.A. & Larigauderie A. 2012: Finding common ground for biodiversity and ecosystem services. BioScience, 62 (5), 503–507.
Rusch M. 2006: Zioła w przedszkolu. Wydawnictwo Jedność, Kielce.

Saan A. 2005: 365 pomysłów na eksperymentowanie. Wydawnictwo Rea, Warszawa.
Saan A. 2006: 365 pomysłów na majsterkowanie. Wydawnictwo Rea, Warszawa.
Sławińska M. 2010: Konstruowanie wiedzy na zajęciach w przedszkolu. Impuls, Kraków.
Soida D. 1993: Bądź Mistrzem – Przyjacielem, czyli edukacja ekologiczna na wesoło. Ośrodek Edukacji Ekologicznej RPK OA PTTK, Kraków.
Solon J., Roo-Zielińska E., Affek A., Kowalska A., Kruczkowska B., Wolski J., Degórski M., Grabińska B., Kołaczkowska E., Regulska E. & Zawiska I. 2017: Świadczenia ekosystemowe w krajobrazie młodoglacjalnym. Ocena potencjału i wykorzystania. Wydawnictwo Akademickie SEDNO, Polska Akademia Nauk, Instytut Geografii i Przestrzennego Zagospodarowania im. Stanisława Leszczyckiego, Warszawa.
Stawiński W. (ed.) 2006: Dydaktyka biologii i ochrony środowiska. Państwowe Wydawnictwo Naukowe, Warszawa.
Sternberg R.J. & Spear-Swerling L. 2003: Jak nauczyć dzieci myślenia. Gdańskie Wydawnictwo Psychologiczne, Gdańsk.
Szafer W., Kulczyński S. & Pawłowski B. 1988: Rośliny Polskie – Cz. I II. Państwowe Wydawnictwo Naukowe, Warszawa.
Taylor B. 1991: Zabawa i nauka. Seria: Baterie i Magnesy, Powietrze i Latanie. Polska Oficyna Wydawnicza BGW, Warszawa.
Telega I. 2014: Kapitał naturalny w koncepcji rozwoju trwałego. Rozprawa doktorska. Uniwersytet Ekonomiczny w Krakowie, Wydział Finansów, Katedra Polityki Przemysłowej i Ekologicznej.
Telega I. 2015: Kapitał naturalny a proces ekologizacji gospodarki. [In:] Kożuch M. (ed.): Ekologizacja gospodarki. Wydawnictwo Fundacja Uniwersytetu Ekonomicznego w Krakowie, Kraków, 31–47.
The General Directorate for Environmental Protection. 2014: Kodeks dobrych praktyk „Ogrodnictwo wobec roślin inwazyjnych obcego pochodzenia". https://rm.coe.int/168048d610, access: 07.01.2020.
The General Directorate for Environmental Protection. 2021: About GDEP. http://www.gdos.gov.pl/eng, access: 29.07.2021.
Tokarska-Guzik B. 2000: Przyrodnicze zagospodarowanie nieużytków miejsko-przemysłowych na przykładzie centrów górniczych Europy. Inżynieria Ekologiczna, 1, 72–80.
Tokarska-Guzik B., Dajdok Z., Zając M., Zając A., Urbisz A., Danielewicz W. & Hołdyński C. 2012: Rośliny obcego pochodzenia w Polsce ze szczególnym uwzględnieniem gatunków inwazyjnych. Generalna Dyrekcja Ochrony Środowiska, Warszawa.
Urbisz A. & Urbisz A. 2004: Atlas i klucz – rośliny zielne i krzewinki Polski – rośliny pospolite, częste. Kubajak, Kraków.
Vopel K.W. 2003: Jak pobudzić kreatywność grupy? Propozycje ćwiczeń i zabaw. Wydawnictwo Jedność, Kielce.
Walter G. 2004: Żywioły w przedszkolu. Seria wydawnicza 4 książek: Ogień, Powietrze, Ziemia, Woda. Wydawnictwo Jedność, Kielce.
Way B. 1997: Drama w wychowaniu dzieci i młodzieży. Wydawnictwo Szkolne i Pedagogiczne, Warszawa.
Woźniak G. 2010: Zróżnicowanie roślinności na zwałach pogórniczych Górnego Śląska [Diversity of vegetation on coal-mine heaps of the Upper Silesia (Poland)]. W. Szafer Institute of Botany, Polish Academy of Sciences, Kraków.
Żytko M. 2002: Kształcenie zintegrowane. Problemy teorii i praktyki. Wydawnictwo Żak, Warszawa.

5.5. Current Reclamation Practices and Their Successfulness

Artur Dyczko
Mineral and Energy Economy Research Institute, Polish Academy of Sciences; J. Wybickiego 7A Str., 31-261 Kraków, Poland; arturdyczko@min-pan.krakow.pl; ORCID: 0000-0001-6387-5339

Karolina Ryś
Institute of Biology, Biotechnology and Environmental Protection, Faculty of Natural Sciences, University of Silesia in Katowice; 28 Jagiellońska Str., 40-032 Katowice, Poland; karorys@o2.pl; ORCID: 0000-0002-3391-5501

Łukasz Radosz
Institute of Biology, Biotechnology and Environmental Protection, Faculty of Natural Sciences, University of Silesia in Katowice; 28 Jagiellońska Str., 40-032 Katowice, Poland; lukasz.radosz@onet.eu; ORCID: 0000-0002-0270-7291

Gabriela Woźniak
Institute of Biology, Biotechnology and Environmental Protection, Faculty of Natural Sciences, University of Silesia in Katowice; 28 Jagiellońska Str., 40-032 Katowice, Poland; gabriela.wozniak@us.edu.pl; ORCID: 0000-0003-1936-2880

ABSTRACT: Mining of natural mineral resources causes transformation of the landscape, vegetation cover, hydrological system, soils, and other environmental elements. Practitioners reclaim the post-excavation sites over a long time to repair the environment. The areas that could not be used for agricultural or forestry purposes need to be reclaimed. The common practice used fertilization of the mineral soil substrate and planting herbaceous plants, shrubs, and trees. Unfortunately, alien plants were frequently planted. Global environmental change shows that these practices were not successful. Environmentalists around the world studied the reasons for the failure of reclamation practices. The study of spontaneous succession, which took place on the same post-mineral excavation sites, has shown that the natural processes are more effective in recovering ecosystem functioning on post-mineral excavation sites. The spontaneous vegetation development and ecosystem functioning on post-mineral excavation sites are non-analogous to any known natural and seminatural habitats. The non-analogous Novel Ecosystems on post-mineral excavation sites are very poorly understood. Adaptation processes are very intense on post-mineral excavation sites. Studies provide evidence that Novel Ecosystems support Ecosystem Services for citizens in the densely populated urban-industrial areas. Understanding the Novel Ecosystems' functioning is crucial for the right actions on the sites that need to be managed to enhance ecosystem functioning (Natural Capital) and Ecosystem Services. Understanding the functioning of the Novel Ecosystems and urban-industrial ecosystems and their proper management is crucial for preventing global change.

KEYWORDS: Natural processes, reclamation management, post-mineral excavation mine heaps, global environmental change, Natural Capital, Novel Ecosystems

INTRODUCTION

Large-scale excavation of mineral resources, including coal mining industry development, leads to significant changes in urban-industrial landscapes. The areas rich in natural resources are mainly influenced by human activity. The period of the most significant expansion in industrial activity in Silesia was recorded during the second half of the nineteenth and twentieth centuries (Jaros 1965; Dulias 2010; Krzysztofik *et al.* 2012; Fagiewicz 2014; Forman 2014; Kicki & Sobczyk 2016).

Underground and opencast mining results in severe landscape transformations in hydrology, vegetation cover, and other biotic and abiotic aspects of the natural environment (Prach *et al.* 2014). In some cases, the loss of ecosystem functions (Szumacher & Pabjanek 2017) is caused by habitat transformation. Defragmentation occurs temporarily until the robust adaptation processes start to act. The previous mining activity left mine buildings, coal heaps, sedimentation pools, mine shafts, open-pits, coal mine drainage, and other surface relief changes, which can be seen even in the centers of the cities in many regions (Rayfield *et al.* 2005; Domański 2009; DeLong *et al.* 2012; Anawar *et al.* 2013; Tarolli & Sofia 2016; Limpitlaw *et al.* 2017).

Coal mine companies conduct mining of coal as an energy source. Some coal mines obtain coking coal. Apart from iron ore, this coal is an important raw material for producing steel, which remains one of the critical materials used in construction, producing various types of machinery, ships, cars, or just everyday objects. Its demand all over the world is systematically growing. The production of one ton of steel requires the use of approx. 800 kg of coal. Technologies that could replace coke in its output are highly unprofitable, and there is no indication that this could change in the coming years.

The exploitation of coal resources causes environmental transformation, including changes to the atmosphere, hydrosphere, and lithosphere. The salinity of surface waters results from drainage of the rock mass. In the landscape, significant amounts of side material are heaped and continuous and discontinuous deformations (direct impacts) and the related changes, e.g., changes in water conditions or lowering of the area and changes in its morphology. The listed examples do not close the catalog of changes caused by the broadly understood mining activity.

The aim of this chapter is to present the current practices of the reclamation of post-mineral mining areas, and the necessary changes in the approach according to the newest environmental knowledge and experience.

THE CURRENT APPROACH TO POST-MINERAL EXCAVATION SITES

An average of about 16 million Mg of coal is mined annually. Along with the coal exploitation, many side-effect mineral materials are excavated (on average about 11 million Mg annually). During excavation, the side-effect mineral materials constitute the mineral material on the heaps. The composition of mineral material on the heaps is determined by the geological and mining conditions occurring in the mines and the high-quality requirements for coal production. Sedimentary rocks with variable proportions are present on the heaps, depending on the geology of the deposit, its exploitation levels, and the group of geological layers (Fig. 1). Thus, in terms of petrography, post-mining heaps gather mineral materials which are:

- sandstones of various grain sizes and clay (carbonate) binder;
- clay shales, which are clay fraction rock with an admixture of mudstone.

Most of the post-mining heaps are disposed of on the Earth's surface. Only a tiny part of them are used in underground workings. According to legal requirements reclamation should be the combination of activities which minimize adverse environmental effects of surface mining and mined lands in order to return them to practical usefulness. End uses may be agriculture, woodland, or residential and commercial development. Some reclamation components include practices that control erosion and sedimentation, stabilize slopes, and avoid and repair impacts on wildlife habitat. The final step is usually topsoil replacement and revegetation with suitable plant species. Reclamation is often phased to be concurrent throughout the life of the mining project.

Figure 1. The differences in particle size of the mineral material collected on heaps do not prevent plants from colonizing the sites in the first year after heaping (Photo: Gabriela Woźniak)

Due to individual technical, economic, and social conditions, each mine has a different concept of heap management on the surface. Still, it is always carried out following local spatial development plans. After the work related to incorporating heaps is completed, the stage of technical reclamation (including profiling of slopes) are followed. The biological reclamation takes place, after technical reclamation the final effect of which is called greening, e.g., by planting trees. On some heaps over the years, reclamation has been carried out using various methods. First, after shaping and compacting, a layer of fertile soil is applied, and then a mixture of grasses is fed with artificial fertilizer.

IS THE CURRENT RECLAMATION MANAGEMENT SUCCESSFUL?

Environmental scientists have collected a growing base of ecological knowledge about self-rebuilding natural capital, novel ecosystems, and ecosystem services, which can be synthesized to scope mining-related issues (Tropek & Konvička 2008; 2011; Tropek *et al.* 2012). There is a still-growing body of knowledge of mineral mining's influence on transforming and establishing terrestrial, wetland, and aquatic ecosystems (Fig. 2). In such heavily changed habitats, spontaneous vegetation succession and related soil substratum biogeochemical processes are dynamic and challenging to predict (Fig. 3). Vegetation and ecosystem establishment occur in former mining areas that remain untouched for many years without management (Woźniak 2010; Krzysztofik *et al.* 2020). Environmental scientists have analyzed this spontaneous vegetation succession in former mineral mining areas worldwide (Krzaklewski 1995; Bradshaw 2000; Grodzińska *et al.* 2000; Wiegleb & Felinks 2001; Rostański 2006; Woźniak 2010; Nicia *et al.* 2014; Urbancová *et al.* 2014; Kodir *et al.* 2017). Many scientists have also studied and discussed varied reclamation strategies on mineral mining heaps (Fox *et al.* 1998; Evans & Willgoose 2000; Hameed *et al.* 2000; Jochimsen 2001; Hancock *et al.* 2003; Krzaklewski & Pietrzykowski 2007; Martínez-Ruiz & Marrs 2007).

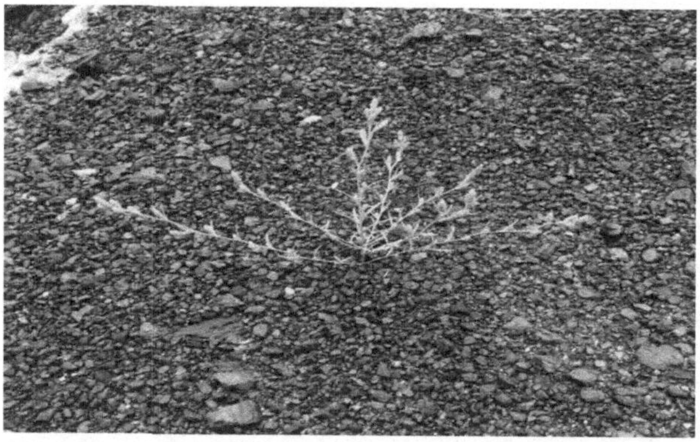

Figure 2. The individual of herbaceous plants which are the first colonizers on the soil substrate of coal mine heaps. No mosses and lichens are observed (Photo: Gabriela Woźniak)

Figure 3. The patchiness of diverse vegetation assemblages – the early stages of primary succession on coal mine heaps (Photo: Gabriela Woźniak)

In non-reclaimed heaps, spontaneous vegetation succession and novel ecosystem establishment have taken place. The harsh habitats conditions, e.g., lack of nutrients, no seed bank, varied soil substratum particle size compositions (Martínez-Ruiz & Fernández-Santos 2005; Zhang & Zhao 2009), and in some mineral excavation sites, the content of heavy metals (e.g., Cabala *et al.* 2006; Błońska *et al.* 2019) do not prevent the establishment of plant communities of widely varying species composition. In the areas set up *de novo* during and after the excavation, mineral resources are effectively colonized by plants with other environmental demands. In one patch, wetland and grassland species grow together, creating an unexpected and unusual species composition, unknown from the natural and seminatural ecosystem (Woźniak 2010). Depending on the bedrock prerequisites and partially on the surrounding vegetation, forest and non-forest ecosystems occur spontaneously (Rahmonov & Oleś 2010; Rahmonov *et al.* 2014; Kompała-Bąba *et al.* 2019).

The post-mining areas (heaps) differ in many habitat and microhabitat parameters. As a result, the organisms (plants) colonizing the sites during the initial stages of succession are characterized by the ecological plasticity of individual species of a specific set of features (height, seed production, biomass, and root features). In the next stage, they develop into phytocenosis and zoocenosis in extreme habitats (Woźniak 2010) and compete for an effectively functioning ecosystem (Rahmonov & Oleś 2010; Rahmonov *et al.* 2013).

The organisms of the first stages of succession produce biomass, mostly plants (the primary producers), which are the starting material for forming soil organic matter (SOM) and soil organic carbon (SOC; Fox *et al.* 1998; Evans & Willgoose 2000). In this way, the relationship between vegetation species composition and soil is significant (Frieswyk *et al.* 2007; Tian *et al.* 2008; Tang *et al.* 2010; Alday *et al.* 2011; Rahmonov *et al.* 2020). The early-successional organisms possess traits that adapt them and give them both morphological and physiological features to colonize and establish in the conditions and habitats of post-mineral excavation sites (Díaz *et al.* 2007; Frieswyk *et al.* 2007).

There are some field studies concerning the influence of post-mineral mining areas on the development and functioning of an ecosystem mosaic and the resulting environments of urban-industrial landscapes. Some of the studies present the results of cognitive models (e.g., former mining "path-centered mosaics") and development mechanisms. Environmental planning (or lack of it) is nevertheless equally crucial in managing this type of area. The paper analyzes two critical elements whose development depends on each other in this particular case. In the first case, the fall of mining led to the abandonment of the land, and only parts found new management methods (Díaz *et al.* 2007; Mokany *et al.* 2008; Domański 2009; Krzysztofik *et al.* 2012; Fagiewicz 2014; Drozdek *et al.* 2017). In the second case, the abandoned post-mineral mining area remained unmanaged. On the non-reclaimed areas, research was undertaken into the development, diversity of vegetation, soil formation, novel ecosystem development, and chemical features of the mineral soil substratum (Rahmonov & Oleś 2010; Woźniak 2010; Rahmonov *et al.* 2013; Kompała-Bąba *et al.* 2019).

Biocenotic systems belong to dynamically changing elements of the geosystem, hence each of them sooner or later will be replaced by another. Changes in the ecosystems are caused by various factors, both natural and anthropogenic (Fig. 4). The ability to change is one of the most important ecosystem features, resulting from the open system of energy and matter in circulation (Rahmonov *et al.* 2010). Such systems include, among others, undeveloped of post-mineral excavation sites and their process of regeneration which depends primarily on the type of bedrock/gangue material and its properties (Burghardt 1994; Oleś *et al.* 2004; Tang *et al.* 2010).

Figure 4. The mosaic of patches of the primary vegetation during the succession process on coal mine heap mineral material (Photo: Gabriela Woźniak)

Figure 5. Vegetation patches dominated by the successful colonizer *Calamagrostis epigejos* on coal mine heap mineral material (Photo: Gabriela Woźniak)

In non-reclaimed sites of mining areas, the vegetation succession is sometimes initiated by, e.g., *Poa compressa*, *Tussilago farfara*, *Calamagrostis epigejos*. The individuals of *Calamagrostis epigejos* show high ecological plasticity and a broad spectrum of habitat colonization (Fig. 5). Its extensive root system is supported by nutrients from the heap soil substratum and water from atmospheric precipitation. The rapid colonization process on coal mine heaps leads to vegetation community assemblages dominated, e.g., by *Calamagrostis epigejos* on the soil substrate surface. The root system of *C. epigejos* inhibits erosive processes. *C. epigejos* spreads vegetatively, contributing to its rapid and extensive spread across large areas (Rebele & Lehmann 2001; Rahmonov 2009; Piekarska-Stachowiak *et al.* 2014). In most heaps and post-industrial regions of Poland and Europe, this species initiates habitat-forming processes by creating thick turf (Zhang & Zhao 2009). Annual biomass production and decomposition of this species lead to an increase in the amount of organic matter on an inferior substrate surface. In subsequent years, it can independently create an organic layer consisting of this species' residues, which enriches the soil and provides nutrients to other species.

On post-mineral excavation sites including post-coal mine heaps regardless of the amount of nutrient elements the small or medium particle sizes promotes the colonization of vegetation (Christensen 2001; Wang *et al.* 2011; Hemkemeyer *et al.* 2018), which is associated with high water holding capacity and aeration. The analyzed area of the substrate is characterized by the features mentioned, hence the formation of the tree dominated (forest) community takes place relatively quickly in different types of post-mineral excavation sites (Fig. 6, 7).

Ecosystems developing on post-mineral excavation sites represent a dynamically changing part of the environment. Each of them will be replaced by another, appropriate to successional stage. Various factors cause changes in ecosystems. The ability to adapt and change is one of the most critical ecosystem features. The adaptation processes of living organisms in extreme habitat conditions of the post-mineral excavation sites are underestimated. It is the organism's response resulting from the open system of energy and matter in circulation (Rahmonov & Oleś 2010).

In the former mining areas, multiway reclamation mostly takes place. The multiway direction in vegetation development applies to both opencast mines and coal mine heaps (Burghardt 1994; Tang *et al.* 2010; Alday *et al.* 2011).

Figure 6. The early establishment of single tree species seedlings during the successional process on coal mine heap mineral material (Photo: Gabriela Woźniak)

Figure 7. The establishment of abundant tree species seedlings during the successional process on coal mine heap mineral material (Photo: Gabriela Woźniak)

Natural succession occurs on non-reclaimed fragments, and the features of rock material storage significantly impact its process (Martínez-Ruiz & Fernández-Santos 2005; Rostański 2006; Kompała-Bąba *et al.* 2019; Rahmonov *et al.* 2020). Therefore, succession can evolve in different directions in each area following the features of the geochemical traits of the soil substratum of the heap (Fig. 8). Their common feature is the sequence of meadow, grassland, wetland, and ruderal species at the initial succession stages in most post-coal mining areas in Europe (Prach 2003; Prach *et al.* 2013; Błońska *et al.* 2019).

Figure 8. The open conditions of post-mining sites are available for species that have run away from their usual habitats, like *Centaurea cyanus* (Photo: Gabriela Woźniak)

CONCLUSION

The legislation prescription concerning the reclamation practices is not updated according to the newest knowledge achieved by environmental sciences. Humans need to observe, understand, follow, and enhance the power and magnitude of natural adaptive, multilevel processes when facing the global environmental change forces.

REFERENCES

Alday J.G., Marrs R.H. & Martínez-Ruiz C. 2011: Vegetation succession on reclaimed coal wastes in Spain: the influence of soil and environmental factors. Applied Vegetation Science, 14, 84–94.

Anawar H.M., Canha N., Santa-Regina I. & Freitas M.C. 2013: Adaptation, tolerance, and evolution of plant species in a pyrite mine in response to contamination level and properties of mine tailings: sustainable rehabilitation. Journal of Soils and Sediments, 13, 730–741.

Błońska A., Kompała-Bąba A., Sierka E., Bierza W., Magurno F., Besenyei L., Ryś K. & Woźniak G. 2019: Diversity of vegetation dominated by selected grass species on coal-mine spoil heaps in terms of reclamation of post-industrial areas. Journal of Ecological Engineering, 20 (2), 209–217.

Bradshaw A. 2000: The use of natural processes in reclamation – advantages and difficulties. Landscape and Urban Planning, 51 (2–4), 89–100.

Burghardt W. 1994: Soils in urban and industrial environments. Zeitschrift für Pflanzenernährung und Bodenkund, 157 (3), 214–250.

Cabala J., Idziak A.F. & Szymala A. 2006: Physicochemical and geophysical investigation of soils from former coal mining terrains in southern Poland. [In:] Cardu M., Ciccu R., Lovera E. & Michelotti E. (eds.): Proceedings of the 15th Mining Planing and Equipment Section, Torino, Italy, 543–548.

Christensen B.T. 2001: Physical fractionation of soil and structural and functional complexity in organic matter turnover. European Journal of Soil Science, 52 (3), 345–353.

DeLong C., Skousen J. & Pena-Yewtukhiw E. 2012: Bulk density of rocky soils in forestry reclamation. Soil Science Society of America Journal, 76 (5), 1810–1815.
Díaz S., Lavorel S., de Bello F., Quétier F., Grigulis K. & Robson T.M. 2007: Incorporating plant functional diversity effects in ecosystem service assessments. Proceedings of the National Academy of Sciences of the United States of America, 104, 20684–20689.
Domański B. 2009: Post-industrial area regeneration – Specificity of challenges and instruments. [In:] Jarczewski W. (ed.): Spatial Aspects of Revitalization of Downtowns, Housing Estates, Post-Industrial, Post-Military and Old-Railway Areas. Revitalization of Polish Cities. Institute of Urban and Regional Development, Kraków, 125–138.
Drozdek M.E., Greinert A., Kostecki J., Tokarska-Osyczka A. & Wasylewicz R. 2017: Thematic parks in urban post-industrial areas. Acta Scientiarum Polonorum, Architectura, 16 (1), 65–77.
Dulias R. 2010: Landscape planning in areas of sand extraction in the Silesian Upland, Poland. Landscape and Urban Planning, 95 (3), 91–104.
Evans K.G. & Willgoose G.R. 2000: Post-mining landform evolution modelling: 2. Effects of vegetation and surface ripping. Earth Surface Processes and Landforms, 25 (8), 803–823.
Fagiewicz K. 2014: Spatial processes of landscape transformation in mining areas (case study of opencast lignite mines in Morzysław, Niesłusz, Gosławice). Polish Journal of Environmental Studies, 23 (4), 1123–1136.
Forman R.T.T. 2014: Urban Ecology: Science of Cities. Cambridge University Press, Cambridge.
Fox H.R., Moore H.M. & McIntosh A.D. (eds.) 1998: Land Reclamation: Achieving Sustainable Benefits. In 4th International Conference, Proceedings of the International Association of Land Reclamationists, Nottingham, The Netherlands, Balkema: Dordrecht.
Frieswyk C.B., Johnston C.A. & Zedler J.B. 2007: Identifying and characterizing dominant plants as an indicator of community condition. Journal of Great Lakes Research, 33, 125–135.
Grodzińska K., Korzeniak U., Szarek-Łukaszewska G. & Godzik B. 2000: Colonization of zinc mine spoils in southern Poland – preliminary studies on vegetation, seed rain and seed bank. Fragmenta Floristica et Geobotanica, 45 (1–2), 123–145.
Hameed A., Jaabir M. & Sankaran S. 2000: Geoenvironmental reclamation – Paddy cultivation in mercury polluted lands. [In:] Paithankar A.G., Jha P.K. & Agarwal R.K. (eds.): Geoenvironmental Reclamation. International Symposium. Oxford & IBH Publishing Co. Pvt. Ltd., New Delhi, 79–85.
Hancock G.R., Loch R.J. & Willgoose G.R. 2003: The design of post-mining landscapes using geomorphic principles. Earth Surface Processes and Landforms, 28 (10), 1097–1110.
Hemkemeyer M., Dohrmann A.B., Christensen B.T. & Tebbe C.C. 2018: Bacterial preferences for specific soil particle size fractions revealed by community analyses. Frontiers in Microbiology, 9, 149.
Jaros J. 1965: History of Coal Mining in Upper Silesian Coal Basin. Ossolineum, Wrocław.
Jochimsen M.E. 2001: Vegetation development and species assemblages in a long-term reclamation project on mine spoil. Ecological Engineering, 17 (2–3), 187–198.
Kicki J. & Sobczyk E.J. (eds.) 2016: Prawne, techniczne, ekonomiczne i środowiskowe aspekty gospodarki skałą płonną w kopalniach węgla kamiennego. Wydawnictwo IGSMiE PAN, Kraków.
Kodir A., Hartono D.M., Haeruman H. & Mansur I. 2017: Integrated post mining landscape for sustainable land use: A case study in South Sumatera, Indonesia. Sustainable Environment Research, 27, 203–213.
Kompała-Bąba A., Bierza W., Błońska A., Sierka E., Magurno F., Chmura D., Besenyei L., Radosz Ł. & Woźniak G. 2019: Vegetation diversity on coal mine spoil heaps – how important is the texture of the soil substrate? Biologia, 74 (4), 419–436.
Krzaklewski W. & Pietrzykowski M. 2007: Diagnoza siedlisk na terenach pogórniczych rekultywowanych dla leśnictwa, ze szczególnym uwzględnieniem metody fitosocjologiczno-glebowej. Sylwan, 1, 51–57.
Krzaklewski W. 1995: The method of directed succession in reclamation work. Postępy Techniki w Leśnictwie, 56, 1–9.
Krzysztofik R., Dulias R., Kantor-Pietraga I., Spórna T. & Dragan W. 2020: Paths of urban planning in a post-mining area. A case study of a former sandpit in southern Poland. Land Use Policy, 99, 104801.
Krzysztofik R., Runge J. & Kantor-Pietraga I. 2012: Paths of environmental and economic reclamation: the case of post-mining brownfields. Polish Journal of Environmental Studies, 21 (1), 219–223.
Limpitlaw D., Alsum A., & Neale D. 2017: Calculating ecological footprints for mining companies – an introduction to the methodology and an assessment of the benefits. The Journal of the Southern African Institute of Mining and Metallurgy, 117 (1), 13–18.

Martínez-Ruiz C. & Fernández-Santos B. 2005: Natural revegetation on topsoiled mining-spoils according to the exposure. Acta Oecologica, 28 (3), 231–238.
Martínez-Ruiz C. & Marrs R.H. 2007: Some factors affecting successional change on uranium mine wastes: Insights for ecological restoration. Applied Vegetation Science, 10 (3), 333–342.
Mokany K., Ash J. & Roxburgh S. 2008: Functional identity is more important than diversity in influencing ecosystem processes in a temperate native grassland. Journal of Ecology, 96, 884–893.
Nicia P., Bejger R., Błońska A., Zadrożny P. & Gawlik A. 2014: Characteristics of the habitat conditions of ash-alder carr (*Fraxinio-Alnetum*) in the Błedowskie Swamp. Journal of Food, Agriculture & Environment, 12 (2), 1227–1232.
Oleś W., Rahmonov O., Rzętała M., Pytel S. & Malik I. 2004: The ways of industrial wastelands management in the Upper Silesian Region. Ekologia (Bratislava), 23 (Suppl. 1), 244–251.
Piekarska-Stachowiak A., Szary M., Ziemer B., Besenyei L. & Woźniak G. 2014: An application of the plant functional group concept to restoration practice on coal mine spoil heaps. Ecological Research, 29, 843–853.
Prach K. 2003: Spontaneous succession in Central-European man-made habitats: What information can be used in restoration practice? Applied Vegetation Science, 6 (2), 125–129.
Prach K., Lencová K., Řehounková K., Dvořáková H., Jírová A., Konvalinková P., Mudrák O., Novák J. & Trnková R. 2013: Spontaneous vegetation succession at different central European mining sites: a comparison across seres. Environmental Science and Pollution Research, 20 (11), 7680–7685.
Prach K., Řehounková K., Lencová K., Jírová A., Konvalinková P., Mudrák O., Študent V., Vaněček Z., Tichý L., Petřík P., Šmilauer P. & Pyšek P. 2014: Vegetation succession in restoration of disturbed sites in Central Europe: the direction of succession and species richness across 19 seres. Applied Vegetation Science, 17 (2), 193–200.
Rahmonov O. & Oleś W. 2010: Vegetation succession over an area of a medieval ecological disaster. The case of the Błędów desert, Poland. Erdkunde, 64 (3), 241–255.
Rahmonov O. 2009: The chemical composition of plant litter of black locust (*Robinia pseudoacacia* L.) and its ecological role in sandy ecosystems. Acta Ecologica Sinica, 29 (4), 237–243.
Rahmonov O., Czylok A., Orczewska A., Majgier L. & Parusel T. 2014: Chemical composition of the leaves of *Reynoutria japonica* Houtt. and soil features in polluted areas. Central European Journal of Biology, 9 (3), 320–330.
Rahmonov O., Krzysztofik R., Środek D. & Smolarek-Lach J. 2020: Vegetation- and environmental changes on non-reclaimed spoil heaps in Southern Poland. Biology, 9 (7), 164.
Rahmonov O., Parusel T., Szymczak A. 2010. The development of ecological systems in the area transformed by human impact (settling ponds of "Jan Kanty" black coal mine). [In:] Rahmonov O. (ed.): Anthropogenic Aspects of Landscape Transformations, 6, 88–95.
Rahmonov O., Snytko V.A., Szczypek T. & Parusel T. 2013: Vegetation development on post-industrial territories of the Silesian Upland (Southern Poland). Geography and Natural Resources, 34 (1), 96–103.
Rayfield B., Anand M. & Laurence S. 2005: Assessing simple versus complex restoration strategies for industrially disturbed forests. Restoration Ecology, 13 (4), 639–650.
Rebele F. & Lehmann C. 2001: Biological Flora of Central Europe: *Calamagrostis epigejos* (L.) Roth. Flora, 196, 325–344.
Rostański A. 2006: Spontaniczne kształtowanie się pokrywy roślinnej na zwałowiskach po górnictwie węgla kamiennego na Górnym Śląsku [Spontaneous plant cover on colliery spoil heaps in Upper Silesia (Southern Poland)]. University of Silesia, Katowice.
Szumacher I. & Pabjanek P. 2017: Temporal changes in ecosystem services in European cities in the continental biogeographical region in the period from 1990–2012. Sustainability, 9 (4), 665.
Tang B., Liu S.R., Cai D.X., Lu L.H., He R.M., Gao Y.X. & Di W.Z. 2010: Soil physical and chemical characteristics under different vegetation restoration patterns in China south subtropical area. Chinese Journal of Applied Ecology, 21 (10), 2479–2486.
Tarolli P. & Sofia G. 2016: Human topographic signatures and derived geomorphic processes across landscapes. Geomorphology, 255, 140–161.
Tian J.-Q., Zhou Z.-Y., Bao B. & Sun J.-X. 2008: Variations of soil particle size distribution with land-use types and influences on soil organic carbon and nitrogen. Chinese Journal of Plant Ecology, 32 (3), 601–610.
Tropek R. & Konvicka M. 2008: Can quarries supplement rare xeric habitats in a piedmont region? Spiders of the Blansky les Mts, Czech Republic. Land Degradation & Development, 19 (1), 104–114.

Tropek R. & Konvička M. 2011: Should restoration damage rare biotopes? Biological Conservation, 144 (5), 1299.

Tropek R., Kadlec T., Hejda M., Kocarek P., Skuhrovec J., Malenovsky I., Vodka S., Spitzer L., Banar P. & Konvicka M. 2012: Technical reclamations are wasting the conservation potential of post-mining sites. A case study of black coal spoil dumps. Ecological Engineering, 43, 13–18.

Urbancová L., Lacková E., Kvíčala M., Čecháková L., Čechák J. & Stalmachová B. 2014: Plant communities on brownfield sites in Upper Silesia (Czech Republic). Carpathian Journal of Earth and Environmental Sciences, 9 (2), 171–177.

Wang L.Y., Han Y.Z., Zhang C.L. & Pei Z.Y. 2011: Reclaimed soil properties and weathered gangue change characteristics under various vegetation types on gangue pile. Acta Ecologica Sinicia, 31 (21), 6429–6441.

Wiegleb G. & Felinks B. 2001: Primary succession in post-mining landscapes of Lower Lusatia – chance or necessity. Ecological Engineering, 17 (2–3), 199–217.

Woźniak G. 2010: Zróżnicowanie roślinności na zwałach pogórniczych Górnego Śląska [Diversity of vegetation on coal-mine heaps of the Upper Silesia (Poland)]. W. Szafer Institute of Botany, Polish Academy of Sciences, Kraków.

Zhang J.Y. & Zhao H.L. 2009: Changes in soil particles fraction and their effects on stability of soil-vegetation system in restoration processes of degraded sandy grassland. Ecology Environment Science, 18 (4), 1395–1401.

Author Index